BioComputing: Computer Tools for Biologists

Stuart M. Brown, PhD
Research Computing Resource
NYU Medical Center
550 First Avenue
New York, NY 10016
(browns02@med.nyu.edu)

Library of Congress Cataloging-in-Publication Data

BioComputing : computer tools for biologists / edited by Stuart M.
Brown.
p. cm.
Includes bibliographical references and index.
ISBN 1-881299-17-1
1. Bioinformatics. I. Brown, Stuart M., 1962-
QH324.2.B54 2003
570'.285--dc21
2002156397

ISBN 1-881299-17-1

Printed in the United States of America

9 8 7 6 5 4 3 2 1

Eaton Publishing
One Research Drive
P.O. Box 1070
Westborough, MA 01581-6070
www.BioTechniques.com

James D. Emerton: *Publisher*
Christine Emerton: *Managing Editor*
Sandy Lamont: *Production Manager*
Anne Hann: *Asst. Production Manager*
Robin Davis: *Marketing Projects Coordinator*
Leigh Linsley: *Sales and Marketing Assistant*

CRANbury

BioComputing: Computer Tools for Biologists

Edited by

Stuart M. Brown
Research Computing Resource
NYU Medical Center
New York, NY, USA

BioTechniques® Press
Eaton Publishing
Part of the Informa Life Sciences Group

CONTENTS

Introduction

For many years, computational biology has been the neglected stepchild of molecular biology and computer science, especially when it comes to publication. Many papers have been found to be "lacking broad interest" in the most prestigious biology journals, yet "not sufficiently novel" for the highly abstract journals such as *CABIOS/Bioinformatics*. *BioTechniques* has often provided a home for these quirky papers that present innovative and useful computing methods for biologists without really breaking new conceptual ground in information theory.

This policy of welcoming papers that are not easily categorized has led *BioTechniques* to be the leading journal for a number of interesting sub-fields such as the use of spreadsheet or word processor macros and scripting languages for biology applications. *BioTechniques* has also published many useful mini-reviews of various categories of bioinformatics tools such as similarity searching, phylogenetics, gene expression, and PCR primer design. These short papers are generally designed to provide a quick introduction to a variety of online or freeware computing tools, with the objective of giving the lab-based biologist some practical help, rather than providing a comprehensive overview of the state of the field.

This Update volume provides a selection of the best papers in the area of biocomputing from *BioTechniques*, together with updates from the authors and the occasional comment from the editors. Many of these papers have held up surprisingly well over the years, earning many citations and still providing useful tools and insights to the researcher. Unlike with many journal articles, the authors of these updates have enthusiastically continued to create new versions of their software and provide support to users many years after the original publication.

Furthermore, many of these papers provide simple programs and open source code that can be used as models to study, copy, and modify for investigators starting out in the area of developing new computer programs to meet their needs—and *BioTechniques* stands ready to publish the papers describing your new software contrivances.

This book contains a collection of papers, previously published in *BioTechniques*, spanning the years 1996–2002, which nicely encapsulates the Internet bubble. We have also included a few papers from the dawn of the Internet era, which might seem a bit naïve today (e.g., how to use a Web server to organize a laboratory's collection of protein and DNA sequences). In fact, these simple technologies are still useful and should not be lost amidst the noise created by the much more complex (and expensive) software that has expanded on these concepts—past the point of practicality in many cases. We have also included a few papers on the use of Internet-based bioinformatics services. While fee-based biocomputing services over the Internet have not fared well in the past few years, free services such as Blast and PubMed at the NCBI and the UCSC Genome Browser have become essential tools for most biologists. Yet most users of these services can learn a great deal from the articles in this volume about how to design better queries and to more carefully interpret their results so that they can make more effective use of these Web sites.

I want to extend my thanks to Bob Horton for writing the introduction to the Simple Scripting Languages section.

Stuart M. Brown
February 2003

Section I

Spreadsheet and Word Processor Macros

Most high-throughput sequence analysis for genome projects, EST pipelines, Web-based similarity search engines, etc., operate on computers using a UNIX operating system. Most new bioinformatics algorithms are also developed first as UNIX software. However, most laboratory biologists work on Microsoft® Windows® or Macintosh® computers and they have little or no knowledge of UNIX. There are also a large number of DNA/protein sequence analysis software packages available for Windows and Macintosh computers, but many of these are expensive and complex commercial products that offer many more tools than the biologist needs to accomplish daily tasks such as translation, restriction analysis, protein hydrophobicity plots, etc. These pre-packaged programs operate as a black box—you can only use their existing commands and interface with no possibility of modifying the program according to your own unique interests and needs.

Gerry Shaw initiated the idea of using the simple programming tools built into Microsoft Word and Excel to create small programs to manipulate DNA and protein sequences that could then be distributed for free to other scientists as macros. Many other authors have been inspired by Shaw's papers to expand on the algorithms and improve the interfaces for these macro programs, and the bulk of these papers have appeared in the pages of *BioTechniques*.

In addition to being free and easy to use, macro programs have several other advantages. First, they are cross platform: any computer that can run Microsoft Office can use these macros. Second, macros are fundamentally open source. Anyone can look at the equations that are implemented in each step and modify them to suit their own needs. This substantially lowers the traditional barrier (or activation energy or learning curve) that separates the biologist from the programmer.

CHAPTER 1

Protein Sequence Interpretation Using a Spreadsheet Program

Gerry Shaw

University of Florida College of Medicine, Gainesville, FL, USA

Reprinted from BioTechniques 19:978-983 (December 1995)

ABSTRACT

Modern spreadsheet programs are inexpensive, easy to use and have a host of powerful mathematical and graphical capabilities. Here I show how such programs can be used in the interpretation of protein sequences. The methods described allow researchers to produce many types of sequence interpretation plots, which can then be printed out or conveniently incorporated into documents. It is also easy to manipulate the parameters used in sequence interpretation without any knowledge of computer programming.

INTRODUCTION

As the number of entries in GenBank®, EMBL and PIR® databases continues to rise exponentially, the interpretation of protein sequences becomes evermore important. Protein sequence interpretation includes the production of diagrams showing the distribution of charge, hydrophobicity, hydrophilicity, antigenicity and propensity to form different types of structure along the sequence. Although numerous computer programs are available for these kinds of analyses, problems are frequently encountered. First, programs tend to be incompatible between different types of computer systems, thus diagrams produced might not be easily incorporated into texts, instead requiring time-consuming and inconvenient photographic reproduction. Second, the user is limited to output formats which were defined by the original programmers and which are usually not easily altered. Third, it is often difficult to find out exactly how these programs work. The programmers may have used upgraded algorithms that are significantly different from those originally published. Frequently the original publications describing a method do not include all relevant details and also may not describe a unique algorithm.

One solution to these problems is to write your own sequence interpretation software, taking advantage of the considerable capabilities built into such graphical user interfaces

BioComputing: Computer Tools for Biologists
Edited by Stuart M. Brown

(GUI) as Windows™ and Apple Macintosh® System 7. However this is a fairly daunting prospect even for an experienced programmer, since the vast majority of the programming is unrelated to the essential algorithm but concerned with making the various elements of the GUI function correctly. There is however a simple answer to this problem, which requires no understanding of the construction of GUI programs or any other programming experience. In recent years, spreadsheet programs have become more and more widely used and versatile. Although they are easy to use, they can perform complex mathematical data manipulations and plot out the results in hundreds of different styles, producing graphs that can be printed out or directly incorporated into other documents. I therefore decided to examine the feasibility of using such programs for the interpretation of protein sequences.

MATERIALS AND METHODS

The spreadsheet program I chose was the Windows Version 5.0 of Microsoft® Excel™ (Redmond, WA, USA)—mostly because I got a free copy with a new PC. This program can produce graphs that can be readily inserted into documents produced by word processor programs such as Microsoft Word™ or into figures produced by drawing programs such as Microsoft PowerPoint™. Several other companies produce similar spreadsheet programs with similar capabilities, and functionally identical versions of Excel are available for Macintosh and other computers. Most sequence interpretation programs translate each amino acid into a numerical value, which reflects some characteristic property such as charge, hydrophobicity or likelihood of being found in a particular structural context. These values are averaged over a "window" of an appropriate number of amino acids down the sequence, and the resulting averages are plotted out. The positions of peaks and troughs on the printout are interpreted in terms of the parameter being examined. Such calculations can be readily performed in Excel. As a very simple example, Figure 1 shows the first 40 rows of the spreadsheet used to calculate the charge profile of the major neurofilament light chain (NF-L) subunit as illustrated in Figure 2. The human NF-L sequence was loaded into the A column of the Excel spreadsheet. This could be manually typed in, but this approach is tedious and error-prone. A better alternative is to download the relevant entry from a sequence database, load it into Microsoft Word or a similar word processor and to format the amino acid sequence for copying into Excel. If the sequence contains numbers and spaces between each amino acid, the numbers can be edited out, being careful to leave a blank space following each amino acid. The "Convert Text to Table" option in the "Table" menu of Word can then be used to format the sequence as a table one character wide by using the blank spaces as delimiters, which turn the sequence into a column, each row of which contains one character. If the database sequence contains only amino acid characters, a simple Word macro can be constructed to automatically add new-line characters following each amino acid, or a new-line character (typing the Enter key) can be inserted manually after each amino acid. The end result in each case is a vertical column with one amino acid in each row, which can be directly copied and pasted into the first row of an Excel spreadsheet.

The next step, as illustrated in Figure 1, is to transform the sequence data into numerical values. In a simple example, the charge profile of a molecule can be examined by assigning a charge of -1 to each aspartic and glutamic acid residue, +1 to each lysine and arginine and +0.5 to each histidine. To transform aspartic acid to a score of +1, all that is required is to make an "Aspartic" column in the spreadsheet as shown in Figure 1 and insert the equation =IF(A2="D",-1,0) into the first cell in this column (B2). This equation gets the amino acid character in cell A2 and compares it to D, the single letter code for aspartic acid. If this amino acid is an aspartic acid, the B2 cell gets a value of -1, if not, a value of 0. If the sequence was loaded into the spreadsheet in the three-letter code, the same result can be obtained simply by substituting "Asp" for "D". The C2, D2, E2 and F2 cells contain similar equations to assign the appropriate values to glutamic acid, lysine, arginine and histidine. The G2 cell then adds the results from each of these five columns, which can be performed using the command =ADD(B2:F2). Since it is also useful to know the charge density irrespective of its positive or negative nature, I also included column H, which determines the absolute amount of charge. This can be done by inserting the equation =ABS(G2), which places the absolute value of the G2 cell into the H2 cell. The equations in the first line of the spreadsheet can be repeated down the length of the sequence in the spreadsheet by selecting the first line, copying it and pasting over the relevant columns in the rest of the spreadsheet. This requires only one copy/paste operation for the entire spreadsheet, since Excel automatically changes the cell addresses in each of the copied equations to the values appropriate for their new positions. The next step is to count the values in the G and H columns over a window of some desired size; in this

	A	B	C	D	E	F	G	H	I	J	K
1	Sequence	Aspartic	Glutamic	Lysine	Arginine	Histidine	Sum	Absolute	Charge	Total Charge	
2	M	0	0	0	0	0	0	0			
3	S	0	0	0	0	0	0	0			
4	S	0	0	0	0	0	0	0			
5	F	0	0	0	0	0	0	0			
6	S	0	0	0	0	0	0	0			
7	Y	0	0	0	0	0	0	0			
8	E	0	-1	0	0	0	-1	1			
9	P	0	0	0	0	0	0	0			
10	Y	0	0	0	0	0	0	0			
11	Y	0	0	0	0	0	0	0			
12	S	0	0	0	0	0	0	0			
13	T	0	0	0	0	0	0	0			
14	S	0	0	0	0	0	0	0			
15	Y	0	0	0	0	0	0	0	2.5	6.5	
16	K	0	0	1	0	0	1	1	3.5	7.5	
17	R	0	0	0	1	0	1	1	3.5	7.5	
18	R	0	0	0	1	0	1	1	3.5	7.5	
19	Y	0	0	0	0	0	0	0	3.5	7.5	
20	V	0	0	0	0	0	0	0	3.5	7.5	
21	E	0	-1	0	0	0	-1	1	3.5	7.5	
22	T	0	0	0	0	0	0	0	4.5	6.5	
23	P	0	0	0	0	0	0	0	5.5	7.5	
24	R	0	0	0	1	0	1	1	5.5	7.5	
25	V	0	0	0	0	0	0	0	5.5	7.5	
26	H	0	0	0	0	0.5	0.5	0.5	5.5	7.5	
27	I	0	0	0	0	0	0	0	5.5	7.5	
28	S	0	0	0	0	0	0	0	5.5	7.5	
29	V	0	0	0	0	0	0	0	5.5	7.5	
30	R	0	0	0	1	0	1	1	4.5	6.5	
31	S	0	0	0	0	0	0	0	3.5	5.5	
32	G	0	0	0	0	0	0	0	2.5	4.5	
33	Y	0	0	0	0	0	0	0	2.5	4.5	
34	S	0	0	0	0	0	0	0	2.5	4.5	
35	T	0	0	0	0	0	0	0	3.5	3.5	
36	A	0	0	0	0	0	0	0	3.5	3.5	
37	R	0	0	0	1	0	1	1	3.5	3.5	
38	S	0	0	0	0	0	0	0	2.5	2.5	
39	A	0	0	0	0	0	0	0	2.5	2.5	
40	Y	0	0	0	0	0	0	0	3	3	

Figure 1. Top 40 rows of the Excel spreadsheet used to construct Figure 2.

example, an arbitrarily selected 28 amino acids. This is performed using the SUM command. Since this window is a running average, the center of which is at amino acid 14, I placed the equation =SUM(G2:G29) in cell J15. The total charge can be computed in a similar manner by putting =SUM(J2:J29) in cell K15. These equations can be simultaneously copied and pasted into all the cells from H15 to 14 cells from the end of the sequence. All that remains is to plot out the data using the "Chart Wizard" option of Excel (Figure 2). These data are plotted directly from the spreadsheet I and J columns, although Excel allows the data to be smoothed using a variety of different mathematical procedures. This final graph can be sized in any desired fashion and printed out or copied and pasted directly into a Word document or other Windows program. In this case it has been sized to exactly match a scaled domain map of NF-L made with Microsoft PowerPoint. NF-L consists of four domains, a globular "head" of about 90 amino acids, an α-helical coiled-coil "rod" of about 310 amino acids (interrupted by two short non-helical linkers), a roughly 42-amino acid "tail a" sequence and finally a 106-amino acid "E segment" (8). As can be seen, the N-terminal head region of NF-L is rather basic, the α-helical rod region is highly charged but slightly acidic in overall charge, the "tail a" region is mildly basic, and the glutamic acid rich E segment is very acidic. This simple figure is quite informative and graphically shows the distinctly different charge properties of the four domains of NF-L.

The example given above can be significantly streamlined by entering the single command =IF(A2="D",-1,IF(A2="E", -1,IF(A2="K",1,IF(A2="R",1,IF(A2="H",0.5,0))))) in the B2 cell and then copying and pasting this equation into all the remaining cells in this column. This statement assigns the values described above to each of the 5 amino acids or, if the amino acid is not one of these, a value of 0, and therefore also makes the SUM

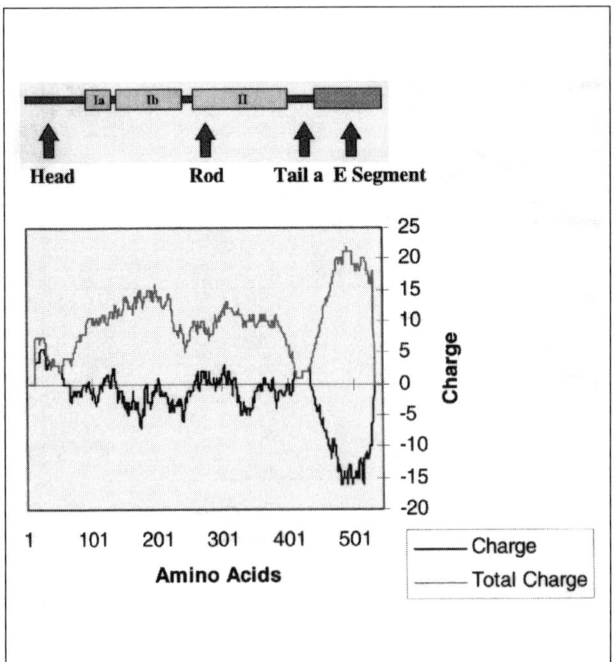

Figure 2. Charge density plotted along the sequence of human NF-L. Upper part is a scale diagram of the different types of sequence found in this molecule. The N-terminal head is about 90 amino acids and is probably a compact β-structure. The rod region is 310 amino acids and contains three segments of α-helical coiled-coil with the characteristic heptad repeat, which are named coils Ia, Ib and II (see Reference 8 and references therein for details). The two major breaks in the coiled-coil regions are indicated. "Tail a" is a short relatively uncharged region followed by the glutamic acid rich E segment. The graph shows the average charge (lower line) and total charge (upper line) of this molecule over 28 amino acid windows plotted to the same scale as the diagram above. The different charge characteristics of the various protein domains are clearly visible.

Table 1. Parameters Used by Various Protein Sequence Interpretation Programs

	[a]Argos	[b]Kyte-Doolittle	[c]von Heijne	[d]GES	[e]Hopp-Woods	[f]Welling
Ala	2.75	1.8	-4.2	-1.6	-0.5	0.115
Cys	1.91	2.5	-6.3	-2.0	-1.0	-0.120
Asp	-4.92	-3.5	31	9.2	3.0	0.065
Glu	-3.54	-3.5	24.7	8.2	3.0	-0.071
Phe	2.99	2.8	-14.2	-3.7	-2.5	-0.141
Gly	0.86	-0.4	0.0	-1.0	0.0	-0.184
His	-2.48	-3.2	14.3	3.0	-0.5	0.312
Ile	4.1	4.5	-10.5	-3.1	-1.8	-0.292
Lys	-2.81	-3.9	17.6	8.8	3.0	0.206
Leu	4.06	3.8	-10.1	-2.8	-1.8	0.075
Met	2.79	1.9	-11.3	-3.4	-1.3	-0.385
Asn	-3.39	-3.5	12.2	4.8	0.2	-0.077
Pro	-2.31	-1.6	13.9	0.2	0.0	-0.053
Gln	-2.47	-3.5	10.1	4.1	0.2	-0.011
Arg	-4.3	-4.5	47.3	12.3	3.0	0.058
Ser	-1.39	-0.8	6.3	-0.6	0.3	-0.026
Thr	-0.44	-0.7	3.8	-1.2	-0.4	-0.045
Val	3.99	4.2	-8.4	-2.6	-1.5	-0.013
Trp	0.61	-0.9	-8.4	-1.9	-3.4	-0.114
Tyr	-0.78	-1.3	4.7	0.7	-2.3	0.013

The original references give details on the interpretation of the plots. Briefly, peaks are interpreted as particularly hydrophobic regions, such as membrane spanning α-helices, using the data from Argos et al. and Kyte-Doolittle. Troughs in plots made using von Heijne and GES parameters are hydrophobic and likely to represent leader sequences or membrane spanning domains. Regions predicted to represent antigenic sites are represented as peaks using the Hopp-Woods and Welling parameters. An appropriate window size to highlight membrane spanning domains is 20–22 amino acids, the length of one membrane spanning α-helix. Since linear antibody recognition sites are generally about 6–7 amino acids long, windows of this size have been used for this kind of analysis. For further discussion of these and other methods of sequence interpretation, see Reference 11.

[a]Values published by Argos et al. (numbers derived from sum of the 5 weighted physical parameters in Table 2 from Reference 1).

[b]Values published by Kyte and Doolittle (data extracted from "Hydropathy index" of Table 2 from Reference 6).

[c]Values published by von Heijne (numbers are derived from "Total" column of Table 1 from Reference 10).

[d]Values published by Engelman et al. (usually known as GES for Goldman, Engelman and Steitz, data from "Water-Oil" column of Table 1 of Reference 3).

[e]Antigenicity after Hopp and Woods (numbers from "Hydrophilicity value" column of Table 1 from Reference 5).

[f]Antigenicity after Welling et al. (numbers from "Antigenicity value" of Table 1 from Reference 12).

command unnecessary. In Excel, a maximum of seven IF loops can be nested in a single cell in this manner. Numerical values therefore can be assigned to all 20 amino acids using three sets of nested IF loops, although a separate column for each amino acid as shown in the example might be preferable for clarity and ease of modification. Published parameters for various measures of hydrophobicity and antigenicity are given in Table 1 and associated references (1,3,5,6,10,12).

The widely used Chou and Fasman algorithms for predicting α-helix, β-sheet and β-turn (2) are somewhat more complex than those described above, but could also be implemented in Excel without difficulty. To more severely test this approach, I attempted to implement the analysis program described by Lupas et al. (7). This widely used and successful program finds polypeptide segments likely to form α-helical coiled-coils but uses a relatively complex algorithm. The defining sequence characteristic of α-helical coiled-coils is the so-called heptad repeat, a loosely defined pattern based on 4 or more contiguous 7 amino acid peptides, the first and fourth amino acid of each of which tend to be hydrophobic, many of the remainder being charged. Lupas and coworkers constructed a library of known α-helical coiled-coil sequences and determined the relative frequency at which each amino acid was found at each of the 7 heptad positions in this library. This table (Table 1 in Reference 7) was used both in the original paper and in this study to give each amino acid 7 scores simultaneously, since it is not known a priori in which of the 7 possible positions the particular amino acid is found. The next amino acid in the sequence is similarly given 7 scores. However this amino acid is one position further down each of the 7 possible α-helical coiled-coils than the previous one, so that the score associated with the first amino acid in a possible heptad repeat must be linked to the score for the second position for the second amino acid and so on. Each sequence therefore produces 7 separate columns of data. The first 28 scores of each column are multiplied together, and then the 28th root (the geometric mean) is calculated. This is repeated for the scores associated with amino acids 2–29 and so on for each 28-amino acid window for each column until the end of the sequence is reached. The largest of the seven 28th roots associated with each of the 7 windows starting at a particular amino acid is determined, and these numbers are then processed to give a measure of probability of α-helical coiled-coil (pCC) formation from 0–1.0, and these results are plotted out. Perhaps surprisingly it was relatively easy to implement this algorithm in Excel, and the plots produced are essentially identical to those produced by Lupas et al. The top of Figure 3 shows a pCC plot obtained from the sequence of the β1 subunit of human trimeric G proteins (Gβ1), which reveals a high pCC at the N terminus. The profile is identical to that obtained by Lupas et al. and was the basis for their prediction that this N-terminal region defined an α-helical coiled-coil. There is now good experimental evidence that this region of Gβ forms a very stable α-helical coiled-coil by dimerization with a complementary α-helical coiled-coil region in the Gγ subunit (4). The bottom of Figure 3 shows a similar plot of rat embryonic skeletal muscle myosin II. This is also very similar to that published by Lupas et al. showing a long C-terminal region with high pCC. This region corresponds to the extended α-helical coiled-coil tail that is well documented from physical studies and visible in the electron microscope. The "hinge" region noted in electron microscopical studies probably corresponds to the region of reduced pCC at about 1150 amino acids, which is also clearly visible both here and in the figure reproduced by Lupas et al. The first approximately 800 amino acids correspond to the N-terminal globular head region, which contains some long α-helices but no α-helical coiled-coils. The sequence in this case was 1939 amino acids long but was processed in only a few seconds in Excel by an IBM®-compatible PC with 16 Mb of RAM and a 66 MHz 486 DX processor.

RESULTS AND DISCUSSION

The methods described here allow the individual scientist to implement and modify published methods for the interpretation of protein sequence data without expert help and should also be valuable for teaching purposes. The graphical output is easy to display, modify, manipulate and incorporate into documents. The approach is versatile, since one single program can perform the role of a variety of more specialized programs and also perform the more usual spreadsheet functions. The basic approach should be applicable to other methods of interpreting protein and also nucleic acid sequences. Other calculations such as the molecular weight or amino acid composition of proteins can easily be performed. Since Excel and similar programs can be run on essentially all available computer systems, the approach should also be widely applicable. Finally, and perhaps most importantly, this approach allows the researcher to understand exactly what published programs are doing and to alter such programs in any desired manner. This approach therefore brings the advantages of computer programming to a wider group of scientists without the considerable investment of time normally required. Detailed descriptions of how to construct spreadsheets to do the various analyses described here can be found in the file SPREAD.DOC, which can be obtained by anonymous file transfer protocol (FTP)

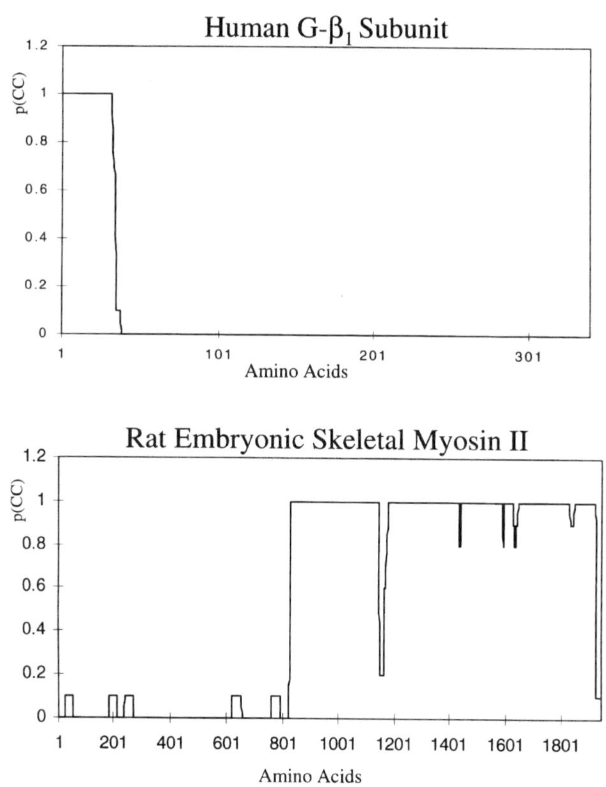

Figure 3. Diagram of coiled-coil propensity (pCC) along the sequences of human Gβ1 and myosin II.

from BIOTECH.UFL.EDU in the PUB/SHAW subdirectory. A second file, AA.DOC, shows how amino acid composition data can also be displayed in a manner essentially identical to that described previously (9), but again in a format compatible with Windows-based programs.

ACKNOWLEDGMENTS

This work was supported by NIH Grant NS22695 and a grant from the Alzheimer's Association.

REFERENCES

1. **Argos, P., J.K.M. Rao and P.A. Hargrave.** 1982. Structural prediction of membrane-bound proteins. Eur. J. Biochem. *128*:565-575.
2. **Chou, P.Y. and G.D. Fasman.** 1978. Empirical predictions of protein conformations. Annu. Rev. Biochem. *48*:251-276.
3. **Engelman, D.M., T.A. Steitz and A. Goldman.** 1986. Identifying nonpolar transbilayer helices in amino acid sequences of membrane proteins. Annu. Rev. Biophys. Biophys. Chem. *15*:321-353.
4. **Garritsen, A., P.J.M. Van Galen and W.F. Simonds.** 1993. The N-terminal coiled-coil domain of β is essential for γ association: A model for G-protein βγ subunit interaction. Proc. Natl. Acad. Sci. USA *90*:7706-7710.
5. **Hopp, T.P. and K.R. Woods.** 1981. Prediction of protein antigenic determinants from amino acid sequences. Proc. Natl. Acad. Sci. USA *78*:3824-3828.
6. **Kyte, J. and R.F. Doolittle.** 1982. A simple method for displaying the hydropathic character of a protein. J. Mol. Biol. *157*:105-132.
7. **Lupas, A., M. Van Dyke and J. Stock.** 1991. Predicting coiled coils from protein sequences. Science *252*:1162-1164.
8. **Shaw, G.** 1991. Neurofilament proteins, p. 185-214. *In* R.D. Burgoyne (Ed.), The Neuronal Cytoskeleton. Alan Liss, New York.
9. **Shaw, G.** 1992. Convenient methods for the display, comparison and interpretation of amino acid composition data. BioTechniques *12*:886-891.
10. **von Heijne, G.** 1981. On the hydrophobic nature of signal sequences. Eur. J. Biochem. *116*:419-422.
11. **von Heijne, G.** 1987. Sequence Analysis in Molecular Biology. Academic Press, London.
12. **Welling, G.W., W.J. Weijer, R. Van der Zee and S. Welling-Wester.** 1985. Prediction of sequential antigenic regions in proteins. FEBS Lett. *188*:215-218.

Update to: Protein Sequence Interpretation Using a Spreadsheet Program

Gerry Shaw

University of Florida College of Medicine, Gainesville, FL, USA (shaw@mbi.ufl.edu)

There has considerable interest in the original paper, at least as judged by the e-mails and letters I have received. There were many suggestions for improvements and extensions to the basic idea, many of which are now included in updated example spreadsheets posted on the Internet at (http://www.mbi.ufl.edu/~shaw/comp.htm). The interest was also enough to warrant an invitation to write a book chapter on the topic (11), and several other groups have published reports apparently stimulated by the first paper (1–4,7–10,13). The approach, therefore, seems to have been a popular one, and one wonders how many laboratories are doing at least some of their protein and nucleic acid sequence analysis using Microsoft® Excel® rather than purchasing more specialized and often very expensive software. The author has also used these spreadsheets for teaching purposes with some success. The students were able to perform quite complex sequence analysis, to see how the basic algorithms worked, and were also able to easily modify the algorithms and immediately see the consequences of these modifications. Some of the more enthusiastic were soon developing and implementing their own algorithms for analyzing protein and nucleic acids in novel ways.

One problem with the original paper was that the method of preparing a sequence for inclusion into the spreadsheet was rather cumbersome. The sequence was processed to form a column one amino acid or base wide, which could be done using the table formatting feature built into Microsoft Word®. However one of my correspondents suggested a better approach that will allow the user to paste in a sequence obtained from GenBank® into the A2 cell of the first sheet of a spreadsheet. This approach works very reliably and is implemented in most of the newer spreadsheets downloadable from the Internet site. The A1 cell can be used to enter the name of the sequence. The entire sequence in uppercase characters is then pasted in the A2 cell. This can be a single or multiple lines of sequence as copied directly from online database. The B, C, D, and E columns contain various equations for removing spaces from the sequence entry and for counting the number of amino acids or bases in the sequence entry. Finally, a complex equation usually in the F column uses the information in these previous columns to select sequentially one amino acid or base at a time from the sequence as entered. This equation is, typically, as found in the F1 cell of Aacomp.xls spreadsheet:

```
=MID(INDEX(B$1:B$100,MATCH(ROWS(F$1:F1)-1,D$1:D$101,1)+1),ROWS
(F$1:F1)-INDEX(D$1:D$101,MATCH(ROWS(F$1:F1)-1,D$1:D$101,1)),1)
```

This extracts the first amino acid or base out of the A2 cell and places it in cell F1. If this equation is copied into cell F2, the variables are automatically incremented to take the second amino acid or base out the sequence entry and so on down the F column. Fortunately, this code can be downloaded from the Internet and does not have to be typed in by the user. Descriptions about how each individual spreadsheet works is given on the Internet site.

Data from sequence databases often include numbers, tab, newline, and space characters that should be removed before the data is entered into most of these spreadsheets.

This can be done manually, although all such characters can also be removed with the reformat macro also included on the Web site listed above. This is part of a suite of macros written in the Visual Basic programming language for use in Microsoft Word, which allow surprisingly complex manipulations of protein and nucleic acid sequences and which was also published in a *BioTechniques* report (12).

As might be expected over an interval of more than 7 years, the author's e-mail address has changed; it is now (shaw@mbi.ufl.edu). Also the Web site linked to these various spreadsheets and other programs has changed, it is now at (http://www.mbi.ufl.edu/~shaw/comp.htm). This Web site now includes links allowing the downloading of a spreadsheet, which calculates charge along a proteins sequence exactly as described in the original paper, called Charged.xls. The other spreadsheets use the more advanced method of configuring the sequence as described above. A quite sophisticated spreadsheet called Aacomp.xls calculates the amino acid composition of protein sequences, compares them to the average composition of various databases, calculates the isoelectric point, and plots out the data in several different ways. Another spreadsheet called Coilcoil.xls implements the Lupas et al. algorithm (5) for the prediction of peptide sequences likely to form α-helical coiled coils. Another spreadsheet called Memb.xls uses a variety of scoring methods to look for membrane spanning α-helices, leader sequences, and antigenic regions. A workable implementation of the Chou and Fasman algorithm is found in the Chofas.xls spreadsheet. A simple example spreadsheet, DNA.xls, shows how GC-rich regions and CpG islands in nucleic acid sequences can be identified and plotted out. Finally, the Dotplot.xls spreadsheet is an implementation of the dotplot method of looking for sequence repeats, including inverted repeats in nucleic acids (6).

REFERENCES

1. **Anbazhagan, R. and E. Gabrielson.** 1999. Spreadsheet-based program for alignment of overlapping DNA sequences. BioTechniques *26*:1180-1185.
2. **Anbazhagan, R., J.G. Herman, K. Enika, and E. Gabrielson.** 2001. Spreadsheet-based program for the analysis of DNA methylation. BioTechniques *30*:110-114.
3. **Delamarche, C.** 2000. Color and graphic display (CGD): programs for multiple sequence alignment analysis in spreadsheet software. BioTechniques *29*:100-107.
4. **Han, B. and A.H. Tashjian, Jr.** 1998. User-friendly and versatile software for analysis of protein hydrophobicity. BioTechniques *25*:256-263.
5. **Lupas, A., M.V. Dyke, and J. Stock.** 1991. Predicting coiled coils from protein sequences. Science *252*:1162-1164.
6. **Maizel, Jr., J.V. and R.P. Lenk.** 1981. Enhanced graphic matrix analysis of nucleic acid and protein sequences. Proc. Natl. Acad. Sci. USA *78*:7665-7669.
7. **McEwan, N.R. and D. Gatherer.** 1998. Adaptation of standard spreadsheet software for the analysis of DNA sequences. BioTechniques *24*:131-138.
8. **Muller, P.Y., L. Barella, and A.R. Miserez.** 2001. RepGene: a spreadsheet template for the management of reporter gene assays. BioTechniques *30*:1294-1298.
9. **Muller, P.Y., E. Studer, and A.R. Miserez.** 2001. Molecular Biocomputing Suite: a word processor add-in for the analysis and manipulation of nucleic acid and protein sequence data. BioTechniques *31*:1306-1313.
10. **Schageman, J.J., M. Basit, T.D. Gallardo, H.R. Garner, and R.V. Shohet.** 2002. MarC-V: a spreadsheet-based tool for analysis, normalization, and visualization of single cDNA microarray experiments. BioTechniques *32*:338-344.
11. **Shaw, G.** 1998. Spreadsheets in molecular biology, p. 203-228. *In* G. Filby (Ed.), Spreadsheets in Science and Engineering. Springer, Berlin.
12. **Shaw, G.** 2000. Useful Microsoft Word macros for molecular biologists and protein chemists [in process citation]. BioTechniques *28*:1198-1201.
13. **Strain, S.R. and J.G. Chmielewski.** 2001. ROCK: a spreadsheet-based program for the generation and analysis of random oligonucleotide primers used in PCR. BioTechniques *30*:1286-1291.

CHAPTER 2

Useful Microsoft® Word™ Macros for Molecular Biologists and Protein Chemists

Gerry Shaw

University of Florida, Gainesville, FL, USA

Reprinted from BioTechniques 28:1198-1201 (June 2000)

ABSTRACT

Biologists today make extensive use of word processing programs for the production of research reports, literature reviews and grant proposals. Frequently, such programs become the default platform for viewing and the later publication of protein and nucleic acid sequence data. Thus, researchers often switch between their word processor and more specialized programs designed to analyze protein and nucleic acid sequences. It would be more convenient to perform these simple sequence analyses using the word processor without switching to another program. The focus here is on the use of the Visual Basic® programming language, which is built into all recent versions of Microsoft® Word™ to generate surprisingly complex and useful macros that can conveniently analyze several important features of protein and nucleic acid sequences. The standard Word interface can also be easily modified to display and run these macros from a pull-down menu. Several examples of this approach are provided.

INTRODUCTION

Most biological researchers constantly switch between word processors and the more specialized programs that are designed for various types of protein and nucleic sequence analysis. It would be more convenient and efficient to perform simple types of protein and nucleic acid sequence analysis without having to leave the word processing program. As demonstrated here, it is relatively easy to customize standard versions of Microsoft® Word™ versions 6.0, 97 or 2000 to perform a variety of useful manipulations that would normally require a separate suite of programs. These Word versions include Visual Basic®, a powerful programming language that allows the production of surprisingly complex and useful macros.

BioComputing: Computer Tools for Biologists
Edited by Stuart M. Brown

Several examples of such macros, of particular use to protein chemists and molecular biologists, are given here. These macros work equally well on Macintosh computers, IBM PCs and compatibles. They can also be used in Microsoft Office® programs such as Excel® and PowerPoint®, which also support the Visual Basic programming language.

Table 1. Macro

```
Sub Nucweight()
'
'Nucweight Macro
'Macro recorded 07/02/99 by Gerry Shaw
'This version for DNA sequences: for RNA MW for
   A = 329.2, U = 306.1, G = 345.2,
   C= 305.2
'
   X = Len(Selection.Text)
   For Y = 1 To X
   Select Case Mid$(Selection.Text, Y, 1)
      Case "A"
      MW = MW + 313.2
      Case "T"
      MW = MW + 304.2
      Case "G"
      MW = MW + 329.2
      Case "C"
      MW = MW + 289.2
      Case Else
      Z = Z + 1
   End Select
   Next
   MW = MW + 18
   If (MW > 18) Then MsgBox ("Selection
includes " & X - Z &" bases, Molecular Weight=
" & MW &" Daltons")
   If (MW = 18) Then MsgBox ("No sequence
selected")
End Sub
```

MATERIALS AND METHODS

Microsoft Word version 6.0 was used initially, and work was continued using Visual Basic programming with Word 97 and Word 2000 when they became available. The Visual Basic macros described here run identically on all three versions; the compiler is built into these versions, and the macros can be constructed with no additional purchases. Releases of Word before version 6.0 use Word Basic, a now obsolete macro language. Although the language is clearly related to Visual Basic, it can be regarded as a distinct and incompatible programming language, and the macros described here must be modified to run on such older versions. However, versions of some of the programs listed here have been written in Word Basic and are available from the author.

RESULTS

A simple example of a useful Visual Basic macro is shown here. Scientists frequently need to know the molecular weight of a sequence or part of a sequence. It would be convenient if the user could select a segment of nucleic acid or protein sequence with the mouse and press a button, tab or hotkey to determine the exact calculated molecular weight of the selection. Table 1 shows a simple Visual Basic macro that performs this function for

DNA sequences and is called NucWeight.

Even those with limited programming experience can probably follow this program. Visual Basic programs start with **Sub**, followed by the name of the macro **(Nucweight()** in this example), and end with **End Sub**. Lines beginning with a single apostrophe **(')** are remarks, with text included for those who wish to understand or modify the program. In the example, the lines beginning with an apostrophe give information about when the program was constructed and by whom. They also give information indicating how to modify the program to read RNA sequences. Because such text is completely ignored by the Visual Basic compiler, it can have any desired content.

The first significant line of computer code in the program is **X = Len(Selection.Text)**, which simply finds the length of the selected text and assigns this number to the variable X. (In all cases, the user will select text using the computer mouse.) The code **For Y = 1 To X** is a loop that starts at 1 and ends at **X**, the number of characters selected by the mouse. The line **Select Case Mid$(Selection.Text, Y, 1)** simply takes one character from the selected text each time the loop is executed, starting with the first character and going sequentially to the end of the selection.

The following lines use the **Case** command to compare each selected character with A, T, G or C. **Case A** means that if the character extracted from the selected text is A, then the following lines of code up to the next **Case** are executed. Here, 313.2 Da, the molecular weight of adenine deoxyribose monophosphate minus the weight of one water molecule, is added to **MW**, a variable that stores the molecular weight of the selected sequence. The following lines add the appropriate molecular weight amount in exactly the same way if the selection contains T, G or C bases.

The final **Case** entry, **Case Else**, deals with characters in the selected text that are not A, T, G or C. If a character in the selected text is not A, T, G or C, the code **Z = Z + 1** increases the variable **Z** by one. Since **X** is the total length of the selected text and **Z** is the number of non A, T, G and C characters, **X–Z** corresponds to the number of bases in the selected text, a calculation actually performed later on.

The lines **End Select** end the part of the macro related to the **Select Case** function, and **Next** indicates the end of the **For** loop. The molecular weights used in this part of the program are bases within DNA sequences bound to deoxyribose and phosphate and lack one water molecule when compared to base deoxyribose monophosphates free in solution. However a nucleic acid sequence has a free N- and C-terminus that, together, contain one water molecule. Therefore, the weight of one water molecule must be added to any nucleic acid sequence to get the correct molecular weight. For this reason, the line **MW = MW + 18** is added.

Finally, the results are printed out using the message box function, **If (MW > 18) then MsgBox ("Selection includes " & X-Z & " bases, Molecular Weight= " & MW & "Daltons").** This line prints out the number of bases and the calculated molecular weight of the selected sequence, if at least one A, T, G or C has been detected. The format for the **MsgBox()** function is simply that text to be printed out is in quotation marks, variables are entered directly and the two types of information are separated by the **&** character. The next line, **If (MW = 18) Then MsgBox ("No sequence selected"),** prints out the message that no sequence has been selected if either no text was selected or if the selected text contains no A, T, G or C characters. In either case, **MW** remains at 18.

The final line, **End Sub**, indicates the end of the source code for this macro. This simple program completely ignores any numbers, spaces, table or other formatting characters in the selected text, and in fact, ignores all characters except uppercase A, T, G and C. Database entries or raw sequence data can therefore be examined without reformatting to remove numbers and spaces. To make the program read lowercase characters, simply change the case of the letters searched for by the **Case** function. Alternately, Word has a command in the Format menu that can change the case of selected text from lower to uppercase. To make the macro determine the molecular weight of RNA sequences, it is only necessary to add 16 to the relevant molecular weights associated with A, G and C to account for the extra oxygen atom, to substitute **Case** "**U**" for **Case** "**T**" and finally, to enter 306.1 for the molecular weight of each U base.

To incorporate this program into your personal Word program, go to the Tools menu, select the Macro option and then Macros from the next menu. When another menu appears, type Nucweight (or whatever name you choose) in the Macro Name box. Press the Create button, which takes you to the built-in Visual Basic editor/compiler. Now, either type in the program or paste it in if you obtained it electronically from the Internet so that the lines not starting with the apostrophe are the same as those shown in Figure 1.

After checking the text, go back to Word, open a nucleic acid sequence and select part of it with your mouse. Now go back to the Visual Basic editor and run the macro by pressing the right arrowhead symbol. If the program has been entered correctly, a window displaying the number of bases and the molecular weight of the selected sequence should appear, similar to the one shown for a different macro in Figure 1. If there is a mistake in the program code, the Visual Basic compiler will highlight the line containing the error, which can then be corrected. After exiting the Visual Basic compiler, the program can be run at any time by going to the Tools menu, selecting Macro and from there, selecting Macros. Next, select Nucweight from the list and press the run button.

Though somewhat inconvenient because it requires several keystrokes to activate the macro, this procedure is workable. However, it is much more convenient to modify the menus of Word to include a new pull-down menu for this and other macros. To do this on an IBM PC running Word, select Tools, then Customize and then the Commands tab. Finally, select New Menu, and the text New Menu will appear to the right of the selection

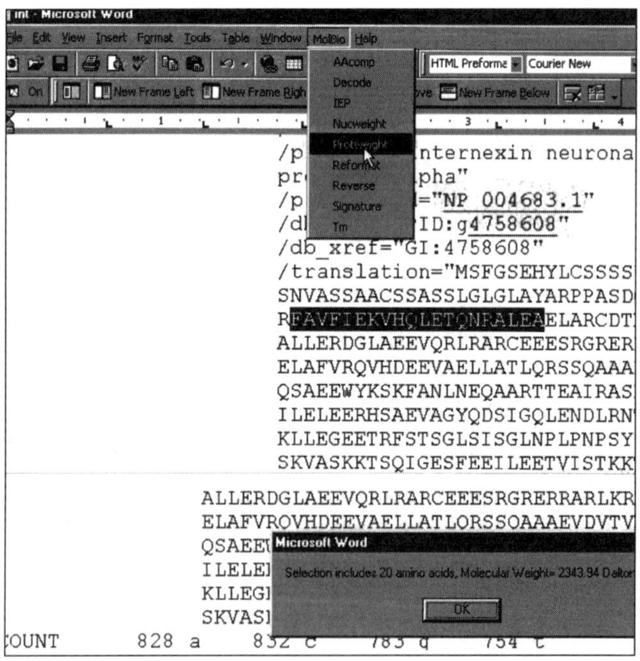

Figure 1. Example of a Visual Basic macro at work in a customized version of Word. The top shows part of the Microsoft Word interface with a new menu item, MolBio, included as described in the text. This menu contains all of the described macros arranged in alphabetical order. In this case, a peptide sequence has been selected with the mouse (outlined in black), and the Protweight macro has also been selected. The lower part of the figure shows the result of this selection, indicating the number of amino acids selected and the exact molecular weight of this sequence.

box. This object can be selected with the mouse and dragged to the desired place on the Word interface where it will appear under the name New Menu. You can change the name by pressing the Modify Selection button and selecting the New Menu name in the properties window with the right mouse button.

As Figure 1 shows, I made a new menu item called MolBio and placed it next to the Help menu, which also includes tabs for the other macros described later. On a Macintosh computer, the procedure is essentially the same except that the New Menu item is added to the various icons below the pull-down menus, rather than among the pull-down menus as shown in Figure 1. To add individual macros to the MolBio menu, go to Tools, select Customize and then Select Macros from the Commands menu. A list of macros will appear and those selected can be dragged and dropped to the MolBio menu.

The macros may have longer names than are convenient for your use—the Nucweight macro was listed as Normal.Autoexec.Nucweight on my PC. However, when this macro is dragged to the MolBio menu, you can select the Modify selection tab and use the right mouse button to change the macro name to Nucweight.

The source codes for this and several other macros can be downloaded from the software library at www.biotechniques.com/ or from www.ufbi.ufl.edu/~shaw/molbio.htm. These can be loaded as a text file into Word, copied and simply pasted into the Visual Basic editor to simultaneously generate all of these macros. The macros can then be incorporated into the menu system as described above.

The first of these other macros, Aacomp, counts the number of each amino acid in a selected sequence and prints out the molar percentage of each. The Decode macro translates the first 24 bases of a selected nucleic acid sequence into the corresponding 7–8 amino acids of protein sequence in all three forward reading frames. It can be easily modified to translate longer regions of nucleic acid sequence, simply by increasing the size of the **outputarray(24)** function (details are included as comments in the appropriate file).

A macro called IEP determines the isoelectric point of a peptide or entire protein selected using the mouse. This program works by first counting how many potentially ionizable (proton-accepting) amino acid residues are present in the selected text and then calculating the total charge of the sequence at pH 2.0. If the charge is positive, as will be the case for almost all peptides, the program then calculates the total charge at pH 2.1, then pH 2.2 and so on, until the charge becomes either zero or negative. The pH at which this occurs is printed on the screen and is a good estimate of the isoelectric point of the sequence. This information predicts where a particular protein will appear on 2-D gels and when it will elute from ion exchange columns.

The Protweight macro is similar to the Nucweight macro described in detail earlier, but it calculates the molecular weight of selected protein sequence, and is shown in operation in Figure 1. The program is useful for rapidly determining the expected SDS-PAGE mobility of recombinant proteins and protein fragments and is particularly useful when studying results from mass spectroscopy experiments. The Reformat macro will remove all numbers, spaces, periods, hyphens and tab- and line-feed characters from a selected sequence. It is useful for preparing database sequences for manipulation in Word.

The Reverse macro will print the antisense sequence of the 3′ 30 bases of selected nucleic acid sequence on the screen. It is printed in the 5′ to 3′ orientation and is useful for the construction of 3′ oligonucleotide primers for PCR and other purposes. The Signature macro simply prints out the author's name, affiliation, address, telephone, fax and e-mail information, saving much time when writing letters.

Finally the Tm macro calculates the length, GC content and melting temperature of selected nucleotide sequences. The program makes use of two different methods for calculating the melting temperature (6). The first is appropriate for longer nucleic acid sequences and is from the work of McCarthy et al. (4); the second, from Itakura and coworkers (7), is appropriate for sequences up to 18 bases. Selecting which method to use is determined by the number of bases in the selected nucleic acid sequence. This macro allows the design of compatible oligonucleotides for sequencing and PCR applications.

DISCUSSION

The methods presented here are simple examples of what can be accomplished using Visual Basic combined with Microsoft Word. Obviously, with an understanding of the Visual Basic programming language, these macros can be modified to perform other useful functions, and macros for other purposes can also be designed. This approach has several obvious advantages over other methods. First, it is free for the majority of us who already have Microsoft Word or a similar program. Second, with a little knowledge of Visual Basic, the user can see exactly what the program is doing and easily modify it—not usually the case with commercial packages. Third, the macros run in Microsoft Word exactly the same on IBM, IBM-compatible PCs and Macintosh computers, so that they should be useful for collaborations or teaching across these two platforms. The macros described here can be downloaded from the software library at www.biotechniques.com/ or from www.ufbi.ufl.edu/~shaw/molbio.htm.

Because the entire suite of the most recent Microsoft Office programs uses the Visual Basic programming language, it is also possible to use similar macros for Excel, PowerPoint and other Microsoft Office programs. With more sophisticated programming, it should be possible to automatically generate graphs and figures from sequence data in Word documents. In fact, previous papers show how it is possible to use Microsoft Excel to perform quite complex types of protein and nucleic acid sequence analysis and plot out the results (1,2,4,5). Hopefully, this article will stimulate others to create and make available programs that use Microsoft Office's versatile suite of programs.

ACKNOWLEDGMENT

This work was supported in part by a grant from the American Heart Association.

REFERENCES

1. **Anbazhagan, R. and E. Gabrielson.** 1999. Spreadsheet-based program for alignment of overlapping DNA sequences. BioTechniques *26*:1180-1185.
2. **Han, B. and A.H. Tashjian, Jr.** 1998. User-friendly and versatile software for analysis of protein hydrophobicity. BioTechniques *25*:256-263.
3. **McConaughy, B.L., C.D. Laird and B.J. McCarthy.** 1969. Nucleic acid reassociation in formamide. Biochemistry *8*:3289-3295.
4. **McEwan, N.R. and D. Gatherer.** 1998. Adaptation of standard spreadsheet software for the analysis of DNA sequences. BioTechniques *24*:131-138.
5. **Shaw, G.** 1996. Protein sequence interpretation using a spreadsheet program. BioTechniques *19*:978-983.
6. **Wahl, G.M., S.L. Berger and A.R. Kimmel.** 1987. Molecular hybridization of immobilized nucleic acids: theoretical concepts and practical considerations. Methods Enzymol. *152*:399-407.
7. **Wallace, R. B., J. Shaffer, R. F. Murphy, J. Bonner, T. Hirose and K. Itakura.** 1979. Hybridization of synthetic oligodeoxyribonucleotides to phi chi 174 DNA: the effect of single base pair mismatch. Nucleic Acids Res. *6*:3543-3557.

Update to: Useful Microsoft® Word™ Macros for Molecular Biologists and Protein Chemists

Gerry Shaw

University of Florida, Gainesville, FL, USA (shaw@mbi.ufl.edu)

This paper built on my interest in using the Microsoft Office® program suite to analyze various features of amino acid and protein sequences. As with a previous paper (2), also included in this volume, there was quite a lot of e-mail and other correspondence and presumably therefore some interest in the method. Some trivial mistakes in the original macros were pointed out by various correspondents and have been corrected in the versions posted on the Internet, and some other macros not discussed in the paper are also posted on the site (http://www.mbi.ufl.edu/~shaw/comp.htm). My e-mail address is now shaw@mbi.ufl.edu.

Some other workers have written papers apparently stimulated by the paper to produce Microsoft® Word® Macros which perform various other functions. For example, Singal and Grimes recently described the use of a macro to analyze cytosine methylation reaction on CpG sequences (3). This analysis is quite complex because methylated cytosine does not react with bisulphite, while unmethylated cytosine is transformed into uracil. This means that each and every CpG may potentially but not necessarily be transformed into a UpG. Since the state of methylation of a CpG on the forward strand may or may not be the same as that on the reverse strand, the two strands may be no longer complementary following the bisulphite reaction. The macro also shows restriction sites that include the CpG sequences, which can be used to cleave a PCR product to see if the relevant CpG is present (indicating methylation at that position) or not (indicating modification by bisulphite and hence lack of methylation). The authors will supply this to interested parties by email (srgymes@prysm.net).

Muller et al. described a whole suite of programs that perform the functions outlined in my original paper and many others as well (1). This suite of programs runs from a menu bar that is easy to install and has some significant advantages over my version. For example, it is possible to paste the results of calculations directly into a Word document. This can be downloaded in the file BIOCOMPUTINGSOFTWARE.ZIP from the BioTechniques software library at http://www.biotechniques.com. It is also possible to edit or alter the versions of the macros downloaded from this site, so the user can modify or customize them in any desired fashion.

References

1. **Muller, P.Y., E. Studer and A.R. Miserez.** 2001. Molecular Biocomputing Suite: a word processor add-in for the analysis and manipulation of nucleic acid and protein sequence data. BioTechniques *31*:1306-1313.
2. **Shaw, G.** 1996. Protein sequence interpretation using a spreadsheet program. BioTechniques *19*:978-983.
3. **Singal, R. and S.R. Grimes.** 2001. Microsoft Word macro for analysis of cytosine methylation by the bisulfite deamination reaction. BioTechniques *30*:116-120.

CHAPTER 3

Adaptation of Standard Spreadsheet Software for the Analysis of DNA Sequences

Neil R. McEwan and Derek Gatherer[1]

Scottish Agricultural College, Aberdeen, Scotland and
[1]John Moores University, Liverpool, England, UK

Reprinted from BioTechniques 24:131-138 (January 1998)

ABSTRACT

This paper presents use of spreadsheet software to derive various statistics on nucleotide and codon distribution and frequency from gene sequence data, including χ^2 tests and their derivatives, and proportion of nucleotides in the third position. The basic principles can be easily extended to other more complex functions. In addition, it can be used to translate a nucleic acid sequence into a protein sequence or to produce a codon usage table. This adaptation permits sequence analysis without expensive dedicated software and easy modification of the analysis according to the user's requirements. It is also ideal for introducing biological science students to the programming potential of spreadsheets.

INTRODUCTION

The ever increasing number of DNA sequences being deposited in various databases necessitates the easy availability of simple computer programs for sequence analysis. Frequently, a problem arises in finding a program able to perform the exact type of analysis required by the experimenter. To most molecular biologists, untrained in computer programming languages, the prospect of having to write a new program or even alter an existing one is a daunting task. A solution to this problem may lie in use of the common spreadsheet program such as Microsoft® Excel®, as suggested by Shaw (9), whose original paper adapted Excel for the analysis of protein sequences. Here, we present an Excel adaptation that can be used for the analysis of DNA sequences, generating a range of statistics on nucleotide and codon usage, in addition to performing translation. Further modification of the program to perform more complex functions requires only a familiarity with Excel.

BioComputing: Computer Tools for Biologists
Edited by Stuart M. Brown

MATERIALS AND METHODS

Computer Programs Used

The programs used in this work are those found already installed in a wide range of IBM®-compatible personal computers—Netscape to allow access to the Internet, Microsoft Word™ Version 6.0 (Microsoft, Redmond, WA, USA) as a word-processing program and Microsoft Excel Version 5.0 as a spreadsheet program. However, other Internet viewers, spreadsheet packages and word-processing documents can be used instead.

Source of DNA Sequences

DNA sequences for input can be accessed and downloaded directly from databases available on the Internet or can be copied directly from sequence information generated by an automated sequencer. DNA sequence information for testing the program was accessed from the GenBank® database (**http://www2.ncbi.nlm.nih.gov/cgi-bin/genbank**). Sequences from GenBank were accessed using Netscape and copied into a word-processing document (Windows® Version 6.0 of Microsoft Word) by "*Copy*"ing and "*Paste*"ing.

Formatting the Sequence for Analysis into a Format for the Excel Spreadsheet

Once in the Word document, all numbers can be removed by selecting the *Replace* function within the *Edit* menu on the toolbar. This was done by selecting the *Any Digit* option from the *Special* option in the *Replace* function, while leaving the *Replace With* option blank, followed by selecting the *Replace All* button. All spaces within the text were removed in a similar manner. For GenBank sequences downloaded in their entirety as Word files, all the descriptive information must also be removed. In addition, all *Paragraph Mark* symbols at the end of lines should be removed, by using the *Special* button within the *Replace* dialogue box. All untranslated 5′ and 3′ sequences must also be excised. Most GenBank entries give details as to the precise locations of the translational start and stop codons and also splice junctions where relevant.

A comma should be inserted before each nucleotide by using the *Replace* function within the *Edit* menu on the toolbar to replace 'A' with 'A,' etc. This should result in the type of appearance shown in Figure 1. The sequence is then put into a table format with three letters per row. This is done by highlighting the text using the mouse and selecting the *Convert Text to Table* option from the *Table* option on the menu

a,t,g,g,g,a,c,t,g,g,a,g,g,a,c,g,a,g,c,a,a,a,a,g,a,t,g,c,t,t,a,c,c,g,a,a,t,c,c,g,g,a,g,a,t,c,c,t,g,a,g,g,a,g,g,a,g,g,a,a,g,a,g,g,a,a,g,a,g
,g,a,g,g,a,a,t,t,a,g,t,g,g,a,t,c,c,c,c,t,a,a,c,a,a,c,a,g,t,g,a,g,a,g,a,g,c,a,a,t,g,c,g,a,g,c,a,g,t,t,g,g,a,g,a,a,a,t,g,t,g,t,a,a,a,g,g,c,c,c
,g,g,g,a,g,c,g,g,c,t,a,g,a,g,c,t,c,t,g,t,g,a,t,g,a,g,c,g,t,g,a,t,t,c,c,t,c,t,c,g,a,t,c,a,c,a,t,a,c,a,g,a,a,g,a,g,g,a,t,t,g,c,a,c,g,g,a,g,g,a,
g,c,t,c,t,t,t,g,a,c,t,t,c,t,t,g,c,a,t,g,c,g,a,g,g,g,a,c,c,a,t,t,g,c,g,t,g,g,c,c,c,a,c,a,a,a,c,t,c,t,t,t,a,a,c,a,a,c,t,t,g,a,a,a,t,a,a,

Figure 1. A typical example of the first 275 nucleotides of a DNA sequence after it has been edited to remove all numbers, and a comma has been inserted in front of each letter.

bar. From this menu, set the number of letters per row to three by typing "3" in the *Number of Columns* box. At this point, the *Separate Text* indicator should be on *Commas*, but if necessary, this should be selected. Then select *OK*. The resulting 3-column table can be highlighted, copied and pasted directly into Excel. Figure 2 gives an example. The 3-column table is the preferred format for those calculations, which are position-specific, e.g., the %A at the first position, etc. as described below. For this type of calculation, the material is ready to be pasted into the spreadsheet program.

However, for calculation of codon usage and derivatives of these, the three columns must be converted to a single-column table. This may be done by highlighting the table and selecting the *Convert Table to Text* function from the *Table* menu. The *Other* option should be selected, and in the box to the right of this, enter a space. Selecting *OK* will produce text where there are three letters per row, and they are separated by spaces. The spaces should be removed by the *Replace* function as already described. This will result in the sequence taking the form as shown in Figure 3.

EXAMPLES OF POTENTIAL APPLICATIONS

Generally, we have found it best to insert descriptive information into the first row (cells A1, B1, C1 etc.) and enter all sequence information below commencing in the second row (cells A2, B2, C2 etc.). In addition, all calculations are based on a sequence of 1000 codons (finishing in row 1001), but altering the range of the number of codons counted can be done by changing the equation to count either a greater or lesser number of cells. This principle can also be applied to all of the applications listed below.

a	t	g
g	g	a
c	t	g
g	a	g
g	a	c
g	a	g
c	a	a
a	a	g
a	t	g
c	t	t
a	c	c
g	a	a
t	c	c
g	g	a
g	a	t
c	c	t
g	a	g
g	a	g
g	a	g
g	a	a
g	a	g
g	a	a
g	a	g
g	a	g
g	a	a

Figure 2. The first 25 codons of the sequence shown in Figure 1, after the *Convert Text to Table* function has generated a 3-column table. This may be copied directly into Excel and is the preferred format for position-specific calculations, e.g., %C in the 2nd position, etc.

Translating DNA Sequences

This uses a translation table that is entered in the spreadsheet as two adjacent columns, e.g., columns R and S. Column R contains the 64 different trinucleotide sequences (e.g., TTT, TTC etc.). Column S contains the appropriate amino acid encoding this sequence (e.g., Phe, Leu etc. or K, L etc.). The sequence, in single-column format (see Figure 3), is entered on the left of the spreadsheet, e.g., in column A. The other two alternate reading frames can be calculated at this time. This is done by removing the first nucleotide of the original DNA sequence for frame two or the first two nucleotides for frame three. The truncated sequences should be treated in the manner described above and can be inserted into columns B and C, respectively. The three reading frames can be translated in columns E, F and G, respectively. If the first triplet of the first reading frame is in cell A2, then the command to translate it in cell E2 is:

```
=VLOOKUP(A2,$R$2:$S$65,2, FALSE)
```

where the translation table covers rows 2–65 in columns R and S. This command should then be copied in all cells in column E:

```
=VLOOKUP(A3,$R$2:$S$65,2,FALSE)
=VLOOKUP(A4,$R$2:$S$65,2,FALSE)
=VLOOKUP(A5,$R$2:$S$65,2,FALSE)
```

and column E then copied in its entirety into columns F and G:

```
=VLOOKUP(B2,$R$2:$S$65,2,FALSE)
=VLOOKUP(B3,$R$2:$S$65,2,FALSE)
=VLOOKUP(B4,$R$2:$S$65,2,FALSE)
:
:
:
=VLOOKUP(C2,$R$2:$S$65,2,FALSE)
=VLOOKUP(C3,$R$2:$S$65,2,FALSE)
=VLOOKUP(C4,$R$2:$S$65,2,FALSE)
```

This results in the contents of the cell of interest from column A being compared with the 64 cells in column R. When an identical match is made between two cells, then the corresponding cell in column E is assigned the value of the cell in column S. This allows a

atg
gga
ctg
gag
gac
gag
caa
aag
atg
ctt

Figure 3. The first 10 codons of the same sequence after creating single columns of three letters. This is the preferred format for entry to Excel for calculations of codon usage and derivatives thereof.

sequence to be translated. In the case of translations being performed for all three reading frames, this takes the form shown in Figure 4.

Counting Stop Codons

In an appropriate cell, for instance I2, enter the command:

=COUNTIF($E$2:$E$1001,"stop")

This can be "pasted" into cells J2 and K2 to display the number of stop codons per reading frame. It will check each of the first 1000 codons to look for STOP codons (TAG, TGA or TAA). The first reading frame should have a single stop codon, but several may appear in the two others. This step is non-essential, but it may give an indication of any errors that have occurred in the process of copying and pasting the data into Excel.

Creating a Codon Usage Table

The codon usage table is an essential element in the analysis of codon bias (8). This method uses the single-column format (Figure 3). The frequency of each codon in the open reading frame (ORF) (and the others if necessary) can be calculated using the COUNTIF function, and the codon usage table displayed in columns M and N as shown below:

Cell	Entry
M2	AAA
N2	=COUNTIF($A$2:$A$1001,M2)
M3	AAC
N3	=COUNTIF($A$2:$A$1001,M3)

Completing the codon set in column M and copying the above commands in column N generates a codon usage table for the first reading frame, for the first 1000 codons. Triplet usage tables for other reading frames can be generated by copying column N in its entirety into columns O and P and modifying it accordingly. For example, in cell P2, the command used would read:

=COUNTIF($C$2:$C$1001,M2) etc.

At first glance, this process may seem rather laborious, as it appears to require entering data in 128 different cells. However, it can be performed more rapidly in the case of column N by entering the value in cell N2 as described and then highlighting both it and the remaining cells to be filled. From the *Edit* option on the toolbar, select *Fill*, and from the submenu, select *Down*. This will complete the cells more rapidly.

Met	Trp	Gly
Gly	Asp	Thr
Leu	Trp	Gly
Glu	Arg	Gly
Asp	Thr	Arg
Glu	Ser	Ala
Gln	Lys	Lys
Lys	Arg	Asp
Met	Cys	Ala
Leu	Leu	Tyr
Thr	Pro	Arg
Glu	Asn	Ile
Ser	Pro	Arg
Gly	Glu	Arg
Asp	Ile	Ser

Figure 4. Translation of the first 15 codons of the above sequence in the 3 reading frames.

Calculating the Frequency of Each Nucleotide in Each Position

There are two methods that can be used to calculate the frequency of a nucleotide in a given position. The strategy to be used depends on which type of further calculations will be performed.

Calculating the frequency using 3-column calculations. The simplest method of performing this calculation is using 3-column sequence input (see Figure 2), and it can be entered in columns U, V and W. This is generally the format used where the result is an end in itself, and no additional calculations will be performed on the data.

Using this method, to calculate the frequency of adenine residues in the first position, enter the command:

```
=COUNTIF(U2:U1001,"a")/COUNTA(U2:U1001)
```

in a chosen cell. The COUNTIF numerator in this function counts the number of adenine residues in column U, and the COUNTA denominator counts all residues. To count the proportion of cytosine, the command is

```
=COUNTIF (U2:U1001,"c")/COUNTA(U2:U1001)
```

likewise repeated for guanine (g) and thymine (t). For the second position, the commands are the same except that the array U2:U1001 becomes V2:V1001 and likewise becomes W2:W1001 for the third position. This generates four values of nucleotide frequency for each position, and these can be combined to generate other statistics. For instance, the frequency of adenine and the frequency of guanine in the third position can be combined to generate the total purine frequency in the third position etc. The total frequency of nucleotides in the reading frame can be calculated by expanding the counted array. For instance, the command:

```
=COUNTIF(U2:W1001,"a")/COUNTA(U2:W1001)
```

gives the total frequency of adenine residues in all three positions.

GC3s is the total proportion of G and C in the third position (7) and can be calculated using the command:

```
=(COUNTIF(W2:W1001,"c")+COUNTIF(W2:W1001,"g"))/COUNTA(W2:W1001)
```

Calculating the frequency using single-column calculations. The method shown in the section above is the most simple means of calculating the frequency of nucleotides in a given position, but it is not easy to apply this method to some of the calculations described below. In the case of calculations used to work out further values, such as the expected frequency of a given codon (see *Calculating Expected Values of Codons*), it is better to use the method described in Figure 5. Briefly, this method sums the number of times a given codon occurs where the nucleotide of interest is contained in the position of interest. For example, the abundance of the A in position 1 would be the sum of all the codons that have A in their first position (i.e., AAA, AAC, AAG, AAT, ACA.....etc). To calculate the abundance of each nucleotide across the gene, this is just a simple matter of summing its occurrence in positions 1, 2 and 3. From here, to calculate the frequency of a nucleotide, it is simply a matter of dividing the total by the number of times that the nucleotide could possibly occur (e.g., for positional calculations, it would be a case of dividing the total number of A nucleotides in position 1, AXX, by the total number of nucleotides present in all of the first positions, the sum of AXX, CXX, GXX and TXX).

Calculating Expected Values of Codons

Derivation of codon bias indices requires that observed codon frequencies be compared with expected frequencies. The expected frequencies can be entered next to the observed, for instance, in column Q. There are three methods of deriving expected frequencies, as follows:

Elementary calculations based on assuming 25% frequency of each nucleotide. The simplest, if most naive, method of calculating expected values for each codon is to assume that each nucleotide occurs with a frequency of 0.25 [i.e., it assumes that the reading frame has DNA arranged in a random manner (5)]. In this case, there is an equal likelihood of each codon, and the expected value is therefore calculated by entering the command:

=SUM(N2:N65)/64

in all cells of column Q. Although this is the easiest method, it takes no account of the amino acid composition of the gene and that there can only be one stop codon per ORF.

Calculations based on the frequency of each nucleotide across the gene. If the frequency of each nucleotide is known, a value can be given for expected trinucleotide frequency based upon this. Figure 5 and the Section *Calculating the Frequency Using Single-Column Calculations* demonstrate the best method to perform the calculations to determine the frequency of each nucleotide.

Thus, the expected abundance of the trinucleotide ATG would be the frequency of A, multiplied by the frequency of T, multiplied by the frequency of G, multiplied by the total number of codons.

Calculations based on the amino acid frequency. The above method does not take account of the amino acid sequence. In the case of 2-fold synonymous groups such as lysine (AAA and AAG), the expected values would be calculated as follows (based on the functions described in the Section *Creating a Codon Usage Table*):

=M2/(M2+M4) and =M4/(M2+M4)

This assumes that the expected value of each of the two codons used is 50% of the total value of number of times that the amino acid is present. These equations could be filled into the cells AA2 and AA4 for AAA and AAG, respectively, as described in the section

Cell	Entry	Comments
M67	=SUM(N2:N17)	Adds the number of "A" nucleotides in position 1
M68	=SUM(N18:N33)	Adds the number of "C" nucleotides in position 1
:		
:		
:		
M75	N2+N6+N10+N14+N18+N22+ N26+N30+N34+N38+N42+N46 +N50+N54+N58+N62	Adds the number of "A" nucleotides in position 3
:		
:		
:		

Figure 5. Some examples of the entries into cells as a means of calculating the abundance of each nucleotide at a given position.

above. In a similar way, the equation for the expected values for asparagine encoding codons would be:

=M3/(M3+M5) and =M5/(M3+M5)

and these would be inserted into cells AA3 and AA5 for AAC and AAT, respectively.

This is the method used for calculating expected codon frequency when deriving Scaled chi-square (10).

Calculations for a coding region on the basis of frequency of the third nucleotide. A more accurate method of calculating expected codon usage is to bias expected values to reflect the composition of the third nucleotide in all of the codons. For example, the expected values of AAA and AAG would now become:

Cell AA2 =(M2/(M2+M4))×(M75/(M75+M77))

Cell AA4 =(M4/(M2+M4))×(M77/(M75+M77))

where M2+M4 is the total number of lysine residues, and M75/(M75+M77) is the proportion of A nucleotides used in position 3, relative to the total codons using either A or G in position 3 (see Figure 5).

Other frequencies can be calculated in a similar manner (e.g., AAC):

Cell AA3 =(M2/(M2+M4))×(M76/(M76+M78))

Calculation of the Difference Between Observed and Expected Values and Chi Squares

Differences between expected and observed values can be calculated by inserting the following equations in column AB:

Cell AB2 =(M2-AA2)
Cell AB3 =(M3-AA3)
Cell AB4 =(M4-AA4)

Chi-squared values for each codon can be calculated by filling in the following equations to column AC:

Cell AC2 =((M2-AA2)^2)/AA2
Cell AC3 =((M3-AA3)^2)/AA3
Cell AC4 =((M4-AA4)^2)/AA4

The chi-squared total can be calculated by using the following equation:

Cell AC66 =SUM(AC2:AC65)

This is then divided by twice the total number of codons to generate Scaled chi-square (10), using for instance the command:

Cell AC67 =AC66/((COUNTA(A2:A1001))×2)

Calculating Frequency of Other Sequences

Recently, the importance of also studying sequences that span codons was highlighted by the observation of J.M. Smith and N.H. Smith (6) that the sequence AG-G (where "-" denotes the division between two consecutive codons) was selected against. Other selective constraints on sequences spanning two codons are at present unidentified. The sequence AG-G appears as a triplet in the second reading frame, so its frequency can be calculated by applying the techniques described above for calculating trinucleotide sequences listed in column A of the single-column calculations to column B trinucleotide sequences.

Microsoft Excel can be used to calculate the relative frequency of any given sequence within the gene, in any reading frame. The same method is used as for triplet sequences; however, the column number is varied, and there are more reading frames. For instance, when examining the frequency of occurrence of hexanucleotides, there are six reading frames, and a 6-column table must initially be generated. This can be compressed to a single column as above.

Educational Use of Spreadsheet Programs

The use of the personal computer (PC) as an educational tool in biological science courses is almost as old as the PC itself (2). Although customization of spreadsheet programs is not computer programming in the strict sense, a case can be made that it is a transferable skill in that science graduates will encounter spreadsheet programs and more general use of PCs in many of their subsequent careers. Programs of the type presented here, in combination with access to sequence databases through Netscape, can be used to introduce undergraduates to basic keyboard and spreadsheet skills, to assist in Internet familiarization and to introduce various concepts in bioinformatics, such as the nucleotide and codon usage statistics described above. Once students are familiar with the mechanism by which the program produces third-nucleotide usage in codons (GC3s) (7) etc., they may be given the opportunity to derive other sequence-based statistics, such as effective codon number (11) or P2 (4). Students could quickly begin to adapt the programs to perform sequence analyses of their own choice, leading to what Belbin and Belbin (1) have described as "discovery learning", which has the advantage of greater long-term retention than more traditional teacher-centered delivery.

CONCLUSIONS

The methods described here can be used to allow relatively easy analysis of DNA sequences without the need for expensive commercial software. They are relatively simple to design and can be customized to the requirements of a single experimenter, without the need for any computer programming per se. In addition to the simple nucleotide and chi-squared calculations described here, we have used this type of spreadsheet for analyzing DNA for a number of more complex statistics, e.g., effective codon number (11) and intrinsic codon deviation index (3).

REFERENCES

1. **Belbin, E. and M. Belbin.** 1969. The Discovery Method in Training. HMSO, London.
2. **Bryce, C.F.A.** 1981. Computer-based learning in biochemistry, p. 13. *In* C.F.A. Bryce (Ed.), Biochemical Education. Croom Helm, London.
3. **Freire-Picos, M.A., M.I. Gonzalez-Siso, E. Rodriguez-Belmonte, A.M. Rodriguez-Torres, E. Ramil and M.E. Cerdan.** 1994. Codon usage in *Kluyveromyces lactis* and in yeast cytochrome c-encoding genes. Gene *139*:43-49.
4. **Gouy, M. and C. Gautier.** 1982. Codon usage in bacteria—correlation with gene expressivity. Nucleic Acids Res. *10*:7055-7074.
5. **Konopka, A.K.** 1994. Sequences and codes: fundamentals of biomolecular cryptology, p. 119-174. *In* D.W. Smith (Ed.), Biocomputing: Informatics and Genome Projects. Academic Press, San Diego.

6.**Maynard Smith, J. and N.H. Smith.** 1996. Site-specific codon bias in bacteria. Genetics *142*:1037-1043.
7.**Sharp, P.M. and K.M. Devine.** 1989. Codon usage and gene expression level in *Dictyostelium discoideum*: highly expressed genes do 'prefer' optimal codons. Nucleic Acids Res. *17*:5029-5039.
8.**Sharp, P.M., M. Stenico, J.F. Peden and A.T. Lloyd.** 1993. Codon usage: mutational bias, translational selection, or both? Biochem. Soc. Trans. *21*:835-841.
9.**Shaw, G.** 1995. Protein sequence interpretation using a spreadsheet program. BioTechniques *19*:978-983.
10.**Shields, D.C. and P.M. Sharp.** 1987. Synonymous codon usage in *Bacillus subtilis* reflects both translational selection and mutational biases. Nucleic Acids Res. *15*:8023-8040.
11.**Wright, F.** 1990. The 'effective number of codons' used in a gene. Gene *87*:23-29.

Update to: Adaptation of Standard Spreadsheet Software for the Analysis of DNA Sequences

Neil R. McEwan[1] and Derek Gatherer[2]

[1]Rowett Research Institute, Aberdeen, Scotland and [2]RiboTargets, Great Abington, Cambridge, England (nrm@rri.sari.ac.uk)

Our original paper (3) had its origin in our work on codon bias in a variety of unusual microorganisms, principally of agricultural or agrobiotech interest. At that time (mid-1997), there were few freestanding codon bias applications available as academic free-

```
1.  From the commands at the top of the screen select the
    Tools option.
2.  From the choices which scroll down select the Macro
    option.
3.  A second choices menu will then appear to the right of the
    first scroll. From this menu select the Record New Macro
    option.
4.  From the box which appears insert the required name in the
    Macro name box.
5.  Click on the OK box.
6.  A new icon will appear on the screen. The icon should have
    the word "Stop" written within the text at the top of the
    box. There should also be two icons displayed within the
    box: "■" and "|| ●"
7.  Carry out the actions which you wish to be "recorded" in
    the macro (e.g. removing numbers or blank spaces, addition
    a tab sign before a particular letter - as described in
    the original paper, etc.
8.  Press the "■" option on the Stop recording icon. This
    should stop the recording of the macro, and Word will now
    save the newly created macro to its list of macros.
9.  As a final precaution it is worth checking that the new
    macro has been saved, this can be done by selecting the
    Tools option again.
10. From the choices which scroll down select the Macro
    option.
11. A second choices menu will then appear to the right of the
    first scroll. From this menu select the Macros option.
12. The name of the newly created macro should feature in the
    large box on the left of the block of options which
    appears.
```

Figure 1. Running a macro in Microsoft Word.

```
Sub Trinucleotides()
'
' Trinucleotides Macro
' Macro recorded 19/01/02 by Neil McEwan
'
    Selection.HomeKey Unit:=wdStory
    Selection.Find.ClearFormatting
    Selection.Find.Replacement.ClearFormatting
    With Selection.Find
        .Text = " "
        .Replacement.Text = ""
        .Forward = True
        .Wrap = wdFindContinue
        .Format = False
        .MatchCase = False
        .MatchWholeWord = False
        .MatchWildcards = False
        .MatchSoundsLike = False
        .MatchAllWordForms = False
    End With
    Selection.Find.Execute Replace:=wdReplaceAll
    Selection.Find.ClearFormatting
    Selection.Find.Replacement.ClearFormatting
    With Selection.Find
        .Text = "^p"
        .Replacement.Text = ""
        .Forward = True
        .Wrap = wdFindContinue
        .Format = False
        .MatchCase = False
        .MatchWholeWord = False
        .MatchWildcards = False
        .MatchSoundsLike = False
        .MatchAllWordForms = False
    End With
    Selection.Find.Execute Replace:=wdReplaceAll
    Selection.Find.ClearFormatting
    Selection.Find.Replacement.ClearFormatting
    With Selection.Find
        .Text = "^#"
        .Replacement.Text = ""
        .Forward = True
        .Wrap = wdFindContinue
        .Format = False
        .MatchCase = False
        .MatchWholeWord = False
        .MatchWildcards = False
        .MatchSoundsLike = False
        .MatchAllWordForms = False
    End With
    Selection.Find.Execute Replace:=wdReplaceAll
    Selection.Find.ClearFormatting
    Selection.Find.Replacement.ClearFormatting
    With Selection.Find
```

Figure 2. The text of a typical macro.

```
        .Text = "a"
        .Replacement.Text = "^ta"
        .Forward = True
        .Wrap = wdFindContinue
        .Format = False
        .MatchCase = False
        .MatchWholeWord = False
        .MatchWildcards = False
        .MatchSoundsLike = False
        .MatchAllWordForms = False
    End With
    Selection.Find.Execute Replace:=wdReplaceAll
    With Selection.Find
        .Text = "c"
        .Replacement.Text = "^tc"
        .Forward = True
        .Wrap = wdFindContinue
        .Format = False
        .MatchCase = False
        .MatchWholeWord = False
        .MatchWildcards = False
        .MatchSoundsLike = False
        .MatchAllWordForms = False
    End With
    Selection.Find.Execute Replace:=wdReplaceAll
    With Selection.Find
        .Text = "g"
        .Replacement.Text = "^tg"
        .Forward = True
        .Wrap = wdFindContinue
        .Format = False
        .MatchCase = False
        .MatchWholeWord = False
        .MatchWildcards = False
        .MatchSoundsLike = False
        .MatchAllWordForms = False
    End With
    Selection.Find.Execute Replace:=wdReplaceAll
    With Selection.Find
        .Text = "t"
        .Replacement.Text = "^tt"
        .Forward = True
        .Wrap = wdFindContinue
        .Format = False
        .MatchCase = False
        .MatchWholeWord = False
        .MatchWildcards = False
        .MatchSoundsLike = False
        .MatchAllWordForms = False
    End With
    Selection.Find.Execute Replace:=wdReplaceAll
    With Selection.Find
        .Text = "n"
```

Figure 2. The text of a typical macro (Continued).

ware, and most of those proved to be difficult to implement on our systems. We had been experimenting with the use of the excellent high-throughput API (Applications Programming Interface) provided by the BioPerl project (http://www.bioperl.org), but also found it, at that time, to have little functionality directable towards codon bias analysis. The inspiration to use spreadsheets came from a previous article in *BioTechniques*, which dealt with protein sequence interpretation using a spreadsheet program (8). Reading Shaw's paper, it became apparent to us that spreadsheets were ideal in cases where: (*i*) the quantity of sequence to be analyzed is limited; (*ii*) all the analysis is performed synchronously (i.e., there are no iterative procedures or recursive methods); (*iii*) graphical output is

```
        .Replacement.Text = "^tn"
        .Forward = True
        .Wrap = wdFindContinue
        .Format = False
        .MatchCase = False
        .MatchWholeWord = False
        .MatchWildcards = False
        .MatchSoundsLike = False
        .MatchAllWordForms = False
    End With
    Selection.Find.Execute Replace:=wdReplaceAll
    Selection.Delete Unit:=wdCharacter, Count:=1
    Selection.WholeStory
    Selection.ConvertToTable Separator:=wdSeparateByTabs,
NumColumns:=3, _
        NumRows:=100, Format:=wdTableFormatNone, ApplyBorders:=True,
ApplyShading _
        :=True, ApplyFont:=True, ApplyColor:=True,
ApplyHeadingRows:=True, _
        ApplyLastRow:=False, ApplyFirstColumn:=True,
ApplyLastColumn:=False, _
        AutoFit:=False
    Selection.Rows.ConvertToText
Separator:=wdSeparateByDefaultListSeparator
    Selection.HomeKey Unit:=wdStory
    Selection.Find.ClearFormatting
    Selection.Find.Replacement.ClearFormatting
    With Selection.Find
        .Text = " "
        .Replacement.Text = ""
        .Forward = True
        .Wrap = wdFindContinue
        .Format = False
        .MatchCase = False
        .MatchWholeWord = False
        .MatchWildcards = False
        .MatchSoundsLike = False
        .MatchAllWordForms = False
    End With
    Selection.Find.Execute Replace:=wdReplaceAll
End Sub
```

Figure 2. The text of a typical macro (Continued).

desired; and (*iv*) the user desires to make small changes to the algorithm and observe the results instantaneously. The last point is particularly important for codon bias in our rare microorganisms, in which unconventional genetic codes may be operating, and there are a plethora of often only subtly variant indices for measurement of codon bias.

Since publication of our paper, we have been pleased to come across reference to it in a variety of contexts, where statistical analysis is being performed on single genes (micro-bioinformatics, as we call it). The fields in which it has been used include primer design (11), sequencing (2), sequence alignment (1), evolutionary biology (10), and also in our own agrobiotech work (4–7). We are also pleased to see further work in this area by our original inspiration, Gerry Shaw (9).

However, when applying the spreadsheet analysis to larger datasets, particularly in the case of applying this work to a large number of putative genes or open reading frames (ORFs) (e.g., from part of a genome sequencing project), it became clear that the process was becoming rather repetitive and tedious. To this effect, the analytical process was customized into a macro that could run within the original Microsoft® Word® document. The user simply "records" the tasks in a step-wise manner, and Word does the rest. A description of how to run a macro within word is shown in Figure 1.

The typical text of a simple macro viewed in Visual Basic is shown in Figure 2 and can be relatively easily changed (see following paragraph). This program: (*i*) returns the cursor to the start of the text; (*ii*) removes all the numbers, blank spaces, and end stops from the text; (*iii*) places a tab before letters A, C, G, T, and N, although additional letters could be added; (*iv*) removes the tab from in front of the first letter; (*v*) forms a table that has three characters per line; (*vi*) converts the newly formed table back into text, where the letters on each line are separated by blank space; and (*vii*) removes the final space from between the letters on each line. In summary, this program can be used to take a piece of DNA sequence data copied into a Word document from the GenBank® database and converts it into blocks of codon-sized pieces.

As stated above, having recorded a macro, making changes to it is a relatively simple matter. Figure 3 demonstrates how to access a Visual Basic editing facility that can be used to edit any part of the macro. Having made the necessary alterations to the file, it can either be saved as a new macro or saved using the same name. The macro is simply a block of different subroutines, in which the commands being carried out are actually fairly similar to those that could be used to describe them in normal English.

```
1.   From the commands at the top of the screen select the
     Tools option.
2.   From the choices which scroll down select the Macro
     option.
3.   A second choices menu will then appear to the right of the
     first scroll.  From this menu select the the Macros
     option.
4.   From the box which appears select the insert the Edit
     option.
5.   The contents of the program should now appear in a Visual
     Basic format on the right of the screen.  At this point
     any commands can be changed from within the program.
```

Figure 3. Preparing to edit a macro program.

While we have continued to use spreadsheets for relatively small problems, we do not suggest that the use of Excel® spreadsheets is the ideal method to tackle larger scale analysis, as this is really the job that should be approached using programs written in languages such as Perl or C++. However, for simple and routine tasks, the combined use of Word and Excel is highly efficient, and the programs are easy to write. It also has the added bonus of allowing scientists trained as molecular biologists or geneticists to carry out relatively simple bioinformatical analysis without having to take the considerable time necessary to learn computer programming languages simply to tackle a relatively small-scale problem and thereby avoiding an approach that may feel akin to using a sledgehammer to crack a walnut.

REFERENCES

1. **Anbazhagan, R. and E. Gabrielson.** 1999. Spreadsheet-based program for alignment of overlapping DNA sequences. BioTechniques *26*:1180-1185.
2. **Dolnik, V.** 1999. DNA sequencing by capillary electrophoresis [review]. J. Biochem. Biophys. Methods *41*:103-119.
3. **McEwan, N.R. and D. Gatherer.** 1998. Adaptation of standard spreadsheet software for the analysis of DNA sequences. BioTechniques *24*:131-136.
4. **McEwan, N.R. and D. Gatherer.** 1999. Codon indices as a predictor of gene functionality in a *Frankia* operon. Can. J. Bot. *77*:1287-1292.
5. **McEwan, N.R., S.C.P. Eschenlauer, R.E. Calza, R.J. Wallace, and C.J. Newbold.** 1999. Protozoal sequences may reveal additional isoforms of the 14-3-3 protein family. Protist *150*:257-264.
6. **McEwan, N.R., S.C.P. Eschenlauer, R.E. Calza, R.J. Wallace, and C.J. Newbold.** 2000a. The 3′ untranslated region of messages in the rumen protozoan *Entodinium caudatum*. Protist *151*:139-146.
7. **McEwan, N.R., D. Gatherer, S.C.P. Eschenlauer, F.M. McIntosh, R.E. Calza, R.J. Wallace, and C.J. Newbold.** 2000b. An unusual codon usage pattern in the ciliate family Ophryoscolecidae and its implications for determining the source of cloned DNA. Anaerobe *6*:21-28.
8. **Shaw, G.** 1995. Protein sequence interpretation using a spreadsheet program. BioTechniques *19*:978-983.
9. **Shaw, G.** 2000. Useful Microsoft Word macros for molecular biologists and protein chemists. BioTechniques *28*:1198-1201.
10. **Smutzer, G. and L.L. Chamberlin.** 1998. The translational start sites of jawless and cartilaginous fish genes. Mol. Mar. Biol. Biotechnol. *7*:303-311.
11. **Strain, S.R. and J.G. Chmielewski.** 2001. ROCK: a spreadsheet-based program for the generation and analysis of random oligonucleotide primers used in PCR. BioTechniques *30*:1286-1291.

CHAPTER 4

Word-Processor Macro for Restriction Endonuclease Analysis

Nelson Cabrera León

Centro de Ingeniería Genética y Biotecnología, La Habana, Cuba

Reprinted from BioTechniques 27:1228-1231 (December 1999)

ABSTRACT

This paper describes a Microsoft Word 97 macro designed for restriction endonuclease analysis. Selected DNA fragments in the active Word document can be analyzed through a dynamic dialog box that formats the enzyme restriction lists for further analysis. The results can be obtained in a new Word document with the name of the enzymes, number of cuts and positions. This macro has several advantages: the results can be printed in a format suitable for record keeping, no additional programs are required and it is simple to use.

INTRODUCTION

The restriction endonuclease analysis of a DNA sequence is one of the basic tools in molecular biology for the development of cloning strategies. Although numerous computer programs are available for this kind of analysis, some are expensive and require some familiarity with them. Those programs are able to perform diverse tasks, but, in their daily practice, researchers only use a few of them. An additional difficulty of other computer programs is that the user must change the output format to include it in text documents.

DNA sequences, in general, are stored in a format that can be read by many word processors, but in some of them, the user can develop applications (macros) to carry out tedious or time-consuming tasks. In this work, we report the use of a Word 97 macro to make endonuclease restriction analyses.

MATERIALS AND METHODS

Computer Programs Used

The Windows™ version of Microsoft® Word 97™ (Microsoft, Redmond, WA, USA) is the only program used in this work.

BioComputing: Computer Tools for Biologists
Edited by Stuart M. Brown

Table 1. Some Restriction Enzymes Written in a Format Suitable for the Macro

```
BglI, 7, GCC^$^$^$^$^$GGC
HaeII, 5, GGCGCC, GGCGCT, AGCGCC, AGCGCT
HgaI, 10, GACGC^$^$^$^$^$
HgaI, 0, ^$^$^$^$^$^$^$^$^$^$GCGTC
HindIII, 1, AAGCTT
```

Files Used

Restriction Sites Analyzer.dot. The macro is contained in a file called Restriction Sites Analyzer.dot. It has three procedures: AutoExec, AutoExit and ResAnalyzer (the latter executes the restriction analysis). A copy of this file in the StartUp directory of Word 97 (the default path to this folder is C:\Program Files\Microsoft Office\Office\StartUp) is necessary to run this macro properly because it has a procedure (AutoExec) that inserts the Restriction Analysis option in the Tools menu when Word 97 starts. The AutoExit procedure deletes the Restriction Analysis option when the user exits Word.

If users do not want to load the macro the next time that Word starts, they simply exit Word and the Restriction Sites Analyzer.dot file from the StartUp folder.

To view or edit the macro, the user must open the Restriction Sites Analyzer.dot file in the StartUp directory and choose Macro in the Tools menu. Then, the user must select the Macros option, then the AutoExec, AutoExit or ResAnalyzer procedure in the dialog box and then click on the Edit button.

Default Enzymes List.txt. To run this macro, the user needs another file: Default Enzymes List.txt. The user must put a copy of this file in the My documents folder (the default path to this folder is C:\ My documents). This file has an alphabetical list of almost all the restriction endonucleases available commercially and includes the restriction enzyme name for each enzyme, its cut positions and all its recognition sites separated by commas (Table 1). The function of the macro is to use the FIND function to search different recognition sites in a DNA sequence that have been typed in a Word document. The user can use the symbols ^$ to represent any nucleotide because the code is understood by the function. Table 1 shows how the code was written for the *Bgl*I, *Hae*II, *Hga*I and *Hin*dIII enzymes.

Enzymes with two different cut positions must be typed separately (see *Hga*I in Table 1). *Hga*I is a non-palindromic endonuclease, so the sequence that it recognizes in a DNA chain is different from the sequence in the complementary DNA chain. In this case, the user must type the same enzyme twice, one following the other, with its respective cut positions and its recognition sites. The user can edit the Default Enzymes List.txt by adding or deleting restriction enzymes to create new lists for specific purposes. These new lists should be saved using different names with the .txt format extension.

DNA sequences in the Word document should contain no numbers, spaces or paragraph mark symbols, which will result in the sequence written as shown in Figure 1.

To run the macro, the user only needs to select the DNA sequence in the active document, then select Restriction Analysis in the Tools menu and a message box will appear. If the user selects Yes, the macro will use the Default Enzymes List.txt file to execute the restriction analysis; if the user selects No, the Open dialog box appears, which allows the user to select another enzyme list. The user can exit the macro automatically by clicking Cancel.

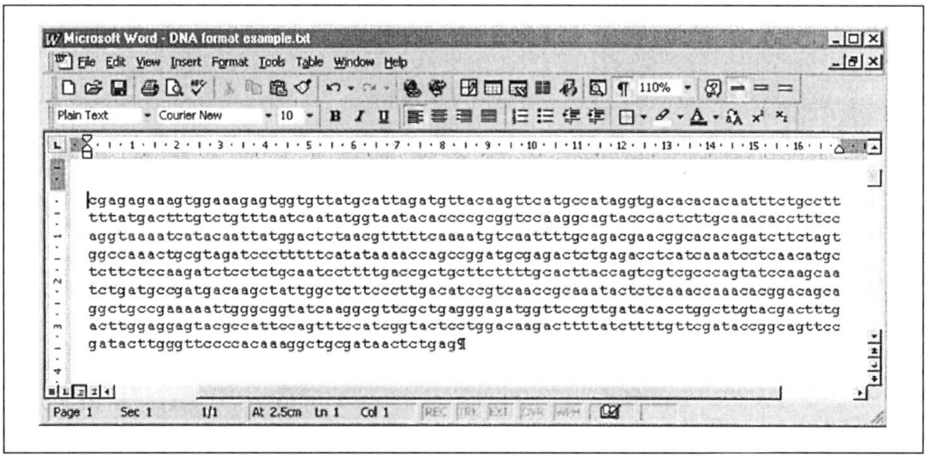

Figure 1. DNA sequence format ready for restriction sites analysis.

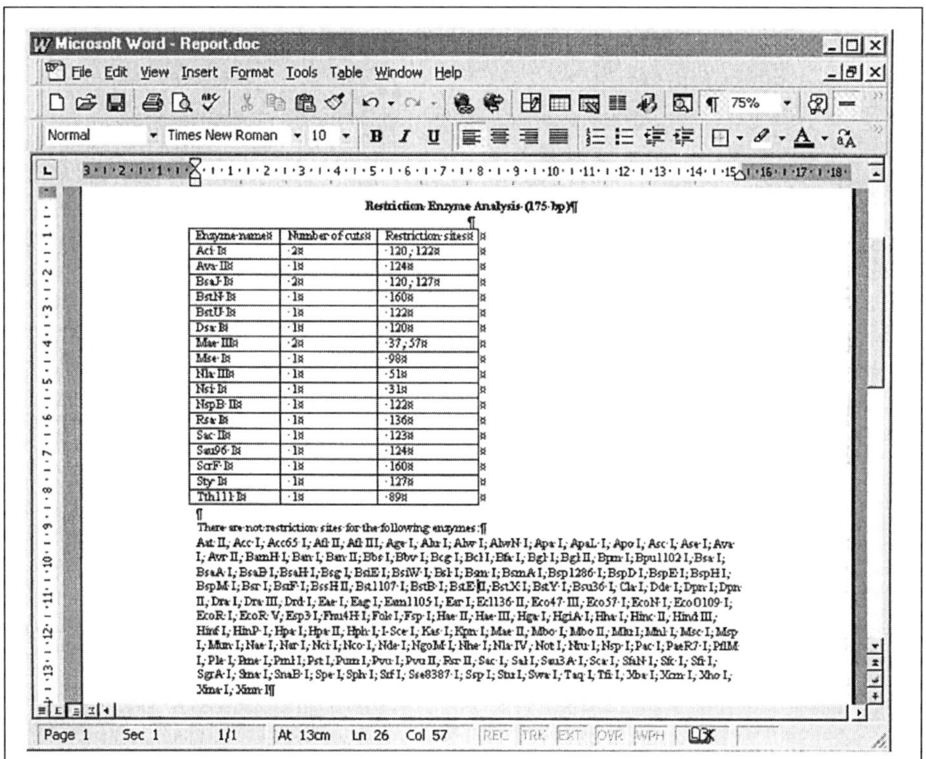

Figure 2. A typical restriction-site analysis report created into a new document.

As the macro runs, a new document is created (Figure 2). This document contains the results of the restriction analysis and shows a table with the name of the enzymes, number of cuts and positions, the DNA length and a list with non-cutter enzymes.

RESULTS AND DISCUSSION

The analysis of the macro was validated for several DNA sequences with BioSOS (1) and Gene Runner v3.02 (Hastings Software) programs with optimal results.

The method described here is a helpful tool for the molecular biologist. It is easy to use and avoids the use of expensive commercial software for the restriction analysis of DNA sequences. Furthermore, the report created is concise and easy to format for record keeping. The Visual Basic for Application code of this macro is a valuable guide for those who want to develop new applications in Word 97.

AVAILABILITY

The files referred to in this article can be downloaded from the **Software Library** on the *BioTechniques* Web site (www.BioTechniques.com).

ACKNOWLEDGMENTS

I thank Tanilo Rivero, Alejandro Soto, Luis Ríos and Dr. Javier Menéndez for their critical reading of the manuscript.

REFERENCES

1. **Bringas, R., R. Ricardo, J. Fernández de Cossio, M.E. Ochagavía, A. Suárez and R. Rodriguez.** 1992. BioSOS: A program package for the analysis of biological sequences in the microcomputer. Biotecnología Aplicada 9:180-185.

Update to: Word-Processor Macro for Restriction Endonuclease Analysis

Nelson Cabrera León

Centro de Ingeniería Genética y Biotecnología, La Habana, Cuba (nakasoni@yahoo.com)

We have detected two areas of troubleshooting for this macro related with (*i*) some Windows™ versions, and (*ii*) using the macro with non-English versions of Microsoft® Word®. Here, we discuss these problems and give some solutions.

TROUBLESHOOTING RELATED WITH WINDOWS VERSIONS

The macro that makes the restriction analysis has, by default, the Default Enzymes List.txt in the path C:\My Documents\Default Enzymes List.txt. Since 1999, other versions of Microsoft® Windows appear, and some, like Windows 2000 or Windows XP, have not a unique My Documents folder for all the users. Instead, they use a My Documents folder for each user. Therefore, the macro cannot find the enzyme database file (Default Enzymes List.txt) in the unique path C:\My Documents\Default Enzymes List.txt.

Solution

This problem has two simple solutions. The first solution is saving the Default Enzymes List.txt file in a folder selected by the user and later, when the macro was started, the user should select the No button when the message Select the enzymes restriction list appears. Then the user can find the previously selected path for the Default Enzymes List.txt file in the Open File dialog box.

The second solution is to edit the code of the macro and change the code that set the path C:\My Documents\Default Enzymes List.txt by default, following these instructions:

1. Open the Restriction Sites Analyzer.dot file.
2. Choose Macro in the Tools menu. At this point, the user must select the Macros… option.
3. Then select the ResAnalyzer procedures in the dialog box and click on the Edit button. This step will open the Microsoft Visual Basic Editor window.
4. The line of code that the user needs to change is:

   ```
   EnzymeList = "c:\My Documents\Default Enzymes List.txt"
   ```

   Here, the user can change this default path for a new one.
5. Save the Restriction Sites Analyzer.dot file.

TROUBLESHOOTING RELATED WITH LANGUAGE VERSIONS OF MS-WORD

This macro was designed using an English version of Microsoft Word 97, so if the user is using another language version of Word, another area of troubleshooting will come up. The macro consist in three procedures named: AutoExec, AutoExit and ResAnalyzer. When Word starts, the AutoExec procedure looks for a menu called Tools. If this menu exists, then AutoExec inserts the Restriction Analysis... option in the Tools menu. So if you are using a non-English version of MS-Word, maybe you could not find the Tools menu, because this menu is named in a different way in this language. For example, in the Spanish version of MS-Word, the Tools menu is named Herramientas instead, so the AutoExec procedure will not find a menu called Tools, and therefore, the macro cannot work properly.

Solution

If you are using a non-English version of Word, you should edit the AutoExec procedure following these steps:

1. Open the Restriction Sites Analyzer.dot file.
2. Choose Macro in the Tools menu. At this point, the user must select the Macros... option.
3. Then select the AutoExec procedures in the dialog box and click on the Edit button. This step will open the Microsoft Visual Basic Editor window.
4. The line of code that the user needs to change is:

   ```
   Set cnt = Application.CommandBars("Menu Bar").Controls("Tools")
   ```

   Here, the user can change the word Tools by the new one. For example, for a Spanish version of Word, this line of code will be:

   ```
   Set cnt = Application.CommandBars("Menu Bar").Controls("Herramientas")
   ```

5. Save the Restriction Sites Analyzer.dot file.

In a similar manner, you should change the code for the AutoExit procedure substituting the word Tools for the corresponding word in this language version of Word.

CHAPTER 5

Molecular BioComputing Suite: A Word Processor Add-In for the Analysis and Manipulation of Nucleic Acid and Protein Sequence Data

Patrick Y. Muller, Etienne Studer, and André R. Miserez

University of Basel, Basel, Switzerland

Reprinted from BioTechniques 31:1306-1313 (December 2001)

ABSTRACT

In all fields of molecular biology, researchers are increasingly challenged by experiments planned and evaluated on the basis of nucleic acid and protein sequence data generally retrieved from public databases. Despite the wide spectrum of available Web-based software tools for sequence analysis, the routine use of these tools has disadvantages, particularly because of the elaborate and heterogeneous ways of data input, output, and storage. Here we present a Visual Basic-encoded Microsoft® Word® Add-In, the Molecular BioComputing Suite *(*MBCS*), available at the* BioTechniques *Software Library (www.BioTechniques.com). The* MBCS *software aims to manage and expedite a wide range of sequence analyses and manipulations using an integrated text editor environment including menu-guided commands. Its independence of sequence formats enables* MBCS *to be used as a pivotal application between other software tools for sequence analysis, manipulation, annotation, and editing.*

INTRODUCTION

As a consequence of the rapid accumulation of sequence information in all fields of molecular biology and biomedicine, researchers increasingly depend on computers to retrieve, analyze, and edit nucleotide and protein sequences electronically. The huge amount of publicly available data sets of complete genomes, the requirement for the routine use of recombinant DNA technology, and the increasing efforts in sequencing of disease genes make it necessary to store and handle nucleotide and protein sequences electronically. Apart from all kinds of sequence alignments (10,21), common tasks in computational molecular biology include the identification of DNA and protein subsequences with specific functions such as cleavage sites of restriction endonucleases (22), consensus binding sites of transcription factors (13), or target sites for posttranslational

BioComputing: Computer Tools for Biologists
Edited by Stuart M. Brown

modifications. As part of such work, DNA sequences frequently need to be reformatted [e.g., in standardized formats such as FASTA (20)], numbered according to the nucleotide positions, or transformed to their complementary, reverse, or antisense counterparts. In addition, DNA and protein sequences often need to be interconverted to each other (i.e., translated or reverse-translated). On the protein level, common physicochemical properties such as molecular weights, isoelectric points (IP), molar absorption coefficients, compositions of amino acids, and their molar percentages are often required. Furthermore, graphical plots of window-averaged biochemical, immunological, and genetic parameters as a function of the individual amino acid residues of a protein have become established procedures in molecular biology (14,29).

Nucleotide and protein sequence analysis and manipulation can be performed using a variety of software tools. These tools include console-based applications (28), Web-based interfaces to server-side PERL or C++ scripts (33), or sophisticated software suites running on UNIX®-like operating systems (8). Although console and Web-based software tools have been introduced increasingly, the routine use of data input, output, and storage in these programs may be inconvenient. Furthermore, performing a series of analyses with different software tools, the sequence data to be analyzed is required in a specific format and frequently needs to be reformatted when proceeding from one analysis tool to another or when the sequence data have to be stored. Although there are several, even Web-based, reformatting tools like, for example, *ReadSeq* (http://bimas.dcrt.nih.gov/molbio/readseq/), switching between different sequence formats usually is cumbersome. In addition, sophisticated software suites providing an integrated environment are often expensive and, thus, beyond the means of small research groups.

Within the last years, several computer applications aiming to provide a convenient way to perform common procedures used in computational molecular biology have been developed (2,3,9,12,17,23–26). Here we present a software application, the *Molecular BioComputing Suite* (*MBCS*) that was implemented in Microsoft® Visual Basic for Applications. *MBCS* operates as an add-in for Microsoft Word® and solves the limitations addressed, including the financial considerations. *MBCS* is an improvement over already existing tools of its kind, as a huge range of operations can be conveniently performed within Microsoft Word with minimal requirements of time. The main objective is to expedite and simplify the analysis and manipulation of nucleic acid and protein sequence data in an integrated text editor environment including menu-guided commands.

MATERIALS AND METHODS

The *MBCS* source code is entirely coded in the object-oriented programming language Visual Basic for Applications and is contained in one module and nine graphical user forms, all integrated in a Microsoft Word add-in file. The *MBCS* software itself comprises a total of 29 individual program components. Their compilation is automatically accomplished with the Visual Basic compiler upon running one of the *MBCS* programs. The Visual Basic compiler is an integrated constituent of Microsoft Visual Basic for Applications and, thus, available on any system where one of the Microsoft Office Applications is installed. The *MBCS* add-in runs within Microsoft Word (version 97 or later for Microsoft Windows® and version 98 or later for Apple® Macintosh®) and, thus, can be operated on any system where this widely used word processor is installed. All common features of the word processor environment can be used for editing, annotation, and storage of sequence data. All programs of the *MBCS* software can be directly executed within Microsoft Word, and the graphical output is generated by a special Microsoft Excel® file (plot.xls). To have permanent access to these programs, it is recommended to copy the *MBCS* application file into the Microsoft Word startup directory (Menu *Tools*, *Preferences*, *File Locations*) or to temporarily activate the file as an add-in (Menu *Tools*, *Templates and Add-Ins*, *Global templates and add-ins*). All programs can then be launched from an additional menu generated by *MBCS* (Figure 1). This menu is added to the Microsoft Word menu bar. Furthermore, to provide rapid access, *MBCS* generates a floatable menu bar containing buttons for the launch of the most frequently used programs (Figure 1). To be able to run *MBCS* in Microsoft Word 2000 and, in the meantime, to maintain macro-virus protection for unknown documents, it is recommended to activate the "Trust all installed add-ins and templates" checkbox (Tools/Macro/Security/Trusted Sources).

For evaluation and validation purposes, the *MBCS* software was subjected to two beta testing phases in one molecular genetics and two unrelated molecular biology laboratories. The *MBCS* file referred to in this article can be downloaded free of charge from the Software Library on the *BioTechniques* Web site (www.BioTechniques.com). Moreover, the unencrypted Visual Basic source code of the module and the graphical user forms can be freely edited and extended to develop further applications.

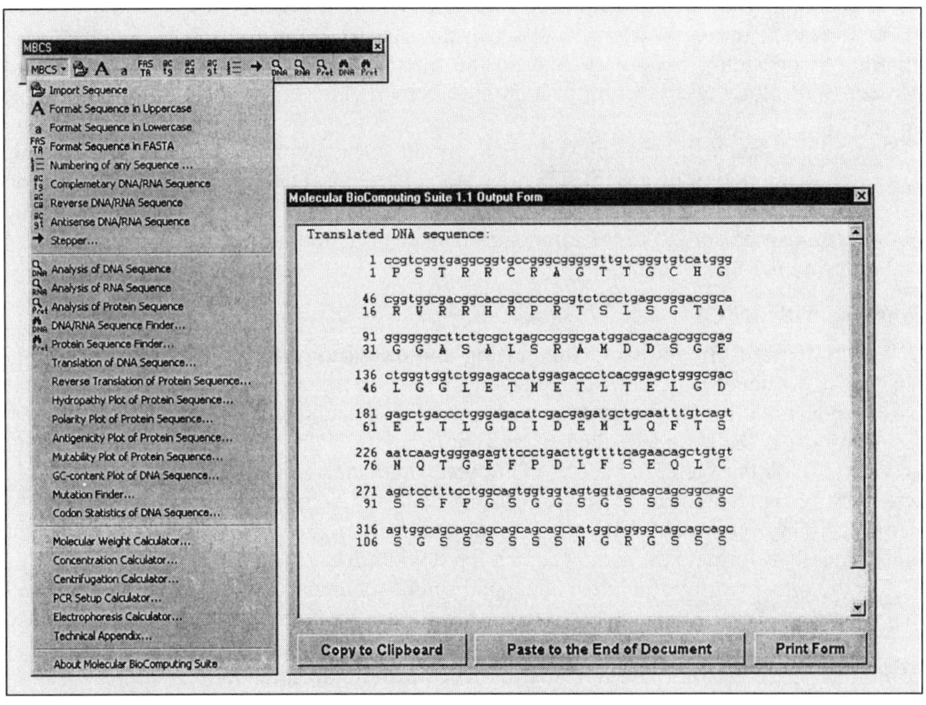

Figure 1. Screen shot of the *MBCS* toolbar and the *MBCS* output form. The floatable toolbar with the *MBCS* menu, subdivided into three sections, allows a rapid and menu-guided launch of all programs of the software. Screen shot of the output form containing output data of the program *Translation of DNA Sequence*.

RESULTS AND DISCUSSION

Structure, Features, and Characteristics of the *MBCS* Software

The individual program components of *MBCS* are divided into three sections. The first section includes utilities for import, manipulation, and formatting of nucleic acid and protein sequences. The second section consists of various programs for the sequence analysis of DNA, RNA, and proteins. The objective of the third section is the calculation of numerous pre- and post-experimental parameters required for common molecular biological methods. The data input requirements for each program of *MBCS* are stated upon launching the respective program.

The data input procedure is basically the same for all programs dealing with the manipulation and analysis of sequence data. First, the sequence stretch to be manipulated or analyzed has to be selected with the mouse pointer; thereafter, the desired program has to be launched, either by selecting an item from the *MBCS* menu or by clicking directly on a button of the menu bar (Figure 1). The sequence stretch selected may contain any characters, digits, or symbols not used in the standard one-letter code for DNA, RNA, or protein sequences. *MBCS* automatically eliminates these characters, digits, and symbols by means of a special program subroutine. Moreover, all programs of *MBCS* are case-insensitive. As a consequence, all kinds of sequence formats are accepted, disregarding whether they contain line breaks, tab stops, spaces, digits, or upper- or lowercase letters. This feature provides highest flexibility and, thus, is useful when working with various sequence formats, including sequences contained in GenBank® files (5) retrieved from the Web. In the case of DNA and RNA sequences, the ambiguous one-letter code according to the International Union of Biochemistry and Molecular Biology (IUB) Commission on Biochemical Nomenclature is supported as well (15).

The output data will be displayed in the output form, a window that opens automatically upon completion of any *MBCS* program (Figure 1). The output form provides the user with the following three options: (*i*) the output data can be directly printed, (*ii*) the output data can be copied to the clipboard, and (*iii*) the output data can be pasted to the end of the active Word document.

Working with *MBCS*

Section 1: tools for sequence formatting and manipulation. Every manipulation and analysis of sequence data requires the desired sequence to be either retrieved from a database or generated by the output of an automatic DNA sequencer. For many applications, GenBank (5) is the sequence database of choice, particularly because of its convenient accessibility on the Web. To handle, manipulate, and analyze a Web-based GenBank file with *MBCS*, the content has to be transferred from the Web browser to Microsoft Word while retaining the layout. This task is managed by the *Import Sequence* program that automatically reformats the sequence in a fixed width font.

For the reformatting of nucleic acid and protein sequences, *MBCS* contains four special programs. Two of them are mainly used to format specific subsections of a sequence in upper- or lowercase characters, a built-in feature of Microsoft Word. This feature is commonly used to annotate and distinguish intron/exon and translated/untranslated regions. The third program for reformatting is intended to convert any kind of nucleic acid and protein sequence into the standardized and widely used FASTA format (20). Another commonly used layout is the numbering of a sequence according to its nucleotide or amino acid positions, which is accomplished by the program *Numbering of Sequence*.

For the conversion of DNA or RNA sequences to their complementary, reversed, or antisense counterparts, three individual programs, named according to these procedures, are provided.

The program *Stepper* offers the opportunity to mark a subsection of a sequence consisting of a user-defined number of nucleotides or amino acids by automatically moving the cursor forward or backward from the actual position, whereas all non-sequence characters are ignored.

Section 2: tools for the analysis of nucleic acid and protein sequence data. The program *Analysis of DNA Sequence* calculates all kinds of physicochemical parameters of the selected DNA sequence. These parameters include overall length, molecular weight of double and single strands, molar percentages of the individual nucleotides, and various melting temperatures. The melting temperatures of oligonucleotides are determined by four different algorithms: (*i*) the popular and simple GC-method (27), (*ii*) the salt-adjusted method of Baldino et al. (4), (*iii*) the advanced thermodynamic nearest-neighbor method (7), and (*iv*) the T_p-method that corresponds to the maximum temperature at which efficient PCR amplification is observed (30).

The program *Analysis of Protein Sequence* calculates all kinds of physicochemical parameters of the selected protein sequence. These parameters include overall length, molecular weight, molar percentages of the individual amino acids, IP according to Bjellqvist et al. (6), and molar absorption coefficient in solution at the 280-nm wavelength (19).

To identify specific DNA or protein subsequences, the *DNA Sequence Finder* and the *Protein Sequence Finder* are provided. They can be used to search for specific DNA or protein sequence stretches, respectively, whereas the ambiguous code for nucleotides and amino acids is supported. In addition to specific amino acids, basic, acidic, nucleophilic, aromatic, and aliphatic side chains can be used as search criteria.

For the inter-conversion of nucleic acid and protein sequences (i.e., translation and reverse translation), the programs *Translation of DNA Sequence* and *Reverse Translation of Protein Sequence* can be used. The output of the program *Reverse Translation of Protein Sequence* generates DNA sequences with synonymous codons by using the ambiguous nucleic acid code.

Established procedures for topological predictions of protein sequences include graphical plots of window-averaged biochemical, immunological, and genetic parameters in function of the individual amino acid residues of a protein. In this context, four programs, named according to the underlying method, can be used for such analyses. The window-averaged hydropathy and the grand average of hydropathy (GRAVY) score are computed according to Kyte and Doolittle (14), the polarity according to Grantham (11), and the antigenicity according to Welling et al. (29). The relative mutability of a specific amino acid corresponds to the reciprocal number of its synonymous codons. The determination of the relative mutability is a useful procedure to predict the effects of potential nucleotide substitution mutations on the protein. Protein regions with a high relative mutability are susceptible to mutations leading to amino acid substitutions. Thus, such protein regions are valuable candidate regions for mutation screening aiming to identify single-nucleotide polymorphisms (SNPs) (1,18) leading to amino acid substitutions. The GC-content plot of a DNA sequence allows the identification of GC-rich sequences and, hence, the selection of appropriate conditions for the sequencing of those templates on the basis of their GC-content (16). The program *Codon Statistics* calculates the individual frequencies of all of the 64 possible codons of a specific DNA sequence and the frequency distributions among synonymous codons. This data provides the basis for the recognition of a possible codon bias frequently observed in the coding region of highly expressed genes because of selective pressure towards maximizing the efficiency of translation (31).

Nowadays, in the post-genomic era, many DNA sequencing applications are directed towards a comparison with already existing DNA sequences—an approach frequently used for mutation detection in human molecular and medical genetics. This laborious process is sped up and simplified by the program *Mutation Finder*. This program performs an alignment of two selected DNA sequences by searching for overlapping sequence motifs. Nucleotide substitutions between the two compared DNA sequences (i.e., nucleotides not matching the template) are displayed in the alignment. The *Mutation Finder* is intended to be used for sequences of up to approximately 500 bases, depending on the performance of the user's computer system.

Section 3: benchtop techniques. This section contains seven programs for benchtop use that all provide a graphical user interface. The objective of these programs is the calculation of various parameters for common molecular biology techniques.

The *Molecular Weight Calculator* determines the molecular weight of any compound on the basis of its chemical formula or atomic composition. The *Concentration Calculator* determines the concentration and further parameters of various kinds of biological samples on the basis of the dilution and the absorbance at specific wavelengths. Thus, this program can be used for the concentration calculation of ssDNA, dsDNA, ssRNA, peptides and proteins, bacterial suspensions, and for the determination of the helix content of proteins based on their circular dichroism measurements (optical ellipticity) (32).

The *Centrifugation Calculator* enables the calculation of the relative centrifugal force, the rotations per minute, or the effective radius of the rotor if any two of these parameters are given.

The calculation of the composition of complex PCR mixtures is accomplished by the *PCR Setup Calculator*. This program calculates all volumes of the individual components that have to be added to the PCR mixture. The input variables for this calculation are the stock concentrations of the individual components and their desired final concentrations in the PCR mixture.

The *Electrophoresis Calculator* is based on the proportional relation between voltage and migration distance and their inverse proportional relation to the migration time when a fragment of given molecular weight is subjected to electrophoresis with otherwise fixed conditions. Applying these relationships, it is possible to calculate voltage, migration distance, or migration time for any electrophoresis once one set of these three parameters has been recorded.

The *Ligation Calculator* enables the calculation of the volumes of the vector and insert stock solutions to be used in the ligation reaction. This is accomplished on the basis of the fragment lengths and the molar ratio of fragment to insert as well as the concentrations of the vector and the insert stock solutions.

Using the *Technical Appendix*, the user has the opportunity to look up the human codon table of the genetic code, the genetically encoded amino acids and their residue molecular weights, the IUB one-letter code for ambiguous nucleotides, the Web addresses of important human SNP databases, and finally the references of all methodical and technical publications used to implement *MBCS*.

CONCLUSION

The *MBCS* software described here is a tool to manage, expedite, and simplify common procedures for sequence analysis and manipulation using an integrated text editor environment including menu-guided commands. The application spectrum of *MBCS* is limited to established procedures using moderate amounts of sequence data and does not intend to replace sophisticated Web- or console-based software suites. Nevertheless, *MBCS* is a useful tool for researchers performing computer applications in the field of molecular biology and biomedicine. Because of its independence of sequence formats, *MBCS* is able to take up a pivotal position between various software tools for sequence analysis, manipulation, annotation, and editing.

ACKNOWLEDGMENTS

The authors thank Harald Janovjak for his expertise and his helpful advice concerning the implementation of object-oriented program components. Moreover, the authors thank the colleagues who assisted them with their valuable suggestions and discussions, especially those who performed the beta testing of the software. This work was supported by the Swiss National Science Foundation grant no. 3200-049125.96/2. A.R.M. is supported by the Swiss Clinicians Opting for Research (SCORE) A grant no. 3231-048896.96 of the Swiss National Science Foundation.

REFERENCES

1. **Altshuler, D., V.J. Pollara, C.R. Cowles, W.J. Van Etten, J. Baldwin, L. Linton, and E.S. Lander.** 2000. An SNP map of the human genome generated by reduced representation shotgun sequencing. Nature *407*:513-516.
2. **Anbazhagan, R. and E. Gabrielson.** 1999. Spreadsheet-based program for alignment of overlapping DNA sequences. BioTechniques *26*:1180-1185.
3. **Anbazhagan, R., J.G. Herman, K. Enika, and E. Gabrielson.** 2001. Spreadsheet-based program for the analysis of DNA methylation. BioTechniques *30*:110-114.
4. **Baldino, F., Jr., M.F. Chesselet, and M.E. Lewis.** 1989. High-resolution in situ hybridization histochemistry. Methods Enzymol. *168*:761-777.
5. **Benson, D.A., M.S. Boguski, D.J. Lipman, and J. Ostell.** 1997. GenBank. Nucleic Acids Res. *25*:1-6.
6. **Bjellqvist, B., B. Basse, E. Olsen, and J.E. Celis.** 1994. Reference points for comparisons of two-dimensional maps of proteins from different human cell types defined in a pH scale where isoelectric points correlate with polypeptide compositions. Electrophoresis *15*:529-539.
7. **Breslauer, K.J., R. Frank, H. Blocker, and L.A. Marky.** 1986. Predicting DNA duplex stability from the base sequence. Proc. Natl. Acad. Sci. USA *83*:3746-3750.
8. **Butler, B.A.** 1998. Sequence analysis using GCG. Methods Biochem. Anal. *39*:74-97.
9. **Cabrera Leon, N.** 1999. Word-processor macro for restriction endonuclease analysis. BioTechniques *27*:1228-1231.
10. **Gaskell, G.J.** 2000. Multiple sequence alignment tools on the Web. BioTechniques *29*:60-62.
11. **Grantham, R.** 1974. Amino acid difference formula to help explain protein evolution. Science *185*:862-864.
12. **Han, B. and A.H. Tashjian, Jr.** 1998. User-friendly and versatile software for analysis of protein hydrophobicity. BioTechniques *25*:256-263.
13. **Heinemeyer, T., X. Chen, H. Karas, A.E. Kel, O.V. Kel, I. Liebich, T. Meinhardt, I. Reuter, F. Schacherer, and E. Wingender.** 1999. Expanding the TRANSFAC database towards an expert system of regulatory molecular mechanisms. Nucleic Acids Res. *27*:318-322.

14. **Kyte, J. and R.F. Doolittle.** 1982. A simple method for displaying the hydropathic character of a protein. J. Mol. Biol. *157*:105-132.
15. **Liébecq, C.** 1992. Biochemical Nomenclature and Related Documents. Portland Press, London.
16. **Motz, M., S. Paabo, and C. Kilger.** 2000. Improved cycle sequencing of GC-rich templates by a combination of nucleotide analogs. BioTechniques *29*:268-270.
17. **Muller, P.Y., L. Barella, and A.R. Miserez.** 2001. RepGene: a spreadsheet template for the management of reporter gene assays. BioTechniques *30*:1294-1298.
18. **O'Dell, S.D., T.R. Gaunt, and I.N. Day.** 2000. SNP genotyping by combination of 192-well MADGE, ARMS and computerized gel image analysis. BioTechniques *29*:500-506.
19. **Pace, C.N., F. Vajdos, L. Fee, G. Grimsley, and T. Gray.** 1995. How to measure and predict the molar absorption coefficient of a protein. Protein Sci. *4*:2411-2423.
20. **Pearson, W.R.** 1994. Using the FASTA program to search protein and DNA sequence databases. Methods Mol. Biol. *24*:307-331.
21. **Phillips, A., D. Janies, and W. Wheeler.** 2000. Multiple sequence alignment in phylogenetic analysis. Mol. Phylogenet. Evol. *16*:317-330.
22. **Roberts, R.J. and D. Macelis.** 1999. REBASE-restriction enzymes and methylases. Nucleic Acids Res. *27*:312-313.
23. **Shaw, G.** 1995. Protein sequence interpretation using a spreadsheet program. BioTechniques *19*:978-983.
24. **Shaw, G.** 2000. Useful Microsoft® Word® macros for molecular biologists and protein chemists. BioTechniques *28*:1198-1201.
25. **Singal, R. and S.R. Grimes.** 2001. Microsoft® Word® macro for analysis of cytosine methylation by the bisulfite deamination reaction. BioTechniques *30*:116-120.
26. **Stowe, R.P. and D.L. Pierson.** 1996. Spreadsheet macro for setting up PCR assay tubes. BioTechniques *20*:1088-1089.
27. **Suggs, S.V., R.B. Wallace, T. Hirose, E.H. Kawashima, and K. Itakura.** 1981. Use of synthetic oligonucleotides as hybridization probes: isolation of cloned cDNA sequences for human beta 2-microglobulin. Proc. Natl. Acad. Sci. USA *78*:6613-6617.
28. **Thompson, J.D., D.G. Higgins, and T.J. Gibson.** 1994. CLUSTAL W: improving the sensitivity of progressive multiple sequence alignment through sequence weighting, position-specific gap penalties and weight matrix choice. Nucleic Acids Res. *22*:4673-4680.
29. **Welling, G.W., W.J. Weijer, R. van der Zee, and S. Welling-Wester.** 1985. Prediction of sequential antigenic regions in proteins. FEBS Lett. *188*:215-218.
30. **Wu, D.Y., L. Ugozzoli, B.K. Pal, J. Qian, and R.B. Wallace.** 1991. The effect of temperature and oligonucleotide primer length on the specificity and efficiency of amplification by the polymerase chain reaction. DNA Cell Biol. *10*:233-238.
31. **Xia, X.** 1996. Maximizing transcription efficiency causes codon usage bias. Genetics *144*:1309-1320.
32. **Yang, J.T., C.S. Wu, and H.M. Martinez.** 1986. Calculation of protein conformation from circular dichroism. Methods Enzymol. *130*:208-269.
33. **Zhang, J. and T.L. Madden.** 1997. PowerBLAST: a new network BLAST application for interactive or automated sequence analysis and annotation. Genome Res. *7*:649-656.

Update to: *Molecular BioComputing Suite*: A Word Processor Add-In for the Analysis and Manipulation of Nucleic Acid and Protein Sequence Data

Patrick Y. Muller, Etienne Studer, and André R. Miserez

University of Basel, Basel, Switzerland
(andre-r.miserez@unibas.ch, patrick.muller@unibas.ch)

In the current update of the *Molecular BioComputing Suite* (*MBCS*) (release 1.2), a number of improvements suggested by the user community have been realized. All bugs reported by users running the software on various platforms have been fixed, and the code of the program Mutation Finder has been considerably optimized for running speed. The present release of *MBCS* has been adapted and tested to run also on Microsoft® Windows™ XP platforms, including Microsoft® Office XP. Furthermore, the layout of the *MBCS* menu has been redesigned in a more structured way. The Visual Basic source code of the module and the graphical user forms is still unencrypted and, thus, can be modified and extended to meet the demands of advanced users. Continuing the philosophy to provide the software to the scientific community, the updated *MBCS* file (release 1.2) can be downloaded free of charge from the Software Library on the *BioTechniques* Web site.

CHAPTER 6

User-Friendly and Versatile Software for Analysis of Protein Hydrophobicity

Bomie Han and Armen H. Tashjian, Jr.

Harvard Medical School, and Harvard School of Public Health, Boston, MA, USA

Reprinted from BioTechniques 25:256-263 (August 1998)

ABSTRACT

We have developed a simple and flexible program to analyze regional hydrophobicity of a protein from its amino acid (aa) sequence. This program runs as a Microsoft® Excel® document into which aa sequence can be copied from any Windows®-compatible or Macintosh® word processor. The program returns the hydrophobicity index of each aa residue and other analyses that can be used to predict transmembrane domains, amphiphilic alpha helices and putative antigenic epitopes in a protein using established algorithms. The program can also be easily modified to test user-defined algorithms and to accommodate non-conventional aa residues or ambiguities in the aa sequence. Simple modification of the program allows direct use of nucleic acid sequence information for various analyses. Graphic visualization of the results is readily achieved using the graphics function of Microsoft Excel. Alternatively, the data can be imported into other graphics software for preparation of publication-quality figures. By running as a document in Microsoft Excel, which can be found in virtually all personal computers, this program provides easy access even to the computer novice.

INTRODUCTION

Analysis of the hydrophobicity pattern of a protein deduced from its amino acid (aa) sequence can give valuable information on overall structure and transmembrane topology (7). The presence of a single transmembrane domain in receptor tyrosine kinases (10) as well as overall transmembrane topology of G protein-coupled receptors (9) or ion channels (2) can often be predicted from such analyses.

The most widely used approach is that of Kyte and Doolittle (7). According to this method, a numeric value indicating hydrophobicity (or hydropathy index) is assigned to each aa residue, and then window-average hydropathy indices are obtained for groups of

BioComputing: Computer Tools for Biologists
Edited by Stuart M. Brown

7 or 9 consecutive aa residues. This window is moved across the entire sequence (thus, a running window average is achieved) to estimate the hydrophobicity at each residue in the protein. Transmembrane domains can be predicted if sufficiently long stretches (20–30 aa residues) of hydrophobic amino acids are found.

Variations on this approach have been proposed. These include changes in the window size, averaging of the hydropathy indices with a weighting factor to detect amphiphilicity of alpha helices (6) and prediction of immunogenic epitopes in a protein (5).

Although the mathematics involved in these analyses is simple, the actual process is often cumbersome due to the large number of aa residues in any given protein (a few hundred in most cases). Computer programs to perform these analyses can be found in a number of different commercial software packages including, for example, Lasergene (DNASTAR, Madison, WI, USA). Alternatively, many of these programs can be accessed through a networked mainframe computer. Unfortunately, commercial software packages and maintaining an account with a mainframe computer can be expensive. Individual researchers have published programs to perform specific analyses for their interests (for example, Reference 3), which can either be downloaded electronically or obtained free from the authors. However, these programs have limited application and platform compatibility.

A spreadsheet program has been published previously (8) that runs on Microsoft® Excel® (Microsoft, Redmond, WA, USA). Such programs provide inexpensive and easy access to various algorithms for protein sequence analysis; however, this particular program is cumbersome to use because loading aa sequence information is labor-intensive. The aa sequence information, downloaded from a sequence database, is formatted appropriately using a word-processor program for importing into the spreadsheet program. The formatting of the sequence data involves inserting a single blank space after each amino acid and converting the text to a one-character-wide table using blank space as delimiters.

We have, therefore, developed a user-friendly and versatile approach that runs on the Microsoft Excel spreadsheet program to analyze protein sequence information. To use this program, aa sequence information is simply copied from any Windows®-compatible or Macintosh® word processor directly into this program. The program automatically converts aa sequence written in horizontal lines into a vertical column and returns analysis results. In addition, the program is easy to modify so that any published algorithms or user-defined algorithm can be easily incorporated and tested.

RESULTS AND DISCUSSION

Overview of the Program

Figure 1 shows the user interface of the program (or the screen output) using the aa sequence of a sample protein, thyrotropin-releasing hormone (TRH) receptor, which belongs to the G protein-coupled seven transmembrane receptor superfamily (11). The user has only to provide the aa sequence of a protein in column A; the rest of Figure 1 is program output. It is likely that the user already has written the aa sequence in a word-processor file, which can be copied into column A on multiple lines starting from row 3 (cell A3). Alternatively, the aa sequence can be written directly into column A. In the example shown in Figure 1, aa sequence was written using the single-letter code; however, both triple-letter code and nucleic acid sequence can be used with simple modifications of the program (see below). In column B, the program returns the aa sequence after eliminating spaces between aa codes in column A. Column C gives the number of aa residues in each line in column B. Column D gives a cumulative total number of aa residues in column B, starting from the first line. The information in columns C and D is useful to confirm that the input sequence is correct. The next four columns give: *(i)* E, the triple-letter code for each aa; *(ii)* F, the single-letter code for each aa; *(iii)* G, the hydropathy index (7); and *(iv)* H, the hydrophilicity index for each residue (5). Columns I and J are reserved for modifications of the program to use nucleic acid sequence data as described below (see Flexibility of the Program). Columns K and L return the residue number and single-letter code in a vertical linear sequence. Columns M and N return the hydropathy index and hydrophilicity index of each aa residue, respectively. These values are obtained by the computer from the information given in columns E–H. Columns O, P and Q give the average value for the hydropathy indices for 7 neighboring residues (H_7)

	A	B	C	D	E	F	G	H	I	J	K	L	M	N	O	P	Q
1	Input Sequence	Trimmed Seq.	#AA	#TAA	TL	SL	HIKD	HIHW	NA	AA	R#	AA	HIKD	HIHW	H7	Ha	H Ag
2			0.0	0	Arg	R	-4.5	3.0			1	M	1.9	-1.3			0.5
3	M E N E	MENETVSELNQTELP	15.0	15	Asp	D	-3.5	3.0			2	E	-3.5	3.0			0.8
4	P Q V A	PQVAVALEYQVVTIL	15.0	30	Glu	E	-3.5	3.0			3	N	-3.5	0.2			0.8
5	L V V V	LVVVICGLGIVGNIM	15.0	45	Lys	K	-3.9	3.0			4	E	-3.5	3.0	-0.8		0.4
6	V V L V	VVLVVMRTKHMRTAT	15.0	60	Ser	S	-0.8	0.3			5	T	-0.7	-0.4	-1.6		0.0
7	N C Y L	NCYLVSLAVADLMVL	15.0	75	Asn	N	-3.5	0.2			6	V	4.2	-1.5	-0.6		0.1
8	V A A G	VAAGLPNITDSIYGS	15.0	90	Gln	Q	-3.5	0.2			7	S	-0.8	0.3	-0.6		0.3
9	W V Y G	WVYGYVGCLCITYLQ	15.0	105	Gly	G	-0.4	0.0			8	E	-3.5	3.0	-0.6		0.7
10	Y L G I	YLGINASSCSITAFT	15.0	120	Pro	P	-1.6	0.0			9	L	3.8	-1.8	-0.6	-0.8	-0.1
11	I E R Y	IERYIAICHPIKAQF	15.0	135	Thr	T	-0.7	-0.4			10	N	-3.5	0.2	-1.7	-0.5	0.2
12	L C T F	LCTFSRAKKIIIFVW	15.0	150	Ala	A	1.8	-0.5			11	Q	-3.5	0.2	-1.0	-0.8	0.2
13	A F T S	AFTSIYCMLWFFLLD	15.0	165	His	H	-3.2	-0.5			12	T	-0.7	-0.4	-0.7	-0.7	0.2
14	L N I S	LNISTYKDAIVISCG	15.0	180	Cys	C	2.5	-1.0			13	E	-3.5	3.0	-1.5	0.4	0.0
15	Y K I S	YKISRNYYSPIYLMD	15.0	195	Met	M	1.9	-1.3			14	L	3.8	-1.8	-1.5	0.7	-0.6
16	F G V F	FGVFYVMPMILATVL	15.0	210	Val	V	4.2	-1.5			15	P	-1.6	0.0	-0.4	-0.3	-0.6
17	Y G F I	YGFIARILFLNPIPS	15.0	225	Ile	I	4.5	-1.8			16	P	-1.6	0.0	-0.1	-0.7	-0.6
18	D P K E	DPKENSKTWKNDSTH	15.0	240	Leu	L	3.8	-1.8			17	Q	-3.5	0.2	1.0	-0.2	-0.9
19	Q N K N	QNKNMNLNTTNRCFN	15.0	255	Tyr	Y	-1.3	-2.3			18	V	4.2	-1.5	0.8	0.3	-0.5
20	S T V S	STVSSRKQVTKMLAV	15.0	270	Phe	F	2.8	-2.5			19	A	1.8	-0.5	1.5	0.7	-0.6
21	V V I L	VVILFALLWMPYRTL	15.0	285	Trp	W	-0.9	-3.4			20	V	4.2	-1.5	1.3	-0.4	-0.5
22	V V V N	VVVNSFLSSPFQENW	15.0	300							21	A	1.8	-0.5	1.6	0.6	-0.5
23	F L L F	FLLFCRICIYLNSAI	15.0	315							22	L	3.8	-1.8	0.5	2.1	-0.7
24	N P V I	NPVIYNLMSQKFRAA	15.0	330							23	E	-3.5	3.0	0.8	1.5	-0.4
25	F R K L	FRKLCNCKQKPTEKA	15.0	345							24	Y	-1.3	-2.3	0.8	0.8	-1.2
26	A N Y S	ANYSVALNYSVIKES	15.0	360							25	Q	-3.5	0.2	0.5	1.8	-1.1
27	D R F S	DRFSTELDDITVTDT	15.0	375							26	V	4.2	-1.5	0.6	2.9	-1.5
28	Y V S T	YVSTTKVSFDDTCLA	15.0	390							27	V	4.2	-1.5	1.6	2.5	-1.5
29	S E K N	SEKNGPSSCTYGYSL	15.0	405							28	T	-0.7	-0.4	2.3	1.7	-1.5
30	T A K Q	TAKQEKI	7.0	412							29	I	4.5	-1.8	3.4	1.9	-1.7

Figure 1. User interface of the program. The screen output of the program is shown with the aa sequence of the rat TRH receptor as an example. The TRH receptor has 412 aa residues; however, the sample output shows the analysis for only a part of the protein (N-terminal 29 residues). Amino acid sequence in column A is provided by the user, and the rest of the figure is the computer output. Column A, aa sequence of rat TRH receptor, which was copied from a word-processor text file that contained 3 spaces between single-letter codes. Only the first 5 residues in each line are shown; column B, spaces between single-letter codes for aa sequence in column A were removed by the program; column C, number of aa residues in each line in column B; column D, cumulative number of aa residues in column B beginning with the first row; column E, triple-letter code for the standard 20 amino acids; column F, single-letter code for each aa in column E; column G, hydropathy index [by Kyte and Doolittle (7)] of each aa residue in column E; column H, hydrophilicity index [by Hopp and Woods (5)] of each aa residue in column E; columns I and J, reserved for other applications; column K, residue number of the sample protein; column L, aa residue number of the sample protein in column K; column M, hydropathy index of each aa residue in column L; column N, hydrophilicity index of each aa residue in column L; columns O, P and Q are for H_7, H_α and H_{Ag}, respectively, for each residue in column L. See text (Overview of the Program) for definition of H_7, H_α and H_{Ag}.

(7), the alpha helicity index (H_α) (6) and the antigenicity (H_{Ag}) (5), respectively. The first two values (H_7 and H_α) are obtained using the hydropathy index by Kyte and Doolittle (7) given in column M, and the last value (H_{Ag}) is obtained using the hydrophilicity index by Hopp and Woods (5) given in column N.

I. For single-letter code sequence in multiple lines

Cell Address	Formula
B2	=SUBSTITUTE($A2," ","")
C2	=LEN($B2)
D2	=SUM(C$2:C2)
K2	=ROWS(K$2:K2)
L2	=MID(INDEX(B$2:B$101,MATCH(ROWS(L$2:L2)-1,D$2:D$101,1)+1),ROWS(L$2:L2)-INDEX(D$2:D$101,MATCH(ROWS(L$2:L2)-1,D$2:D$101,1)),1)
M2	=INDEX(G$2:G$26,MATCH($L2,$F$2:$F$26,0))
N2	=INDEX(H$2:H$26,MATCH($L2,$F$2:$F$26,0))
O5	=AVERAGE(M2:M8)
P10	=(M2+M18+M3+M17+(M5+M15)/4+M6+M14+(M7+M13)*3/4+(M9+M11)/2+M10)/10
Q2	=AVERAGE(N2:N7)

II. When the whole sequence in single-letter code can be put in one cell (A2)

Cell Address	Formula
L2	=MID($B$2,ROWS($L$2:L2),1)

III. For triple-letter code sequence in multiple lines

Cell Address	Formula
C2	=LEN($B2)/3
L2	=INDEX($F$2:$F$26,MATCH(MID(INDEX(B$2:B$101,MATCH(ROWS(L$2:L2)-1,D$2:D$101,1)+1),(ROWS(L$2:L2)-INDEX(D$2:D$101,MATCH(ROWS(L$2:L2)-1,D$2:D$101,1)))*3-2,3),$E$2:$E$26,0))

IV. When the whole sequence in triple-letter code can be put in one cell (A2)

Cell Address	Formula
C2	=LEN($B2)/3
L2	=INDEX($F$2:$F$26,MATCH(MID($B$2,ROWS(L$2:L2)*3-2,3),$E$2:$E$26,0))

Figure 2. The program as written in a Microsoft Excel document. The program shown is for the first cell of each column. Once the program is written for the first cell, highlight it by double-clicking the mouse, and then copy it into the remaining cells in the same column. The complete program shown in I is for the aa sequence written in single-letter code in multiple lines. For an aa sequence written in single-letter code in a single line (II), in triple-letter code in multiple lines (III) or in triple-letter code in a single line (IV), replace the formulae in I with those in II, III and IV. Note that the first set of quotation marks within the formula for cell B2 contains a single space, while the second set contains no character. This formula removes spaces from sequence input in cell A2.

Writing the Program

To write the program use the following steps. Open a new document in the Microsoft Excel. Type in row 1 and columns E–H, as shown in Figure 1, into the Microsoft Excel document. Leave Column A empty for sequence input. Columns B–D and columns K–Q in Figure 1 are program outputs, and Figure 2-I shows the actual program for only the first cell of each column (row 2 except for columns O and P). Type in the formula for each cell as shown in Figure 2-I. Save the program with an appropriate name (for example, HPANAL1.XLS). Modifications of the program are explained in Flexibility of the Program and are shown in Figure 2-II, III and IV and in Figure 4. Alternatively, examples of each program shown in this paper are available in the Software Library section of the BioTechniques Web site (http://www.BioTechniques.com) or can be obtained from the authors on a floppy disk either in PC or Macintosh format.

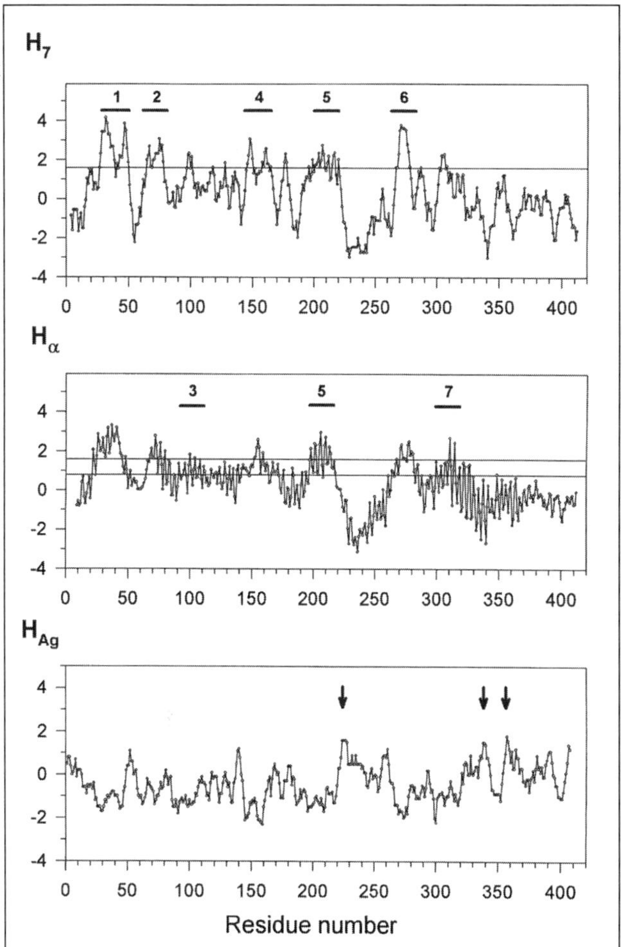

Figure 3. Analysis of the rat TRH receptor. The full length (412 residues) aa sequence of the rat TRH receptor was analyzed using this program and the graphic output was prepared using SigmaPlot. See text (Overview of the Program) for definition of H_7, H_α and H_{Ag}. In H_7 and H_α, values above the horizontal line at 1.6 are considered hydrophobic. In H_α, values below the horizontal line at 0.8 are considered hydrophilic. In H_{Ag}, regions with high values are likely to be antigenic. Bars in the plot of H_7 and H_α indicate predicted transmembrane helices, and arrows in the plot of H_{Ag} predict regions of high antigenicity.

Using the Program

Amino acid sequence written in the single-letter code is “copied” from any Windows-compatible word processor and “pasted” into column A on multiple lincs starting from cell A3. It is essential that the sequence starts from cell A3 and that cell A2 be left empty. Up to 255 characters, including space, can be written in one line, which is the maximum number allowed in one cell in the Microsoft Excel document. To use triple-letter code or nucleic acid sequence, see Flexibility of the Program. It is not necessary to remove spaces within or between lines; however, non-sequence characters (such as residue number) should be eliminated. The sequence information in column A should not exceed 100 lines. If more than 100 lines are needed, see Flexibility of the Program.

If the text file containing the sequence was not prepared using a Windows-based word processor, there is no need to worry. Most Windows word processors can read documents created by non-Windows or DOS text files. Once the sequence is read by a Windows word processor, then it can be "copied" and "pasted" into the program. Alternatively, the sequence can be written directly into the program.

To analyze the hydrophobicity of the protein, simply copy the formulae in the top row of the columns B–D and K–Q into the rest of the cells in the same column using the copy and paste functions of Excel. Figure 1 shows the result you get. Give it a separate name (for example, TRHR.XLS) so you can recognize which protein you analyzed, and save it using the *save as* function of Microsoft Excel.

For columns B–D, the formulae are copied down the column to the same row as the end of the sequence in column A. For example, when the aa sequence of a protein is writ-

Cell Address	Formula
C2	=LEN($B2)/3
L2	=INDEX($J$2:$J$101,MATCH(MID(INDEX(B$2:B$101,MATCH(ROWS(L$2:L2)-1,D$2:D$101,1)+1),(ROWS(L$2:L2)-INDEX(D$2:D$101,MATCH(ROWS(L$2:L2)-1,D$2:D$101,1)))*3-2,3),$I$2:$I$101,0))

	I	J
1	NA	AA
2	TTT	F
3	TTC	F
4	TTA	L
5	TTG	L
6	TCT	S
7	TCC	S
8	TCA	S
9	TCG	S
10	TAT	Y
11	TAC	Y
12	TAA	#
13	TAG	#
14	TGT	C
15	TGC	C
16	TGA	#
17	TGG	W

	I	J
18	CTT	L
19	CTC	L
20	CTA	L
21	CTG	L
22	CCT	P
23	CCC	P
24	CCA	P
25	CCG	P
26	CAT	H
27	CAC	H
28	CAA	Q
29	CAG	Q
30	CGT	R
31	CGC	R
32	CGA	R
33	CGG	R

	I	J
34	ATT	I
35	ATC	I
36	ATA	I
37	ATG	M
38	ACT	T
39	ACC	T
40	ACA	T
41	ACG	T
42	AAT	N
43	AAC	N
44	AAA	K
45	AAG	K
46	AGT	S
47	AGC	S
48	AGA	R
49	AGG	R

	I	J
50	GTT	V
51	GTC	V
52	GTA	V
53	GTG	V
54	GCT	A
55	GCC	A
56	GCA	A
57	GCG	A
58	GAT	D
59	GAC	D
60	GAA	E
61	GAG	E
62	GGT	G
63	GGC	G
64	GGA	G
65	GGG	G

Figure 4. Use of nucleic acid sequence data. Replace the formulae in Figure 2-I with those shown here. In addition, type in the nucleic acid to aa conversion table into columns I and J as shown here. # indicates stop codon.

ten row 30 in column A, then the formulae in columns B–D should be copied down to the row 30. For columns K–N, the formulae are copied down the column until the residue # (output of column K) is the same as the last residue number of the protein (or the output in the last row of column D). For example, when the protein contains 412 aa residues, the formulae in columns K–N are copied down to the 412th line (row 413). Column O (H_7) starts from the 4th residue and ends 4 residues before the end of the sequence, because the H_7 value assigned to position *i* is the average value of hydropathy indices of seven consecutive aa residues centered at residue *i* (7). H_7 for the N-terminal first three residues and the C-terminal last three residues are meaningless. Similarly, column P (H_α) begins with the 9th residue and ends 9 residues before the end of the sequence, because the H_α value for position *i* is a weighted average of hydropathy indices of seventeen consecutive residues centered at residue *i* (6). Column Q (H_{Ag}) begins with the first residue but ends 6 residues before the end of the sequence, because the H_{Ag} value for position *i* is an average of hydrophilicity of six consecutive residues beginning with the residue *i* (5).

In practice, the formula in each column can be copied further down the column into empty rows without affecting the results. In other words, the length of each column B–D can be longer than that of column A, and the length of each column K–Q can be longer than the total number of aa residues in the protein. Output in the inappropriate cells will be either null or error messages ("#N/A") will be obtained from the Excel program. Therefore, instead of trying to figure out the exact length for each column, we simply make sure that the entire length of the protein is covered in each column and then delete, if necessary, inappropriate cells containing error messages.

Once the analysis is complete, you can print out the result, visualize it using the graphics function of Microsoft Excel or copy and paste the result into other graphics software to draw a figure. Figure 3 gives the output for the rat TRH receptor as an example. The sequence was analyzed by this program, and the graphs were created using SigmaPlot® Version 3.0 (SPSS, Chicago, IL, USA) after copying and pasting columns K–Q into it.

For interpretation of the results, refer to the original papers and recent reviews describing the methods (4–7). Briefly, a transmembrane helix is predicted if $H_7 \geq 1.6$ for a stretch of aa residues whose length is long enough to transverse the lipid bilayer (>19 aa residues) (7). For example, H_7 plot of the rat TRH receptor (Figure 3, Top) predicts clearly 5 transmembrane helices (TM 1, 2, 4, 5 and 6). Two transmembrane helices (TM 3 and 7) are not obvious from this analysis; although, the TRH receptor belongs to the G protein-coupled receptor family, which is believed to have 7 transmembrane helices (1,9). Transmembrane segments may not necessarily satisfy $H_7 \geq 1.6$ if they are surrounded by other hydrophobic or amphiphilic helices within the lipid bilayer. Indeed, for G protein-coupled receptors in general, transmembrane helices 3 and 7 appear to be amphiphilic, and their hydrophilic surfaces are predicted to be largely shielded from the lipid bilayer by the other transmembrane helices (1). Analysis of amphiphilicity in conjunction with the analysis of hydrophobicity may significantly improve prediction of transmembrane helices. A transmembrane amphiphilic alpha helix is predicted if H_α oscillates between hydrophobic (≥1.6) and hydrophilic (≤0.8) values with a period of 3–4 residues (6). Transmembrane helices 3 and 7 for the rat TRH receptor are now more readily appreciable from the plot of H_α (Figure 3, Middle). Analysis of the hydrophilicity profile can also be used to predict segments of synthetic peptides to be used for antibody preparation. Regions of a protein that are exposed at the surface are more likely to be antigenic than the regions in the core of a protein (3,5). In addition, antibodies against surface epitopes are more likely to be useful in most applications. Such regions are predicted to have high H_{Ag} values (3,5).

Flexibility of the Program

When the aa sequence is given in multiple lines, it is essential to leave cell A2 empty. However, if the entire sequence is short enough to fit in one line (<256 aa residues), the sequence can be placed in a single cell A2, and the lengthy formula for cell L2 in Figure 2-I can be replaced with a formula given in Figure 2-II.

The program was written for aa sequence in the single-letter code. If your sequence file is written in the triple-letter code, modifications of the columns C and L, as shown in Figure 2-III, are sufficient to handle the triple-letter code. One can mix small and capital letters, because the program does not distinguish between them. Again, in rare cases when the entire sequence written in triple-letter code can fit in one cell (cell A2), the formula in cell L2 can be shortened as shown in Figure 2-IV.

The current program allows up to 100 lines in column A for sequence input. If more than 100 lines are needed to input a very long aa sequence, change B$101 and D$101 in the formulae, for cell L2 in Figure 2, to B$201 and D$201, respectively, to allow 200 lines.

If the sequence file was written using abbreviations for the aa residues different from the standard abbreviations shown in column E, modify the abbreviations in column E of the program accordingly. If aa residues not provided in column E are found in your protein, simply assign appropriate hydrophobicity values to them (if known) and add them to the table in columns E–H. Up to 5 additional aa residues can be added to the table without any modification of the program. If more space is needed, modify the formula for cell M2 and N2 in Figure 2. Change $E$26, $F$26, $G$26 and $H$26 to $E$31, $F$31, $G$31 and $H$31, respectively, to allow 5 additional aa residues. Ambiguities in the sequence can be handled efficiently using nonstandard user-defined aa codes.

To use the nucleic acid sequence, replace the formulae in columns C and L in Figure 2 with those shown in Figure 4, and type in the nucleic acid-to-amino acid conversion table into columns I and J, as shown in Figure 4. The nucleic acid sequence is written in column A in multiple lines starting from row 3 (cell A3), and the aa sequence is returned by the program in column L. By converting the nucleic acid sequence into the aa sequence, this program can be used to search the longest open reading frame. It is essential that the number of bases in each line in column A be an integer multiple of 3. This can be easily checked by examining the number of aa residues in each line in column C. Numbers in column C should be integral. For this reason, it is important to set the display of the number in column C to show the first decimal point.

Users can define their own functions to analyze the hydrophobicity information in many other ways. For example, one can test the effect of variation of window size in obtaining hydropathy index average (column O). By using mathematical formulae and copy and paste function provided in Microsoft Excel, virtually any type of manipulation of the hydrophobicity information of individual aa residues in the protein can be achieved using this program cost free.

ACKNOWLEDGMENTS

This investigation was supported in part by a Research Grant from the National Institute of Diabetes, Digestive and Kidney Diseases (No. DK 11011).

REFERENCES

1. **Baldwin, J.M.** 1993. The probable arrangement of the helices in G protein-coupled receptors. EMBO J. *12*:1693-1703.
2. **Catterall, W.A.** 1994. Molecular properties of a superfamily of plasma-membrane cation channels. Curr. Opin. Cell Biol. *6*:607-615.
3. **Hopp, T.P.** 1995. Evidence from sequence information that the interleukin-1 receptor is a transmembrane GTPase. Protein Sci. *4*:1851-1859.
4. **Hopp, T.P.** 1993. Retrospective: 12 years of antigenic determinant predictions, and more. Pept. Res. *6*:183-190.
5. **Hopp, T.P. and K.R. Woods.** 1981. Prediction of protein antigenic determinants from amino acid sequences. Proc. Natl. Acad. Sci. USA *78*:3824-3828.
6. **Jahnig, F.** 1990. Structure predictions of membrane proteins are not that bad. Trends Biochem. Sci. *15*:93-95.
7. **Kyte, J. and R.F. Doolittle.** 1982. A simple method for displaying the hydropathic character of a protein. J. Mol. Biol. *157*:105-132.
8. **Shaw, G.** 1995. Protein sequence interpretation using a spreadsheet program. BioTechniques *19*:978-983.
9. **Strader, C.D., T.M. Fong, M.R. Tota, D. Underwood and R.A.F. Dixon.** 1994. Structure and function of G protein-coupled receptors. Annu. Rev. Biochem. *63*:101-132.
10. **van der Geer, P., T. Hunter and R.A. Lindberg.** 1994. Receptor protein-tyrosine kinases and their signal transduction pathways. Annu. Rev. Cell Biol. *10*:251-337.
11. **Zhao, D., J. Yang, K.E. Jones, C. Gerald, Y. Suzuki, P.G. Hogan, W.W. Chin and A.H. Tashjian, Jr.** 1992. Molecular cloning of a complementary deoxyribonucleic acid encoding the thyrotropin-releasing hormone receptor and regulation of its messenger ribonucleic acid in rat GH cells. Endocrinology *130*:3529-3536.

CHAPTER 7

Color and Graphic Display (CGD): Programs for Multiple Sequence Alignment Analysis in Spreadsheet Software

Christian Delamarche

Université de Rennes 1, CNRS UMR 6026, Rennes, France

Reprinted from BioTechniques 29:100-107 (July 2000)

ABSTRACT

Interpretation of multiple sequence alignments is of major interest for the prediction of functional and structural domains in proteins or for the organization of related sequences in families and subfamilies. However, a necessity for the bench scientist is the use of outstanding programs in a friendly computing environment. This paper describes Color and Graphic Display (CGD), a set of modules that runs as part of the Microsoft® Excel® spreadsheet to color and analyze multiple sequence alignments. Discussed here are the main functions of CGD and the use of the program to highlight residues of importance in a water channel family. Although CGD was created for protein sequences, most of the modules are compatible with DNA sequences.

INTRODUCTION

Many commercial or freeware programs for alignment display and analysis are available (http://mercury.ebi.ac.uk/biocat/Alignment_editing_and_display.html) (11,20). Unfortunately, many of these programs are not useful for biology students or researchers because they offer a limited choice of methods and optional parameters (input and output formats, graphical types), or they may require hours of practice in unfamiliar environments, for example, using the UNIX® operating system or Adobe® PostScript® files.

Becoming familiar with a given piece of software can be time consuming. Thus, a non-expert in sequence analysis would prefer to use multipurpose software. Spreadsheet software is useful, widely used by biologists and provides many statistical tools, sorting functions and chart types. The spreadsheet software, Microsoft® Excel®, has been used to calculate enzyme kinetic parameters (18), to color amino acids (14) and to predict structural domains in proteins (2,6,13,30). This software has been used to develop Color and

BioComputing: Computer Tools for Biologists
Edited by Stuart M. Brown

Graphic Display (CGD), a package of 14 modules dedicated to multiple sequence alignment analysis.

For this paper, CGD was applied to the prediction of biological data in major intrinsic protein (MIP) channels, proteins that mediate flows of water or solutes (glycerol and other small uncharged molecules) across biological membranes (10). A topological model of an MIP channel consists of a homotetramer, in which each monomer is a pore and composed of six transmembrane α–helices connected by five loops. The internal loop B and the external loop E protrude into the pore, resulting in an hourglass shape (1,19). In this paper, sequence analysis is used to predict residues implicated in the selectivity of solutes by MIP channels.

SYSTEMS AND METHODS

Overview of CGD

The programs are written in Visual Basic® language (Microsoft). The present version of CGD (CGD.xla) has been tested on IBM PC-compatible computers with Excel 5.0, 95 and 97. (The extra-module logo.xla is not supported by Excel 95.) CGD does not create the alignments but is compatible with the widely used programs PILEUP, CLUSTAL, TREEALIGN, MULTALIGN, MAP, PIMA and BLOCKS (5).

First, the user copies the multiple alignment from a common text editor or from an HTML output file and pastes it into a cell of an Excel worksheet. When running the first module of CGD, the letters (amino acids or bases) are automatically redistributed into multiple cells. CGD uses conventional Excel dialog boxes such as color palette or format cell. Specific dialog boxes are also displayed in order to choose between algorithms or to classify amino acids into groups. In modules using computations, the results are presented in tables. The data can be plotted using the Chart Wizard of Excel, or they can be modified by applying user-defined equations.

The main modules available in CGD are briefly described here. CGD can be freely obtained from the author (cdelam@univ-rennes1.fr) or downloaded from the software library at BioTechniques at http://www.BioTechniques.com.

The MIP Family

A set of 42 MIP protein sequences was extracted from GenBank®, EMBL, TREMBL and SWISSPROT databases, which can be found at http://www.ncbi.nlm.nih.gov/ or http://www.ebi.ac.uk/embl/ or http://www.epasy.ch/sprot/, respectively. The sequences were partitioned into two functional subfamilies, one for water transport (AQP, aquaporin subgroup) and the other for solute transport (GlpF, glycerol facilitors and mixed channels). AQP subgroup: Y07625 (NLM1), P93004 (PIP3), M84344 (TIPG), P43285 (WC1A), X68293 (WC1B), X69294 (WC1C), P43286 (WC2A), P43287 (WC2B), P30302 (WC2C) and U39485 (TIPD) from *A. thaliana*; X97159 (AQPcic) from *C. viridis*; U38664 (AqpZ) from *E. coli*; Q39935, Q39957 and Q39958 from *H. annuus*; M77829 (AQP1), D31846 (AQP2), D63412 (AQP4), P55064 (AQP5), Q13520 (AQP6) and P30301 (MIP26) from *H. sapiens*; L36095 (MIPA), Q40266 (MIPB) and Q40260 (MIPE) from *M. crystallinum*; L24754 (AQPA) from *R. esculenta*; X70257 (AQP1), D13906 (AQP2), U14007 (AQP4), U16245 (AQP5); and P56405 (AQP8) from *R. norvegicus*.

GlpF subgroup: P18156 (GlpF) from *B. subtilis*; M55990 (GlpF) from *E. coli*; P44826 from *H. influenzae*; D25280 (AQP3) and AB006190 (AQP7L) from *H. sapiens*; M58315 (YDP1) from *L. lactis*; U49666 from *P. aeruginosa*; L35108 (AQP3), P56403 (AQP7) and AF016406 (AQP9) from *R. norvegicus*; P23900 (FPS1) from *S. cerevisae*; and P37451 (PDFU) from *S. thyphimurium*.

The sequences were aligned with three programs: PILEUP within the GCG package (8), CLUSTALW 1.7 (31) and PROTOMAT (17). Each time, the global alignment of the 42 sequences was separated (using Excel's cut-and-paste function) into two distinct subalignments that correspond to sequences of the two functional subfamilies (30 AQP, 12 GlpF). Then, both the global alignment and each subalignment were analyzed with the CGD tools. Key positions were retrieved from regions giving consistent results with the different alignment programs.

RESULTS AND DISCUSSION

Coloring

Coloring identical or similar residues in multiple alignments is useful to predict transmembrane regions, derive consensus sequences, find periodic repeats of patterns or simply to compare a new sequence with other members of the family. In CGD, there are two ways to color the alignments. The first way colors residues that are identical or similar in all sequences; the second way colors residues using any reference sequence selected by the user. A dialog box is displayed to assign amino acids to groups of similarity. Cell patterns and letter fonts are selected from the color palette of Excel (Figure 1).

Hydrophobicity and Similarity Plots

Hydrophobicity profiles are popular tools for predicting the topology of membrane proteins directly from their sequence. CGD proposes three scales of hydrophobicity (9,22,26) and two options. The first option assigns the hydropathy values to a single sequence selected by the user, a method independent of the alignment. In the second option, CGD averages the individual hydropathy indices at each position of the alignment. This method, based on the information given by the alignment, is more accurate because there is more information in a set of related sequences than in a single sequence (25).

One module of CGD calculates an average similarity score at each position along the multiple alignments. The score is the arithmetic mean of the distance values between all amino acid pairs. In an alignment of *n* sequences, the number of pairwise distances is as follows:

$$\binom{n}{2} = \frac{n!}{2(n-2)!}$$

and the distance values are taken from one of these three distance matrices: Blosum62 (15), Gribskov (12) and PAM250 (7).

Using the table of results created by CGD, hydrophobicity and similarity plots can be easily constructed for any window size, for the complete alignment or for some part of it. Figure 2 superimposes the average hydrophobicity and similarity plots for the alignment of 42 MIP sequences. This permits a rapid visualization of the alignment. The figure shows that loop B and E are highly conserved in the MIP family. Low values on the similarity plot indicate that these regions are poorly conserved. Negative values on the hydrophobicity plot correspond to extra amino acids (insertions) that characterized the GlpF subgroup in loop D and at the end of loop E (10,21).

Information Quantification

A module of CGD determines the frequency of occurrence $f(x, L)$ of each symbol x at each position L of the alignment. From this frequency table, it is easy to calculate the information content along the alignment, with the help of the worksheet functions of Excel. A nice method, based on the Shannon information theory (29), has been introduced by Schneider (28):

$$I_{seq}(L) = \log_2 S - H(L) - e(n) \text{ where } H(L) = \sum_{X=A}^{S} f(x, L) \log_2[f(x, L)]$$

$I_{seq}(L)$, measured in bits, is the information at position L, S is the number of symbols (4 for nucleotides or 20 for amino acids), $\log_2 S$ is the maximum value of uncertainty, n is the number of sequences, $H(L)$ is the uncertainty at position L and $e(n) = (S - 1)/(2\, n \ln 2)$ is a correction factor for finite-size samples. A sequence logo is a graphical representation of

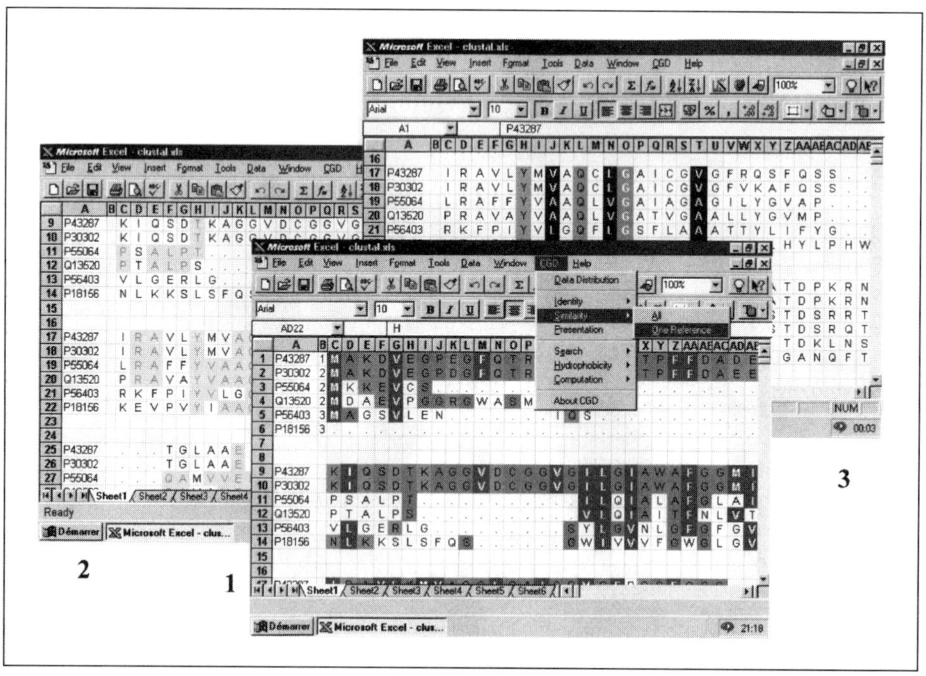

Figure 1. Three examples of multiple alignment presentations with CGD. (1) Similarities with sequence P43287 according to structural properties of the amino acids, (A, C, G, P, S, T, W, Y), (R, N, D, Q, E, H, K), (I, L, M, F, V). Numbers typed in the first block of column B permit the manual indexing of each sequence for computing the residue frequencies (see the weighting method). (2) Identities with sequence P55064. (3) Similarities between all sequences according to chemical properties of the amino acids, (D, E), (H, R, K), (C, S, T), (N, Q), (M), (A, I, L, V), (F, W, Y), (P), (G). **(See color plate A1.)**

$I_{seq}(L)$, in which the letters of the sequences are stacked on top of each other, at each position of the alignment (27), as follows:

$$\text{Height of letter } x \text{ at position } L = f(x, L)\, I_{seq}(L)$$

The size of each letter grows with the increasing frequency of the letter, and the stack height of letters represents the total information at that position. CGD includes an Excel worksheet, dedicated to protein logos (only Excel 97 and further versions). One modification in the computing of $I_{seq}(L)$, was to add the information given by gaps that are introduced in sequence alignments. Although gaps do not really exist, their presence in multiple alignments can reveal insertion/deletion events important for the structure or the function of the molecule. Therefore, in CGD, the symbol S represents 21 equally probable symbols (20 amino acids and gaps). This method reduces the stack height at positions containing many gaps. Logos are more subtle than consensus sequences because they associate a mathematical formulation of sequence conservation with a colored representation of amino acid classes. For example, the height of Q at position 350 of the MIP logo (Figure 3) indicates a low contribution of glutamine in that position of the pattern. Indeed, only one MIP sequence matches with the pattern Q-x-N-P-A-V-T.

Sequence Weighting

In a multiple alignment, the results of a counting method—and subsequently, the predictions based on it—can be biased by redundancies in the source of sequences. For example, the information contained in the orthologous sequences will dominate if each

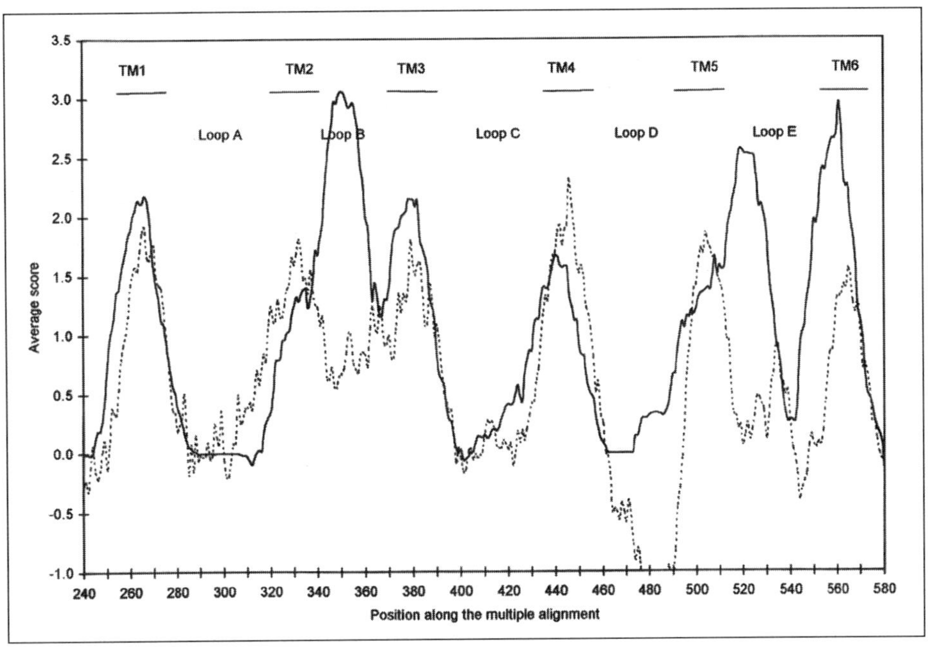

Figure 2. Average similarity (continuous curve) and hydropathy (dotted curve) plots of the MIP family proteins. CGD program was used with the similarity matrix of Gribskov (12) and hydropathy scale of Kyte and Doolittle (22), with a window size of 19 residues. For convenience, the N- and C-termini are not shown.

sequence is considered independently in a multiple alignment composed of many sequences from one evolutionary group and few sequences from another group. It is possible to improve the performance of a counting method in assigning weights to sequences (3,16). CGD provides a simple and original method to represent the heterogeneity in a set of aligned sequences. The user indexes the sequences into subclasses (evolutionary kingdoms, structural/functional particularities, clusters in a dendrogram and so forth). All sequences with the same attribute are considered as one. The result is that, when counting one residue at any position, CGD divides it by the number of sequences included in the corresponding subclass. In the example of Figure 1.1, if the counting method is applied without correction, the results for the first and second column of the alignment are respectively 5 M, 1 gap and then 3 A, 1 K, 1 D, 1 gap (total of each column = 6). When the sequences are classified in three groups as indicated in column B, the new results of residue counting are respectively 2.5 M, 0.5 gap and then 1.83 A, 0.33 D, 0.33 K, 0.5 gap (total of each column = 3).

Filtering Methods and Statistical Analysis

The statistical methods and database functionality of Excel can be applied directly to the results generated by CGD. An example is shown in Table 1. First, the weighting method described earlier was applied to classify MIP sequences into six kingdoms (mammalian, amphibian, insect, plant, yeast and bacteria) to calculate the "corrected frequency" of each residue at each position of the AQP and GlpF subalignments. Then, the amino acids were grouped and the advanced filter command of Excel was used to select posi-

Figure 3. Logo obtained from the 42 MIP sequence alignment in region of loop B. Colors are assigned according to the Blockmaker criteria (17): red for negative (D, E), blue for positive (H, K, R), light grey for polar uncharged (C, S, T), green for amide (N, Q), yellow for methionine (M), black for hydrophobic (A, I, L, V), orange for aromatic (F, W, Y), purple for proline (P) and grey for glycine (G). **(See color plate A2.)**

Table 1. Sequence Specificity of Each Functional Subgroup of the MIP Family

		AQP subgroup									GlpF subgroup									
pos.	loc.	DE	HRK	CST	NQ	M	AILV	FWY	P	G	DE	HRK	CST	NQ	M	AILV	FWY	P	G	r
259	T 1	100.0	0.0	0.0	0.0	0.0	0.0	0.0	0.0	0.0	100.0	0.0	0.0	0.0	0.0	0.0	0.0	0.0	0.0	1.00
267	T 1	0.0	0.0	0.0	0.0	0.0	100.0	0.0	0.0	0.0	0.0	0.0	0.0	0.0	13.3	86.7	0.0	0.0	0.0	0.99
326	T 2	0.0	0.0	0.0	0.0	0.0	98.2	0.0	1.8	0.0	0.0	0.0	0.0	0.0	0.0	100.0	0.0	0.0	0.0	1.00
328	T 2	0.0	0.0	0.0	0.0	3.6	83.9	12.5	0.0	0.0	0.0	0.0	0.0	0.0	0.0	88.9	11.1	0.0	0.0	1.00
334	T 2	0.0	0.0	1.8	0.0	0.0	98.2	0.0	0.0	0.0	0.0	0.0	0.0	0.0	0.0	100.0	0.0	0.0	0.0	1.00
346	Lp B	0.0	0.0	1.8	0.0	0.0	98.2	0.0	0.0	0.0	0.0	0.0	0.0	0.0	0.0	100.0	0.0	0.0	0.0	1.00
347	Lp B	0.0	0.0	100.0	0.0	0.0	0.0	0.0	0.0	0.0	0.0	0.0	100.0	0.0	0.0	0.0	0.0	0.0	0.0	1.00
348	Lp B	0.0	0.0	0.0	0.0	0.0	0.0	0.0	0.0	100.0	0.0	0.0	5.6	0.0	0.0	0.0	0.0	0.0	94.4	1.00
350	Lp B	0.0	100.0	0.0	0.0	0.0	0.0	0.0	0.0	0.0	0.0	94.4	0.0	5.6	0.0	0.0	0.0	0.0	0.0	1.00
352	Lp B	0.0	0.0	0.0	100.0	0.0	0.0	0.0	0.0	0.0	0.0	0.0	0.0	100.0	0.0	0.0	0.0	0.0	0.0	1.00
353	Lp B	0.0	0.0	0.0	0.0	0.0	0.0	0.0	100.0	0.0	0.0	0.0	0.0	0.0	0.0	6.7	0.0	93.3	0.0	1.00
355	Lp B	0.0	0.0	0.0	0.0	0.0	100.0	0.0	0.0	0.0	0.0	0.0	0.0	0.0	0.0	94.4	5.6	0.0	0.0	1.00
356	Lp B	0.0	0.0	100.0	0.0	0.0	0.0	0.0	0.0	0.0	0.0	0.0	100.0	0.0	0.0	0.0	0.0	0.0	0.0	1.00
370	T 3	0.0	0.0	3.6	0.0	0.0	90.1	0.0	0.0	6.3	0.0	0.0	0.0	0.0	0.0	80.0	20.0	0.0	0.0	0.96
373	T 3	0.0	0.0	0.0	0.0	0.0	0.0	100.0	0.0	0.0	0.0	0.0	0.0	0.0	0.0	0.0	100.0	0.0	0.0	1.00
375	T 3	0.0	0.0	1.8	0.0	0.0	98.2	0.0	0.0	0.0	0.0	0.0	5.6	0.0	0.0	87.8	0.0	0.0	6.7	1.00
377	T 3	0.0	0.0	0.0	100.0	0.0	0.0	0.0	0.0	0.0	0.0	0.0	0.0	100.0	0.0	0.0	0.0	0.0	0.0	1.00
379	T 3	0.0	0.0	0.0	0.0	0.0	98.2	1.8	0.0	0.0	0.0	0.0	0.0	0.0	0.0	100.0	0.0	0.0	0.0	1.00
380	T 3	0.0	0.0	0.0	0.0	0.0	0.0	0.0	0.0	100.0	0.0	0.0	0.0	0.0	0.0	0.0	0.0	0.0	100.0	1.00
387	T 3	0.0	0.0	1.3	0.0	0.0	93.8	5.0	0.0	0.0	0.0	0.0	20.0	0.0	0.0	80.0	0.0	0.0	0.0	0.97
388	T 3	0.0	1.3	0.0	0.0	0.0	98.7	0.0	0.0	0.0	0.0	0.0	6.7	0.0	0.0	93.3	0.0	0.0	0.0	1.00
392	Lp C	0.0	0.0	60.8	10.0	4.3	23.6	1.2	0.0	0.0	0.0	0.0	0.0	0.0	0.0	0.0	100.0	0.0	0.0	-0.18
436	T 4	100.0	0.0	0.0	0.0	0.0	0.0	0.0	0.0	0.0	80.0	0.0	0.0	20.0	0.0	0.0	0.0	0.0	0.0	0.97
505	T 5	0.0	0.0	0.0	0.0	0.0	98.2	1.8	0.0	0.0	0.0	0.0	0.0	0.0	0.0	100.0	0.0	0.0	0.0	1.00
509	T 5	0.0	90.7	0.0	0.0	0.0	9.3	0.0	0.0	0.0	0.0	0.0	6.7	33.3	0.0	5.6	0.0	0.0	54.4	-0.20
516	Lp E	0.0	0.0	100.0	0.0	0.0	0.0	0.0	0.0	0.0	0.0	0.0	100.0	0.0	0.0	0.0	0.0	0.0	0.0	1.00
519	Lp E	0.0	0.0	62.0	0.0	0.0	0.0	0.0	0.0	38.0	0.0	0.0	0.0	0.0	0.0	94.4	0.0	0.0	5.6	-0.14
521	Lp E	0.0	0.0	0.0	100.0	0.0	0.0	0.0	0.0	0.0	0.0	0.0	0.0	100.0	0.0	0.0	0.0	0.0	0.0	1.00
523	Lp E	0.0	0.0	0.0	0.0	0.0	98.7	0.0	0.0	1.3	0.0	0.0	13.3	0.0	0.0	86.7	0.0	0.0	0.0	0.99
524	Lp E	0.0	98.7	0.0	0.0	0.0	1.3	0.0	0.0	0.0	0.0	100.0	0.0	0.0	0.0	0.0	0.0	0.0	0.0	1.00
525	Lp E	0.0	0.0	96.9	0.0	0.0	3.1	0.0	0.0	0.0	100.0	0.0	0.0	0.0	0.0	0.0	0.0	0.0	0.0	-0.11
529	Lp E	0.0	0.0	2.5	0.0	0.0	97.5	0.0	0.0	0.0	0.0	100.0	0.0	0.0	0.0	0.0	0.0	0.0	0.0	-0.11
555	T 6	0.0	0.0	0.0	0.0	0.0	0.0	100.0	0.0	0.0	0.0	0.0	0.0	0.0	0.0	0.0	0.0	100.0	0.0	-0.11
556	T 6	0.0	0.0	0.0	0.0	0.0	1.3	98.7	0.0	0.0	0.0	0.0	0.0	0.0	38.9	61.1	0.0	0.0	0.0	-0.15
559	T 6	0.0	0.0	0.0	0.0	0.0	0.0	0.0	100.0	0.0	0.0	0.0	0.0	0.0	0.0	0.0	0.0	100.0	0.0	1.00
561	T 6	0.0	0.0	0.0	0.0	1.8	98.2	0.0	0.0	0.0	0.0	0.0	0.0	0.0	0.0	100.0	0.0	0.0	0.0	1.00
562	T 6	0.0	0.0	0.0	0.0	0.0	1.8	0.0	0.0	98.2	0.0	0.0	0.0	0.0	0.0	5.6	0.0	0.0	94.4	1.00
570	T 6	0.0	12.5	0.0	0.0	0.0	1.8	85.7	0.0	0.0	0.0	0.0	0.0	0.0	0.0	0.0	100.0	0.0	0.0	0.99

Column 1, position in the global multiple alignment; column 2, probable localization in a transmembrane or a loop region; columns 3–20, frequency of each amino acid group, expressed in percentage; column 21, correlation coefficient. Shaded lines point out amino acid groups that could modulate the selectivity of MIP channels.

tions in the multiple alignment that follow given criteria. A position was considered prominent in the MIP family when the conservation of one amino acid group was ≥80% in the two subfamilies, and a position was selective when the percentage of one amino acid group was ≥80% in one subfamily and ≥20% in the other (grey lines in Table 1).

The Pearson's correlation coefficient method, directly available in Excel, permits a statistical comparison of the distribution of the amino acid groups for the selected positions. The r values range from -1 to +1. Dissimilar positions between the subgroups have a negative r score (the groups are mutually determined, but they vary in the opposite sense). From Table 1, it is easy to describe the two major functional types of MIP channels from a limited number of key positions (gray areas). According to these predictions, we have recently shown that positions 555 and 556 (Table 1) participate in the selection of the transported molecule, water or glycerol (23,24). From prominent positions in the multiple alignment (table 1), it is also possible to define a new concept of signature sequence in which amino acids in brackets have common properties. For example, the signature [CST]-G-x-[HRK]-x-[NQ]-P-x-[AILV]-[CST], defined from positions 347–356, can be translated in a natural language: polar uncharged, glycine, any, positive, any, amide, proline, any, hydrophobic, polar uncharged. Such a description can help the structural interpretations. The new signature retrieves 203 MIP sequences in the SWISS-PROT and TREMBL databases at http://www.expasy.ch/tools/scnpsit2.html. The official signature of the MIP family, PS00221 in the PROSITE database (4), which is [HNQA]-x-N-P-[STA]-[LIVMF]-[ST]-[LIVMF]-[GSTAFY], retrieves 214 sequences. However, this sequence signature is less informative because it is only a consensus with no meaning in terms of the conservation of biological properties along the sequence.

CONCLUSION

CGD, software for multiple sequence alignment analysis, runs on personal computers equipped with Excel. The methods are easy to understand, and at any time, the user can follow and access the processing of its data. Many statistical methods are available in Excel, and users can select the most versatile presentation for slides or paper-quality figures, choosing between a large variety of charts, styles and colors. Users can then implement their own algorithms, taking advantage of the high programming capabilities of Visual Basic.

ACKNOWLEDGMENTS

I would like to thank Drs. Jean-Jacques Lebrun and Suad Ali (McGill University, Montréal) for software testing. I thank Prof. J. Gouranton and Dr. D. Thomas (Université de Rennes 1) for their support and for stimulating discussions during the development of this work.

REFERENCES

1. **Agre, P., M. Bonhivers and M.J. Borgnia.** 1998. The aquaporins, blueprints for cellular plumbing systems. J. Biol. Chem. *273*:14659-14662.
2. **Aloy, P., J. Cedano, B. Oliva, F.X. Aviles and E. Querol.** 1997. 'TransMem' : a neural network implemented in Excel spreadsheets for predicting transmembrane domains of proteins. Comput. Appl. Biosci. *13*:231-234.
3. **Altschul, S.F., R.J. Carroll and D.J. Lipman.** 1989. Weights for data related by a tree. J. Mol. Biol. *207*:647-653.
4. **Bairoch, A., P. Bucher and K. Hofmann.** 1996. The PROSITE database, its status in 1995. Nucleic Acids Res. *24*:189-196.
5. **Briffeuil, P., G. Baudoux, C. Lambert, X. De Bolle, C. Vinals, E. Feytmans and E. Depiereux.** 1998. Comparative analysis of seven multiple protein sequence alignment servers: clues to enhance reliability of predictions. Bioinformatics *14*:357-366.
6. **Clotet, J., J. Cedano and E. Querol.** 1994. An Excel spreadsheet computer program combining algorithms for prediction of protein structural characteristics. Comput. Appl. Biosci. *10*:495-500.
7. **Dayhoff, M.O., R.M. Schwartz and B.C. Orcutt.** 1978. A model of evolutionary changes in proteins, p. 345-358. *In* M.O. Dayhoff (Ed.), Atlas of Protein Sequence and Structure; vol. 5, suppl. 3. National Biochemical Research Foundation, Washington, DC.
8. **Devereux, J., P. Haeberli and O. Smithies.** 1984. A comprehensive set of sequence analysis programs for the VAX. Nucleic Acids Res. *12*:387-395.
9. **Eisenberg, D. and A.D. McLachlan.** 1986. Solvation energy in protein folding and binding. Nature *319*:199-

203.
10.**Froger, A., B. Tallur, D. Thomas and C. Delamarche.** 1998. Prediction of functional residues in water channel and related proteins. Protein Sci. *7*:1458-1468.
11.**Gotoh, O.** 1999. Multiple sequence alignment: algorithms and applications. Adv. Biophys. *36*:159-206.
12.**Gribskov, M. and R.R. Burgess.** 1986. Sigma factors from E. coli, B. subtilis, phage SPO1, and phage T4 are homologous proteins. Nucleic Acids Res. *14*:6745-6763.
13.**Han, B. and A.H. Tashjian.** 1998. User-friendly and versatile software for analysis of protein hydrophobicity. BioTechniques *25*:256-263.
14.**Haygood, M.G.** 1993. Spreadsheet macros for coloring sequence alignments. BioTechniques *15*:1084-1088.
15.**Henikoff, S. and J.G. Henikoff.** 1992. Amino acid substitution matrices from protein blocks. Proc. Natl. Acad. Sci. USA *89*:10915-10919.
16.**Henikoff, S. and J.G. Henikoff.** 1994. Position-based sequence weights. J. Mol. Biol. *243*:574-578.
17.**Henikoff, S., J.G. Henikoff, W.J. Alford and S. Pietrokovski.** 1995. Automated construction and graphical presentation of protein blocks from unaligned sequences. Gene *163*:GC17-GC26.
18.**Hernandez, A. and M.T. Ruiz.** 1998. An EXCEL template for calculation of enzyme kinetic parameters by non-linear regression. Bioinformatics *14*:227-228.
19.**Heymann, J.B., P. Agre and A. Engel.** 1998. Progress on the structure and function of Aquaporin 1. J. Struct. Biol. *121*:191-206.
20.**Hofmann, K.** 1998. Protein classification and functional assignment, p. 18-21. Trends Guide to Bioinformatics. Elsevier Science, New York.
21.**Ishibashi, K. and S. Sasaki.** 1998. The dichotomy of MIP family suggests two separate origins of water channels. News Physiol. Sci. *13*:137-142.
22.**Kyte, J. and R.F. Doolittle.** 1982. A simple method for displaying the hydropathic character of a protein. J. Mol. Biol. *157*:105-132.
23.**Lagrée, V., A. Froger, S. Deschamps, I. Pellerin, C. Delamarche, G. Bonnec, J. Gouranton, D. Thomas and J.F. Hubert.** 1998. Oligomerization state of water channels and glycerol facilitators. J. Biol. Chem. *273*:33949-33953.
24.**Lagrée, V., A. Froger, S. Deschamps, J.F. Hubert, C. Delamarche, G. Bonnec, D. Thomas, J. Gouranton and I. Pellerin.** 1999. Switch from an aquaporin to glycerol channel by two amino acids. J. Biol. Chem. *274*:6817-6819.
25.**Persson, B. and P. Argos.** 1994. Prediction of transmembrane segments in proteins utilising multiple sequence alignments. J. Mol. Biol. *237*:182-192.
26.**Rao, J.K.M. and P. Argos.** 1986. A conformational preference parameter to predict helices in integral membrane proteins. Biochim. Biophys. Acta *869*:197-214.
27.**Schneider, T.D. and R.M. Stephens.** 1990. Sequence logos: a new way to display consensus sequences. Nucleic. Acids Res. *18*:6097-6100.
28.**Schneider, T.D., G.D. Stormo, L. Gold and A. Ehrenfeucht.** 1986. Information content of binding sites on nucleotide sequences. J. Mol. Biol. *188*:415-431.
29.**Shannon, C.E.** 1948. A mathematical theory of communication. Bell System Tech. J. *27*:379-423.
30.**Shaw, G.** 1995. Protein sequence interpretation using a spreadsheet program. BioTechniques *19*:978-983.
31.**Thompson, J.D., D.G. Higgins and T.J. Gibson.** 1994. CLUSTAL W: improving the sensitivity of progressive multiple sequence alignment through sequence weighting, positions-specific gap penalties and weight matrix choice. Nucleic Acids Res. *22*:4673-4680.

Update to: Color and Graphic Display (CGD): Programs for Multiple Sequence Alignment Analysis in Spreadsheet Software

Christian Delamarche and Grégory Ranchy

Université de Rennes 1, UMR CNRS 6026, Rennes, France
(christian.delamarche@univ_rennes1.fr)

The last version of Color and Graphic Display (CGD) can be obtained by e-mail at (christian.delamarche@univ-rennes1.fr).

WHAT IS CGD?

CGD offers a set of Add-Ins for the spreadsheet software Microsoft® Excel®. The Add-Ins provide many tools to assist biologists in the interpretation of multiple sequence alignments of nucleotides or amino acids. It is an integrated and useful solution to perform analysis of large alignments.

NEW FEATURE IN THE VERSION 1.1

The Program Interface

Many dialog boxes and algorithms have been changed to speed up the analysis and to guide the user in the choice of parameters. Here lays the main CGD tool bar (Figure 1), indicating that a new method is available to automatically parse out any multiple sequence alignment to Excel.

Tables and Matrices

Twelve tables of amino acid properties and twelve score matrices are available for secondary structure predictions and similarity profiles (Figure 2). When using these scales,

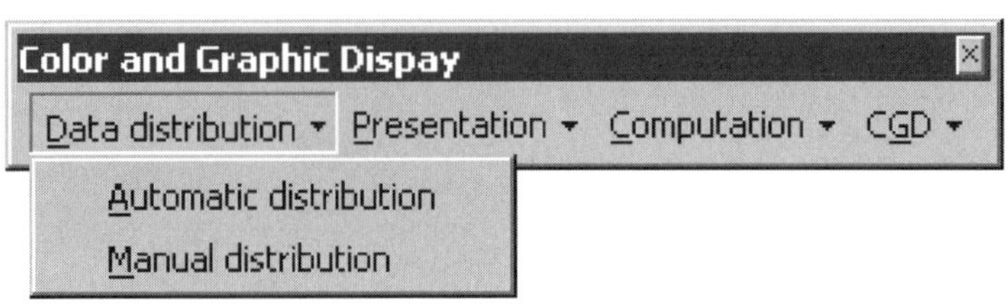

Figure 1. The main CGD tool bar.

the properties of one sequence or the averaged values of hydrophobicity and/or similarity at each position of an alignment are copied in a new Excel worksheet. A dialog box is used to fix the window width when necessary.

Counting Methods

There are now four methods to calculate the distribution of the letters in a multiple alignment. The new modules are named COUNTING 3 and COUNTING 4. These modules are useful when the sequences are indexed into subclasses (see the sequence weighting method). COUNTING 3 gives the mean of each letter (amino acids or nucleotides) for each subclass. The COUNTING 4 module gives a frequency table of each letter at each position of the alignment for each subclass.

Logos

This option has been updated. One file named logo50.xls is now included in the CGD package. This can be useful to automatically create protein sequence logos. On a logo, the horizontal axis shows the positions in the alignment, and the vertical axis represents the amount of information at each position (in bits). The new max-default for logos in CGD is 50 columns. When the user pastes the results of a counting table (COUNTING 2 and COUNTING 4), a sequence logo is immediately created. If the user has a good working knowledge of Excel, it is possible to create customized logos for proteins by using the ChartWizard and pictures of amino acids.

Tools

When working on large multiple alignments, it is difficult to know the position of a particular letter in a sequence or the position of this letter in the alignment. Four tools are available to give the relationships between alignment positions, letters, and sequence names.

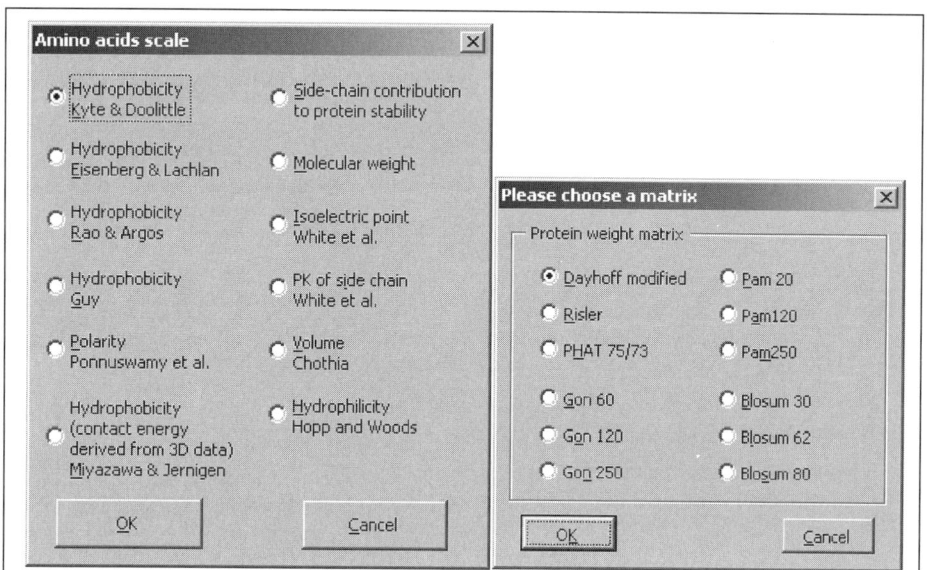

Figure 2. Tables of amino acid properties and score matrices.

Spreadsheet Macros for Coloring Sequence Alignments

Margo G. Haygood

Scripps Institution of Oceanography, La Jolla, CA, USA

Reprinted from BioTechniques 15:1084-1089 (December 1993)

ABSTRACT

This article describes a set of Microsoft® Excel™ macros designed to color amino acid and nucleotide sequence alignments for review and preparation of visual aids. The colored alignments can then be modified to emphasize features of interest. Procedures for importing and coloring sequences are described. The macro file adds a new menu to the menu bar containing sequence-related commands to enable users unfamiliar with Excel to use the macros more readily. The macros were designed for use with Macintosh® computers but will also run with the DOS version of Excel.

INTRODUCTION

Computer algorithms that aid in producing sequence alignments are very helpful, but alignments should always be reviewed visually. The powerful pattern recognition capabilities of the human brain are greatly enhanced by the use of color, and display of sequence alignments in color is beneficial. I present here a technique for coloring alignments (Figure 1) for review, documentation and preparation of visual aids that uses inexpensive, widely available software. Sequences can also be displayed in color by some commercial sequence analysis programs, as well as noncommercial programs such as the UNIX-based Genetic Data Environment. Custom-designed fonts with colored backgrounds can also be used with text editors. The technique presented here has the advantages of low cost, minimal hardware requirements and ease of output. Related techniques can be found in References 7–10. It was developed for use on Macintosh® computers with the spreadsheet program Microsoft® Excel™ 4.0 (Microsoft, Redmond, WA, USA) and has been used in our laboratory for the past year as a convenient adjunct to other computer resources. The macros are portable to the DOS version of Excel.

The macros are contained in a file called color sequences. When the file is opened in Excel, a new menu (**Sequence**) appears on the menu bar, containing sequence-related

BioComputing: Computer Tools for Biologists
Edited by Stuart M. Brown

commands. The first group of commands, **Parse Sequence** and **Parse Column**, are used in importing sequences into Excel. The second group, **Color Nucleotides** and **Color Amino Acids**, adds colors to the sequence. The third group of commands are used in highlighting regions of interest by removing or dimming colors in other regions. **Color Amino Acids** colors most amino acids a different color (Figure 1). The color palette was designed with reference to the Dayhoff log-odds matrix (3), which weights amino acids according to their observed frequency of replacement in proteins and has been shown to be an effective means of detecting relationships (4). Amino acids that are heavily weighted are given the brightest colors, those that are more frequently replaced have less arresting colors; related amino acids are given similar colors. Since Excel's color palette is limited to 16 colors, a few amino acids with the lowest weights are colored gray.

Color Nucleotides colors nucleotide sequence data (Figure 1). The color palette gives purines one pair of related colors and pyrimidines another to reflect the typical higher frequency of transitions over transversions in nucleotide substitutions, and colors Gs and Cs more intensely to communicate their stronger hydrogen bonds. The colored worksheets can be printed on paper or transparencies in color or black and white, or output as slides. An overview of the procedures appears in Figure 2.

IMPORTING SEQUENCES INTO EXCEL

See Figure 3. To use the macros, open the file color sequences. In order to avoid cluttering the screen, the macro sheet is formatted as hidden and thus is not visible. To work with sequences in Excel, each character of the sequence must be in a separate cell. When Excel imports data from text files, tabs are used to separate data into cells, and hard returns separate data into rows. Data that are not formatted appropriately with tabs must be "parsed." Data can be parsed manually using the **Parse** command on the **Data** menu (see below). Parsing routines for some common sequence types are found on the **Sequence** menu. Sequence files can be opened in Excel if they are saved as text files or

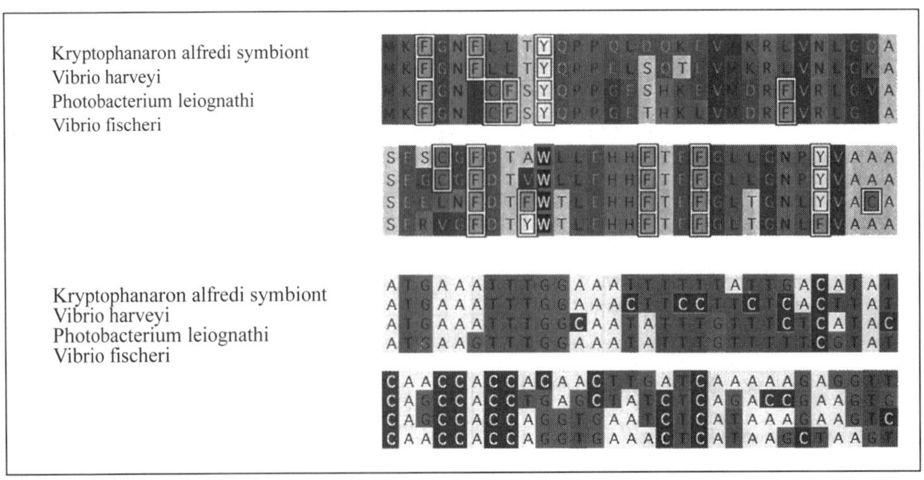

Figure 1. Examples of colored sequences. Upper panel: Partial alignment of *luxA* (bacterial luciferase alpha subunit) amino acid sequences. Lower panel: Partial alignment of *luxA* nucleotide sequences. References for sequences: (1, 2, 5, 6). **(See color plate A2.)**

copied and pasted into Excel from a word processing program. Individual sequences should be parsed in separate worksheets and then combined by copying and pasting. For all parsing operations make sure that there are no blank rows above the data.

Excel worksheets are limited to 256 columns, and cells can contain a maximum of 255 characters. Line length in the file to be imported should be ≤250 characters. If the user tries to import a file with a longer line length, the extra characters will be lost, so always work with a copy of the file. For editing purposes, the longer the line length the better. For preparation of figures and slides, shorter line lengths are preferable. It is best if the line length is the same throughout the file (except for the last line). If not, the first line must be ≥ all of the others in length in order for parsing commands to work properly because the parse commands on the **Sequence** menu work by counting the number of characters in the first cell of the column containing sequence and placing a single character into a cell that number of times. The user can adjust the line length as follows: In a word processing program, use a non-proportionally spaced font (e.g., Courier), strip hard returns, increase the margin, select a line and do a character count, and adjust the margin as needed. When the line length has the desired value, save as "text with line breaks," which will add hard returns at the end of each line.

Some characters cause problems and should be replaced. Spaces in the data matrix are removed during parsing and thus should not be used as a gap character. Spaces that break all sequences into blocks are all right since the alignment will not be changed when they are removed. Likewise, hyphens have a special meaning in Excel and cannot be used as a gap character. Hyphens or spaces used as alignment gaps should be replaced with another character, such as underscore, before importing into Excel. Tabs are recognized as separating data into cells and may need to be removed.

The command **Parse Sequence** on the **Sequence** menu is designed to parse sequences that consist of a column of names or numbers, followed by one or more spaces, followed by sequences all beginning at the same position, as in the following example:

```
Name1         XXXXXXXXXXXXXXXXXXX
Another_name  YYYYYYYYYYYYYYYYYYY
Third_name    ZZZZZZZZZZZZZZZZZZZ
              XXXXXXXXXXXXXXXXXXX
              YYYYYYYYYYYYYYYYYYY
              ZZZZZZZZZZZZZZZZZZZ
```

Parse Sequence places the names in one column and then distributes the sequence one character to a cell. It recognizes the name column by one or more spaces between the name and the sequence, so if the names have spaces within them, they must be removed. Nexus (PAUP) and PHYLIP files can be imported with **Parse Sequence**. Open the file in a text editor and select the data matrix only (names and sequences), copy and paste into a new worksheet in the upper left corner and run **Parse Sequence**. To import GenBank® files, select the numbers and sequence only in a text editor, then copy and paste into a new worksheet in the upper left corner and choose **Parse Sequence**. Then select the first column and choose **Insert** (on the **Edit** menu) to create a new column. Enter the name of the sequence in the top left cell. To import a file containing only the sequence itself, check to see that the line length is ≤250 (sequence-only files usually do not contain hard returns), then copy and paste into a new worksheet as above and choose **Parse Column**. Insert a new column as above and enter the name of the sequence in the top left cell.

For other sequence formats some experimentation may be necessary, keeping in mind the guidelines above. On the **Sequence** menu, **Parse Sequence** puts everything in front of the first space into a cell and then distributes all of the remaining characters into individual cells; **Parse Column** distributes all of the characters in the line into individual cells. Tabs can be used to force data into different cells, if, for example, names are different lengths and are not filled out by spaces to ensure that sequence data begins at the same position for all lines. The command **Parse** on the **Data** menu can be used to parse data manually. Select the range of cells to be parsed (only one column). Choose **Parse** on the **Data** menu. A dialog box appears showing the contents of the first cell, with square brackets showing where cells will be separated. Type brackets in the appropriate places. Brackets can be removed with the Clear button. Because it is tedious to separate all of the sequence characters with brackets, it is convenient to use **Parse** on the **Data** menu to separate the sequence data from other items and then use **Parse Column** on the **Sequence** menu to distribute the sequence into cells. The phylogenetic analysis program MacClade (Sinauer Associates, Sunderland, MA, USA) imports a variety of data types and saves in spreadsheet format ("simple table"), providing another option for easily transferring sequence data to Excel.

The **Parse Sequence** and **Parse Column** commands on the **Sequence** menu bring up the Save As dialog box after processing the data. This is to encourage the user to save under a different name from the original text file and to ensure that the file is in Excel normal format. If it remains in text format, color information will be lost. If data is imported by another means, be sure to **Save As** (**File** menu) and choose Options: normal format.

COLORING SEQUENCES

See Figure 4. To color a sequence, choose **Color Nucleotides** or **Color Amino Acids** as appropriate from the **Sequence** menu. The command first brings up the Save As dialog box to encourage the user to save under a different name from the data file. The color commands save periodically and will overwrite the original file unless this is done. Choose Cancel in the Save As dialog box if the user does not wish to save under another name. The user is asked to select the range containing the sequence to be colored (not including names, etc.) and then click the resume button. The resume button is a small box with an icon on it that appears in the upper left of the window at

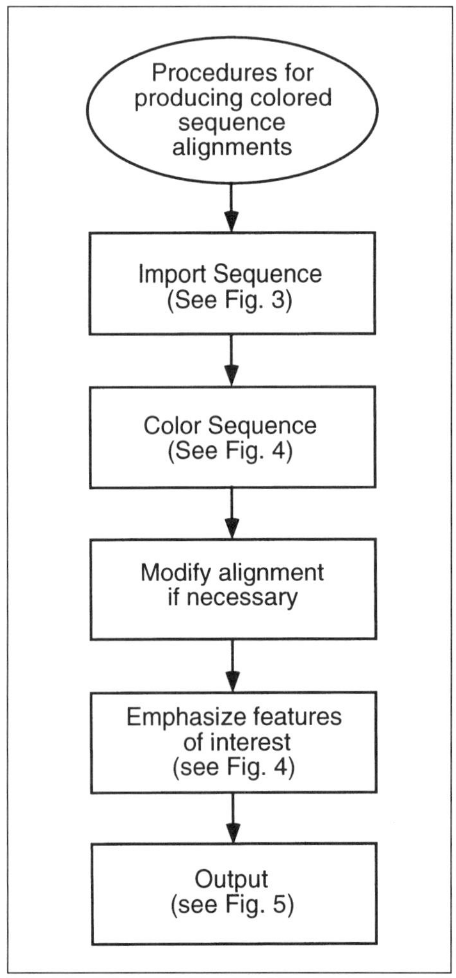

Figure 2. Overview of procedures for coloring sequences.

this point in the process. To increase the speed of the macro, the screen does not redraw, but the row and column numbers of the cell it is working on are displayed in the status line at the bottom left corner of the screen and the macro beeps when finished. Because changes are held in memory, the macro will slow down on large files unless the file is frequently saved. For this reason, the macro saves the file after every 100 cells. It also saves at the end. The macro does not run in background but can be interrupted with Command-period at any time (choose Halt in the dialog box that appears), and the remaining sequence can be selected and colored later. Remember that cells processed since the last save have not been saved if the macro is interrupted; therefore, it is a good idea to save immediately after aborting the macro.

EDITING ALIGNMENTS

See Figure 4. Editing alignments is easier in a spreadsheet than in a word processing program, since vertical blocks of sequence can be shifted by insertion or deletion or cut and pasted as a block. For example, to insert an alignment gap into a group of sequences, select the first cells after the gap in the sequences affected (it is not necessary to select the whole line), choose **Insert** on the **Edit** menu and choose Shift Cells Right in the dialog box that appears. The selected sequences will shift right one cell. Sequences can be shifted by more than one position by selecting the appropriate number of cells. Multiple noncontiguous selections can be made by selecting a cell and holding down the command key and selecting the next cell. A section of sequence can be moved, for example, from the end of one block of sequence down to the beginning of the next, by selecting the cells, choosing **Cut** from the **Edit** menu, moving to the beginning of the next block, selecting the first cell and choosing **Insert** from the **Edit** menu. Do not use **Paste**, or you will overwrite data. To minimize the need for this operation, use the maximum possible line length. To move a whole

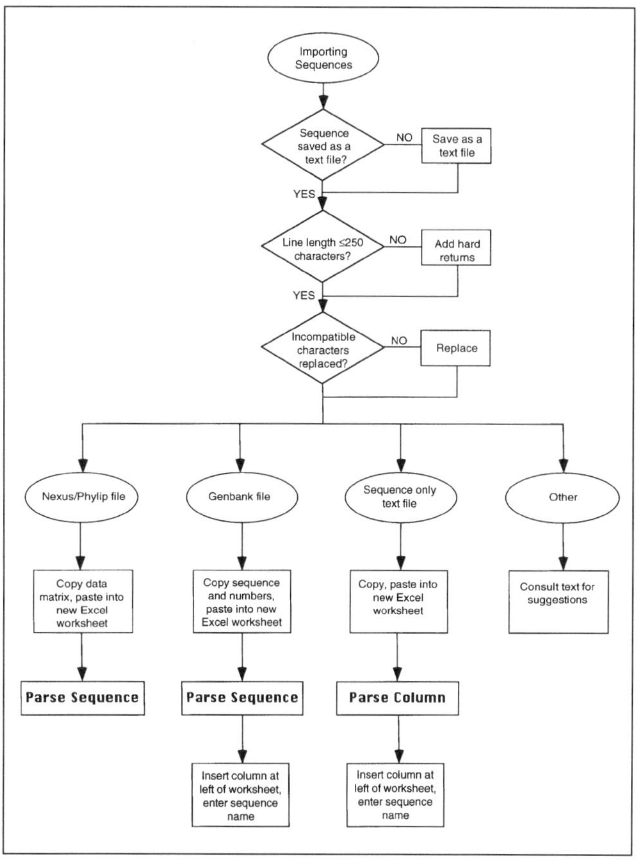

Figure 3. Flowchart of procedures for importing data.

row, for example, if creating an alignment from individual sequences, select the row by clicking on the row number at the left, choose **Cut** from the **Edit** menu, select the row above which you want the cut row to appear and then choose **Insert** from the **Edit** menu. Excel files can be saved as text, tabs stripped in a word processing program and then the data can be used for further analysis.

EMPHASIZING FEATURES OF INTEREST

See Figure 4. After sequences have been colored and alignments evaluated and adjusted if necessary, it may be desirable to emphasize certain features for use by the researcher or for preparing visual aids. The commands **Auto Clear/Dim**, **Clear**, **Dim** and **Undim** are used for this purpose. **Clear** removes colors, but does not affect the sequence data, **Dim** adds a shaded pattern to deemphasize cells and **Undim** reverses **Dim**. To automatically modify constant or variable regions use **Auto Clear/Dim**. For example, to highlight variable regions by clearing constant regions, use **Auto Clear/Dim** and choose Constant and Clear in the dialog box. The command will not work with noncontiguous selections. The user may modify any region of choice with **Clear** or **Dim**. Select the region to modify (noncontiguous selections are all right) and choose **Clear** or **Dim** (keyboard shortcuts are shown on the menu). **Undim** reverses the **Dim** command. **Clear** can be reversed by running **Color Nucleotide** or **Color Amino Acids** on the cleared cells. Do not include blank cells, otherwise they will be shaded (**Dim**) or black (**Undim**). This can be corrected with **Clear**, which will return them to blank. For example, to produce an alignment in which only phylogenetically informative sites are colored, run **Auto Clear/Dim** and choose Constant and Clear, and then clear the remaining variable but uninformative sites manually with **Clear**.

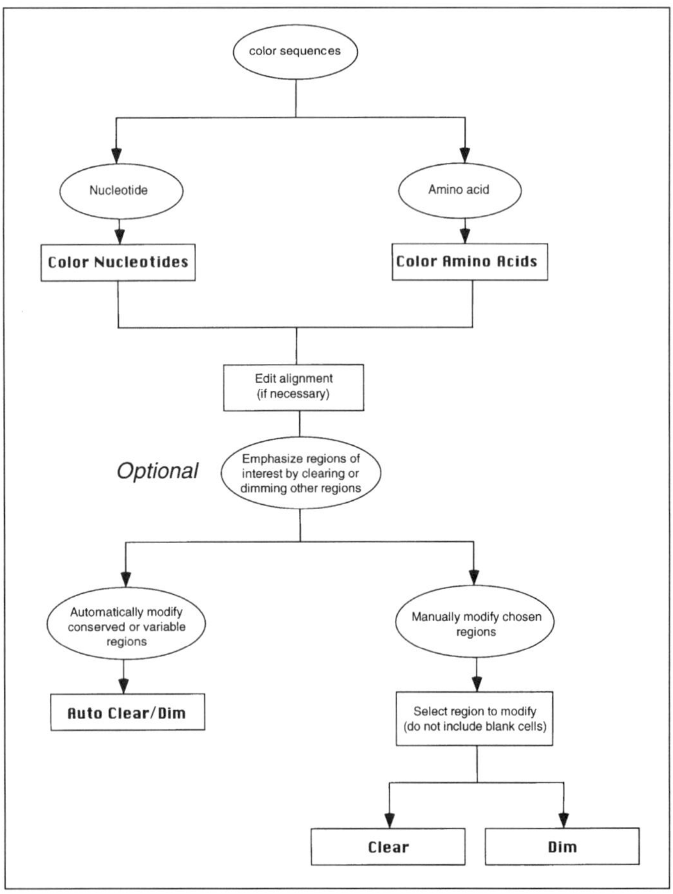

Figure 4. Flowchart of procedures for coloring and modifying sequence display.

OUTPUT

See Figure 5. The colored alignments can be printed quite satisfactorily on an inexpensive color inkjet printer as either paper or transparencies. To obtain a quick copy for documentation of the alignment used for a particular analysis, the colored worksheet can be printed to a black and white printer, in which case the colors are replaced by patterns. The patterns used are arbitrary and not as informative as colors, particularly for amino acids. I recommend this approach, however, because it allows the colored worksheet to be printed directly without modification. If a patterned worksheet that corresponds more closely to the colors is desired, another macro "pattern_aas" is included in the macro file but not listed on the **Sequence** menu. This works in the same way as **Color Amino Acids**, but since it removes colors, it must be used on a copy of the colored worksheet. It can be accessed by choosing **Run** on the **Macro** menu, and selecting "–'color sequences'!pattern_aas" from the list. For preparation of slides, the desired region of the worksheet is selected and copied using **Copy as Picture** (hold down the shift key while pulling down the **Edit** menu to see this choice and choose As Shown On Screen) and pasted into a drawing program. It is important to allocate plenty of memory to the drawing program because the file is large due to the numerous objects. It can then be saved as a PICT file and output to a film recorder or sent to a service bureau. Small alignments can be photographed from the monitor.

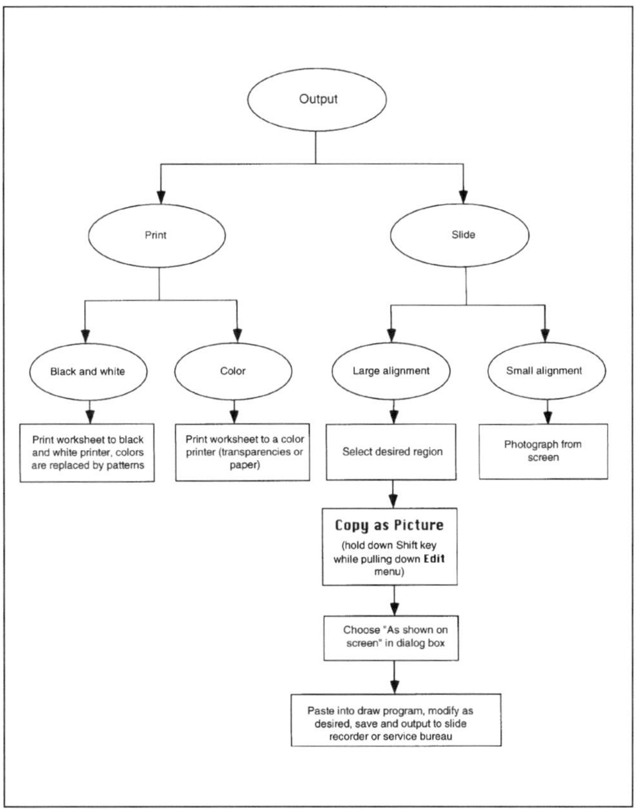

Figure 5. Flowchart of output procedures.

MODIFYING THE MACROS

The macro file is formatted as an international macro, so it can be used with international systems (the text messages will appear in English). The macro sheet is formatted as hidden to prevent accidental modification. To view and edit the macro sheet for specialized purposes, choose **Unhide** from the **Windows** menu. The macros are annotated to facilitate modification. For example, some of the ALERT messages that are no longer needed by an experienced user can be removed.

ACKNOWLEDGMENTS

I thank Ron Caspi for DOS testing and anonymous reviewers for suggestions for improving the macros. This work was supported by the U.S. Office of Naval Research.

REFERENCES

1.**Baldwin, T.O., J.H. Devine, R.C. Heckel, J.-W. Lin and G.S. Shadel.** 1989. The complete nucleotide sequence of the *lux* region of *Vibrio fischeri* and the *lux* ABN region of *Photobacterium leiognathi* and the mechanism of control of bacterial luminescence. Biolumin. Chemilumin. *4*:326-341.

2.**Cohn, D.H., A.J. Mileham, M.I. Simon, K.H. Nealson, S.K. Rausch, D. Bonam and T.O. Baldwin.** 1985. Nucleotide sequence of the *lux A* gene of *Vibrio harveyi* and the complete amino acid sequence of the α subunit of bacterial luciferase. J. Biol. Chem. *260*:6139-6146.

3.**Dayhoff, M.O.** 1978. A model of evolutionary change in proteins. Matrices for detecting distant relationships, p. 345-358. *In* M.O. Dayhoff (Ed.), Atlas of Protein Sequence and Structure. National Biomedical Research Foundation, Washington, D.C.

4.**Feng, D.F., M.S. Johnson and R.F. Doolittle.** 1985. Aligning amino acid sequences: comparison of commonly used methods. J. Mol. Evol. *21*:112-125.

5.**Haygood, M.G.** 1990. Relationship of the luminous bacterial symbiont of the Caribbean flashlight fish, *Kryptophanaron alfredi* (family Anomalopidae) to other luminous bacteria based on bacterial luciferase (*luxA*) genes. Arch. Microbiol. *154*:496-503.

6.**Illarionov, B.A., M. Protopopova, V.A. Karginov, N.P. Mertvetsov and J.I. Gitelson.** 1988. Nucleotide sequence of part of the *Photobacterium leiognathi lux* region. Nucleic Acids Res. *16*:9855.

7.**Schneider, T.D. and R.M. Stephens.** 1990. "Sequence Logos": A new way to display consensus sequences. Nucleic Acids. Res. *18*:6097-6100.

8.**Shankarappa, B.** 1993. Transformation of sequence data into geometric symbols. BioTechniques. *14*:990-996.

9.**Taylor, W.R.** 1986. The classification of amino acid conservation. J. Theor. Biol. *119*:205-218.

10.**Zimmerman, P.A., M.L. Spell, J. Rawls and T.R. Unnasch.** 1991. Transformation of DNA sequence data into geometric symbols. BioTechniques *11*:50-52.

CHAPTER 9

ROCK: A Spreadsheet-Based Program for the Generation and Analysis of Random Oligonucleotide Primers Used in PCR

Steven R. Strain and Jerry G. Chmielewski

Slippery Rock University of Pennsylvania, Slippery Rock, PA, USA

Reprinted from BioTechniques 30:1286-1291 (June 2001)

ABSTRACT

Oligonucleotide primers used to amplify target DNA regions via PCR should meet certain design criteria to maximize the potential for efficient priming. The Random Oligonucleotide Construction Kit (ROCK), a spreadsheet-based program that runs under Microsoft® Excel® 97 or later version for Microsoft Windows™, was developed to facilitate the design of efficient random oligonucleotide primers. Primer sequences are generated that meet user-defined criteria with regard to G+C content, size of a 3′ GC clamp, maximum intramolecular/intermolecular complementation potential, and maximum intersequence similarity. The user can analyze the intramolecular/intermolecular complementation potential of program-generated primer sequences or of sequences entered manually. The latter may contain any of the standard nucleotide symbols, including ambiguous bases. Primer sequence length, GC%, individual base composition, molecular weight, approximate melting temperature, and mass/volume/concentration relationships can be determined for any sequence generated by ROCK or entered manually.

INTRODUCTION

Randomly amplified polymorphic DNA (RAPD) marker analysis is a PCR-based method in which DNA fragments are amplified through the use of short oligonucleotide primers that are not directed toward any particular target sequence. The amplified fragments produced by a single primer are generally assumed to be neutral markers, and the use of many different primers, each amplifying different regions of the target DNA, permits investigators to rapidly collect large amounts of polymorphism data for subsequent analyses. Since their development, RAPD markers have been used extensively to analyze

BioComputing: Computer Tools for Biologists
Edited by Stuart M. Brown

diversity and genetic structure within and between populations. Regardless of a number of technical and analytical problems, RAPD analysis has been shown to be useful, especially at the intraspecific level (3,10).

Despite the acronym RAPD, the short primers used to amplify regions of the target DNA are not, or should not be, truly random. As with any PCR-based technique, whether specifically directed toward a particular target DNA sequence or not, there are certain primer design criteria that should be satisfied to increase the probability that the amplification reaction will proceed efficiently. Primer design considerations shown (7) to play a critical role include: (*i*) the G+C content of the primer sequence, (*ii*) the presence of a GC clamp at the 3′ end of the primer sequence, and (*iii*) the elimination or minimization of dimer and/or hairpin-forming potential. Given the design considerations specified above, commercially available sets of random primers may not be optimal. Furthermore, commercial oligonucleotide synthesis services often provide moderately greater amounts of primers at costs comparable to these sets. Of greatest significance, though, is the potential for increasing the probability of successful priming in reactions through attention to primer design parameters. In response to these concerns, the Random Oligonucleotide Construction Kit (ROCK) was designed to simplify the generation of random oligonucleotide primers that satisfy design criteria specified by the user.

MATERIALS AND METHODS

Obtaining and Starting the Program

ROCK is a macro-based program designed primarily to facilitate the generation and analysis of random oligonucleotide primers, although many of its functions will work with user-entered oligonucleotide sequences. The macros, written using the features of the Visual Basic for Applications language built into Microsoft® Excel® 97 or later version for Microsoft Windows™, work well with either of these applications running under the Windows 95 or later Windows version operating system. To obtain ROCK, download the spreadsheet file (ROCK.xls) from the Web site http://www.sru.edu/depts/artsci/bio/ROCK.htm.

ROCK may be started using any of the following methods: (*i*) start the Excel application, choose **File | Open** and select ROCK.xls, (*ii*) open ROCK.xls by double-clicking the file in Windows Explorer® or My Computer®, or (*iii*) click on **Start | Run** from the Task Bar on the Windows desktop and select ROCK.xls. Choose **Enable Macros** in the dialog box. A new menu item called **ROCK** will be added to the Excel menu bar. The functions required to generate and analyze random primers are located in this menu.

Generating Parameterized Random Sequences

Initialization of ROCK creates the Scratch worksheet, which may be used for temporary storage of sequences or for the manual entry of sequences to be analyzed. Choose **ROCK | Primers | New sheet** to create a primer generation sheet. ROCK sets the name of the worksheet to the user-supplied name with the identifier "[Primer]" appended.

Primer design criteria appear in cells C1–C7, and settings for display of generated primers appear in cells F1–F5. These values may be altered for the current primer generation sheet by choosing **Rock | Options | Active primer sheet...**. Cells I1–I7 display statistics associated with the primer generation process. The function of most generation parameters are self-evident, but the *Quantity = "Good" Primers* and *Show Details* options warrant explanation. For the former, if the *Yes* option is selected, then random sequences are generated until the number of acceptable sequences equals the value of *Quantity* (cell C1); otherwise, *Quantity* determines the total number of sequences to be generated regardless of their acceptability. For the latter, if the Yes option is selected, then all generated sequences are displayed; otherwise, only acceptable sequences are displayed. Click **OK** to update the parameter values or **Cancel** to discard any changes. Default parameters may be changed by selecting **ROCK | Options | Default....**

To generate random primer sequences, choose **ROCK | Generate primers**. The specified number of random sequences is generated by the Excel pseudo-random number generator. User-defined, critical primer design considerations including the GC% content, the presence of a 3′ GC clamp, the number of contiguous residues potentially capable of intramolecular or intermolecular complementation and the similarity to previously generated, acceptable sequences are used by ROCK to determine the acceptability of each generated sequence. Generated sequences are displayed in column B on the worksheet (Figure 1). Sequences that meet each of the four primer design criteria are indicated with an "X" in column C. An "X" in any of the columns D–G indicates the sequence did not meet one or more of the primer design criteria specified.

The **ROCK | Primers | Clear primer list** menu item deletes previously generated primers and resets the statistics associated with the generation sheet. This option is useful if the user wishes to generate a new list of primers without creating a new generation sheet.

Analysis of Intramolecular/Intermolecular Complementation Potential

ROCK | Dimer analysis | Restricted complementation or **ROCK | Dimer analysis | Unrestricted complementation** assesses the potential for complementation within a

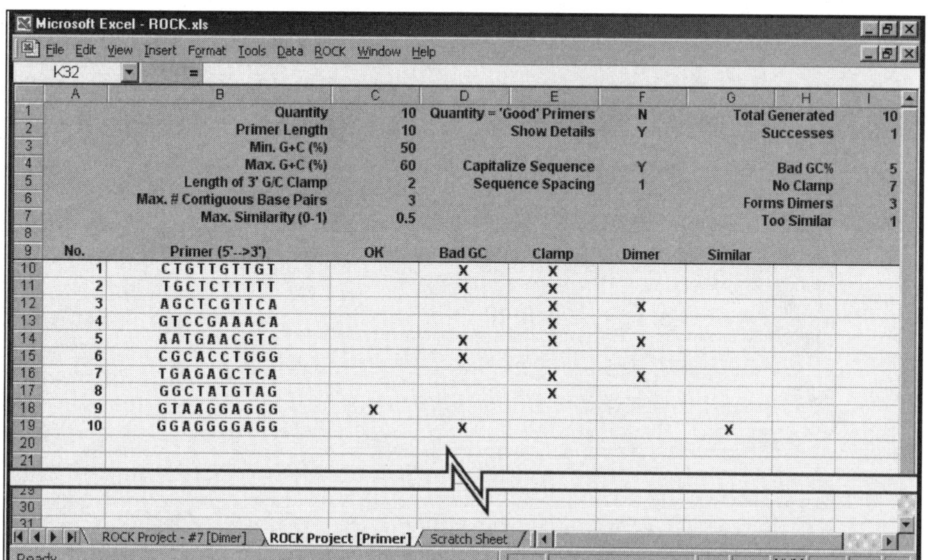

Figure 1. Primer generation worksheet. The 10 primer sequences shown in cells B10–B19 were generated by the program using the parameters displayed in the shaded area at the top. Note that even for short sequences of only 10 bases, only one (no. 9) of the 10 sequences met all of the design criteria.

primer sequence or between two primers of the same sequence for generated or manually entered sequences. Select the sequence to be analyzed and choose one of the menu items above. The restricted and unrestricted complementation options differ in the way that complementation of ambiguous bases in the sequence is determined and give different results only if the sequence contains such bases. The complementation rules used by ROCK for restricted and unrestricted complementation may be obtained by manually entering the sequence ACGTUIRYMKSWHBVDN on the Scratch sheet and selecting one of the above menu items. If the sequence to be analyzed was generated by ROCK, the user will be prompted to input the maximum number of contiguous bases involved in forming complementary pairs considered to be acceptable. The default value in the input dialog box is set to the *Max. # Contiguous Base Pairs* parameter (cell C6 in Figure 1) on the primer generation sheet. ROCK will create a new dimer analysis sheet with a name derived from the name of the primer generation sheet, the number of the selected primer and the identifier "[Dimer]", and will perform the analysis (Figure 2). The procedure for analyzing sequences entered manually on the Scratch worksheet is identical, with these exceptions: (*i*) the user will be prompted to supply a name for the dimer analysis sheet, and (*ii*) there is no default value for the acceptable maximum number of contiguous bases forming pairs.

Regions potentially capable of intramolecular/intermolecular complementation are represented by contiguous diagonal stretches of complementation from the upper left to the lower right in the square matrix. Complementary regions are color-coded by ROCK according to their length, relative to the maximum allowed contiguous base pairs value specified. Green indicates regions that do not exceed the maximum value. Yellow, orange, and red, respectively, are used to indicate regions that are one base, from two to four bases, and greater than four bases over the maximum value. The ranges associated with each of the four colors can be modified by choosing **ROCK | Options | Default....**

The **ROCK | Dimer analysis** submenu (*i*) alternates the color-coding scheme between color and grayscale (**Color <-> B/W**), (*ii*) changes the complementation type (**Change**

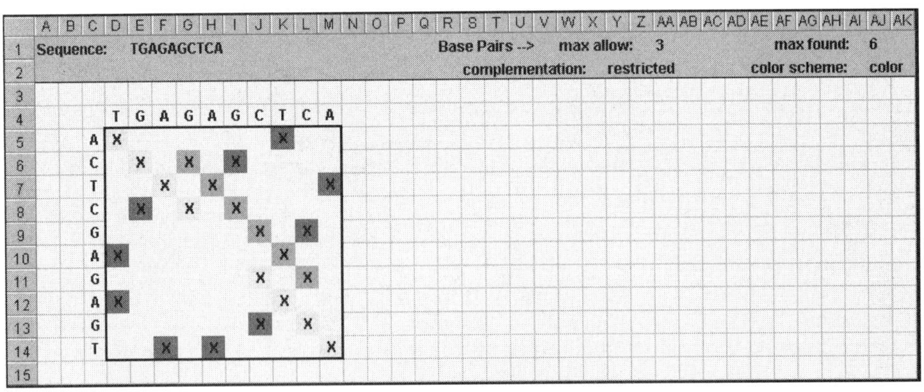

Figure 2. Dimer analysis worksheet. The primer sequence 5′-TGAGAGCTCA-3′ (sequence no. 7 in Figure 1) is shown following the analysis for intramolecular/intermolecular complementarity. ROCK constructs a square matrix with the sequence across the top (5′ to 3′) and down the left side (3′ to 5′). At each row-column intersection, an "X" indicates complementation. In this example, three contiguous complementary base pairs was the maximum allowed (cell AA1). Two regions (yellow) composed of four contiguous pairs were detected near the termini of the sequence, and a single region (red) of six contiguous pairs was found near the middle of the sequence. **(See color plate A3.)**

complementation type), or (*iii*) changes the maximum number of contiguous bases involved in forming complementary pairs (**Change max. pairing allowed**). The default color-coding scheme (color versus grayscale) of a dimer analysis sheet can also be altered by choosing **ROCK | Options | Default....**

Obtaining Sequence Information

Base composition, physical properties [molecular weight and melting temperature (T_m)], and sample calculations for primer sequences may be obtained by selecting the sequences of interest and choosing **ROCK | Primers | Information**. A dialog box containing the selected sequences will be displayed (Figure 3). Information is displayed in the remaining portions of the dialog box by selecting a sequence.

Items in the Generation Results section of the dialog box are applicable only for generated primer sequences. The Base Composition section shows the length of the currently selected sequence and enumerates each of the different bases in the sequence. Because ROCK is capable of dealing with user-entered sequences containing ambiguous bases, the GC% of sequences is calculated as [G + C + S + 0.5 * (N + M + Y + K + R) + 0.67 * (B + V) + 0.33 * (D + H) / (sequence length)] * 100%, where each base code corresponds to its frequency of occurrence in the sequence. The molecular weight of the primer is calculated as the sum of the individual molecular weights of the component bases (adjusted for the effects of dehydration during synthesis) minus 62 (2,11). The final term adjusts for the lack of both a 5′ phosphate group and a 3′ dehydration reaction. The approximate T_m

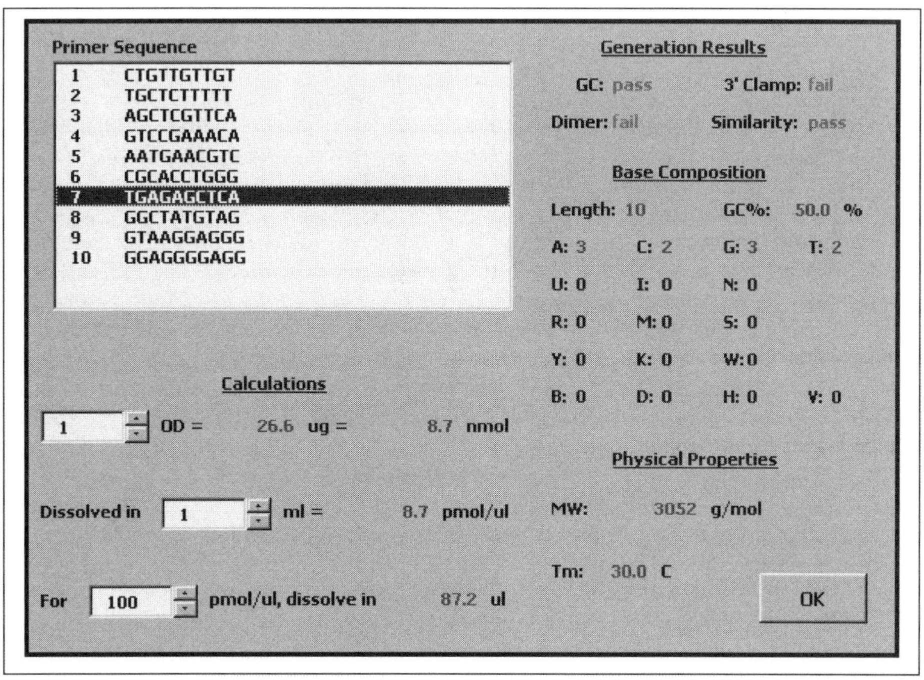

Figure 3. Dialog box displaying sequence information. The list of primer sequences in the upper left contains the sequences displayed on the primer generation sheet in Figure 1. Information is currently being shown for sequence no. 7. For sequences entered by the user, the individual items under Generation Results would be labeled as "N/A". See text for details regarding the calculation of values.

of the primer in °C is calculated depending on the length of the sequence (Application Note; MWG Biotech, High Point, NC, USA). For sequences of 15 or fewer bases, T_m is calculated as [2 * (1 - GC%) * (sequence length)] + [4 * (GC%) * (sequence length)]; for sequences greater than 15 bases in length, T_m is calculated as 69.3 + (41 * GC%) - [650 / (sequence length)]. For both equations, GC% is expressed as a ratio (i.e., 0.6 rather than 60%). In the Calculations section, the results of calculations are shown, which are useful in reconstituting lyophilized primers to a specific concentration after synthesis. Each result is an example of a mass/volume/concentration conversion, with the exception of the calculation of nmol of primer. The nmol of primer is calculated as (A_{260} * 100) / (extinction coefficient of the primer), where the denominator is the sum of the extinction coefficients of the component bases of the sequence (2,11).

Deleting Primer Generation and Dimer Analysis Sheets

The **ROCK | Delete** submenu is used to permanently delete ROCK-generated worksheets. Worksheets created by the user should be deleted using the built-in Excel commands. The currently active worksheet is deleted with **Delete active sheet**. Deletion of multiple worksheets is facilitated with **Delete...**, which lists ROCK-generated worksheets. Single or multiple worksheets may be selected from the list for deletion; hold down the *Shift* or *Ctrl* key to select a range or several individual worksheets, respectively. If the box beneath the worksheet list is checked, then deletion of a primer generation worksheet will automatically delete all associated dimer analysis worksheets.

RESULTS AND DISCUSSION

ROCK automates the generation of random primer sequences while simultaneously ensuring that the sequences satisfy user-defined parameters. One of the principal advantages of the program is the rapidity with which random primers can be generated and analyzed. For example, more than 150 10-mers were generated to find 10 primers with the design criteria specified in Figure 1. Even so, a 75-MHz Pentium® computer required an average of only 9 s to generate 10 acceptable sequences. In addition, ROCK permits the user to generate sequences satisfying a specified set of design parameters, thus increasing the likelihood that the primers will perform well in RAPD analyses by ensuring that the GC% of the primer matches the average GC% of the target DNA as closely as possible, that the 3′ end of the primer has a minimum number of G or C residues, and that the sequence does not have a significant potential for intramolecular/intermolecular complementation.

One of the PCR parameters cited as a cause of artifactual banding patterns is the primer concentration (3,6). ROCK permits the analysis of intramolecular/intermolecular complementation potential of primer sequences because primers involved in either of these interactions are unable to anneal to the template DNA, lowering the effective concentration of primer in the reaction mixture. ROCK also offers the user the option of analyzing intramolecular/intermolecular complementation potential with unrestricted complementation. Companies offering custom synthesis of oligonucleotides generally accept ambiguous base specifications in the sequence. In practice, primers synthesized to these specifications are mixtures containing equimolar concentrations of sequences, in which the position containing the ambiguous base is replaced by one of the cognate-specific residues. It is possible that certain of the primers in the mixture will possess intramolecular/intermolecular complementation potential, again reducing the effective primer concentration in the reaction. The conventional complementation rules for ambiguous bases (i.e., those used by ROCK in restricted complementation) would generally not detect this, as these rules state that ambiguous bases only complement other ambiguous bases. In contrast, ROCK is able to detect this situation by treating an ambiguous base as the specific cognate bases; complementation by any of the specific residues represented by an ambiguous code is considered to be positive complementation.

The use of computer spreadsheets for the rapid completion of a variety of tasks in molecular biology has become quite common (1,4,5,8,9). The advantage of these spreadsheet-based applications is clear: analyses or calculations that can be done by hand but are tedious, time consuming, and error prone may be completed quickly with the assistance of Excel running on the nearly ubiquitous personal computer. A number of utilities exist that can be used to select primers for PCR applications requiring the use of primers that are directed toward a particular target sequence. However, ROCK specifically addresses the task of generating sequences for use in RAPD protocols where the primers are not chosen to complement specific target sequences. Furthermore, ROCK's ability to handle user-defined sequences containing any of the standard nucleotide bases extends its usefulness beyond the simple generation of primers.

ACKNOWLEDGMENTS

The authors express their appreciation to Dr. Jack R. Layne, Jr. for providing valuable editorial comments on the manuscript.

REFERENCES

1. **Anbazhagan, R. and E. Gabrielson.** 1999. Spreadsheet-based program for alignment of overlapping DNA sequences. BioTechniques *26*:1180-1185.
2. **Dawson, R.M.C., D.C. Elliot, W.H. Elliot, and K.M. Jones.** 1986. Data for Biochemical Research, 3rd ed. Oxford University Press, New York.
3. **Grosberg, R.K., D.R. Levitan, and B.B. Cameron.** 1996. Characterization of genetic structure and genealogies using RAPD-PCR markers: a random primer for the novice and nervous, p. 67-100. *In* J.D. Ferraris and S.R. Palumbi (Eds.), Molecular Zoology: Advances, Strategies, and Protocols. Wiley-Liss, New York.
4. **Han, B. and A.H. Tashjian, Jr.** 1998. User-friendly and versatile software for the analysis of protein hydrophobicity. BioTechniques *25*:256-263.
5. **McEwan, N.R. and D. Gatherer.** 1998. Adaptation of standard spreadsheet software for the analysis of DNA sequences. BioTechniques *24*:131-138.
6. **Muralidharan, K. and E.K. Wakeland.** 1993. Concentration of primer and template qualitatively affects products in random amplified polymorphic DNA PCR. BioTechniques *14*:362-363.
7. **Sharrocks, A.D.** 1994. The design of primers for PCR, p. 5-11. *In* H.G. Griffin and A.M. Griffin (Eds.), PCR Technology: Current Innovations. CRC Press, Boca Raton, FL.
8. **Shaw, G.** 1995. Protein sequence interpretation using a spreadsheet program. BioTechniques *19*:978-983.
9. **Stowe, R.P. and D.L. Pierson.** 1996. Spreadsheet macro for setting up PCR assay tubes. BioTechniques *20*:1088-1089.
10. **Williams, J.G.K., A.R. Kubelik, K.J. Livak, J.A. Rafalski, and S.V. Tingey.** 1990. DNA polymorphisms amplified by arbitrary primers are useful as genetic markers. Nucleic Acids Res. *18*:6531-6535.
11. **Windholz, M. (Ed.).** 1983. The Merck Index, 10th ed. Merck & Company, Rahway, NJ.

Update to: ROCK: A Spreadsheet-Based Program for the Generation and Analysis of Random Oligonucleotide Primers Used in PCR

Steven R. Strain and Jerry G. Chmielewski

Slippery Rock University of Pennsylvania, Slippery Rock, PA, USA (steven.strain@sru.edu)

INTRODUCTION

Since being made available to the scientific community in 2001, the Random Oligonucleotide Construction Kit (ROCK) has been downloaded by numerous individuals, and user comments have been positive. However, based upon our own experience with ROCK, we have identified a potentially limiting aspect of the original program for some users with regard to the ability to integrate ROCK-generated data with other Microsoft® Excel® data files. To facilitate the integration and organization of research project-related data, including ROCK-generated primers, into a single Excel workbook, we are releasing ROCK version 1.2, which functions as an Excel add-in component.

The original version of ROCK requires that the generation and analysis of primers be done with a copy of the ROCK workbook opened in Excel. Integration of other Excel-based data with ROCK-generated data can only be accomplished by entering the data initially into a ROCK workbook or by copying existing data sheets in other workbooks into a ROCK workbook. However, the first option is not possible for preexisting data, and the second is laborious and potentially prone to error.

In contrast to the original stand-alone version of ROCK, the ROCK add-in component behaves as an extension of Excel's normal capabilities rather than being associated with a specific workbook. Consequently, the generation and analysis of primers by ROCK are available to any open workbook within Excel. In addition, once the ROCK add-in is installed in Excel, it will always be available without requiring that a ROCK workbook be opened. As the code for the ROCK program is no longer stored in each workbook containing ROCK-generated data, an added advantage to using the add-in version of ROCK is a substantial decrease in the file size of these workbooks, a significant consideration when transporting files.

INSTALLING AND USING THE ROCK ADD-IN

It is strongly recommended that the ROCK add-in file (Rock.xla) be copied to a non-removable hard drive, as Excel will need to access the file each time the program is started. To install the ROCK add-in for the first time, choose the Tools/Add-Ins item on the Excel main menu bar to open the Add-In Manager dialog box. Click the Browse button, select the Rock.xla file from the directory where it was copied, and click the OK button to install the ROCK add-in in Excel. A new menu item called ROCK will be added to the Excel main menu bar. Unless the ROCK add-in is uninstalled by manually deselecting the ROCK checkbox in the Add-In Manager dialog box, it will be available each time Excel starts.

The use of the add-in version of ROCK is essentially identical to the original standalone version. However, this version of ROCK introduces what we have termed a "ROCK-enabled" workbook: an Excel workbook which contains the hidden worksheets required for the proper operation of ROCK. ROCK is designed to work when ROCK-enabled workbooks are active. When non-ROCK-enabled workbooks are active, most of the items in the ROCK menu will be unavailable with three notable exceptions: (*i*) a workbook that is not ROCK-enabled can be made ROCK-enabled by choosing ROCK/Primers/New sheet; (*ii*) sequence information can be obtained for valid sequences in a non-ROCK-enabled workbook by choosing ROCK/Primers/Information; and (*iii*) a new ROCK-enabled workbook can be created by choosing ROCK/Primers/New workbook. The last menu item is new to the add-in version of ROCK.

AVAILABILITY OF ROCK AND ADDITIONAL UTILITIES

Due to an ongoing redesign of our institution's Web infrastructure, the Web address cited in the original ROCK article may no longer be valid. To avoid future difficulties with regard to obtaining ROCK, we have chosen to distribute the ROCK add-in version on compact disc read-only memory (CD-ROM). Users interested in the ROCK add-in should contact the authors to obtain a copy. In addition to the ROCK add-in, the CD-ROM will contain a utility to convert workbooks created with the original version of ROCK to a ROCK-enabled workbook format for use with the ROCK add-in, and an add-in that will serve as a programming interface to many of the basic functions used by ROCK. The latter may be of interest to users who design their own Excel-based applications.

CHAPTER 10

Spreadsheet Software for Thermodynamic Melting Point Prediction of Oligonucleotide Hybridization with and without Mismatches

Ekkehard Schütz and Nicolas von Ahsen

Georg-August-University, Göttingen, Germany

Reprinted from BioTechniques 27:1218-1224 (December 1999)

ABSTRACT

The use of thermodynamic parameters for the calculation of oligonucleotide duplex stability provides the best estimates of oligonucleotide melting temperatures (T_m). Such estimates can be used for evidence-based design of molecular biological experiments in which oligonucleotide melting behavior is a critical issue, such as temperature or denaturing gradient gel electrophoreses, Southern blotting or hybridization probe assays on the LightCycler™. We have developed a user friendly program for T_m calculation of matched and mismatched probes using the spreadsheet software Microsoft Excel®. The most recently published values for entropy and enthalpy of Watson-Crick pairs are used, and salt and oligonucleotide concentrations are considered. The 5′ and 3′ end stability is calculated for the estimation of primer specificity. In addition, the influence of all possible mutations under a given probe can be calculated automatically. The experimental evaluation of predicted T_m with the LightCycler, based on 14 hybridization probes for different gene loci, showed an excellent fit between measured results and values predicted with the thermodynamic model in 14 matched, 25 single mismatched and 8 two-point mismatched assays ($r=0.98$; $Sy.x= 0.90$; $y=1.01\times -0.38$). This program is extremely useful for the design of oligonucleotide probes because the use of probes that do not discriminate with a reasonable T_m difference between wild-type and mutation can be avoided in advance.

BioComputing: Computer Tools for Biologists
Edited by Stuart M. Brown

INTRODUCTION

The chemical- or temperature-driven melting behavior of short oligonucleotides that are hybridized to PCR-amplified DNA is the basic principle of several mutation detection techniques. These techniques include temperature or denaturing gradient gel electrophoreses (TGGE/DGGE) (17), DNA blotting (7) and the hybridization probe assay on the LightCycler™ (Roche Molecular Biochemicals) (8). They are only possible because the stability of dsDNA oligonucleotides is grossly influenced by mismatches under a specifically designed synthetic probe (18). During the last decade, several publications that discuss the influence of mismatches on dsDNA stability, based on thermodynamic principles, have offered insight into the underlying physico-chemical mechanisms (1–4,10). The stability of a single base pair binding is influenced by the surrounding base pairs (nearest neighbor). The thermodynamic parameters for every matched or mismatched base pair are experimentally defined (1–4,10). Such definition allows the prediction of oligonucleotide duplex stability with or without mismatches. For longer sequences (>50 bp), alternative methods based on length and GC content give reliable estimates of the melting temperature (T_m) (18), but only for duplex stability without mismatches. However, mismatches have only a minor influence on the T_m of longer dsDNAs.

For the enthalpy and entropy of matched pairs, unified values for the prediction of DNA duplex stability are available that provide a reliable calculation of the T_m (12). The T_m is the temperature at which 50% of the oligonucleotides in solution are in a duplex formation. Because entropy depends on salt and oligonucleotide concentration, the T_m is also influenced by these parameters. Therefore, they must be considered by the use of experimentally derived correction equations (12).

We have used these parameters to develop a user friendly program for the T_m calculation using the Excel® spreadsheet software (Microsoft, Redmond, WA, USA). It offers a sequence that can be pasted from any source into the input table. The mismatch must be defined together with sodium buffer equivalents and DNA concentration to calculate the T_m of both matched and mismatched oligonucleotide probes. A function to calculate the influence of any possible single base pair mismatch under the probe is also included. Nearly every laboratory will be able to meet the program's hardware and software needs. The program is fully compatible with the Windows environment. Copy and paste into every program running under Windows is possible (e.g., into e-mail oligonucleotide custom synthesis order programs). This program reliably calculates estimates of hybridization probes' T_m when compared with experimental data from genotyping assays used in our laboratory and is of particular interest for the design of such hybridization-probe mutation detection systems.

MATERIALS AND METHODS

Every standard PC equipped with Microsoft Windows (NT4.0, WIN95 or WIN98) or a Macintosh® computer (Apple Computer, Cupertino, CA, USA) with Microsoft Excel Ver. 5.0 and later can run the program. For the calculation of enthalpy (ΔH^0) and entropy (ΔS^0), we used the unified parameters derived from six independent laboratories (12). Parameters for the stability of mismatched duplexes were taken from publications of the same group (1–4,10). The Appendix gives the equations that are used for the calculations in this program. Oligonucleotides (PCR primers and fluorescein probes) were custom synthesized (MWG-Biotech, Germany). LC-Red640-N-hydroxysuccimide ester (Roche

Molecular Biochemicals, Mannheim, Germany) was linked with the respective oligonucleotide by using an aminolinker purchased from Glen Research (Sterling, VA, USA) and was purified by HPLC. Experimental T_m values were determined with the LightCycler using standard techniques in hybridization probe format (16). The actual T_m was calculated by the LightCycler software (Ver. 3.0) as the maximum of the first derivative of the fluorescence as opposed to the temperature function. Experiments were run in triplicates and the CV was <1% for experiments repeated on different days.

Program Overview

The Excel spreadsheet consists of three parts: an input table, an output table for all possible mismatches under the probe and the program itself. After loading, the program adds a new menu item to the standard menu that incorporates four program functions (Figure 1): *(i)* calculation of the T_m of self-complementary oligonucleotides; *(ii)* calculation of the T_m of non-self-complementary oligonucleotides; *(iii)* generation of a spreadsheet with the T_m of any possible one-point mutation; *(iv)* calculation of buffer strength in sodium equivalents.

The calculation of oligonucleotide T_m takes <1 s for the program items 1 and 2. The time for calculation of all possible single-point mismatch T_m under the probe (program 3) depends on the length of the probe because 3× (x-2) sequences are considered (where x is the length of the probe). For example, it takes approximately 5 s to generate the output spreadsheet for a 30-mer probe on a PC equipped with an AMD K6-350 that runs on Windows NT 4.0 and Microsoft Excel 7.0.

Users can type the sequence (only nucleotides A,T,G and C are allowed) into the input table or paste it from other sources, such as the GenBank® database (**http://www2. ncbi. nlm.nih.gov/cgi-bin/genbank**). The sequence must be free of unnecessary characters, such as numbers and spaces. Two sequences are always required as input. The length of the input sequence is restricted to <256 bp because the model is not useful for longer DNA sequences. The oligonucleotide concentration and the concentration of sodium equivalents must be entered in the respective cells because these parameters also influence the T_m. For user convenience, a sodium equivalent calculator computes the monovalent ion concentration on user input of Mg^{++} and K^+ concentration (6) of the used buffers; this number is then pasted into the respective cells of the input table. Until the influence of Mg^{++} on DNA duplex stability is fully investigated, we recommend this formula. However, Mg^{++} concentrations higher than approximately 8 mM will not be properly accounted for. When dimethyl sulfoxide (DMSO) is included in the PCRs, its influence on T_m is considered when a value >0% is given in the respective cell.

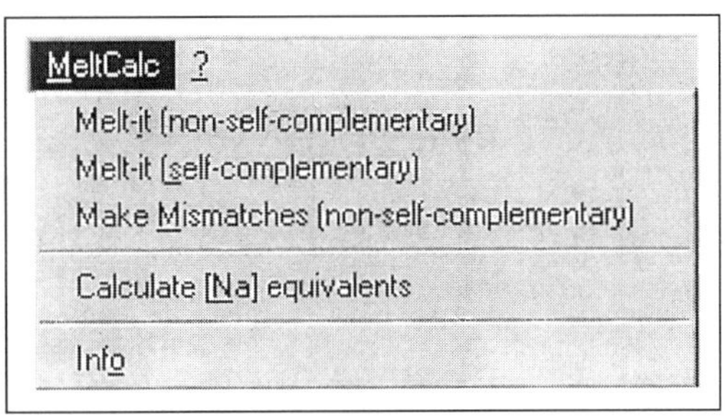

Figure 1. Menu for the calculation of oligonucleotide thermodynamics.

After these steps are completed, the user can start the calculation by choosing the desired function from the "MeltCalc" menu. When starting the T_m calculation "Melt-it" (self-complementary or non-self-complementary), the given sequences are separated into single nearest-neighbor pairs, and the thermodynamic values of the resulting pairs are taken from the internal table. If the second sequence contains a mismatch, then the respective mismatch parameters are used. Total thermodynamic values for the sequences are then calculated according to the equations given in the Appendix. The program writes back T_m values for the matched and mismatched oligonucleotide and the difference in T_m between the given wild-type and mutation (mismatch). In addition, GC%, the free energy (ΔG^0) of the 5′ and 3′ pentamer end of the oligonucleotide are also calculated. The pentamer stability is useful for the estimation of primer performance because it has been shown that primers with high priming efficiency are somewhat more stable at their 3′ end (11). The system calculates enthalpy and entropy values under standard conditions (1 mol/L NaCl, 100 mol/L oligonucleotide). For the entropy, the values after correction of the given salt concentration are calculated and shown. Because it might be favorable to probe the antisense strand for a mutation detection caused by a higher T_m difference between matched and unmatched sequence, the same outputs just described are given for this strand together with the respective sequence. Figure 2 shows the calculation of duplex thermodynamics of an oligonucleotide with a mismatch.

Additionally, the program will calculate the influence of every possible mutation that can occur under a given probe. This is helpful to estimate the detection specificity of a mutation. When the user selects this menu item, the system calculates the T_m for every possible single point mismatch between the perfect matched and the mismatched sequence.

Oligoconc. [µM]	0.1	DMSO [%]	5	Unified parameters					
Na eq. [mM]	300								
Length [bp]	19	Tm	ΔTm	ΔG-5(5′-end)	ΔG-5(3′-end)	GC[%]	ΔH	ΔS1	ΔSmod
Sequence (perfect match)	cggcgattgtcgcaccagt	64.5	1.7	-8.6	-6.0	63.2	-155.0	-412.2	-420.2
Sequence (mismatch)	cggcgcttgtcgcaccagt	62.8		-8.6	-6.0	68.4	-140.9	-372.9	-380.9
Reverse sequence (pm)	actggtgcgacaatcgccg	64.5	5.5	-6.0	-8.6	63.2	-155.0	-412.2	-420.2
Reverse sequence (mm)	actggtgcgacaagcgccg	59.1		-6.0	-8.6	68.4	-138.5	-370.4	-378.4

Figure 2. Input spreadsheet after calculation of thermodynamic values (example). Cells for user input are marked with dark letters on white background. Note the difference in ΔT_m when the detection probe is sited on the antisense strand.

Mismatch	Sequence	Reverse Sequence	ΔG_{37}	TMmm	ΔTm
(perfect match)	cggcgattgtcgcaccagt	actggtgcgacaatcgccg	-27.09	64.5	
c6	cggcgCttgtcgcaccagt	actggtgcgacaaGcgccg	-25.14	62.8	1.7
c10	cggcgattgCcgcaccagt	actggtgcgGcaatcgccg	-24.86	62.2	2.3
c14	cggcgattgtcgcCccagt	actggGgcgacaatcgccg	-24.69	62.0	2.5
c3	cgCcgattgtcgcaccagt	actggtgcgacaatcgGcg	-25.23	61.2	3.3
c8	cggcgatCgtcgcaccagt	actggtgcgacGatcgccg	-24.77	61.0	3.5
c7	cggcgaCtgtcgcaccagt	actggtgcgacaGtcgccg	-24.43	60.7	3.8
c18	cggcgattgtcgcaccaCt	aGtggtgcgacaatcgccg	-24.63	60.6	3.9
t18	cggcgattgtcgcaccaTt	aAtggtgcgacaatcgccg	-24.21	60.5	4.0
g6	cggcgGttgtcgcaccagt	actggtgcgacaaCcgccg	-23.33	60.4	4.1
t3	cgTcgattgtcgcaccagt	actggtgcgacaatcgAcg	-23.78	60.0	4.5
t6	cggcgTttgtcgcaccagt	actggtgcgacaaAcgccg	-24.13	59.9	4.6
c17	cggcgattgtcgcaccCgt	acGggtgcgacaatcgccg	-24.22	59.9	4.7
c9	cggcgattCtcgcaccagt	actggtgcgaGaatcgccg	-23.89	59.6	4.9
g7	cggcgaGtgtcgcaccagt	actggtgcgacaCtcgccg	-23.84	59.4	5.1
a8	cggcgatAgtcgcaccagt	actggtgcgacTatcgccg	-24.08	59.4	5.1
t9	cggcgattTtcgcaccagt	actggtgcgaAaatcgccg	-23.32	59.1	5.4
a3	cgAcgattgtcgcaccagt	actggtgcgacaatcgTcg	-23.52	59.1	5.5
a2	cAgcgattgtcgcaccagt	actggtgcgacaatcgcTg	-23.87	59.0	5.5

Figure 3. Output after calculation of all possible mismatches under a given probe (example; truncated).

Results are sorted by the T_m difference with the perfect match and are written into a separate spreadsheet together with the respective sequence. The position of the mismatch in the sequence is indicated by an uppercase letter together with the base that was changed. The complementary reversed sequence is also displayed in an additional row (Figure 3).

Because no thermodynamic parameters for dangling ends, double and terminal mismatches are available as yet, these are not calculated by the program. The same restrictions apply for double mismatches as direct neighbors, with the exception of those pairs (e.g., GG/**TT** or GT/**TG**) for which respective values are known (1). Such calculations are only possible with the HyTher program of SantaLucia's group, available at **http://jsl1.chem.wayne.edu** that uses as yet unpublished thermodynamic data. For all other situations, the presented spreadsheet application calculates results equivalent to the HyTher program.

RESULTS

We have generated eight different single base pair mutations by site-directed mutagenesis under a wild-type and a mutation complementary probe that we used for the detection of the apolipoprotein-B3500 mutation (14). These mutations were used with two different probes for experimental validation of the calculated T_m values together with 12 other hybridization-probe sets that are currently used in our laboratory for genotyping of various gene loci (15,16 and unpublished data). A total of 14 matched and 33 unmatched hybridization experiments were performed in triplicate. The annealing oligonucleotide concentration is needed for T_m calculation according to [Equation 3]. We have used the final concentration of the hybridization probe in each assay (0.1 and 0.2 µmol/L) for parameterization of the model. If DMSO was present in the PCRs the lowering influence on T_m, which is 0.6°C/% DMSO (9), was considered. The results in Figure 4 show excellent agreement of calculated and measured T_m, for both wild-type experiments and for those with mismatches (n=47; r=0.98, Sy/x= 0.90). The T_m of the perfectly matched and the mismatched hybridization was predicted with an accuracy of ±2.4°C (±2 SD). Figure 4 shows a Bland-Altman difference plot (5) of T_m values predicted with the thermodynamic model vs. those values derived from LightCycler experiments plotted against the mean of both.

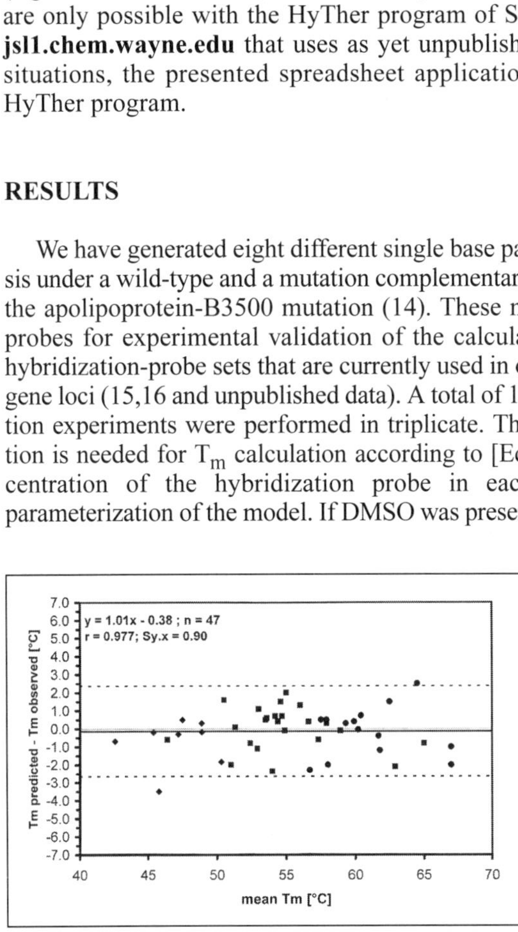

Figure 4. Difference plot for predicted and observed T_m derived from 14 different LightCycler hybridization-probe assays vs. the mean T_m of both. Probe lengths were 17–29 bp, GC content was between 24% and 63%. Melting temperatures from 14 matched (●), 25 different single mismatches (■) and 8 two-point mismatches (◆) evaluated in 14 hybridization-probe assays are shown (r=0.977). The solid line (—) represents the mean difference of predicted and observed T_m (-0.13°C), and (- - - -) shows ±2 SD (2.4°C). Regression estimates are derived from principle component analysis of calculated vs. measured T_m values. Correlation coefficient represents Pearson's r.

DISCUSSION

An easy-to-use spreadsheet program is presented that calculates the T_m of oligonucleotide duplexes on the basis of thermodynamic parameters for nearest-neighbor pairs. Recently published values for enthalpy and entropy allow a precise prediction of duplex stability and the T_m with an error of approximately 2°C (1–4,10,13). Our data show that this holds true for hybridization-probe experiments after enzymatic amplification of genomic DNA with an estimated error of 2.4°C, with appropriate parameterization of the thermodynamic model for matched, one-point and two-point mismatched oligonucleotide sequences.

The application of this program for the design of hybridization assays on the LightCycler is of special interest when selecting a probe/anchor pair that gives the best discrimination (difference in T_m between wild-type and mutation). In our experience, a minimal difference of 1.5°C is necessary, and at least 3°C is advisable for the safe detection of heterozygosity. Furthermore, it estimates the discriminating power of such assays for the detection of a mutation of interest and other mutations that may be present in the region covered by the probe by calculating the T_m of all possible mutations.

A more detailed discussion about hybridization-probe assays on the LightCycler and their discrimination power was published recently (14). Because dye labeled probes (ca. $3.80 per assay) are expensive, added to the cost for technician time in setting up a suitable and reliable genotyping assay, the use of this thermodynamic approach before probe design and synthesis may save time and money. The program can be used in almost every scientific laboratory that works with T_m-based molecular biological assays.

From a theoretical point of view, such evidence-based design of probes is also applicable to DNA blotting, microarray experiments or for the estimation of suitable stringency conditions (sodium concentration) that give reliable and reproducible results for respective experiments. How valid the model is for other hybridization assays can not be concluded from our data. However, other assay systems (e.g., TaqMan) should also profit from a probe design that generates a maximum difference between perfectly matched and mismatched sequences. In addition, the "make all mismatches" feature may be helpful to calculate how specific a mutation can be detected by a certain T_m shift. We encourage scientists working with hybridization assays to use this program and to report their experience to the scientific community.

The program is available at **http://server1.medikc. med.uni-goettingen.de/meltcalc. htm**, which can also be accessed through a link from the **Software Library** on the *BioTechniques* Web site (**www.biotechniques.com**), or by e-mail upon request from the authors.

Formulas and Equations

The enthalpy (ΔH^0) and entropy (ΔS^0) and free energy ΔG^0 of an oligonucleotide duplex is calculated as the sum of the entire nearest neighbors plus helix initiation and symmetry terms:

$$\Delta H^0_{total} = \Sigma i \Delta H^0\ (i) + \Delta H^0\ (5'init) + \Delta H^0\ (3'init) + \Delta H^0\ (symmetry) \quad \text{[Eq. 1]}$$

For ΔS^0 and ΔG^0, the same formula applies. The ΔH^0, ΔS^0 and ΔG^0 values for nearest neighbors for matched and mismatched base pairs and those for initiation and symmetry correction were taken from published data (1–4,10,12).

Apart from entropy and enthalpy of the nucleotides in the sequence, the concentration of monovalent ions and oligonucleotides must be taken into account as the data basically provided by Allawi and SantaLucia are derived at 1 mol/L NaCl. Because enthalpy is independent of salt concentration, only entropy values for a given [Na] were calculated according to the formula:

$$\Delta S^0\ [Na] = \Delta S^0\ [1M\ Na] + 0.368 \times N \times \ln\ [Na] \quad \text{[Eq. 2]}$$

where N is the total number of phosphates in the duplex divided by two (12) which usually equals the length minus one.

The T_m of self-complementary oligonucleotides at a given concentration CT in a solution with the sodium equivalence concentration [Na] is calculated as:

$$T_m = \Delta H^0/(\Delta S^0\,[Na] + R \times \ln CT) \qquad \text{[Eq. 3]}$$

where R is the gas constant (1.987 cal/K×mol); for non-self-complementary sequences, CT is the concentration of oligonucleotides divided by 4. If strands are in different concentrations, CT is defined as Ca-Cb/2, where Ca and Cb are the concentrations of either oligonucleotide and Ca > Cb.

Calculation of monovalent cations (6):

$$[\text{Na equivalents}] = [MgCl_2] \times 100 + [KCl]$$

ACKNOWLEDGMENT

We gratefully thank Sandra Hartung and Reiner Andag for their skillful technical assistance and Prof. Dr. V.W. Armstrong for the critical revision of the manuscript.

REFERENCES

1. **Allawi, H.T. and J. SantaLucia, Jr.** 1997. Thermodynamics and NMR of internal GT mismatches in DNA. Biochemistry *36*:10581-10594.
2. **Allawi, H.T. and J. SantaLucia, Jr.** 1998. Nearest-neighbor thermodynamics of internal AC mismatches in DNA: sequence dependence and pH effects. Biochemistry *37*:9435-9444.
3. **Allawi, H.T. and J. SantaLucia, Jr.** 1998. Nearest-neighbor thermodynamic parameters for internal GA mismatches in DNA. Biochemistry *37*:2170-2179.
4. **Allawi, H.T. and J. SantaLucia, Jr.** 1998. Thermodynamics of internal CT mismatches in DNA. Nucleic Acids Res. *26*:2694-2701.
5. **Bland, J.M. and D.G. Altman.** 1986. Statistical methods for assessing agreement between two methods of clinical measurement. Lancet *1*:307-310.
6. **Chester, N. and D.R. Marshak.** 1993. Dimethyl sulfoxide-mediated primer T_m reduction: a method for analyzing the role of renaturation temperature in the polymerase chain reaction. Anal. Biochem. *209*:284-290.
7. **Conner, B.J., A.A. Reyes, C. Morin, K. Itakura, R.L. Teplitz and R.B. Wallace.** 1983. Detection of sickle cell beta S-globin allele by hybridization with synthetic oligonucleotides. Proc. Natl. Acad. Sci. USA *80*:278-282.
8. **Lay, M.J. and C.T. Wittwer.** 1997. Real-time fluorescence genotyping of factor V Leiden during rapid-cycle PCR. Clin. Chem. *43*:2262-2267.
9. **Musielski, H., W. Mann, R. Laue and S. Michel.** 1981. Influence of dimethylsulfoxide on transcription by bacteriophage T3- induced RNA polymerase. Z. Allg. Mikrobiol. *21*:447-456.
10. **Peyret, N., P.A. Seneviratne, H.T. Allawi and J. SantaLucia, Jr.** 1999. Nearest-neighbor thermodynamics and NMR of DNA sequences with internal AA, CC, GG, and TT mismatches. Biochemistry *38*:3468-3477.
11. **Rychlik, W.** 1995. Priming efficiency in PCR. BioTechniques *18*:84-90.
12. **SantaLucia, J., Jr.** 1998. A unified view of polymer, dumbbell, and oligonucleotide DNA nearest-neighbor thermodynamics. Proc. Natl. Acad. Sci. USA *95*:1460-1465.
13. **SantaLucia, J., Jr., H.T. Allawi and P.A. Seneviratne.** 1996. Improved nearest-neighbor parameters for predicting DNA duplex stability. Biochemistry *35*:3555-3562.
14. **von Ahsen, N., M. Oellerich, V.W. Armstrong and E. Schütz.** Application of a thermodynamic nearest-neighbor model to estimate nucleic acid stability and optimize probe design: prediction of melting points of different mutations of apolipoprotein B 3500 and factor V Leiden with a hybridization probe genotyping assay on the LightCycler. Clin. Chem. (In press).
15. **von Ahsen, N., M. Oellerich and E. Schütz.** Using two reporter dyes without interference in a single tube rapid cycle PCR: alpha1-antitrypsin genotyping by multiplex real time fluorescence PCR with the LightCycler. Clin. Chem. (In press).

16. **von Ahsen, N., E. Schütz, V.W. Armstrong and M. Oellerich.** 1999. Rapid detection of prothrombotic mutations of prothrombin (G20210A), factor V (G1691A) and methylenetetrahydrofolate reductase (C677T) by real time fluorescence PCR with the LightCycler. Clin. Chem. *45*:694-696.
17. **Wartell, R.M., S. Hosseini, S. Powell and J. Zhu.** 1998. Detecting single base substitutions, mismatches and bulges in DNA by temperature gradient gel electrophoresis and related methods. J. Chromatogr. A *806*:169-185.
18. **Wetmur, J.G.** 1991. DNA probes: applications of the principles of nucleic acid hybridization. Crit. Rev. Biochem. Mol. Biol. *26*:227-259.

Update to: Spreadsheet Software for Thermodynamic Melting Point Prediction of Oligonucleotide Hybridization with and without Mismatches

Ekkehard Schütz and Nicolas von Ahsen

Georg-August-University, Göttingen, Germany (eschuetz@meltcalc.com)

THERMODYNAMICS OF OLIGONUCLEOTIDE HYBRIDIZATIONS IN PCR: FROM ANALYSIS TO ASSAY DESIGN

In the original article, we describe the use of thermodynamic nearest-neighbor parameters for the calculation of melting temperatures (T_ms) of oligonucleotides in the polymerase chain reaction (PCR) environment. Since the time of publication, the application of this concept has been proven for several single nucleotide polymorphism (SNP) assays (3,5,6). Based on the fact that it could be shown that such calculations can predict the melting point within a small error, the software has been further developed to help scientists design such assays based on thermodynamic T_m prediction in silico (9). This finally lead to a program that will find the most discriminating probe (i.e., the probe with the highest T_m difference between matching probe and mismatch probe) for a defined mismatch and also, if a LightCycler® (Roche Molecular Biochemicals, Indianapolis, IN, USA) is used, an anchor probe for fluorescence energy transfer (FRET). The reliability of the prediction of T_ms has been greatly improved by the availability of thermodynamic parameters for so-called "dangling ends" (1), which are affecting the binding stability of DNA/probe hybrids in all experimental cases in which PCR products are probed with oligonucleotides. The software also accounts for dangling end effects when a hybridization assay is designed automatically. We have used this approach thoroughly for assay design of more than 50 different SNPs in our laboratory, and all probes show the predicted performance. An example of this is an assay system for thiopurine methyltransferase, which comprises eight different SNPs. When using the MeltCalc© software for assay design, we were able to do the complete genotyping in three LightCycler capillaries (3). Furthermore, it is noteworthy that thermodynamic calculations are also helpful to estimate the theoretical error rate of genotyping. Although not that important for most such assays, it clearly has to be considered for SNPs in "hot spots." A more detailed discussion of this can be found in the first publication of a practical approach of this concept to real time PCR (5). On the other hand, by using thermodynamic NN calculations, it is possible to define the limitations of genotyping based on double-stranded (ds)DNA melting, which is important for diagnostic genotyping in humans (8). An informed judgment of the yielded results should be based on a reliable estimate of the underlying possible error rate (3). Themodynamic NN calculations can also be extended to insertion–deletion (I/D) polymorphisms. Even though there is not a defined single mismatch under the probe in that case, it is possible to predict the melting behavior based on the comparison of the stability of a given probe hybridizing to a given sequence with an I/D polymorphism (7). Another important refinement of the basic calculations has been done by us through evaluation of the influence of the typical PCR chemistry on the T_m. Since the enthropy is salt-dependent, and the correction is based on sodium concentration, the calculation of sodium equivalents from bivalent cations, which is essential for most thermostable polymerases, is critical. A review of more than 450 LightCycler assays lead to an algorithm for translating those into sodium equivalents for T_m calculations, which has recently been published (10). In addition to that, we defined the influence of other compounds often present in real time PCR mixtures, such as dimethyl sulfoxide and SYBR® GreenI, which is a dsDNA fluorescence dye. All the aforementioned improvements, based on newly published parameters, have been integrated into the software, thus enabling us the best calculation–prediction of T_ms for oligonucleotides. This lead to a commercial version of the program featuring an assays' design module that calculated probes with the highest possible T_m difference between the wild-type and mutation for SNP analysis. More detailed information can be found on our Website at (http://www.meltcalc.com). At that URL, the original freeware version of MeltCalc is also available and can be downloaded and used by noncommercial users without any restrictions.

In some cases, it may not be possible to find a probe that has a high T_m difference between the wild-type and the mutation of interest. That is based on the type of mismatch and the surrounding sequence (nearest neighbors). In such cases, using "structured probes" may be a solution. The reason is the fact that such probes do not follow a simple 2-state transition (random–coil versus hybridized), but have two transitions (structured–hybridized–random coil). The best known example for such structured probes are "molecular beacons". It has been shown recently that such an approach can increase the T_m difference by several degrees (2). Since this may be a remedy for such critical cases, we intend to implement a feature for the calculation and design of such beacon-type probes into one of the next versions of the MeltCalc software.

Overall, it is possible to predict the melting behavior of oligonucleotides within a very small prediction error, which has been significantly improved in the last years by the availability of new thermodynamic parameters and other important estimates. We have incorporated those into the MeltCalc software and implemented the above mentioned features based on our experience and needs in a genetic diagnostic laboratory. It often takes time to convince the scientific community that, even though experience plays an important role in the laboratory, evidence-based assay design and interpretation is a step forward to better reliability and reproducibility, one example of which has been shown by us (4) recently, and many more can be found in the literature. In conclusion, in silico design of these type of SNP assays based on physical–chemical evidence can significantly shorten the time to success of establishing assays and also minimize the error rate in critical cases.

REFERENCES

1. **Bommarito, S., N. Peyret, and J. Santa Lucia, Jr.** 2000. Thermodynamic parameters for DNA sequences with dangling ends. Nucleic Acids Res. *28*:1929-1934.
2. **Bonnet, G., S. Tyagi, A. Libchaber, and F.R. Kramer.** 1999. Thermodynamic basis of the enhanced specificity of structured DNA probes. Proc. Natl. Acad. Sci. USA *96*:6171-6176.
3. **Schütz, E., N. von Ahsen, and M. Oellerich.** 2000. Genotyping of eight thiopurine methyltransferase mutations: three-color multiplexing, "two-color/shared" anchor, and fluorescence-quenching hybridization probe

assays based on thermodynamic nearest-neighbor probe design. Clin. Chem. *46*:1728-1737.

4. **Schütz, E. and N. von Ahsen.** 2001. Genotyping of the most common thiopurine methyltransferase mutations (TPMT*3) with the LightCycler. Optimization of hybridization probe assays for the detection of mutations causing stable mismatches in DNA duplexes, p. 143-152. *In* Meurer, S., C.T. Wittwer, and K.I. Nakagawara (Eds.), Rapid Cycle Real Time PCR—Methods and Applications. Springer Verlag, Berlin.
5. **von Ahsen, N., M. Oellerich, V.W. Armstrong, and E. Schütz.** 1999. Application of a thermodynamic nearest-neighbor model to estimate nucleic acid stability and optimize probe design: prediction of melting points of multiple mutations of apolipoprotein B-3500 and factor V with a hybridization probe genotyping assay on the LightCycler. Clin. Chem. *45*:2094-2101.
6. **von Ahsen, N., M. Oellerich, and E. Schütz.** 2000. A method for homogeneous color-compensated genotyping of factor V (G1691A) and methylenetetrahydrofolate reductase (C677T) mutations using real-time multiplex fluorescence PCR. Clin. Biochem. *33*:535-539.
7. **von Ahsen, N., M. Oellerich, and E. Schütz.** 2000. DNA base bulge vs unmatched end formation in probe-based diagnostic insertion/deletion genotyping: genotyping the UGT1A1 (TA)(n) polymorphism by real-time fluorescence PCR. Clin. Chem. *46*:1939-1945.
8. **von Ahsen, N., M. Oellerich, and E. Schütz.** 2001. Limitations of genotyping based on amplicon melting temperature. Clin. Chem. *47*:1331-1332.
9. **von Ahsen, N. and E. Schütz.** 2001. Using the nearest neighbor model for the estimation of matched and mismatched hybridization probe melting points and selection of optimal probes on the LightCycler, p. 43-56. *In* Meurer, S., C.T. Wittwer, and K.I. Nakagawara (Eds.), Rapid Cycle Real Time PCR—Methods and Applications. Springer Verlag, Berlin.
10. **von Ahsen, N., C.T. Wittwer and E. Schütz.** 2001. Oligonucleotide melting temperatures under PCR conditions: nearest-neighbor corrections for Mg(2+), deoxynucleotide triphosphate, and dimethyl sulfoxide concentrations with comparison to alternative empirical formulas. Clin. Chem. *47*:1956-1961.

CHAPTER 11

Spreadsheet Macro for Setting Up PCR Assay Tubes

Raymond P. Stowe and Duane L. Pierson[1]

KRUG Life Sciences and [1]National Aeronautics and Space Administration, Lyndon B. Johnson Space Center, Houston, TX, USA

Reprinted from BioTechniques 20:1088-1089 (June 1996)

ABSTRACT

This article describes a Microsoft® Excel™ macro designed to calculate PCR master mix amounts based on variations in the DNA template amounts added to each tube, the total number of PCR assay tubes being set up or both. This macro uses a dynamic dialog box to quickly calculate and display the new component volumes after changes in one or both of the variables. The PCR assay mix protocol can then be printed in a format suitable for record keeping. This macro was designed for use with the Windows™ version of Excel but will also run on Macintosh® computers.

INTRODUCTION

The polymerase chain reaction (PCR) is a powerful tool in molecular biology research. PCR is a simple automated procedure that can be used to amplify (1) or modify a specific target DNA sequence. The reaction often is carried out in 25-, 50- or 100-µL volumes contained in a 0.2-mL reaction tube. The final reaction mixture typically consists of the DNA target (template), four deoxynucleoside triphosphates (dATP, dCTP, dGTP and dTTP), complementary primers, Tris buffer and DNA polymerase. Some protocols also call for other reagents such as dimethyl sulfoxide (DMSO), glycerol or formamide to be included.

PCR protocols generally use master mix tubes. Individually dispensing the PCR reagents to each PCR assay tube is labor-intensive, uses large numbers of pipet tips and may introduce pipetting errors when small volumes are transferred. The use of master mix tubes is less labor-intensive and more cost-effective because fewer pipettings are made, and they also allow larger volumes to be pipetted into assay tubes, thus reducing pipetting errors inherent in transferring small volumes.

Computer programs exist that select optimized PCR primer pairs from DNA sequence files and establish optimal experimental conditions. However, calculations for determining the volumes of PCR components are frequently performed on handheld calculators. This approach is time-consuming and may lead to calculation errors. We present a program that has the advantages of minimal computer requirements and ease of calculations

BioComputing: Computer Tools for Biologists
Edited by Stuart M. Brown

and record keeping. It was developed for use on the IBM® PC with Excel™ for Windows™ 4.0 (Microsoft, Redmond, WA, USA) and has been used in our laboratory for the past several months. The macros also can be run on the Macintosh® version of Excel.

The macro is contained in a file called **pcrmac.xlm**, which is loaded into the xlstart directory within Excel. When Excel is started, a blank screen appears on the monitor instead of the usual unmodified spreadsheet (due to the **Hide** command built into the macro). An important note is that all active (displayed) worksheets must be closed for the macro to run properly. When the macro is started by the command key **<CTRL/A>**, a dialog box appears (Figure 1). The user can input the **Reaction Mixture Volume** by clicking on the appropriate button (25 or 100 µL); the number of **Assay Tubes (#)** and the **Template** amount (µL per reaction tube) are entered from the keypad (entry is not blocked in component columns, but they are not designed as data entry spaces). Clicking on the **OK** button automatically calculates the amounts of each assay component to be added to **Mix-Tube 1** or **Mix-Tube 2**. The amounts of **Mix-Tube 1** and **Mix-Tube 2** to be added to the reaction tubes are presented at the bottom of each column (**Volume [µL/tube]**). The protocol can be printed to obtain a hard copy suitable as a notebook record (Figure 2). The user exits the dialog box by clicking on the **Cancel** button, which automatically hides the file. Quitting Excel produces a dialog box that asks the user if the changes in the macro file are to be saved.

The Macintosh version (**PCRMACInt.XLM**) of the macro can be opened by simply double-clicking on the icon. The **Hide** window functions have been removed, since Macintosh Excel does not have an "xlstart" directory. Upon opening, a minimized window will appear. The macro is started by the command key **<CTRL/A>**, and then by following the instructions above. The macro can be exited by clicking on the close square in the upper left-hand corner.

The assay setup displayed in Figure 2 is based on a method currently used in our laboratory. The PCR reagents are dispensed into the two mix tubes in the following manner: **Mix-Tube 1** contains 10× Tris buffer (Perkin-Elmer, Norwalk, CT, USA), DMSO (5% final conc.) and H_2O (to adjust the final volume); **Mix-Tube 2** contains dNTPs, both primers, *Taq* DNA polymerase (0.625 U/25 µL per assay tube) and water to bring reagents to correct concentrations. We only included 25- and 100-µL buttons, since the user could multiply for a 50-µL reaction volume. The macro can be modified to add/ delete reagents as necessary and combine reagents (without *Taq* or DMSO) into a single mix tube for freezing. Increasing flexibility can also be performed by adjusting reagent (e.g., primers) concentrations to support fixed volumes in the macro. This division (Mix-Tube 1 and 2) is arbitrary so that this may also be applied to other

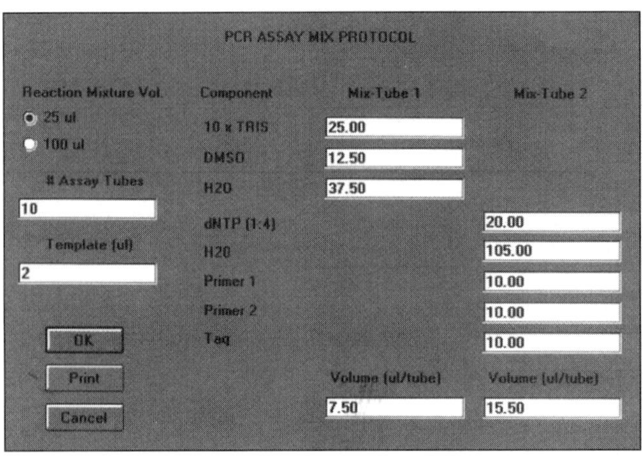

Figure 1. The dialog box for the PCR assay mix protocol.

PCR ASSAY PROTOCOL

Component	1 Assay (ul)	# Assays (ul)
		10
10X Buffer	2.5	25.00
DMSO	1.25	12.50
dNTP	12.5	125
Template	2	20
Primer 1	1	10
Primer 2	1	10
Taq	1	10
	21.25	212.5
H2O	3.75	37.5
	25	250

Mix 1

10x buffer		25.00	
DMSO		12.50	
H2O		37.50	
		75	
Div/by		10	7.50
			ul/assay

Mix 2

dNTP (1:4)		20.00	
H2O		105.00	
		125	
Primers	1	10.00	
	2	10.00	
Taq		10.00	
		155	
Div/by		10	15.50
			ul/assay

Figure 2. The printout of the PCR assay mix components in table form.

configurations (e.g., top and bottom mixes). Additionally, by entering a larger number of tubes than necessary in the **Reaction Mixture Volume**, a "safety factor" can be figured in so that no shortage of reaction mixture occurs.

MODIFYING THE MACRO

The macro file (**pcrmac.xlm**) is formatted as an international macro; the sheet is "hidden" to prevent accidental modification. To view and edit the macro sheet, choose **Unhide** from the **File** menu. After viewing/modifying the macro, choose the **Hide** command to protect against accidental error when running the macro again. The Macintosh version (**PCRMACInt. XLM**) does not require the **Unhide** command. The window can be maximized to view the program codes. The user can then minimize the window and save the changes before exiting the macro.

OBTAINING THE MACRO

The software discussed in this paper is available for downloading from the BioComputing section of the *BioTechnques* World Wide Web site located at www/BioTechniques.com.

ACKNOWLEDGMENTS

We thank Christine Wogan for her editorial work. This work was supported by NASA Grant No. NAS-18492.

REFERENCE

1. **Innis, M.A., D.H. Gelfand, J.J. Sninsky and T.J. White.** 1990. PCR Protocols: A Guide to Methods and Applications, p. 3-12. Academic Press, NewYork.

Update to: Spreadsheet Macro for Setting Up PCR Assay Tubes

Raymond P. Stowe

University of Texas Medical Branch, Galveston, TX, USA (rpstowe@utmb.edu)

It has been fascinating to watch the area of molecular genetics, which has undergone an explosive growth in recent years. Accordingly, there has been a tremendous increase in the number of genomics- and proteonomics-based computer programs; the complexity of these programs has generally mirrored the rapid changes in molecular techniques that generate massive amounts of data.

Many sophisticated computer programs can be purchased that analyze DNA sequences, polymerase chain reaction (PCR) primers, etc., to aid in molecular analyses. A major disadvantage of commercially available software is that it is relatively expensive, and a typical molecular laboratory can only afford a single license. This limits the needs of multiple users in the laboratory (i.e., the creation of a sign-up sheet!). In addition, the software may not be used enough to justify its cost.

On an alternate note, many individual laboratories have written focused applications to handle their biocomputing needs. Many of these programs were written as a Microsoft® Excel® spreadsheet or macro that will run on virtually any personal computer (PC), thus allowing each individual user to analyze their data without delay. To handle the numerous PCR assays in our laboratory, we wrote a spreadsheet macro for rapidly calculating volumes for PCR master mixes (see original article); this macro has not required any major changes since its implementation almost 9 years ago.

Recent advances in molecular biology techniques have resulted in the generation of massive amounts of information from a limited number of samples (i.e., DNA microarrays). In keeping with these advances, many laboratories have written computer programs to analyze the massive amounts of data particular to their needs (1,2). These relatively simple, yet flexible programs are easily modified and may be adapted by students for laboratory use. As molecular biology techniques continue to change and evolve, it will be interesting to speculate where the area of biocomputing will be in the next 10 years.

The computer programs discussed in this update may be located from the *BioTechniques* Web site.

REFERENCES

1. **Conway, T., B. Kraus, D.L. Tucker, D.J. Smalley, A.F. Dorman, and L. McKibben.** 2002. DNA array analysis in a Microsoft Windows environment. BioTechniques *32*:110-119.
2. **Strehlow, D.** 2000. Software for quantitation and visualization of expression array data. BioTechniques *29*:118-121.

Spreadsheet-Based Program for Alignment of Overlapping DNA Sequences

Ramaswamy Anbazhagan and Edward Gabrielson

Johns Hopkins University School of Medicine, Baltimore, MD, USA

Reprinted from BioTechniques 26:1180-1185 (June 1999)

ABSTRACT

Molecular biology laboratories frequently face the challenge of aligning small overlapping DNA sequences derived from a long DNA segment. Here, we present a short program that can be used to adapt Excel® spreadsheets as a tool for aligning DNA sequences, regardless of their orientation. The program runs on any Windows® or Macintosh® operating system computer with Excel 97 or Excel 98. The program is available for use as an Excel file, which can be downloaded from the BioTechniques Web site. Upon execution, the program opens a specially designed customized workbook and is capable of identifying overlapping regions between two sequence fragments and displaying the sequence alignment. It also performs a number of specialized functions such as recognition of restriction enzyme cutting sites and CpG island mapping without costly specialized software.

INTRODUCTION

Sequencing large DNA segments typically requires cutting the DNA into a number of small overlapping fragments, which are then individually cloned and sequenced from both ends. These fragment sequences can be in forward, reverse, complimentary or antiparallel orientations. Identification of the overlapping segments between various fragments and arranging them into a single continuous sequence is a time-consuming process. Furthermore, software available commercially for sequence assembly is relatively expensive. Here, we report the adaptation of a commonly available spreadsheet program, Excel® 97 or Excel 98 (Microsoft, Redmond, WA, USA) for this purpose. The program opens a customized workbook specifically designed for this and is capable of identifying overlapping regions between two sequence fragments and displaying the sequence alignment.

BioComputing: Computer Tools for Biologists
Edited by Stuart M. Brown

MATERIALS AND METHODS

Downloading and Opening the Program

We have created an Excel file with a built-in program that performs all major functions required for sequence alignment. The program works well with Excel 97 for Windows® operating system and with the Excel 98 for Macintosh® (Apple Computer, Cupertino, CA, USA) operating system. We recommend downloading the file (alignwin.xls for Windows and alignmac.xls for Macintosh operating system) from the Software Library on the BioTechniques Web site (http:// www.BioTechniques.com). The program can be started in two ways: (*i*) either click open the downloaded file or (*ii*) start the Excel program and open the downloaded file from within the program using **File/Open**. A dialogue box appears asking to enable or disable macros. Click on the **Enable macros** button. When the program opens, you will notice that a new menu named *Align* is added to the program menu bar. This menu has six menu items, namely *New-Workbook*, *Convert*, *Paste*, *Mark-Align*, *Single-Step* and *Clear-Sheet2*. These menu items can be executed to open a customized workbook and to perform various functions necessary to identify the region of alignment between two overlapping sequences.

Opening a New Workbook and Entering a Sequence

The Excel workbook is organized as a collection of sheets, with each sheet's name appearing on a tab at the bottom of the workbook. Navigation between the sheets is possible by clicking on the tab at the bottom. Refer to the User's Guide for further explanation about the organization and use of Excel workbook and sheets. Go to the **Align** menu and click on the **New-Workbook** menu item. A new workbook is opened with three sheets. The program automatically enters the column headings such as Clone ID, Primer, Count and Sequence in sheet 1, as shown in Figure 1. The user enters the relevant details of the sequence fragments in the corresponding columns manually. The clone identification numbers are entered in column A, and the sequencing primers such as T7, T3 and M13 are entered in column B. If a new insert-specific oligonucleotide is used as a primer for sequencing, the sequence of the primer itself can be entered in column B. The sequences read from the sequencing gel are entered in column D, using the standard abbreviation (a, c, g and t). Ambiguous nucleotides should be designated as "n". Each cell in Excel 97 can hold up to 32 767 characters, and therefore, very long sequences can be entered in these cells.

The program automatically fills up column C with a formula that will count and display the total number of nucleotides of the sequences that are entered in column D by the user. Column C initially displays 0 before the entry of any sequence in column D. When the sequences are entered or copied to column D, the nucleotide count of each sequence is automatically displayed in column C. Enter the sequences using lowercase letters. The

Clone ID	Primer	Count	Sequence
Clone 1	T3	53	acagagatagacagatacagctcgctgacagatcgctccagtacgctgacacg
Clone 2	T7	50	atatatagacacagaacagacagagatagatagctttgcctcgccgttta
Clone 3	M13	54	tacgctgacacgagatagcttcgctcgctcgaagatgcctcgacagtccccgct
Clone 4	T3	51	tcgaagatgcctcgacagtcgccatctcgctacccacatatagaccatgaa

Figure 1. First sheet (sheet 1) of the Excel workbook. Clone identification numbers, primers and sequence information are shown.

program codes are case-sensitive and will function properly only if the sequences are entered in lowercase letters. The nucleotide count in column C is updated automatically whenever any sequence entry is altered in column D.

The second sheet serves as a temporary worksheet for running the macros to align any two sequences. The labels *Clone 1*, *Clone 2* and *Overlap* are automatically displayed by the program in sheet 2 in the appropriate cells. The overlapping fragments, identified by the program, are arranged manually into a single continuous sequence in the third sheet, which is customized by the program for this purpose.

Sequence Alignment

To align the sequences, it is first necessary to search for the overlapping sequences between different clones. During cloning, the fragments can get oriented in any direction, so that the overlapping sequences can be matching in forward, reverse, compliment or antiparallel orientations. Thus, it is necessary to search and align all possible orientations of the fragment sequences. The entire alignment can be performed as a single step or as a series of three steps. We first present the three-step approach to illustrate the strategy used and to demonstrate additional options available.

Select two sequences from sheet 1 to be aligned, and copy them into cells B1 and B3 of sheet 2. Click on the **Align** menu, and click on the **Convert** menu item; this will convert the sequence in cell B1 into a reverse sequence, which is pasted in cell B2. The second sequence pasted in cell B3 is converted into reverse, complimentary and antiparallel orientations and pasted into cells B4, B5 and B6, respectively. The program also finds the longest overlapping region between these two groups of sequences and displays the overlapping sequences in cells B9 and B10. The program is capable of converting and aligning sequences of any length entered in the appropriate cells. The program also can handle ambiguous nucleotides. If the sequence contains an "n", indicating ambiguity, it will be aligned against any of the four nucleotides. If the overlapping region contains any ambiguous nucleotide, then the sequence in B9 and B10 reflects this.

Next, execute the menu item **Paste**, which will paste all sequences in cells B1 through B6 in a column format (starting from cell B13) with each nucleotide occupying a single cell. Executing the menu item **Mark-Align** is the next step; this reads the overlapping sequence from cells B9 and B10 and uses them as search strings to search the vertically pasted sequences, and the overlapping regions are highlighted with a bright pink background. The actual alignment of the two sequences is also displayed in column J, with the overlapping regions highlighted. Figure 2 shows a sample sheet 2 displaying the results of all three procedures. Sequence alignments can be repeated for different sets of sequences. To erase the previous entries, including the alignment, click on the **Clear-Sheet2** menu item in the **Align** menu.

If there is a true overlapping region between the two sequences, the program always identifies that. In most other cases, the program identifies a couple of nucleotides or a single nucleotide at the terminal region as the region of alignment. It is up to the user to decide whether these few nucleotide overlaps reflect real alignment or not. For example, if you are sequencing a known insert with multiple primers, and if you know the position and sequence of the primers used, then you will be able to identify and confirm even if there is an overlap of a few nucleotides as a valid result. If you are sequencing a novel insert, and if you get an overlap of a few nucleotides between two fragments, you might need further sequencing data from different primers to confirm or exclude this result.

In some instances, when the **Mark-Align** menu item is executed, the program will identify and mark more than one region in each sequence as the region of alignment. This

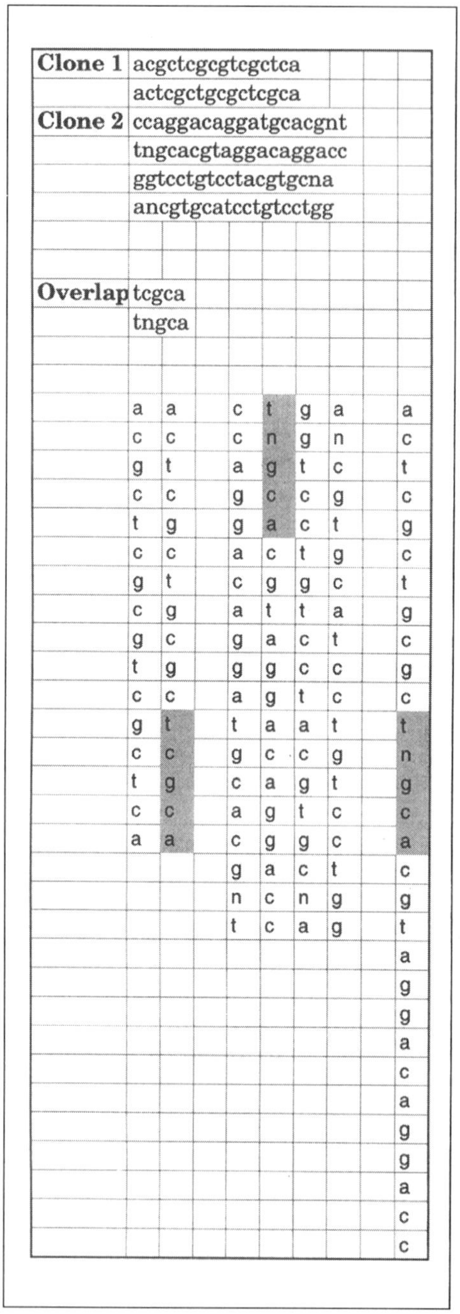

Figure 2. The result of executing the first three macros in sheet 2. The sequences are converted into various orientations and pasted below in a vertical column format. The aligned regions are highlighted, and the alignment is displayed.

might happen when there are internal sequences similar to the terminal region, or when the overlapping region identified has only few or a single nucleotide. If this happens, the program displays the message "Multiple matching regions found. Alignment display not performed" and

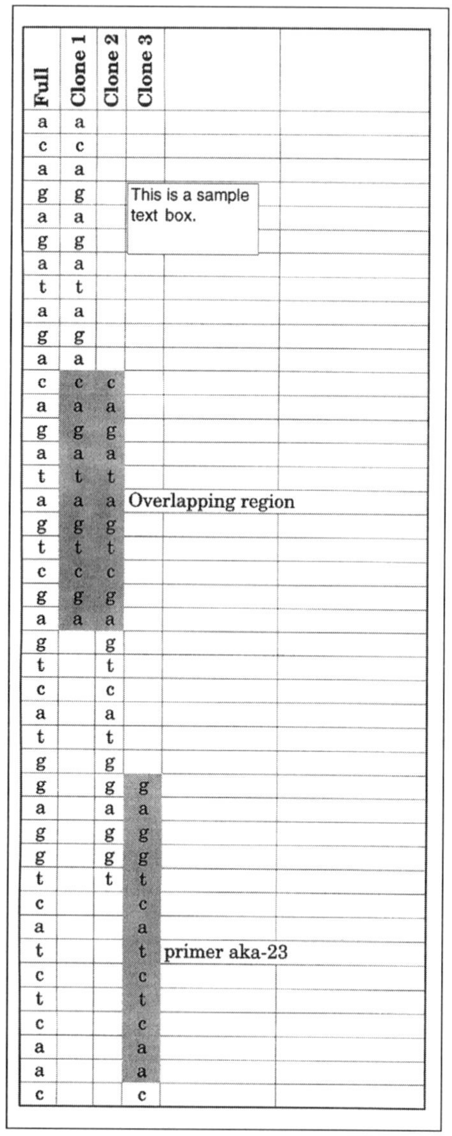

Figure 3. A sample of sheet 3 with arranged sequences.

does not proceed with the display of the alignment. However, all the regions that match with the overlapping sequence will be highlighted.

Arranging the Overlapping Fragments into a Single Continuous Sequence

Although the program identifies the overlapping sequence and also displays the alignment, the actual arrangement of large number of sequence fragments into a single continuous sequence has to be performed manually by the user on sheet 3, as shown in Figure 3. To arrange a continuous sequence, fragment sequences on sheet 2, with highlighted aligned regions, are copied and pasted into sheet 3 manually, with overlapping regions aligned next to one another to show continuity. The first column of sheet 3 is allocated for entering the full continuous sequence, which can also be copied from sheet 2. The clone and primer details are entered in the top cell of each column to keep track of the sequence identities (Figure 3). The third sheet is custom-designed to make it easy to arrange and view the sequences. The program automatically freezes the first column and top rows. If the viewing field is moved in a vertical or horizontal direction, the first column and the top row always remain visible, thus making it easy to follow up the continuity and the identity of the sequences. You can rename and save the workbooks, but the worksheet names should not be altered. We recommend making multiple copies of the downloaded file and using separate files for individual projects. Always keep a copy of the original file for future use.

Other Special Functions

The menu item **Single-Step** combines the functions of the three menu items (**Convert**, **Paste** and **Mark-Align**) and executes them in the same order in a single step. Although the entire program for aligning sequences can be run as a single step, using the three menu items individually allows the program to be used for other purposes. For example, the menu items **Paste** and **Mark-Align** could be used to search for any sequence of interest such as restriction enzyme cutting sites or CpG islands within a given target sequence. For these purposes, enter one or more target sequences in the cells from B1 to B6 in sheet 2. If more than one target sequence is used, enter them in the adjacent cells. Enter the sequence to be searched in cell B9. Now execute **Paste** and **Mark-Align**. The target sequence is pasted in a column format, and the matching regions (to the search sequence entered in cell B9) is highlighted.

While performing sequence alignments, very rarely one might face a situation in which a short fragment is contained within a longer sequence. Because the **Mark-Align** procedure of the program only recognizes overlapping sequence at the ends of sequences, it is not possible to identify this situation. However, if this possibility is suspected, then the following procedure can be used to identify this. Copy the longer sequence to cell B1, and paste the suspected segment in cell B9. Now, if the menu items **Paste** and **Mark-Align** are executed, the shorter segment within the longer sequence is identified.

DISCUSSION

The menu items **Paste** and **Mark-Align** are designed to aid the assembly of sequences on sheet 3 in a column format. We chose the column format for arranging sequences because the number of cells available in a column is much greater than the number available in a row. Excel 97 has 65 536 cells in a column. Thus, it is possible to arrange 65 kb of sequence using a column format.

In addition to assembling and identifying the continuity of the sequence, a number of additional data can be recorded in sheet 3. For example, the position of primers can be conveniently marked by color highlighting the appropriate cells, and the name of the primer can be recorded in the adjacent cell. The 5′ and 3′ ends of the sequencing primers also can be marked as indicated. Notes can be inserted to record additional information (Figure 3).

The two menu items, **Paste** and **Mark-Align**, can also be used to search for and mark amino acid sequences within a given protein sequence. If amino acid and protein sequences (entered in single letter format) are used in place of DNA sequence in the above example, the menu items **Paste** and **Mark-Align** will search and highlight the matching amino acid sequence within a target protein sequence.

The commonly available spreadsheet program Excel has been successfully used in molecular biology for a number of purposes. Shaw (3) described a method of using a spreadsheet program for protein sequence interpretation. McEwan and Gatherer (2) have described adapting a spreadsheet program to translate nucleic acid sequence into a protein sequence and to produce a codon usage table. Han and Tashjian (1) have developed a program to analyze regional hydrophobicity of a protein from its amino acid sequence. We have now described a method of using a spreadsheet program for sequence alignment. In addition to assembling a continuous sequence from small sequences of variable orientation, the program can also be used for a number of other individual needs.

REFERENCES

1. **Han, B. and A.H. Tashjian, Jr.** 1998. User-friendly and versatile software for analysis of protein hydrophobicity. BioTechniques *25*:256-263.
2. **McEwan, N.R. and D. Gatherer.** 1998. Adaptation of standard spreadsheet software for the analysis of DNA sequences. BioTechniques *24*:131-138.
3. **Shaw, G.** 1995. Protein sequence interpretation using a spreadsheet program. BioTechniques *19*:978-983.

Update to: Spreadsheet-Based Program for Alignment of Overlapping DNA Sequences

Ramaswamy Anbazhagan and Edward Gabrielson

Johns Hopkins University School of Medicine, Baltimore, MD, USA (anba@jhmi.edu)

PROGRAM FOR THE ALIGNMENT OF OVERLAPPING SEQUENCES

This is a small Microsoft® Excel® spreadsheet-based program with many useful functions for sequence assembly and analysis. It is very handy in finding out overlapping sequences. It also provides a sequence assembly sheet for assembling multiple sequences into a single continuous one. Primers, restriction sites, and other information can be added in this page. This helps to easily review the assembled sequence with all the relevant annotation. This program can also be used to find out restriction enzyme cutting sites or to search for any sequence motif within a sequence. This program has been successfully used to assemble sequences in small sequencing projects (1). The program is freely available for academic users and can be obtained by sending an e-mail (anba@jhmi.edu) request to the author.

REFERENCE

1. **Anbazhagan, R., J.G. Herman, K. Enika, and E. Gabrielson.** 2001. Spreadsheet-based program for the analysis of DNA methylation. BioTechniques *30*:110-114.

CHAPTER 13

Spreadsheet-Based Program for the Analysis of DNA Methylation

Ramaswamy Anbazhagan, James G. Herman, Kalyan Enika, and Edward Gabrielson

The Johns Hopkins University School of Medicine, Baltimore, MD, USA

Reprinted from BioTechniques 30:110-114 (January 2001)

ABSTRACT

Methylation of DNA in CpG dense regions of gene promoters (CpG islands) is important for transcriptional inactivation of selective genes in normal and neoplastic cells. Here, we present a spreadsheet-based program adapted from Microsoft® Excel® that is useful for identifying CpG islands and for assisting in the laboratory analysis of DNA methylation of these regions. Upon execution of the program, a customized workbook analyzes an entered DNA sequence for the total number and percentage cytosine and guanine nucleotides, the total number and percentage of CpG sites, and a CpG:GpC ratio. The program also displays the distribution of CpG sites in a visual format as well as in two different graphical formats. Finally, the program assists in laboratory studies of DNA methylation that employ bisulfite modification of DNA by displaying methylation-dependent effects of bisulfite treatment on DNA sequences.

INTRODUCTION

While the CpG dinucleotide occurs at less than 10% of its expected frequency in the human genome, many gene promoter regions have high densities of CpG sites. These regions are known as CpG islands, and methylation of cytosines in CpG islands is a common mechanism for transcriptional inactivation of genes on the inactivated X-chromosome in females and of silenced imprinted alleles on autosomal chromosomes. Furthermore, methylation is increasingly recognized as an important mechanism for inactivation of tumor suppressor genes in neoplasia (reviewed by Baylin et al. in Reference 2).

CpG islands are defined as regions of DNA ranging in size from 0.5 to 2 kb that have a C + G content of greater than 60% and a ratio of CpG to GpC of at least 0.6 (3,4). Identification of CpG islands is a critical initial step for studying transcriptional regulation of genes by methylation. The status of CpG islands can then be evaluated in the laboratory using

BioComputing: Computer Tools for Biologists
Edited by Stuart M. Brown

restriction enzymes that are methylation specific, by sequencing bisulfite modified DNA, or by methylation-specific PCR (MSP) (5) of bisulfite-modified DNA. Here, we report the adaptation of a commonly available spreadsheet program, Microsoft® Excel® 2000, as a tool for identifying CpG islands and assisting in the analysis of DNA methylation.

MATERIALS AND METHODS

Downloading and Opening the Program

We have created an Excel file with a built-in program that performs all major functions required for the analysis of CpG islands. The program works with Excel 2000. The program file (CpG-Win) can be downloaded from the Software Library on the Internet at http://128.20.85.49/genomics. After downloading the file, the program can be started in two ways: (*i*) either open the downloaded file or (*ii*) start the Excel program and open the downloaded file from within the program using File/Open. A dialogue box appears, asking to enable or disable macros. Click on the Enable Macros button. When the program opens, you will notice that a new menu named Analyze-CpG is added to the program menu bar. This menu has seven menu items, namely New-Workbook, Format, Count-CpG, Mark-CpG, Bis-Modify, and Make-Graph. These menu items can be executed to perform various functions necessary for the analysis of CpG islands within the DNA sequences.

Opening a New Workbook and Entering a Sequence

The Excel workbook is organized as a collection of sheets, with each sheet's name

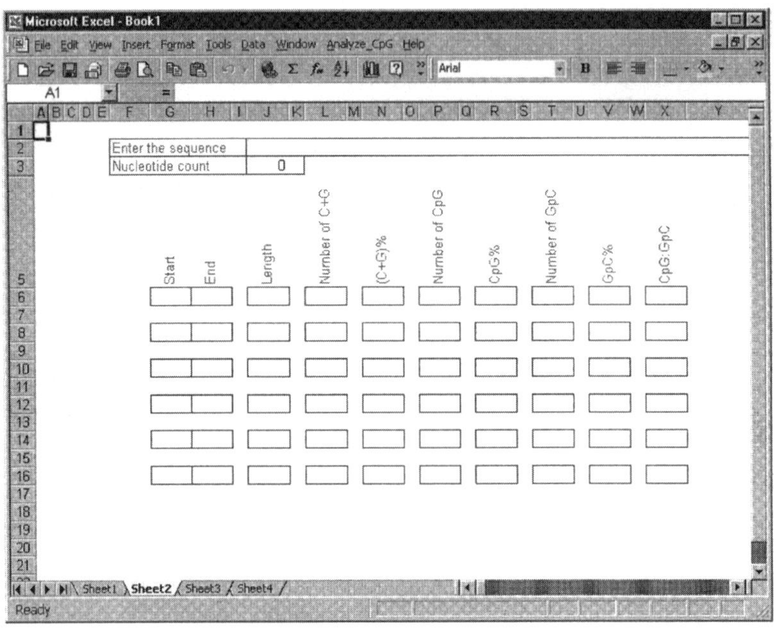

Figure 1. Sheet2 of a new workbook opened by executing the menu item New-Workbook in the CpG-Analyze menu.

appearing on a tab at the bottom of the workbook. Navigation between sheets is possible by clicking on the tab at the bottom. Refer to the Microsoft User's Guide for further explanation about the organization and use of Excel workbook and sheets. To use this program, go to the newly added Analyze-CpG menu and click on the New-Workbook menu item. This command will open a new workbook with four sheets. The program automatically formats the Sheet2 as shown in Figure 1. To analyze a specific DNA sequence, enter the nucleotide sequence in cell A1 of Sheet1. Nucleotides in the sequence may be entered in either uppercase or lowercase. Ambiguous nucleotides may be designated as "n". The cells can hold more than 30 000 characters and can therefore accommodate long sequences. These sequences may be copied and pasted from other documents. GenBank® sequences viewed through Internet browsers can also be directly copied and pasted along with numbers and spaces. If you copy and paste sequences from GenBank, you will notice that the sequence occupies multiple rows in Sheet1. After entering or pasting the sequence in cell A1 of Sheet1, go to the Analyze-CpG menu and click on the Format menu item. This command removes all the numbers and spaces from the sequence (if there are any), concatenates the sequences in all the rows, and also transfers the sequence from cell A1 of sheet1 to cell J2 of Sheet2. The command also activates Sheet2 so that you can see the transferred sequence there.

The program automatically fills up cell J3 of Sheet2 with a formula that will count and display the total number of nucleotides in cell J2 of Sheet2. When a sequence is transferred from Sheet1 to cell J2 of Sheet2, the nucleotide count of that sequence is automatically displayed in cell J3. The nucleotide count in cell J3 is also updated automatically whenever any sequence entry is altered in cell J2.

Analyzing the CpG and GpC Distribution in Different Regions of the Sequence

The next step is to designate which portions of the sequence are to be analyzed for CpG islands. This is done by entering the starting and ending position numbers of the fragments to be analyzed in column G and H, respectively, below the labels Start and End. Up to six fragments can be entered in these columns with the starting and ending position numbers of their nucleotides. After entering these values, go to the Analyze-CpG menu and click on the Count-CpG menu item. After a few seconds delay (depending on the length of the sequence), the analyzed data are displayed in columns J–X as shown in Figure 2.

Column J (labeled "Length") displays the length (number of nucleotides) of each of the fragments entered in columns G and H. Column L (labeled "Number of C+G") displays the total number of cytosine and guanine nucleotides within each of the fragments. The percentage of cytosine and guanine nucleotides within each fragment is displayed in column N (labeled "(C+G)%"). Column P (labeled "Number of CpG") displays the total number of CpG islands within each of these fragments. The percentage of CpG islands within each of these fragments is displayed in column R (labeled "CpG%"). Similarly, the total number and the percentage of GpC base pairs are displayed in columns T and V (labeled "Number of GpC" and "GpC%", respectively). The last column, X, shows the CpG:GpC ratio in the sequence.

Marking the CpG and GpC Base Pairs for Visual Display

Although the various parameters of the CpG islands are analyzed as described above, it does not give a visual impression of the distribution of the CpG and GpC sites within a sequence. The next menu item, Mark-CpG, is meant to perform this function. When this menu item is executed, the program takes the sequence in cell J2 of Sheet2 and displays it in columns B and D (labeled "CpG Marked" and "GpC Marked", respectively) of Sheet3 in a vertical column format with each nucleotide occupying a single cell. The program also highlights the CpG sites in column B and GpC sites in column D with a bright pink background. This highlighting helps to visually appreciate the distribution of the CpG and GpC sites (Figure 3). This visual effect can be better appreciated by clicking on the View menu bar and executing the Zoom function, which are built-in Excel commands. By adjusting the zoom ratio to 10% of the original, an overall view the marked sequence can be seen.

Displaying the Bisulfite-Modified DNA Sequence

MSP and sequencing of bisulfite-modified DNA are commonly employed to analyze methylation of DNA. These techniques both involve treating the DNA with bisulfite, which chemically modifies the unmethylated cytosine residues to uracil. In effect, this modification is reflected as thymine after sequencing or MSP. Bisulfite treatment of DNA therefore results in different products, depending on the methylation status of the DNA before modification.

The next command in our custom menu, Bis-Modify, helps to identify possible nucleotide sequences after bisulfite treatment, depending on prior methylation status. When this menu item is executed, the DNA sequence from cell J2 of Sheet2 is displayed in columns G, H, and I of Sheet3 (labeled "Original", "Unmethylated" and "Methylated",

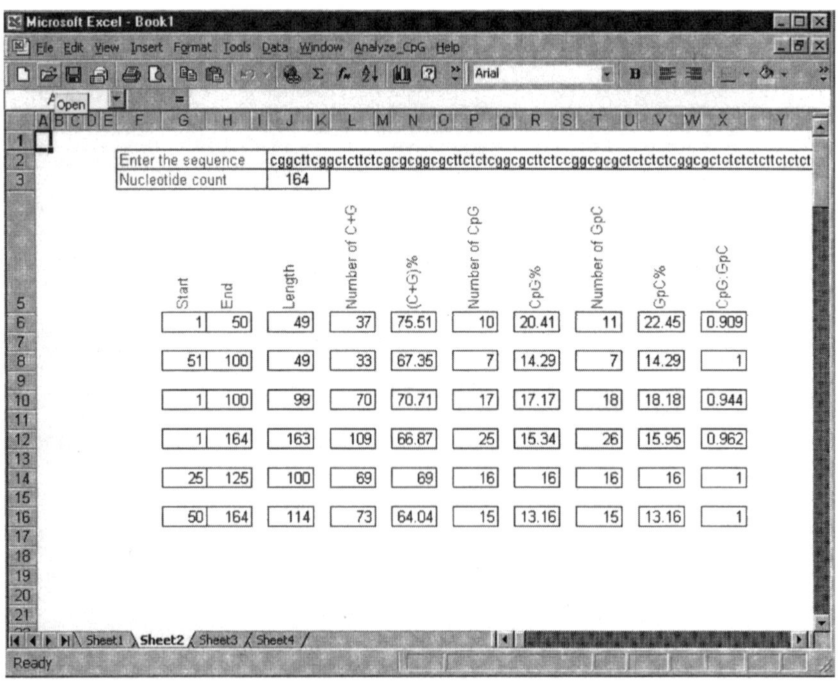

Figure 2. A sample of Sheet2 displaying the results of analysis of a sequence by executing the Count-CpG menu item.

respectively), in a vertical format, with each nucleotide occupying a single cell. The column labeled "Original" displays the original sequence for the purpose of comparison with others. The column labeled "Unmethylated" displays the modified sequence that would result from bisulfite treatment of unmethylated DNA. Note that in this situation, all cytosine nucleotides would be converted to thymine regardless of whether or not they are associated with a part of a CpG island. These modified nucleotides are highlighted with a bright pink background. The next column labeled "Methylated" displays the sequence that would result from bisulfite treatment of DNA with methylated sequences in CpG islands. In this situation, all cytosines not followed by guanine (thus not making a CpG pair) are modified to thymine, while those followed by guanine (thus making a CpG pair) are retained as cytosines. As in the previous case, all the modified nucleotides are highlighted with a bright pink background. Thus, the execution of this menu item displays all three sequences (original, unmethylated, and methylated) side by side with the modified nucleotides being highlighted. This display format facilitates comparison of the sequences and identification of possible sites of methylation. The results of bisulfite modification of the complementary strand are displayed in columns K, L, and M.

To design primers for MSP, it may be desirable to copy these unmodified and modified sequences to another file or to another program used for designing primers. To make it easy to copy these sequences, they are also displayed in a horizontal format with each sequence occupying a single cell as shown in Figure 3. Actually, the menu item Bis-Modify performs this function also. The unmethylated or methylated version of DNA sequence could be easily copied from column O.

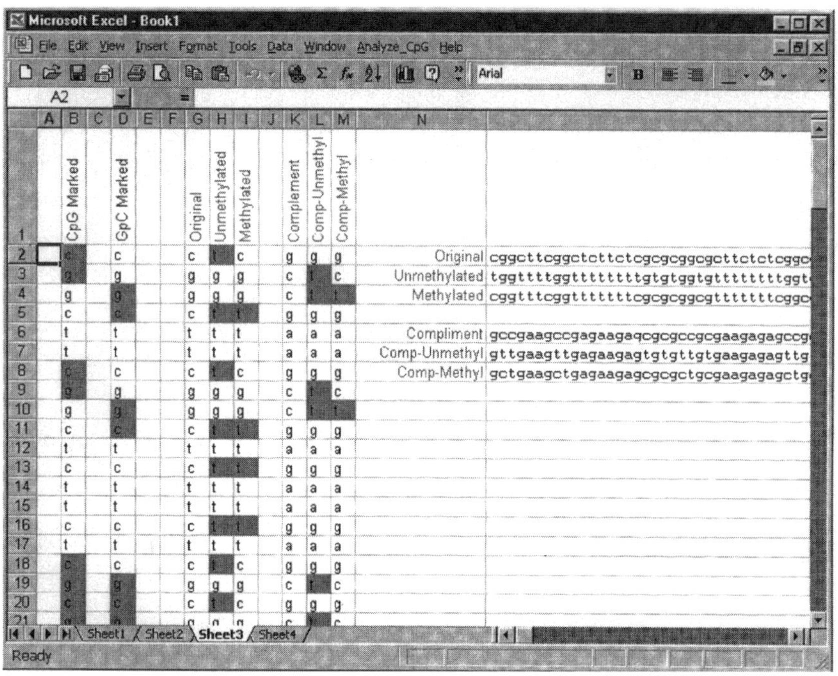

Figure 3. A view of Sheet3 displaying results of executing Mark-CpG and Bis-Modify menu items.

Displaying the CpG and GpC Base Pairs in a Graphical and Pseudographical Formats

The next menu item, Make-Graph, is used to display the distribution of CpG and GpC base pairs in graphical formats. When this command is executed, the program takes the sequence in cell J2 of Sheet2, analyzes the CpG distribution, and plots two graphs. First, the program scans the sequence through a window of 10 bp, computes the CpG frequency, makes a line graph, and displays it in Sheet4 as shown in Figure 4. The Y-axis in this graph represents the exact number of CpG base pairs counted within a window of 10 bp. Then, the program scans the sequence again through a window of 100 bp, computes the CpG frequency, and displays the graph in Sheet4 just below the earlier graph. The Y-axis in this graph represents the percentage of CpG frequency and has a fixed range of 0%–12%. While the first graph will be useful for the analysis of shorter fragments, the second will be very useful to analyze larger fragments.

The program also makes a text string "pseudograph" for CpG and GpC base pairs. In this pseudograph, the CpG or GpC pairs are represented by the "l" symbol, and all other possible pairs of nucleotides are represented by "." symbol. For example, a sequence such as "acgtcgac" is displayed ".l..l.." to represent the distribution of the CpG pairs. These pseudographs are displayed just above the previous graph with labels "CpG base pairs" and "GpC base pairs". This gives another visual tool to appreciate the distribution of the CpG islands in a sequence. The sequence and the results can be saved using the built-in Save As command in Excel. The user can then open a new workbook and continue analyzing new sequences, if necessary.

DISCUSSION

One important feature that we have not included in this program is the ability to search for restriction sites, especially for methylation-specific enzymes within the DNA sequence. We did not include this feature in the current program because our earlier programs, Win-Align and Mac-Align, can perform these functions effectively (1). In summary, we have described a spreadsheet-based program for the analysis of CpG islands. This program has features to copy and paste sequences from the GenBank database and would be very useful for identifying CpG frequency in different regions of DNA. The user can get a numerical output of the CpG frequency in selected regions and appreciate the output visually through a color-shaded vertical display and through graphical and pseudographical formats. The program also outputs the sequence that is expected after bisulfite treatment of DNA. This feature will be especially useful for designing primers for methylation-specific PCR.

REFERENCES

1. **Anbazhagan, R. and E. Gabrielson.** 1999. Spreadsheet-based program for alignment of overlapping DNA sequences. BioTechniques *26*:1180-1185.
2. **Baylin, S.B., J.G. Herman, J.R. Graff, P.M. Vertino, and J.P. Issa.** 1998. Alterations in DNA methylation: a fundamental aspect of neoplasia. Adv. Cancer. Res. *72*:141-196.
3. **Bird, A.P.** 1986. CpG-rich islands and the function of DNA methylation. Nature *321*:209-213.
4. **Gardiner-Garden, M. and M. Frommer.** 1987. CpG islands in vertebrate genomes. J. Mol. Biol. *196*:261-282.
5. **Herman, J.G., J.R. Graff, S. Myohanen, B.D. Nelkin, and S.B. Baylin.** 1996. Methylation-specific PCR: a novel PCR assay for methylation status of CpG islands. Proc. Natl. Acad. Sci. USA *93*:9821-9826.

Update to: Spreadsheet-Based Program for the Analysis of DNA Methylation

Ramaswamy Anbazhagan and Edward Gabrielson

Johns Hopkins University School of Medicine, Baltimore, MD, USA (anba@jhmi.edu)

PROGRAM FOR THE ANALYSIS OF DNA METHYLATION

This Excel spreadsheet-based program is very popular among methylation researchers. It is very useful in mapping CpG islands within DNA sequences. Users can estimate CG content, CpG percentage, and CpG/GpC ratio in selected regions of the DNA sequence. In addition, CpG pairs can also be marked with different colors for easy visualization. Perhaps the most useful function of the program is its ability to convert a DNA sequence into its bisulfite-modified versions. This is very useful for designing primers for methylation-specific polymerase chain reaction and in analyzing the sequences after bisulfite treatment (1). This program has been distributed to more than 80 institutions all over the world, and requests are coming in even now. It is freely available for academic users and can be obtained by sending an e-mail (anba@jhmi.edu) request to the author.

REFERENCE

1. **Anbazhagan, R. and E. Gabrielson.** 1999. Spreadsheet-based program for alignment of overlapping DNA sequences. BioTechniques *26*:1180-1185.

CHAPTER 14

A Computer Program for Modeling the Kinetics of Gene Expression

James L. Hargrove, Martin G. Hulsey, Frederick H. Schmidt, and Elmus G. Beale

The University of Georgia

Reprinted from BioTechniques 8:654-659 (June 1990)

ABSTRACT

Enzyme induction may be modeled on the basis of four, quantifiable processes that control the rates at which specific gene products accumulate and decay. These processes include synthesis of functional mRNA, translation and degradation of mRNA, and degradation of the protein product. We present a simple computer program that permits mathematical simulation of gene expression on the basis of experimentally determined rates of synthesis and degradation. The program was implemented as a spreadsheet using Microsoft Excel™ for Macintosh™ and MS-DOS operating systems and also was adapted for Hypercard™ on the Macintosh. It contains a formula to account for growth of tissue or cell populations. The program predicts amounts of individual mRNAs and proteins (or enzyme activities) in cells as a function of time after a stimulus alters their rates of synthesis or degradation.

INTRODUCTION

A Simple, Mathematical Model of Gene Expression

The rates of degradation of mRNAs and the proteins they encode are thought to control the time course of induction when the rate of transcription changes (3,7,23–25). Because the cellular content of mRNAs and proteins is regulated by mechanisms that control synthesis and stability (1–4,7,11,15,20,29), it is important to understand the different effects produced

Abbreviations and symbols: mRNA, messenger ribonucleic acid; PEP, phosphoenolpyruvate; TyrAT, tyrosine aminotransferase; R_t, mRNA concentration at time specificity by subscript; P_t, protein concentration at specified time; k_{s1}, rate constant for mRNA synthesis; k_{s2}, fractional rate constant for protein synthesis; k_{d1}, rate constant for mRNA degradation; k_{d2}, rate constant for protein degradation.

BioComputing: Computer Tools for Biologists
Edited by Stuart M. Brown

by stimuli that affect each step. A mathematical model with separate "compartments" for mRNA and protein, yields good estimates of actual concentrations of mRNAs and their encoded proteins (14,29). Here we describe a computer program based on this model.

The computer program is based on two differential equations that result from the assumptions outlined in Figure 1; the derivation and limitations to these equations have been described previously (7,14,24,30). To summarize, any mRNA in a population of cells is synthesized from nucleotide precursors at a rate that is proportional to the concentration of active genes, but is independent of the pre-existing pool of mRNA. Thus, biosynthesis of any specific mRNA may be described by a zero-order rate constant (k_{s1}). Each species of mRNA is degraded at a characteristic rate, described by a first- order rate constant, k_{d1}, which equals (ln 2)/half-life. Each of the constants has a fixed value under steady-state conditions, but increases or decreases to a new value as a result of transcriptional activation or other regulatory events. The concentration of each mRNA at any time (R_t) is predicted from these two constants (eq.1). The concentration of the corresponding protein at any time (P_t) is proportional to the concentration of its mRNA and the rate of degradation for that protein, as defined by first-order rate constants for translation and protein degradation (k_{s2} and k_{d2}, respectively). The computer program provides solutions to these two equations:

$$R_t = \frac{k_{s1}}{k_{d1}} - \left(\frac{k_{s1}}{k_{d1}} - R_o\right)e^{-k_{d1}t} \qquad \text{(eq. 1)}$$

$$P_t = \frac{k_{s1}k_{s2}}{k_{d1}k_{d2}} - \left(\frac{k_{s1}k_{s2}}{k_{d1}k_{d2}} - P_o\right)e^{-k_{d2}t} - k_{s2}\left(\frac{k_{s1}}{k_{d1}} - R_o\right)\left(\frac{e^{-k_{d2}t} - e^{-k_{d1}t}}{k_{d1} - k_{d2}}\right) \qquad \text{(eq. 2)}$$

where R_0 and P_0 are the initial, steady-state concentrations of mRNA and protein, respectively. To use the computer program, one must therefore measure or calculate these initial concentrations and values for the four rate constants.

Calculating Rates of Synthesis for mRNAs and Proteins

The units for the rate constants used in the computer program are defined by the user. It is convenient to express the concentrations of mRNA and protein in terms of molar amounts present per cell or per gram of tissue and to define rate constants appropriately. The concentrations of enzymes may be calculated from the activity present per unit of tissue, the catalytic activity per mole of purified enzyme, and the molecular weight of the enzyme (an example is given in the Appendix). However, concentrations of mRNAs are usually reported in terms of the percentages they represent of the total population of mRNA. To convert these percentages to molar concentrations, one must estimate the total amount of mRNA present. Approximate concentrations of DNA, RNA and protein, and other pertinent data for liver are shown in Table 1.

Rates of synthesis and degradation for mRNAs and proteins may be calculated from the steady-state relationships derived from equations 1 and 2. At steady state, the exponential terms become negligible, and the limiting values for mRNA and protein are as follows:

Table 1. Concentrations of Macromolecules and Cell Numbers in Rat Liver[a]

	Typical Values	References
Total cells/g	2.3×10^8	(33)
Hepatocytes/g	1.2×10^8	(33)
DNA Concentration wet weight	1.9–2.2 mg/g	(6)
DNA per cell	8.7 pg	
RNA Concentration (Average, 7.6 mg/g)	5.9–9.3 mg/g	(6,10)
mRNA	76 μg/g	(28)
Rate of mRNA Synthesis[b]	64 μg/g/day	
Protein Concentration	160–170 mg/g	(6)
Rate of Protein Synthesis[c]	39 mg/g/day	

[a]Estimates made by different investigators vary up to twofold. The values do not apply to other organs and species.
[b]The rate of mRNA synthesis was calculated by multiplying the concentration of mRNA (76 μg/g liver) times the rate constant for degradation (0.832 d^{-1}) based on a half-life of 20 h (16).
[c]The rate of protein synthesis was calculated by multiplying the concentration of protein (170 mg/g) times a rate constant for degradation (0.23 d^{-1}) based on a half-life of 3 day (23).

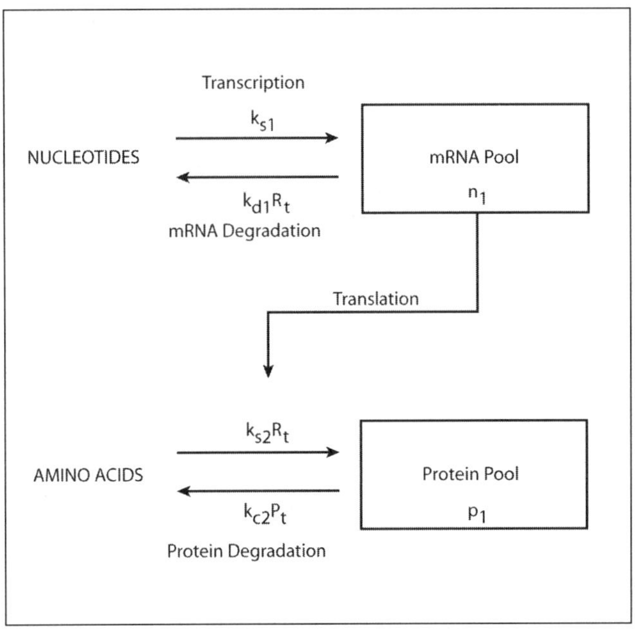

Figure 1. Basis for the mathematical model of gene expression. Each species of mRNA in a population of cells is assumed to be synthesized from a pool of nucleotides at a rate that is independent of the amount of pre-existing mRNA. The rate constant that defines this process (k_{s1}) is therefore zero-order with respect to mRNA concentration. Each protein is synthesized at a rate that is proportional to the concentration of mRNA (R_t) but independent of the pre-existing protein (P_t). Rates of degradation of the mRNA and protein are thought to be first order with respect to the concentrations of these two pools. Equations 1 and 2 (see text) may be derived on the basis of these assumptions (14,31).

$$R_{ss} = \frac{k_{s1}}{k_{d1}} \quad \text{(eq. 3)}$$

$$P_{ss} = \frac{k_{s1}k_{s2}}{k_{d1}k_{d2}} \quad \text{(eq. 4)}$$

At steady state, the fractional rates of degradation and synthesis for total protein (or mRNA) in a tissue are equal, and eq. 3 or eq. 4 may be applied to calculate pool sizes. Because protein in liver has a half-life of ca. 3 days (23), the rate constant of degradation is ca. 0.693/3 days, or 0.23/day (k_d = ln 2/half-life). The rate of protein synthesis per gram of liver is then ca. [0.23/day] × [170 mg/g], or 39 mg/g/day. Assuming that hepatic proteins have average molecular weights of 45 000, this represents 860 nmol of protein/day/g of tissue. Since the half-life of mRNA in liver has been estimated to be 20 h (0.83 days) (16), the rate of synthesis for mRNA in liver is ca. [0.693]/ [0.83 days] × [176 µg of mRNA/g liver], or 64 µg/g/day. The average molecular weight of hepatic mRNA is ca. 640 000 (9), so mRNA is synthesized at a rate of about 100 pmol/g/day in liver.

The preferred method of measuring mRNA concentration is by quantitative hybridization (5). However, because this information is frequently not available the abundance and rate of syntheses of individual mRNAs in liver may be calculated from the proportions they represent of total mRNA. These values are often determined by translating mRNA *in vitro* and immunoprecipitating the product. By multiplying the percentage times the total quantity of mRNA and dividing the result by the molecular weight, one obtains an estimate of the number of moles of each mRNA per unit of tissue. Rates of synthesis are then obtained by multiplying the concentration times the rate constant for degradation (Eq. 3, see Appendix for an example).

The kinetic model also requires an estimate for the rate at which mRNAs are translated, defined by the fractional rate constant, k_{s2}, for which we employed units of pmol of protein synthesized per pmol of mRNA per day. An estimate may be made on the basis of the measured concentrations and rates of synthesis of mRNA and protein in liver (Table 1). Because the rate of protein synthesis equals ca. 860 nmol/g/day, and the mRNA concentration is ca. 120 pmol/g, this yields 7160 pmol of protein/day/pmole mRNA. The rate of ovalbumin synthesis in estrogen-stimulated chick oviduct is ca. 13 200 pmol/day/pmol mRNA (18). The assumption that the molar amplification inherent in protein synthesis is ca. 10 000 in vertebrate tissues provides reasonable agreement between measured rates of transcription and calculated rates of mRNA synthesis (Table 2). The data suggest that a rate of transcription of 100 parts per million is equivalent to ca. 1 pmol of mRNA synthesized per gram of tissue per day.

Values for k_{s2} may be calculated from the estimated concentrations of individual mRNAs and proteins based on equation 4, but one must be cautious that the conditions approximate a steady state. Since proteins are usually much more stable than mRNAs, the mRNAs will generally attain equilibrium values much earlier than the encoded proteins. Indeed, values for ks2 calculated from published data vary from ca. 5000 to over 30 000 (11). Because the high values are not congruent with ones calculated from measured rates of elongation for tyrosine and alanine aminotransferases (26), we have assumed a value of 10 000 for the purposes of this report (Table 2).

Rate Constants for Degradation of mRNAs and Proteins

Rate constants for degradation may be calculated from reported values for half-lives

Table 2. Kinetic Constants for mRNAs and Proteins in Liver[a]

	Tyrosine Amino-transferase	PEP Carboxy-kinase	Malic Enzyme	Alanine Amino-transferase	Serum Albumin[d]
Basal Enzyme Activity[b] (U/g)	2.4	1.5	1.0	10	None
Basal Protein Concentration (pmol/g)	80	800	325	350	—
K_{d2} (days^{-1})	11.1	2.73	0.247	0.23	—
Protein Synthesis (pmol/g/day)	888	2180	81	65	800
Basal mRNA Concentration[c] (pmol/g)	0.089	0.22	0.008	0.008	6.58
K_{d1} (days^{-1})	13.9	24.8	1.39	0.693	0.756
K_{s1} (pmol/g/day)	1.24	5.46	0.011	0.006	4.97
Rate of Transcription (parts per million)	100–1100	500–3600	10–50	—	710–1670

[a]Values for tyrosine aminotransferase and alumine aminotransferase are from references 8, 13, 14, 17 and 26; those for PEP carboxykinase are from 1, 2, 12, 27 and 30; those for malic enzyme are from 4, 28, and 29; and those for albumin are from 10, 19, and 32. Protein concentrations and rates of synthesis were calculated from basal enzyme concentrations and half-lives using the single-compartment model (eq. 3., see text and appendix for methods).

[b]Units of enzyme activity are expressed as μmol of product formed per min at 37°C.

[c]Messenger RNA concentrations and rates of synthesis were calculated with a value for k_{s2} of 10 000 pmol of protein/day/pmol of mRNA and values for k_{d1} calculated from published half-lives using eq. 4 (see text).

[d]Serum albumin is secreted after it is synthesized, so no value of k_{d2} applies. Calculations are based on a rate of synthesis of 11.3% (10).

from the relationship, k_d = ln 2/half-life. If the half-lives are not known, the rate constants for degradation of an mRNA or protein may be calculated from linear transformations of equation 1 (21,25). The two general methods rely upon measurements made during approach to steady state after an inducing stimulus is administered or withdrawn or upon pulse-chase analysis. For inducible mRNAs and proteins, a straightforward estimate can be made by plotting $\Delta C/\Delta t$ vs. C, where ΔC is the change in concentration of mRNA or protein during an interval and C is the concentration at the midpoint of each interval (21). The slope of this line equals $-k_d$. Such approaches have provided estimates for rates of turnover that vary by over 100-fold for specific gene products expressed in the same kinds of cell and subcellular compartments (7,23).

Simulating Enzyme Induction with the Computer Program

Once the initial concentrations and the rate constants for synthesis and degradation of an mRNA and protein have been calculated, the data may be entered into the program to simulate effects of altering synthesis or turnover. We will describe how to implement the program using a spreadsheet (Microsoft Excel™) that is available for MS- DOS and Macintosh™ operating systems (a similar version of the program is available for Hypercard™). To use the spreadsheet, one opens a file named "Open Me First," which begins with instruction for use of the program. At the end of the instructions, an area is provided into which the values for the kinetic constants may be entered (Figure 2, right panel), as well as a calculator to interconnect half-lives and rate constants for degradation (not shown). The program automatically calculates predicted changes in concentrations of mRNA and protein that occur if any of the four rate constants is changed to a new value. The results are listed in specific columns (out of view at the bottom of Figure 2) and may either be exported to a graphics program or plotted immediately using a window linked to a resident graphics package. Output is then plotted and viewed automatically whenever a

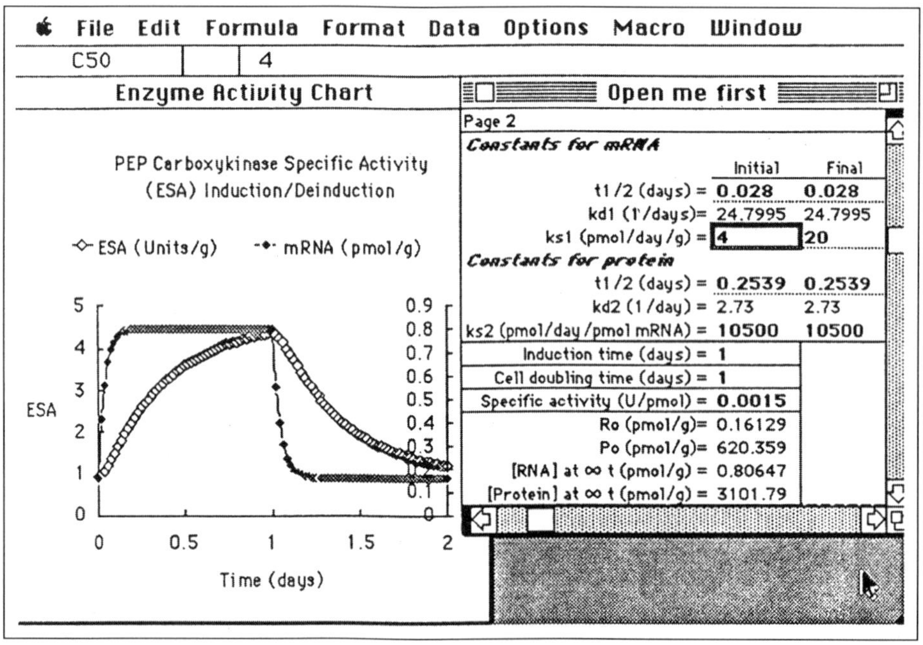

Figure 2. Use of a spreadsheet to simulate gene expression. The equations used for simulating gene expression may be entered into spreadsheets for rapid, convenient modeling. This figure represents a display from the program implemented in Microsoft Excel. The spreadsheet window is opened first (right), after which one opens windows for charting the results. User-adjustable parameters are shown on the spreadsheet in bold type, and values in plain type are calculated automatically. The user enters the four kinetic constants, the periods for induction and reinduction, cell doubling time and specific activity of pure enzyme (the latter two values are optional and require separate charts). The degradation constants may be entered as half-lives, from which the program calculates k_d values. The program simulates one period of induction using the "final" kinetic constants and an equal period of reinduction during which the initial constants are substituted. Simulation may be initiated manually from the "options" menu (top of figure), or may be performed automatically when new values are entered. The chart on the left is scaled automatically, with asymptotic values appearing on the spreadsheet (lower right). The program may easily be customized for other experimental designs.

new rate constant or concentration is entered (left side of Figure 2).

The spreadsheet also provides a means to plot output in terms of enzyme activity, rather than picomoles of protein per unit of tissue. This only requires that the user enter the catalytic activity per picomole of purified protein in the appropriate box; this is multiplied times protein concentration. An example for PEP carboxykinase is shown in Figure 2. Because each chart in Microsoft Excel must retrieve the appropriate values from specified columns in the spreadsheet, we have provided separate charts as templates for plotting protein as pmol/g, enzyme activity per gram, or total per system if growth is occurring (not shown).

The kinetic model predicts that different rates of syntheses and degradation for mRNAs and proteins markedly influence the initial concentrations and time courses of induction. In the simulations shown in Figure 3, transcription was increased 10-fold by specifying a 10-fold increase in k_{s1} at the initial time. This rate was maintained for 24 h, whereupon it was returned to the initial value for the next 24 h. Be cause tyrosine amino-

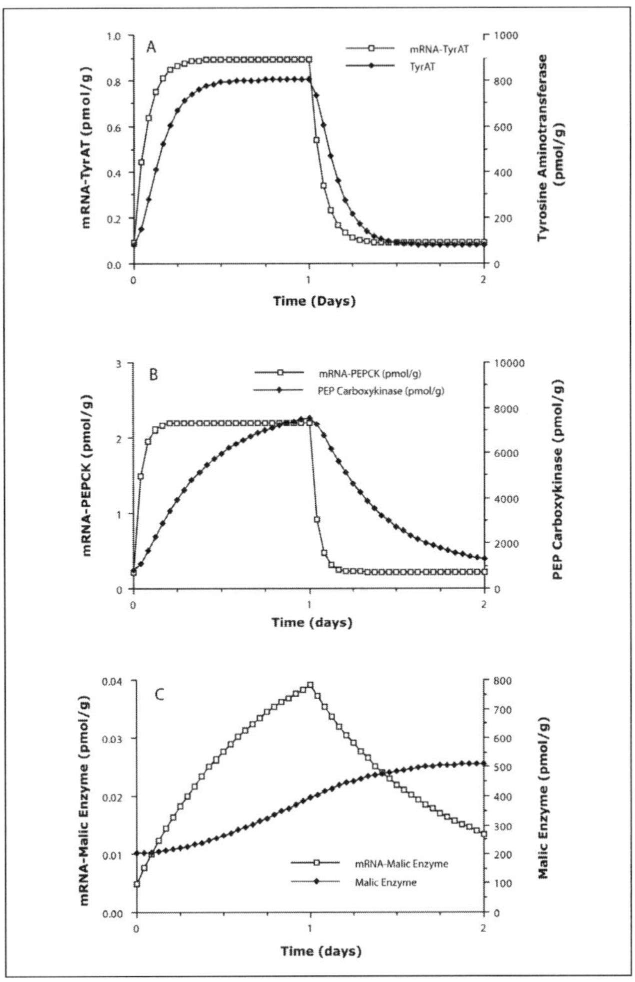

Figure 3. Simulating induction of hepatic enzymes. The kinetic constants and concentrations of several proteins that are expressed in liver were obtained from the literature (Table 2) and used to generate predicted induction curves. The curves correspond to the following: A) tyrosine aminotransferase and its mRNA; B) PEP carboxykinase and its mRNA; and C) malic enzyme and its mRNA. The initial pool sizes were calculated from reported enzyme activities in liver and the steady-state values for mRNA (eq. 3) and protein (eq. 4). Induction was initiated by increasing k_{s1} by 10-fold and maintaining it at that level for 24 h; deinduction was simulated by returning the rate of mRNA synthesis to the initial level after 24 h. Any combination of changes may be entered for a given simulation.

transferase and its mRNA are both labile *in vivo*, their initial concentrations are relatively low, but both attain new steady-state levels rapidly, as dictated by k_{d1} and k_{d2} (eq. 2). In contrast, malic enzyme and its mRNA are relatively stable, so that they do not attain a new steady state within the 24 h period (Figure 3). PEP carboxykinase represents an intermediate case, since its mRNA is relatively unstable (half-life of 0.5–2.4 h), while the protein is more stable than tyrosine aminotransferase (half-life of ca. 5 h).

Note that a delay always occurs between the time required for each mRNA to increase to 50% of its new, steady-state level and the time required for the protein to increase to the same degree. Because the half lives of tyrosine aminotransferase and its mRNA are short (1–2 h), the delay is small (Figure 3). When the proteins are much more stable than the mRNAs, the delay becomes quite long (Figure 3). Because the rate constants for degradation dictate the time course of induction, the time required for synthesis and degradation of relatively unstable mRNAs to attain a new steady state is much shorter than the time needed for the relatively stable protein to reach equilibrium. This relationship is not obvious, but is predicted by the model. Therefore, one use of the computer program is to educate students concerning the relative roles of synthesis and degradation in dictating the time course for expression of gene products.

Even for the small collection of mRNAs and enzymes shown in Table 2, each of the four kinetic constants varies over a range of at least 30-fold. Basal rates of transcription differ by 60-fold, and stimulated values increase this range to over 300-fold. Because the constants can vary independently, the potential range of final protein concentration is over 100 000-fold. Remarkably, the actual concentration of the enzymes shown varies by 10-fold, suggesting a requirement for a minimum amount of each catalyst in liver. It is impressive to compare the similar rates of transcription for albumin and PEP carboxykinase to the large difference in synthesis of the two proteins, which is explained by differential turnover of the two mRNAs.

Correcting for Growth of Cell Populations

The quantity of any mRNA or protein in a tissue depends upon the number of cells and biosynthesis capacity present. The computer program includes a feature for calculating total cell number or quantities of tissue, mRNA and protein on the basis of measured values for cell or tissue doubling times and maximum numbers of cells attained per system. The correction is based on a simple, exponential function:

$$N_t = N_0 \, e^{gt} \qquad \text{(eq. 5)}$$

where N_t equals cell number (or tissue mass) at any time, N_0 equals initial cell number and g equals the doubling time. To use this feature, one enters the doubling time into an area located in the file named "Open Me First." The program then calculates the fold-increase in cell number and multiplies this value by the concentration of mRNA or protein to yield the total quantity of each. This could be valuable in predicting the rate at which gene products are synthesized in bacterial cultures or mammalian cells during exponential growth.

CONCLUSIONS AND PROSPECTS

Mathematical models provide a theoretical framework to aid in interpreting information. In the context of enzyme induction, it is important to understand how rates of synthesis and degradation may be regulated to yield catalysts for metabolic pathways. By providing a quantitative basis for comparing the time course of induction for mRNAs and enzymes, this program for microcomputers should contribute to assimilating information concerning transcription, translation and turnover, which is now available for hundreds of mRNAs and proteins. The formulas used in the calculations can readily be entered into commercially available spreadsheets and adapted to specific applications.

The mathematical model on which this computer program is based is one of the simplest that shows how changes in synthesis or stability of mRNA will affect protein concentration. Its limitations are evident in the lumping of several processes into single kinetic constants (Figure 1) and the assumption that the new values of the constants must be attained instantaneously. Because techniques for measuring the kinetic constants are difficult (15,21,24), predictions based on the program must be interpreted cautiously. Nevertheless, it should be possible to use microcomputers to integrate information about genetic regulation of enzyme concentration with data concerning enzyme kinetics to predict flux of substrates through biochemical pathways.

NOTE: Versions of the computer program described here (in BASIC, HyperCard® or as templates for Microsoft Excel®) may be obtained from BioTechNet. Instructions for downloading the program are provided in the BioTechNet newsletter of this issue. Copies on diskettes may also be obtained along with documentation concerning their use by sending a check or purchase order for $25 (US) to: The University of Georgia Research Foundation, COSMIC Program, 382 E. Broad Street, Athens, GA 30602. Please specify your operating system and diskette format.

ACKNOWLEDGMENTS

This research was supported by U.S.P.H.S. grant 131139329 to J.L.H., grant GM39895 to E.G.B. and an award from the American Diabetes Association to E.G.B.

REFERENCES

1. **Beale, E.G., J.L. Hartley and D.K. Granner.** 1982. $N^6,O^{2'}$-Dibutyryl cyclic AMP and glucose regulate the amount of messenger RNA coding for hepatic phosphoenolpyruvate carboxykinase (GTP). J. Biol. Chem. *757*:2022-2028.
2. **Beale, E.G., C.S. Katzen and D.K. Granner.** 1981. Regulation of rat liver phosphoenolpyruvate carboxykinase (GTP) messenger ribonucleic acid by $N^6,O^{2'}$-dibutyryladenosine 3′, 5′-phosphate. Biochemistry *20*:4878-4883.
3. **Berlin, C.M. and R.T. Schimke.** 1965. Influence of turnover rates on responses of enzymes to cortisone. Mol. Pharmacol. *1*:149-156.
4. **Dozin, B., M.A. Magnuson and V.M. Nikoden.** 1986. Thyroid hormone regulation of malic enzyme synthesis. Dual tissue-specific control. J. Biol. Chem. *261*:10290-10292.
5. **Durnam, D.M. and R.D. Palmiter.** 1983. A practical approach to quantitative specific mRNAs by solution hybridization. Anal. Biochem. *131*:385-393.
6. **Goodman, M.N. and N.B. Ruderman.** 1980. Starvation in the rat. I. Effect of age and obesity on organ weights, RNA, DNA, and Protein. Am. J. Physiol. *239*:E269-E276.
7. **Hargrove, J.L. and F.H. Schmidt.** 1989. The role of mRNA and protein stability in gene expression. FASEB J. *3*:2360-2370.
8. **Hashimoto, S., W. Schmoe and G. Schütz.** 1984. Transcriptional activation of the rat liver tyrosine aminotransferase gene by CAMP. Proc. Nat. Acad. Sci. USA *81*:6637-6641.
9. **Hastie, N.D. and J.O. Bishop.** 1976. The expression of three abundance classes of messenger RNA in mouse tissues, Cell *9*:761-774.
10. **Keller, G.H. and J.M. Taylor.** 1976. Effect of hypophysectomy on the synthesis of rat liver albumin. J. Biol. Chem. *251*:3768-3773.

11.**Kenney, F.T. and K.-L. Lee.** 1982. Turnover of gene products in the control of gene expression. Bioscience *32*:1 81-184.
12.**Lamers, W.H., R.W. Hanson and H.M. Meisner.** 1982. cAMP stimulates transcription of the gene for cytologic phosphoenolpyruvate carboxykinase in rat liver nuclei. Proc. Natl. Acad. Sci. USA *79*:5137-5141.
13.**Lee, K.-L., K.R. Isham, A. Johnson and F.T. Kenney.** 1986. Insulin enhances transcription of the tyrosine aminotransferase gene in rat liver. Arch. Biochem. Biophys. *248*:597-603.
14.**Lee, K.-L., J.R. Reel and F.T. Kenney.** 1970. Regulation of tyrosine α-ketoglutarate transaminase in rat liver. IX. Studies of the mechanisms of hormonal inductions in cultured hematoma cells. J. Biol. Chem. *245*:5806-5811.
15.**McKnight, G.S. and R.D. Palmiter.** 1979. Transcriptional regulation of the ovalbumin and conalbumin genes by steroid hormones in chick oviduct. J. Biol. Chem. *254*:9050-9058.
16.**Moore, R.E., T.L. Goldsworthy and H.C. Pitot.** 1980. Turnover of 3′ polyadenylate-containing RNA in livers from aged, partially hepatectomized, neonatal, and Morris 5123 C hepatoma-bearing rats. Cancer Res. *40*:1449-1457.
17.**Nickol, J.M., K.-L. Lee and F.T. Kenney.** 1978. Changes in hepatic levels of tyrosine aminotransferase messenger RNA during induction by hydrocortisone. J. Biol. Chem. *253*:4009-4015.
18.**Palmiter, R.D.** 1975. Quantitation of parameters that determine the rate of ovalbumin synthesis. Cell *4*:189-197.
19.**Peavy, D.E., J.M. Taylor and L.S. Jefferson.** 1985. Time course of changes in albumin synthesis and mRNA in diabetic and insulin-treated diabetic rats. Am. J. Physiol. *248*:E656-E663.
20.**Petersen, D.D., S.R. Koch and D.K. Granner.** 1989. 3′ conceding region of phosphoenolpyruvate carboxykinase mRNA contains a glucocorticoid-responsive mRNA-stabilizing element. Proc. Natl. Acad. Sci. USA *86*:7800-7804.
21.**Rodgers, J.R., M.L. Johnson and J.M. Rosen.** 1985. Measurement of mRNA half-life as a function of hormonal treatment. Methods Enzymol. *109*:572-580.
22.**Sala-Trepat, J.M., M.J. Savage and J. Bonner.** 1978. Isolation and characterization of poly(adenylic acid)-containing messenger ribonucleic acid from rat liver polysomes. Biochim. Biophys. Acta *519*:173-193.
23.**Schimke, R.T.** 1973. Control of enzyme levels in mammalian tissues. Adv. Enzymol. *37*:135-187.
24.**Schimke, R.T., E.W. Sweeney, and C.M. Berlin.** 1965. The roles of synthesis and degradation in the control of rat liver tryptophan pyrrolase. J. Biol. Chem. *240*:322-331.
25.**Segal, H.L. and Y.S. Kim.** 1963. Glucocorticoid stimulation of the biosynthesis of glutamic-alanine transaminase. Proc. Nat. Acad. Sci. USA *50*:912-918.
26.**Stiles, C.D., K.-L. Lee and F.T. Kenney.** 1976. Differential degradation of messenger RNAS in mammalian cells. Proc. Nat. Acad. Sci. USA *73*:2634-2638.
27.**Tilghman, S.M., R.W. Hanson and F.J. Ballard.** 1976. *In* Gluconeogenesis. Its Regulation in Mammalian Species, p.47-91. R.W. Hanson and M.A. Mehlman (Eds.),Wiley Interference, New York.
28.**Towle, H.C., C.N. Mariash and J.H. Oppenheimer.** 1980. Changes in the hepatic levels of messenger ribonucleic acid for malic enzyme during the induction by thyroid hormone and diet. Biochemistry *19*:579-585.
29.**Towle, H.C., C.N. Mariash, H.L. Schwartz and J.H. Oppenheimer.** 1981. Quantitation of rat liver messenger ribonucleic acid for malic enzyme during induction by thyroid hormone. Biochemistry *20*:3486-3492.
30.**Veneziale, C.M., J-D. Donofrio and H. Nishimura.** 1983. The concentration of P-enolpyruvate carboxykinase protein in murine tissues in diabetes of chemical and genetic origin. J. Biol. Chem. *258*:14257-14262.
31.**Yagil, G.** 1975. Quantitative aspects of protein induction. Curr. Top. Cell. Regul. *9*:183-286.
32.**Yamauchi, A., E. Imam, T. Noguchi, T. Tanaka, S. Yamamoto, H. Mikami, Y. Fukuhara, M. Fujii, Y. Orita and T. Kamada.** 1988. Albumin gene transcription is enhanced in liver of nephritic rats. Am. J. Physiol. *254*:E676-E679.
33.**Zahlten, R.N. and R.W. Stratman.** 1974. The isolation of hormone-sensitive rat hepatocytes by a modified enzymatic technique. Arch. Biochem. *163*:600-608.

APPENDIX

Calculating Constants for Kinetic Modeling

Appropriate kinetic constants for computer-based simulation of enzyme induction may be derived in several ways. As an example, the values used for tyrosine aminotransferase (TyrAT) and its mRNA were calculated as follows: The unit of enzyme activity is expressed as microcodes of product formed per min at 37°C, and purified TyrAT has a specific activity of about 600 units per mg of protein. Livers of fasted rats contain about 2–3 units of enzyme activity/g of tissue, or the equivalent of 4 µg of TyrAT. Since the subunit molecular weight of TyrAT is 50 000 g/mol, this is equal to ca. 80 pmol of enzyme/g liver.

The half-life of TyrAT has been estimated to be 1.5 h *in vivo* (11). The rate constant for degradation (k_{d2}) therefore equals (0.693)(24h/d)/1.5 h, or 11.1 d^{-1}. At steady state, synthesis and degradation are balanced, and the rate of synthesis equals the protein concentration times the rate constant for degradation. Thus, TyrAT is synthesized at a rate of ca. (80 pmol/g liver) (11.1 d^{-1}, or ca. 888 pmol/g/d (11).

The units for k_{s2} (eq. 2) are expressed as pmol of protein synthesized per pmol of mRNA per day. This value may be estimated if the concentration of mRNA is known; translational assays for the mRNA provide minimum values of 0.02% of total mRNA (17). Assuming that the mRNA concentration in liver is 76 µg/g, then the minimum estimate for the mRNA for TyrAT equals about 15 ng/g liver. If the molecular weight of the mRNA for this enzyme is 640 000, this equals .024 pmol/g liver. Thus the maximum value for k_{s2} would be (888 pmol TyrAT/g liver/day)/ (.024 pmol mRNA/g liver), or 37 000. Because this value is much higher than ones calculated from the transit time for ribosomes on this mRNA (26), we used a value for k_{s2} of 10 000 for each of the simulations shown in Figure 3 (see text).

As the mRNA concentration at steady state also equals the ratio between the rate of synthesis and the rate of degradation, the rate of synthesis for this mRNA may be estimated from the half-life of 1.2 h. This yields a k_{d1} of (.693)(24 h/d)/(1.2 h), or 13.86 d^{-1}. Thus, the rate of mRNA synthesis for TyrAT is at least (.024 pmol/g liver)(13.86 d^{-1}), or 0.33 pmol of mRNA/g liver/day.

Update to: A Computer Program for Modeling the Kinetics of Gene Expression

James L. Hargrove

Department of Foods and Nutrition, University of Georgia, Athens, GA (jhargrov@fcs.uga.edu)

Our previous article (6) indicated that a simple and inexpensive method is available to biological scientists by which processes involved in messenger RNA (mRNA) metabolism and protein metabolism may be quantitatively linked to predict the time course of enzyme induction. Indeed, computer-assisted modeling potentially allows changes in gene expression to be coupled to adjustments in phenotype through the use of concepts derived from the fields of Metabolic Control Analysis and risk assessment (2,7,8). This is particularly true when a gene encodes an enzyme that affects flux though a metabolic pathway, but the idea can readily be applied to systems in which the gene product is a receptor or a peptide hormone, such as insulin or leptin.

The approach taken in our work has been to use a technique called compartmental analysis. Here, an individual species of mRNA or the corresponding protein is assumed to exist as a homogeneous population with a quantifiable rate of formation. Degradation of each kind of molecule is assumed to be first-order (proportional to concentration) and to be characterized by a half-life and equivalent first-order decay constant. This approach allows an investigator to focus on the rates of synthesis and degradation that characterize primary transcripts, mRNAs, and proteins as they move to their destinations. One advantage of this method is that it corresponds to the well-established theory of approach to steady state, which has been very useful in pharmacodynamics and other disciplines (1).

The methods described in the original paper (6) served as the basis for a monograph on modeling gene expression (3). Recently, a book was published that includes chapters about gene expression and a CD-ROM with kinetic models, as well as other applications in the health sciences (4). One form of the model has been posted on the Internet (http://www.arches.uga.edu/~jhargrov/) and can be operated using a trial version of STELLA® software (High Performance Systems, Hanover, NH, USA).

It seems plausible that computer-assisted modeling can help investigators generate dynamic models using information that is available from genomic and proteomic databases. One possible means to do this was reviewed in a recent update (5). In principle, much of the information that controls gene expression is located in promoter sequences. This information can be mined to produce dynamic maps and even networks of interacting systems that may help predict consequences of altered gene expression. However, for individual investigators and small groups, the main utility of computer-assisted modeling is to state falsifiable hypotheses in a quantitative manner and then to compare experimental outcomes with theoretical predictions. This capability can help with publication of results and also with grantsmanship. The ability to use general models to predict the time course of gene expression rightfully represents an important aspect of computational biology and the emerging field of bioinformatics.

REFERENCES

1. **Gallaher, E.J.** 1996. Biological system dynamics: from personal discovery to universal application. Simulation *66*:243-257.
2. **Hargrove, J.L.** 1993. Microcomputer-assisted kinetic modeling of gene expression. FASEB J. *7*:1163-1170.
3. **Hargrove, J.L.** 1994. Kinetic Modeling of Gene Expression. R.G. Landes, Austin, TX.
4. **Hargrove, J.L.** 1998. Dynamic Modeling in the Health Sciences, p. 298. Springer Verlag, New York.
5. **Hargrove, J.L.** 2001. Integrating nutrition knowledge: from computer-assisted modeling to nutritional genomics, p. 519-529. *In* Bowman, B.A. and R. Russell (Eds.), Present Knowledge in Nutrition, 8th Ed. ILSI Press, Washington, DC.
6. **Hargrove, J.L., M.G. Hulsey, F.H. Schmidt, and E.G. Beale.** 1990. A computer program for modeling the kinetics of gene expression. BioTechniques *8*:654-659.
7. **Hargrove, J.L., A.O. Summers, and M.G. Hulsey.** 1993. From genotype to phenotype: computer-based modeling of gene expression with STELLA II. BioTechniques *15*:1096-1101.
8. **Kohn M.C., N.J. Walker, A.H. Kim, and C.J. Portier.** 2001. Physiological modeling of a proposed mechanism of enzyme induction by TCDD. Toxicology *162*:193-208.

Section II

Simple Scripting Languages

Scripting for Biologists: Why Biologists Need to Learn Something About Programming

Though few would fail to acknowledge that computer literacy is an essential skill for anyone who works in a laboratory, research biologists are sometimes reluctant to learn programming because they feel it is both overwhelming and outside the scope of their field. If the articles in this section have a common theme, it is that it is indeed practical to learn enough about computer programming to do useful things. Programming is itself a component of computer literacy, and some experience with programming generally makes one a more effective computer user.

Computers are indispensable tools for biotechnology research. For bioinformatics in particular, the computer revolution could not have come at a more auspicious time. It is hard to imagine how the high-throughput sequencing and expression experiments at the leading edge of molecular biology would be possible without the low-cost and ubiquitous personal computer and easily accessible networked databases, or how we would otherwise deal with the huge quantities of information so obtained.

Beyond the now-commonplace sequence analysis chores of molecular biology, many biological specialties make use of mathematical modeling in an attempt to obtain global insights into how various parts fit together to make systems. Individual binding constants, ion permeabilities, substrate concentrations, and myriad other details so painstakingly measured in the laboratory are just so many bland facts until a model connects them into a system; in many cases, it is the computer that brings these models to life.

Furthermore, the PC is now an important communication device, indispensable for researching bibliographic citations, maintaining (and sometimes initiating) collaborations with colleagues, and, to an extent, for publishing results. A research scientist must be a "power user" of the Internet and the PC, as well as of the pipette, the microscope, and the cell culture hood.

In established fields of investigation, there are commonly used programs to do the common sorts of calculations. However, since researchers typically hope to one day find themselves in uncharted territory, the nature of research often requires flexibility in how computers are used. Many of the most flexible programs, which are used for purposes their designers did not explicitly foresee, are themselves "scriptable" ¾ that is, you can write your own little programs to control how these programs work (for example, NIH Image).

Programming languages provide flexible human–computer interfaces. A computer language is a tool invented by human beings to help other humans express problems in a way that can ultimately be translated into the opcodes and binary numbers used by a computer processor. Myriad such languages exist, and more are constantly being invented. Unfortunately, many people see this plurality of languages as an obstacle in itself: How do I choose a language, and how do I know it won't become obsolete by the time I learn it?

While these are legitimate questions, they view the situation from the "glass half empty" perspective. It is also valid to remark that so many languages increase the odds that you may find one suited to your problems and your tastes, and that it is a good thing that the industry is constantly providing new languages designed to make the latest ideas and technologies easier to use.

And if you find it difficult to learn a computer language (which was, after all, designed for the use of mere mortals), how much more difficult will it be to explain the subtleties of your biological system to a hapless programmer? In many cases it is worthwhile to write a simple prototype yourself. The process frequently helps you to define the problem more clearly. Then, even if you find you require a more sophisticated software product than you are able to construct on your own, the prototype can be a useful tool for explaining the problem requirements to programmers.

Another mitigating phenomenon is that computer languages tend to "evolve" from earlier languages: Java, Javascript, and Perl, for example, all inherit much of their syntax from the venerable C programming language. And even the "old" languages themselves may not go away. Most modern operating systems (Windows, Linux, Macintosh OS X, Solaris, etc.) are written in C or an object-oriented derivative of C, as are the programs of the NCBI toolbox, and many other programs where performance is important.

The articles we have chosen illustrate biological applications of several languages. Using a spreadsheet for calculations may be more of a paradigm than a language, but it clearly falls into the same general category of tools that let you express problems in a way that will allow a computer to solve them. Whatever language you choose to start with, you will likely be able to use it for biological problems of your own.

Robert M. Horton

CHAPTER 15

PCR Master Mix Volume Calculators in JavaScript™

Robert M. Horton[1] and Remington J.S. Stone[2]

[1]Attotron Biosensor Corporation, Carson City, NV and
[2]University of California-Davis, Davis, CA, USA

Reprinted from BioTechniques 24:420-422 (March 1998)

When setting up multiple polymerase chain reactions (PCRs), it is often a good idea to make a "master mix" containing all of the common ingredients, which can then be added to each reaction tube in a single pipetting step. Individual ingredients (such as templates for different reactions) can then be added to each tube. Using master mixes saves pipetting steps (thus saves time and money) and increases the accuracy of reactions by avoiding the measurement of extremely small (perhaps sub-microliter) volumes of reagents that need to be added to a single tube.

Calculating the amount of each ingredient to add to a master mix can be a chore, especially if you are doing two-dimensional titrations where varying concentrations of two ingredients are used in different combinations and you want to make one big master mix to contain the ingredients for the sub-masters, which will themselves be further divided. Each nested step must include a "fudge" factor to allow for pipetting errors; this ensures that you don't run short on the last tube. Good calculations using a reasonable fudge factor save reagents more so than using a fudge factor that is easy to calculate. In other words, a quick and dirty approach might be just to add enough extra ingredients for an additional reaction. However, a fudge factor based on a percentage of the total volume is probably more appropriate; one extra reaction might be excessive if you are only making four reactions, and it might not be enough if you are making 96.

Computerized Worksheets

Setting up a worksheet to calculate master mix ingredients is very helpful to keep everything straight. Generic worksheets including the standard amounts of each reagent for one reaction together with a column of blank lines to enter the calculated amounts needed for a set of reactions (including a fudge factor) are very useful and also give you a record to put in your notebook. Computers are, of course, ideal tools for performing such calculations, and a worksheet on a computer gives you the best of both worlds.

A program for performing such calculations using a macro for Microsoft® Excel® has been published (1). This approach has several interesting points from a programming and user's perspective. A macro is a set of commands that are executed inside of another pro-

BioComputing: Computer Tools for Biologists
Edited by Stuart M. Brown

gram; in this case, the macro program runs inside Excel. The remarkable thing about this approach is that it doesn't matter what kind of computer the user has as long as it runs Excel (which is available for both Windows® and Macintosh®).

Unfortunately, a macro for a program like Excel suffers from some drawbacks that plague many pre-World Wide Web (WWW) computing approaches. Users must have Excel (which can be rather expensive, even with an academic discount), obtain the macro (download it or get it on a floppy disk), install it and figure out how to use it. Ironically, such obstacles can cause people to use their multi-hundred-megahertz computers to print out a worksheet, which they fill in using a calculator to crunch the numbers.

Here we present an approach that neatly overcomes many of these difficulties. Rather than writing the program as a macro for Microsoft Excel, we write it as a "macro" for WWW browsers using the JavaScript™ scripting language.

As noted in earlier Internet On-Ramp articles (2,3), JavaScript programs run right within a Web browser (Netscape Navigator or, theoretically, Internet Explorer; both of which are now free). You locate a program just by going to its Web site and run it right on the Web page. To use it offline or to give a copy to friends, download it by saving the Web page. There is no installation;

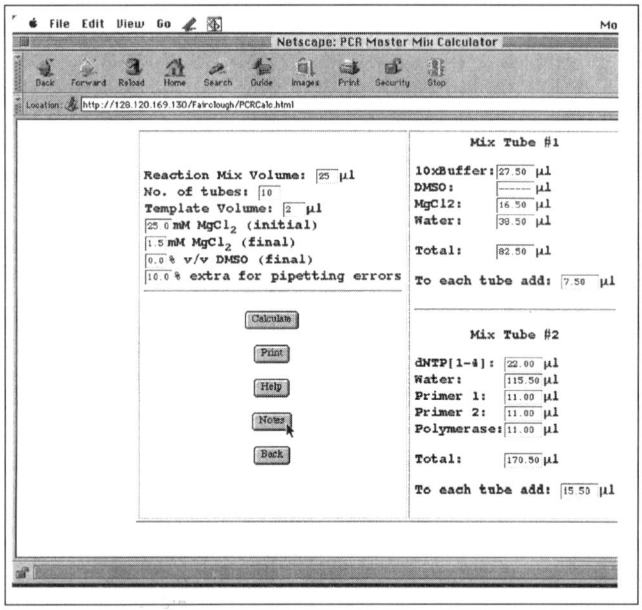

Figure 1. JavaScript PCR master mix volume-calculating program with an interface similar to Excel macro of Reference 1.

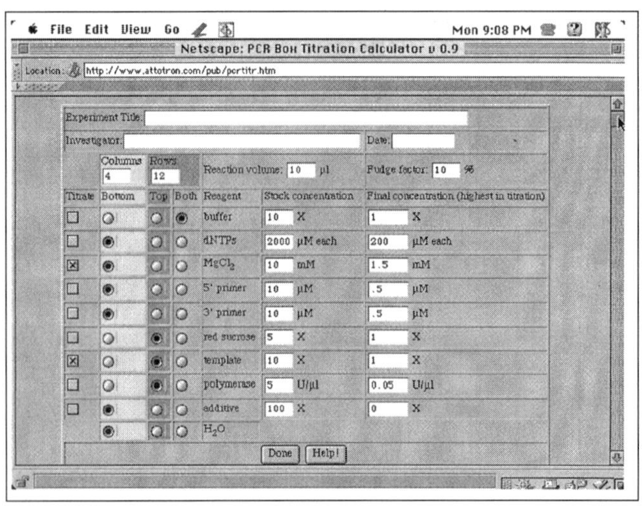

Figure 2. A JavaScript program to calculate reagent volumes for Ampli-Grease hot-start PCR protocol of Reference 4.

just opening the page in your browser makes it ready to run. And since it uses standard Web-page interaction elements like text areas and buttons, the interface should be familiar.

JavaScript Calculators

JavaScript is ideal for making specialty calculators. The programs themselves are fairly simple and small because the difficult work of drawing the user interface is handled by the browser. Thus the script is mostly concerned with the formulas. Here we present two master mix calculators that take different approaches.

Figure 1 shows a JavaScript program with an interface that is vaguely similar to the Excel macro (1). This page can be found at http://128.120.169.130/Fairclough/PCR-Calc.html. Input concentrations, reaction size, number of tubes, etc. in the upper left corner and click on *Calculate*; the volumes of each reagent to include in the master mix appear on the right-hand side. The PRINT button creates a worksheet for your notebook.

The example in Figure 2 (http://www.attotron.com/pub/pcrtitr.htm) implements the "AmpliGrease" hot-start protocol (4) using PCR-compatible Cresol Red loading dye (5). However, many options can be changed to implement a variety of experimental protocols, and it does not necessarily need to be used only for hot start.

The program can calculate reagent volumes for a single reaction (by entering 1 row and 1 column), for one-dimensional titrations or for two-dimensional "box" titrations in which separate ingredients are titrated in the top and bottom mixtures. While the default distribution separates the ingredients according to the protocol described (4), reagents can be reassigned to "top" or "bottom" mixture, or split proportionately between them. To use stock solutions at different concentrations, type over the values in the boxes. A reagent can be omitted by entering a final concentration of zero. You can permanently reset the default values in your own copy of the program by editing the source code; default concentrations are set at the beginning of the program to make them easy to find.

Clicking on *Done* puts the program in action; it asks you for the remaining concentrations if you are doing a titration, and it calculates a worksheet describing how much of each reagent to use in each tube for your experiment, complete with a "fudge factor" to take into account pipetting errors. The program checks your inputs to be sure they meet certain criteria (you must have at least one column, you can only titrate one ingredient in the top mixture and/or one in the bottom, etc.), and it won't let you proceed until it is satisfied. But it will not necessarily complain about bizarre experimental designs.

As with all JavaScript programs, the source code can be examined using your browser's *View Source* option (*View: Page Source* in Navigator 4). You can use this to learn how to create your own specialized calculator or (with due credit to the authors!) modify the code to fit your needs.

REFERENCES

1. **Stowe, R.P. and D.L. Pierson.** 1996. Spreadsheet macro for setting up PCR assay tubes. BioTechniques *20*:1088-1089.
2. **Horton, R.M.** 1996. Java and JavaScript applications in biomedical research. BioTechniques *21*:834-836.
3. **Horton, R.M.** 1997. Interactive internet instruction: web quizzes and a new JavaScript™ QuizMaker. BioTechniques *23*:1048-1050.
4. **Horton, R.M., B.L. Hoppe and B.M. Conti-Tronconi.** 1994. AmpliGrease: "hot start" PCR using petroleum jelly. BioTechniques *16*:42-43.
5. **Hoppe, B.L., B.M. Conti-Tronconi and R.M. Horton.** 1992. Gel-loading dyes compatible with PCR. BioTechniques *12*:679-680.

CHAPTER 16

JavaScript™ Program for Browser-Based Presentations

Robert M. Horton

Attotron Biosensor, Carson City, NV, USA

Reprinted from BioTechniques 26:456-458 (March 1999)

With the increasing availability of computer projection devices and portable computers, it is becoming more common to give presentations directly from a computer, as opposed to using slides. This can be much more convenient than using traditional slides; it saves you the expense and trouble of preparing slides, lets you show animations and allows you to modify your presentation at the last minute.

If you already prepare slides on a computer, there is a very shallow learning curve to switching to a computer projector. Most presentation programs commonly used for preparing slides, such as Microsoft® PowerPoint™ etc., can also be used to show the same presentations on a projector directly.

Here, I describe approaches to giving presentations directly from a Web browser, reading a file from the hard drive without necessarily having a network connection, rather than using a traditional commercial presentation program. There are several reasons to take this approach. Web browser software is free, whereas some of the full-featured presentation packages can be expensive. Web browsers can display a wide variety of types of information, including animated dynamic HyperText Markup Language (HTML) pages, and can be used to create quite sophisticated presentations. They are also ubiquitous; in fact, PowerPoint and similar programs can save presentations in HTML format, so you can use a browser to view or show presentations originally created in those programs.

Several freeware JavaScript™ programs are available to assist in preparing Dynamic HTML presentations. The most sophisticated of these JavaScript presentation editors is probably the one from Netscape (http://developer.netscape.com/docs/technote/javascript/prestemp/prestemp.html), which goes to great length to be both cross-browser (i.e., it works on Microsoft Internet Explorer and Netscape Navigator) and cross-platform compatible. It automatically detects the screen size and adjusts its font sizes accordingly. Unfortunately, it is also a bit complex.

Here, I present a very basic JavaScript presentation program, Presentation Maker, tailored for scientists. It is intended to be very easy to use; once you get started, you can advance to more sophisticated programs as your skills allow and your artistic needs require. Since it is free (as are many of the more sophisticated presentation packages and templates), you do not need to invest anything in it other than (hopefully) a small amount

BioComputing: Computer Tools for Biologists
Edited by Stuart M. Brown

of time. The program can be used on-line (http://www.attotron.com/pub/presmaker.html), or you can download a copy to your own computer to use off-line.

This program uses Dynamic HTML layers. Dynamic HTML loosely refers to the set of scriptable browser objects available in 4th-generation browsers. Unfortunately, the Dynamic HTML in Netscape Navigator is implemented differently on Internet Explorer, and this version of Presentation Maker only runs on Navigator.

Enter your data as plain text, as illustrated in Figure 1, and the program automatically generates a "slide presentation". Separate the data for each slide by a line containing three equals signs, "===". The first line in each slide is automatically used as the title, while subsequent lines are used as text in the body of the slide. A hyphen at the start of a line is converted to a "bullet" mark. To include an image on a slide, simply type or paste the name of a graphics interchange format (GIF) or Joint Pictures Expert Group (JPEG) file into the text area; the program will locate any file names that end in .gif, .jpg or .jpeg and create the proper HTML image tags from them.

An additional feature is specifically designed for scientific presentations: you can place a molecular structure file into a slide just as you do an image. Any file name ending with .pdb (a molecular structure in the Brookhaven Protein Data Bank [PDB] format) is used by the program to create the HTML tag needed to view that file using the Chemscape™ Chime™ molecular structure viewing plug-in (available from http://www.mdli.com/download/). As described in an earlier Internet On-Ramp article (1), this plug-in allows you to view and manipulate a ray-traced image of the molecule and even view it stereoscopically. Figure 2 shows a slide with an image displayed by Chime.

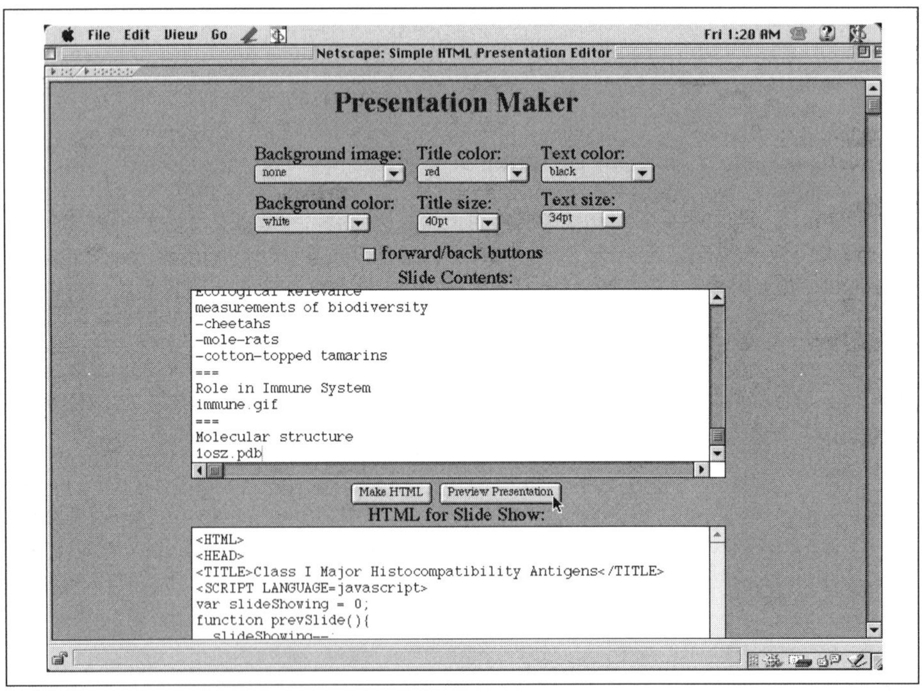

Figure 1. Presentation Maker screen. The text of the slides has been typed into the top text area. Clicking on the **Make HTML** button causes the program to translate this data into a slide show.

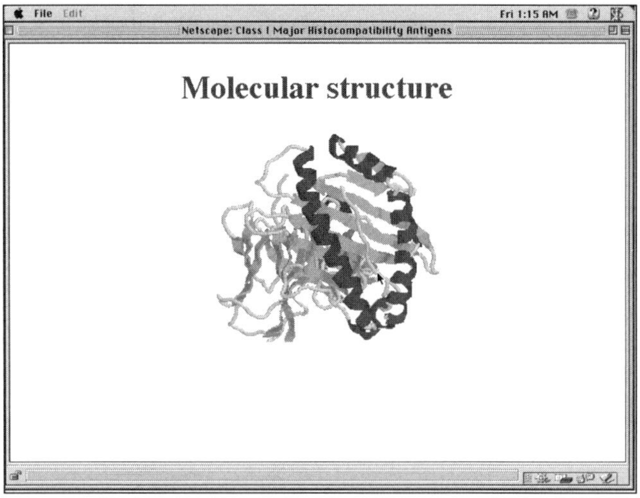

Figure 2. The slide show in action. Presentation Maker automatically converts the names of certain kinds of files into the appropriate HTML codes to display them on a slide. Here, the slide shown displays a PDB molecular structure file of a mouse class I major histocompatibility complex (MHC) antigen (2) using the Chime molecule viewer plug-in.

Formatting is done by selecting options from the pop-down lists at the top of the *Presentation Maker* window.

Once you have entered the text for your slides and changed the formatting options, if you like, click on **Make HTML** to get the program to create the code for your presentation. To preview your presentation in its own browser window, click on **Preview**. The entire presentation uses a single HTML file. Each slide is a different layer, and when you press the space bar to change slides, the program simply makes the next layer visible and hides the rest. Your browser's BACK button will not go back one slide, since all the slides are actually on the same page; use the RETURN key to cycle through the slides backwards.

You can give a presentation directly from the **Preview** window; if you want to save your data to regenerate the presentation later, just copy it and paste it into a text editor or word processor. Alternatively, you can save the HTML for the completed presentation in its own file. Don't use "File:Save As..."; this will not save the JavaScript portion of the presentation correctly. Instead, copy the HTML from the lower text area and paste it into a text editor (Windows® NotePad™ or Macintosh® SimpleText™); if you paste it into a word processor instead (like MS Word™); be sure to save it as "text".

To keep the interface simple, only a limited number of formatting options are available from the pull-down lists. To go beyond these limits, edit the HTML in the lower window, then click on **Preview** again.

It is possible to print your virtual slides to actual 35-mm slides, but it may be challenging to adjust the areas properly. Your graphics department may insist that your presentations be made with traditional presentation program if they are to print it for you.

HTML presentations can have elaborate features such as sounds (http://developer. netscape.com/viewsource/wyner_slideshow2/wyner_slideshow2.html) and transition effects (http://home. earthlink.net/~jhclark/iedhtml.htm). For more sophisticated molecular viewing and display tools, periodically check the following site: (http://www. umass.edu/microbio/rasmol/index.html). Fortunately, since Web pages can be linked together so easily, you can make different parts of a browser-based presentation using different tools. Once you complete your presentation, why not put it on your laboratory Web page?

ACKNOWLEDGMENT

Development of the Presentation Maker software was supported by Grant No. 1 R43 RR13645-01 from the National Center for Research Resources (NCRR).

REFERENCES

1. **Horton, R.M. and R.J.S. Stone.** 1997. Molecules on web pages: the Chime molecule viewer plug-in. BioTechniques *22*:884-886.
2. **Ghendler, Y., M.K. Teng, J.H. Liu, T. Witte, J. Liu, K.S. Kim, P. Kern, H.C. Chang, J.H. Wang and E.L. Reinherz.** 1998. Differential thymic selection outcomes stimulated by focal structural alteration in peptide/major histocompatibility complex ligands. Proc. Natl. Acad. Sci. USA *95*:10061-10066.

CHAPTER 17

Scripting Wizards for Chime™ and RasMol

Robert M. Horton

Attotron Biosensor, Carson City, NV, USA

Reprinted from BioTechniques 26:874-876 (May 1999)

The molecule viewing program RasMol and the Web browser plug-in Chime™ (reviewed in References 1 and 2) share a common scripting language for specifying how structures should be displayed. A "Wizard" is a program that "walks you through" the steps of a task and lets you select options. Here I present a set of **Scripting Wizards** intended to help new users begin using these remarkable visualization tools. They also take advantage of some of the new features introduced in Version 2.0 of Chime. They are written in client-side JavaScript™ and can be found on-line (http://www.attotron.com/pub/cs).

A Web of Wizards

Figure 1 shows how the various Wizard components are related and which other wizards each calls and answers to. The *Chime Scripter* at the top is the starting point for embedding Chime scripts and molecules in Web pages. The **Script Wizard** in the center does the actual assembly of scripts.

The *Chime Scripter* coordinates construction of the HyperText Markup Language (HTML) tags to "embed" molecules and control buttons on a Web page. It presents a window with **Add Molecule** and **Add Button** buttons. Once a molecule tag has been created, buttons can be added to trigger execution of additional scripts.

A script can be added to either a molecule tag (to specify the initial display) or to a button (to present a particular view in response to a user click), so both the **Add Molecule** and **Add Button** wizards have the ability to call the Script Wizard.

Wizards at Work

Figure 2 shows the *Chime Scripter* in action. The window in the background contains the tags and scripts constructed by the various Wizards, and the *Preview* window shows how the molecule looks and behaves on a page. Only the hemoglobin protein backbone was initially displayed; clicking the buttons displayed iron, heme and oxygen, and zoomed to higher magnification.

The various Wizards (not shown for lack of space) are collections of menus. The **Scripting Wizard** presents the major categories of scripting commands in a navigation frame; choosing a category loads a pull-down menu in the *Current Command* frame, where you select a specific command. Commands may take parameters; for example, the

BioComputing: Computer Tools for Biologists
Edited by Stuart M. Brown

colour command changes the color of the currently selected atoms, so the **Script Wizard** presents a sub-window with a list of available colors. Other sub-windows collect options such as numeric values. Most of these sub-windows are fairly simple, and their code is built into the **Scripting Wizard** page.

The most complicated options required by commands are *atomic expressions*, which specify subsets of the atoms. This specification process is complex enough to warrant a separate **Expression Wizard**.

The layout of the **Expression Wizard** is similar to that of the **Scripting Wizard**; a navigation frame lets you to choose types of expressions or predefined subsets of atoms. Sets include atoms of a given element, atoms in

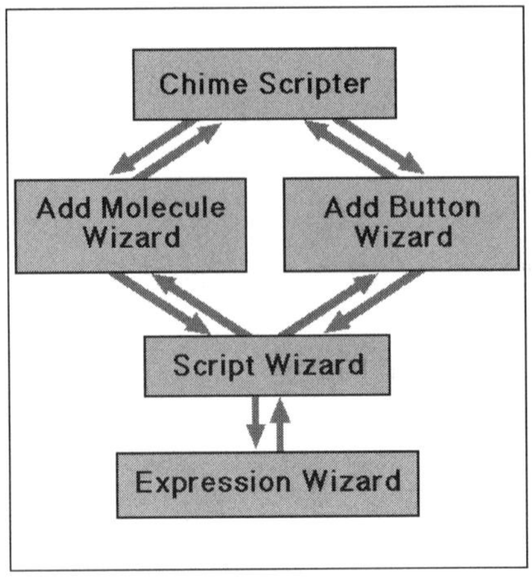

Figure 1. Relationships among the Wizard components.

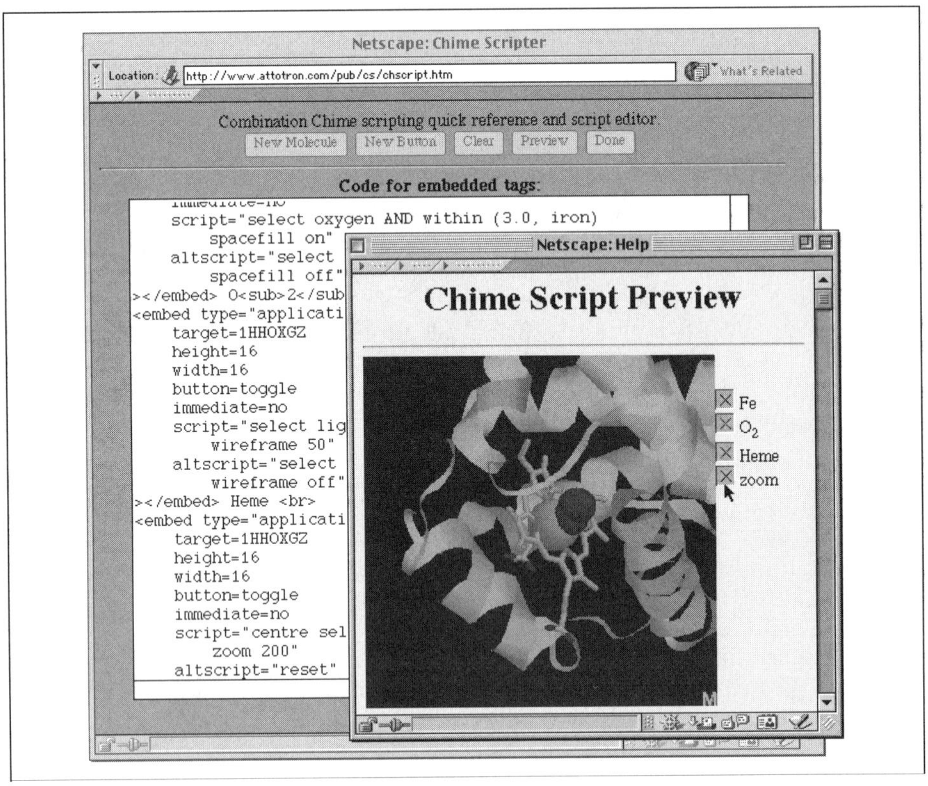

Figure 2. The Chime Scripter (background) and the resulting molecular view. The structure is oxyhemoglobin (3). Clicking the buttons on the right side of the *Preview* window runs the scripts attached to those buttons. The scripts for the **Fe**, **O2** and **Heme** buttons toggle the display of the corresponding atoms or molecules on and off, and the "zoom" button switches between two views at different magnifications. This example is online (http://www.attotron.com/pub/cs/).

particular amino acids or types of amino acids, in DNA, in proteins, in the solvent, in heterogeneous small molecules etc. Choosing a category loads the *Current Term* frame with the appropriate widgets to select specific values.

Since atomic expression terms can be combined into complex expressions using Boolean operators (i.e., "manganese OR magnesium atoms","surface AND NOT hydrophylic"), the **Expression Wizard** lets you negate a term before you add it to the complex expression.

Wizards Within Wizards: Which Window?

Perhaps the most interesting expression is "within", which lets you choose atoms inside a specified radius of atoms specified in a sub-expression. Thus, the term "within(3.0,iron)" selects all atoms within 3.0 Angstroms of an iron atom. ANDing this with oxygen gives us the attached oxygen in oxyhemoglobin; this is a generally useful method for visualizing binding sites. The "within" expression takes an expression as a parameter, which can be another "within". To create such nested "within" expressions, the **Expression Wizard** calls up another copy of itself, until a final (e.g., non-"within") term is encountered. You must complete all the windows in order, starting with the most deeply nested.

The presence of many open windows can be confusing, so be sure to fill out sub-windows from top to bottom. Terms in your script or expression that are set off in exclamation points, such as !colour! or !expression!, are placeholders that should have been replaced by the value returned from a sub-window. If you see an odd exclamatory remnant, it means all sub-windows did not complete correctly or in the proper order.

These Wizards act as both quick reference manual and command template. Once you become familiar with a particular command, you might find it more efficient to simply type it into the Script box. Thus, scripts and HTML code are made directly accessible in the various Wizards. As your abilities grow, you can gradually take over typing the commands you know, and just use Wizards to generate HTML tag templates or to look up less familiar commands. At any point, you can copy scripts or HTML tags from the various Wizards to paste into other files or Web pages (for example, the Chime Presentation Template; http://www.umass.edu/microbio/chime/prsswc/template.htm).

The menu-driven approach taken by these Wizards can be complementary to graphical scripting interfaces currently under development by other groups, which let you do things such as select residues by clicking on them instead of having to know their numbers (see the RasMol Home Page [http://www.umass.edu/microbio/rasmol]). Since a wizard can be called from another browser window, it would be fairly simple to integrate wizards like those described here as accessories to other browser-based tools.

ACKNOWLEDGMENT

Development of the Scripting Wizards was supported by Grant No. 1 R43 RR13645-01 from the National Center for Research Resources.

REFERENCES

1. **Horton, R.M. and R.J.S. Stone.** 1997. An introduction to molecular visualization: seeing in stereo with RasMol. BioTechniques *22*:660-662.
2. **Horton, R.M. and R.J.S. Stone.** 1997. Molecules on web pages: the Chime molecule viewer plug-in. BioTechniques *22*:884-886.
3. **Shaanan, B.** 1983. Structure of human oxyhaemoglobin at 2.1 A resolution. J. Mol. Biol. *171*:31-59.

CHAPTER 18

Biological Sequence Analysis Using Regular Expressions

Robert M. Horton

Attotron Biosensor Corporation, Carson City, NV, USA

Reprinted from BioTechniques 27:76-78 (July 1999)

"Regular expressions", also known as "RegExps" or sometimes as "grep patterns", are a way of specifying patterns of characters within a text file or longer string of characters. Regular expressions can contain various "wild cards" or sets of alternatives, which can be used to include ambiguity in the pattern to be matched. One common use is to perform sophisticated find-and-replace operations.

The fourth-generation Web browsers (Version 4.0 or higher of Netscape Navigator and Internet Explorer) incorporate regular expressions into their JavaScript™ scripting language, bringing the power of these tools to client-side, browser-based scripting. This article illustrates some ways that this technology can be used for analysis of nucleic acid sequences.

Finding Sequence Patterns

We begin with the classic problem of finding a genetic sequence capable of encoding a given peptide. The **Sequence Pattern Finder** program illustrated in Figure 1 uses a two-step process to generate a regular expression representing a degenerate nucleotide sequence that matches any gene sequence capable of encoding the specified peptide. The regular expression is then used to search a nucleic acid sequence pasted in the target box.

After typing in the desired peptide sequence, clicking on **reverse translate** generates an ambiguous nucleic acid sequence in which each single-letter amino acid code character is replaced with the corresponding nucleotide sequence, using the standard genetic code and ambiguity characters (1). For amino acids whose codons can vary at more than one nucleotide position, the genetic representation contains alternative codons. For example, the six codons encoding leucine are represented by either "uur" (uracil, uracil, purine) or "cun" (cytosine, uracil, any base). These alternatives are written concisely as (uur|cun); they are separated by a vertical bar character, used to represent OR, and the limit of what is included in each alternative is set by the parentheses.

The **disambiguate** button replaces each ambiguous code letter with the set of bases it represents; for example, n in the ambiguous sequence is replaced with the character set [acgt]. The degenerate sequence encoding leucine then becomes (uu[ag]|cu[acgu]). The **highlight matches** button searches for the pattern and displays a copy of the target sequence in a new browser window. If a region of the sequence matches the pattern, it is wrapped in Hypertext Markup Language (HTML) **<FONT>** tags to change its color.

BioComputing: Computer Tools for Biologists
Edited by Stuart M. Brown

The two-step reverse translation process makes the tool more flexible; you can type ambiguous codes directly into the second box, without starting from amino acids, or enter your own completed regular expressions in the third box. In fact, regular expressions entered directly are not confined to matching nucleic acid sequences; you can enter peptide sequence patterns to match against a protein sequence pasted in the target box or, similarly, match any text pattern.

Regular expression pattern matching is fundamentally different from sequence alignment and "best-fit" algorithms that show you where the best match is found between probe and target sequences. While the regular expression can contain ambiguities, it will only find exact matches to that ambiguous sequence. However, it has flexibility in other ways that make it an interesting alternative to sequence similarity searching.

For example, the following regular expression matches two *Eco*RI sites in the same reading frame: gaattc([acgt]{3})*gaattc

This regular expression matches codons encoding serine or threonine, separated by any two other codons: ((uc[acgu]|ag[cu])|ac[acgu])([acgt]{3}){2}((uc[acgu]|ag[cu])|ac [acgu])

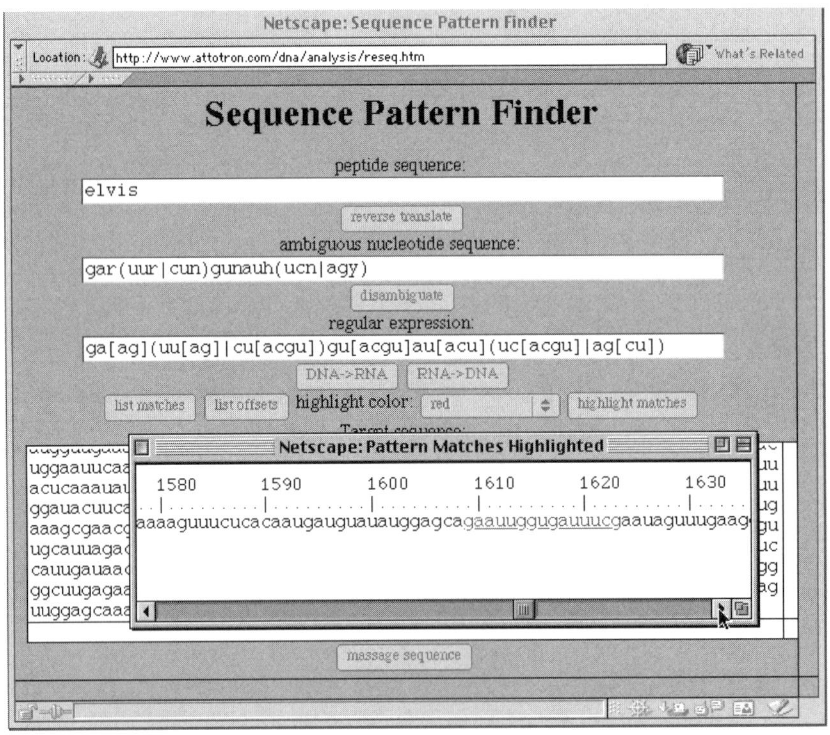

Figure 1. A JavaScript program to locate patterns in a target sequence, using regular expressions. In the top of the window, nucleotide regular expressions are automatically generated from a peptide sequence, using a two-step reverse translation. Three buttons search for the pattern in the target sequence: (*i*) **list matches** shows the actual sequence(s) matching the pattern; (*ii*) **list offsets** shows the number of characters (bases) preceeding each match and (*iii*) **highlight matches** produces a copy of the target sequence, with the portion matching the pattern colored and underlined. This program requires Netscape Navigator Version 4.0 or higher, and can be found at (http://www.attotron.com/dna/analysis/reseq.htm).

For an example of an on-line database supporting searches with regular expressions, see the Yeast Genome Grepmatch page (http://www-genome.stanford.edu/Sacch3D/grepmatch.html).

Massaging Sequences

The **massage sequence** button calls up another JavaScript program in a new window. This *Sequence Massager* uses regular expressions to manipulate a DNA sequence file. Various buttons cause predefined (and hopefully useful) operations to be carried out on whatever sequence (or other text) has been pasted into the target box. Most of these actions are programmed using regular expressions. The **DNA->RNA** and **RNA->DNA** buttons, for instance, simply replace t (or T) with u (or U) and vice versa. Other buttons remove comments, line breaks, spaces etc. Using a slightly more elaborate pattern, a button labeled **remove HTML tags** finds any HTML tags, such as **<H1>** or **<FONT COLOR=red>** and removes them (by replacing them with nothing). The pattern for an HTML tag is defined by the regular expression <[^>]*>, which tells the replace method to look for an opening angle bracket (or less than sign, '<'), followed by any number of characters that are NOT a closing angle bracket (greater than sign, '>'), followed by a closing angle bracket.

Readers with an interest in programming are encouraged to view the source code for these programs (by choosing **View:Page Source** from the top menu in Netscape). The use of regular expressions makes these programs remarkably compact; many of the JavaScript statements using regular expressions for find-and-replace in the *Sequence Massager* are concise enough to be incorporated directly into the HTML button tags, rather than being defined as functions. The heart of the Pattern Finder program (the *findIt()* function) contains only two lines.

The "dialect" of regular expressions used in JavaScript is very similar to that used in the Perl programming language, and JavaScript regular expressions can generally be used in Perl with little or no modification. Perl comes with many Unix/Linux distributions and is available for Macintosh® and Windows® at no cost. Unlike client-side JavaScript, Perl can directly access files on your hard drive, and is more appropriate for such tasks as file searching and manipulation. It is also very well suited for server-side programs. For more information on uses of Perl for biological sequence analysis and manipulation, see the BioPerl home page (**bio.perl.org**) and Reference 2. For in-depth background and tutorials on regular expressions and discussion of various dialects, see Reference 3.

The tools presented here are intentionally simple, and are geared for student use. Since they are written in client-side JavaScript, they can be used in classrooms with or without a network. A variety of related sequence analysis tools can be found on the following Web site (www.attotron.com/dna/analysis).

These tools illustrate some of the power, flexibility, and relative simplicity of using regular expressions for analysis and manipulation of sequence data. This approach to pattern matching in biological sequences can be useful in a wide variety of applications.

ACKNOWLEDGMENT

Development of the JavaScript programs described here was supported by Grant No. 1 R43 RR13645-01 from the National Center for Research Resources.

REFERENCES

1.**Nomenclature Committee of the International Union of Biochemistry (NC-IUB).** Nomenclature for incompletely specified bases in nucleic acid sequences. Recommendations 1984. Eur J. Biochem. 1985 Jul 1;150:1-5.

2.**Stein, L.** 1999. A DNA sequence class in Perl. Dr. Dobb's Journal *300*:50-58.

3.**Friedl, J.E.F.** 1997. Mastering Regular Expressions. O'Reilly Associates, Sebastopol, CA.

CHAPTER 19

A Virtual Molecular Biology Teaching Laboratory

Robert M. Horton[1] and Robert C. Tait[2]

[1]Attotron Biosensor, Carson City, NV, and [2]Biotechnology Consultant, Davis, CA, USA

Reprinted from BioTechniques 27:298-300 (August 1999)

Recombinant DNA technologies and genetic analysis have become such an important aspect of technological culture that the analysis of disease-related genes, production of mammalian hormones and antibodies in bacteria, the use of DNA evidence in criminal proceedings and mammalian cloning all appear routinely as news items. The economic and social impact of molecular genetics has created a demand for training in these areas that extends well beyond the confines of research laboratories (1). Laboratory researchers can expect to be called upon to teach these technologies with increasing frequency to students from an ever broadening spectrum of disciplines. However, the expense and technical demands of maintaining biological strains, enzymes and DNA samples for even simple laboratory exercises can be burdensome and can severely strain educational budgets.

Here we present the restriction analysis module of the Cybertory™ Molecular Biology Laboratory Simulator, which is being developed in an effort to reduce the costs and technical demands associated with presenting training in principles of molecular biotechnology to a broad audience. This program is written in client-side JavaScript™, and runs in the Netscape Navigator web browser (Version 4.03 or higher). It can be run either from the Internet or on stand-alone machines (Macintosh® and Windows®).

Launching the Laboratory

Start the program from the start page (http://www.attotron.com/cybertory/start. htm). Buttons let you choose a work area size. The program is designed to use a 600 × 800 pixel area; however, it can be launched in 640 × 480 mode for display on a low-resolution monitor or television set or in a larger window to cover the desktop.

Experimental parameters such as the available restriction enzymes and DNA sequences are set by placing an "order" with BOBCO, a fictitious Virtual Supply Company. BOBCO has an on-line (or off-line, for that matter) catalog, where one can find technical information about the various "products", such as restriction enzyme recognition sequences, the sizes of the bands in the molecular weight marker, etc. Placing an order sets client-side cookies that remember your choices from one computer session to the next. An order specifies four restriction enzymes and six DNA samples.

BioComputing: Computer Tools for Biologists
Edited by Stuart M. Brown

The BOBCO order form is password-protected, so that you can choose whether or not to let your students change the experimental parameters; give out the password if you want them to order their own enzymes and/or DNA samples. The secret password is Laguna.

Figure 1 illustrates the interface for setting up the reactions. The example shown here uses the "Sickle Cell Anemia Genetic Diagnosis Sample Kit", which includes linear DNA samples (virtual PCR products) from one positive control individual homozygous for the sickle cell mutation (SS), one negative control homozygous wild type (WW) and four individuals of unknown genotype (Pt1–Pt4). The goal of the experiment is to determine these unknown genotypes.

Twelve reagent tubes are arrayed on the left side of the screen, and twelve reaction tubes are arrayed on the right. Clicking on a reagent tube pops its top open, and the micropipettor travels to the selected tube. Click on the volume control wheel to increase volume (right side to increase, left to decrease), or select the volume display with the mouse and type a new value. Draw fluid into the pipet by clicking on the plunger. To move the material to another tube, click on the target tube; its top will pop open, and the pipettor will move to the newly selected tube. Click on the plunger again to dispense the pipet contents.

Click on the tip ejector or the waste bin to eject a used tip. Click on the tip ejector again (or the tip box) to get a new one.

When all reaction ingredients have been added, the Start button of the timer begins the

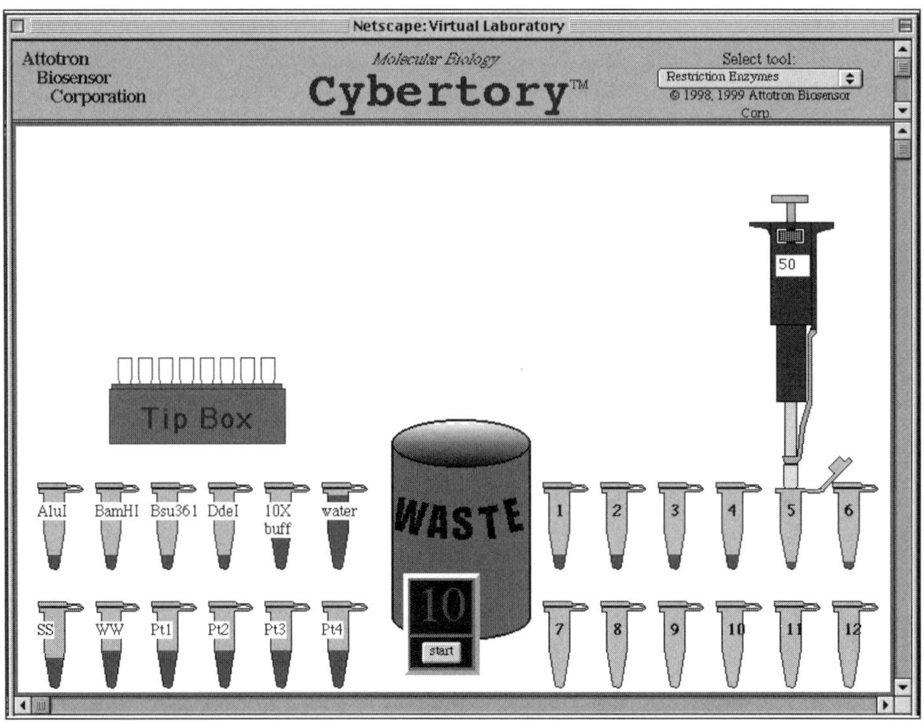

Figure 1. Restriction enzyme reaction setup. Clicking on tubes moves the pipettor, allowing mixing of ingredients from the reagent tubes on the left into the reaction tubes on the right.

incubation. The contents of the tubes will be fully digested with whatever enzymes were added (no partial digests are allowed).

Gel Analysis

To determine the sizes of the restriction fragments, choose *Agarose Gel* from the pull-down menu at the upper right corner of the screen (Figure 2). The **Auto Load Gel** button on the transilluminating gel box causes the twelve reaction tubes to be loaded into the gel wells from left to right, with molecular weight marker in the thirteenth well. Set an appropriate voltage and time on the power supply (BOBCO gels never melt), then click on the **on/off** button to begin electrophoresis. A red light on the power supply and bubbles from the electrodes confirm that the power is on, and the blue loading dyes migrate through the gel. To see the DNA, turn the UV light on and the room lights off.

Measure migration distances by clicking on and dragging the fluorescent ruler. Turn the room lights on briefly to line up the ruler with the bottom of the gel. Plotting migration curves has been left as a student exercise.

Though the catalog has several pre-designed "kits," the experiments themselves are

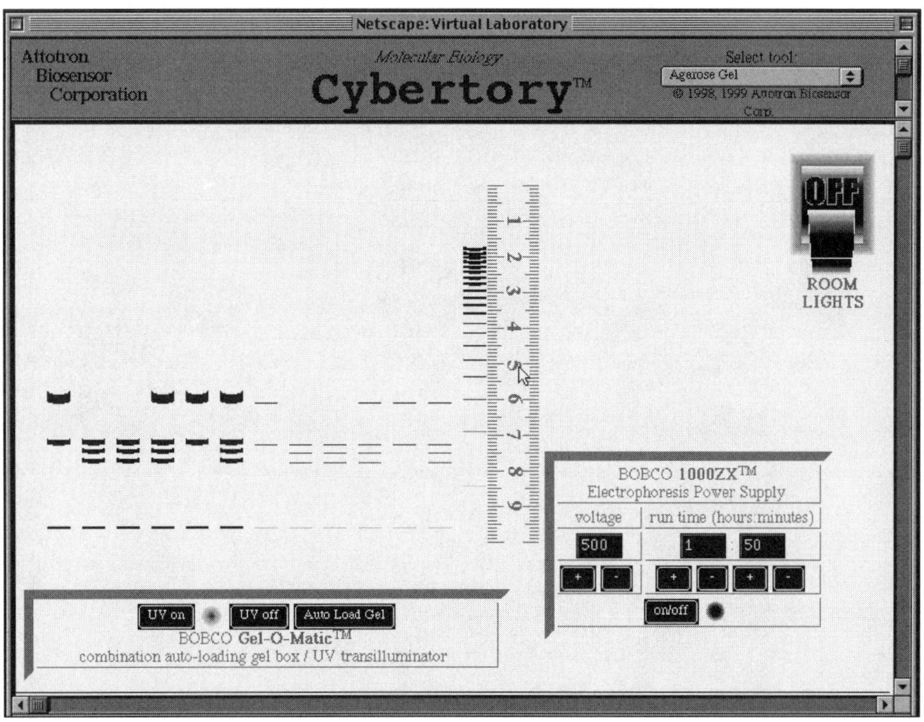

Figure 2. Gel electrophoresis. Six virtual DNA samples representing 771-bp PCR products of human genomic β-globin have been cut with the restriction enzyme *Bsu*36 I. Samples are run in two concentrations. Lanes 1 and 7: homozygous sickle-cell; lanes 2 and 8: homozygous normal; lanes 3–6 and 9–12: patients being tested. The normal gene yields bands of 254, 228, 201 and 88 bp. The mutation destroys a *Bsu*36 I site, giving bands of 429, 254 and 88 bp. Lane 13 contain a 150-bp ladder, with bands from 150 to 3000 bp. Ladder bands are equimolar, so larger bands contain proportionately more DNA. Starting sequences were taken from GenBank® locus HUMHBB using the primer sequences from Reference 2, but are 1 bp shorter than described in Reference 2.

not built into the simulator. The *custom DNA synthesis* option lets you combine any sequence (within your computer's memory limits) with your choice of restriction enzymes to create your own experiments. This may be useful for "dry-labs" with which students can practice before undertaking the more laborious, more expensive, and less reliable real thing.

Molecular biology in the real world demands sophisticated equipment and supplies, technical expertise and manual dexterity. The goal of the virtual molecular biology laboratory is to lower these barriers and facilitate teaching of key molecular genetics concepts.

ACKNOWLEDGMENT

Development of these programs was supported by Grant No. 1 R43 RR13645-01 from the National Center for Research Resources.

REFERENCES

1. **Tait, R.C.** 1997. An Introduction to Molecular Biology. Horizon Scientific Press, Wymondham UK.
2. **Husain SM, P. Kalavathi, and M.P.J.S. Anandaraj.** 1995. Analysis of sickle cell gene using polymerase chain reaction and restriction enzyme Bsu 361. Indian J. Med. Res. *101*:273-276.

CHAPTER 20

An Introduction to Perl: the QuizMaker Revisited

Robert M. Horton[1] and Robert C. Tait[2]

[1]Attotron Biosensor Corporation, Carson City, NV, USA and [2]Biotechnology Consultant, Davis, CA, USA

Reprinted from BioTechniques 27:470-472 (September 1999)

The Practical Extraction and Report Language (PERL) for programming was originally designed to sift through large computer data files (such as a collection of restriction enzyme recognition sites), extract information of interest and output it in a useful format (e.g., a file format used by a new program).

It turns out that the abilities of this language are well suited to the demands of computing on the Internet. Perl shines in creating "reports" in Web page format and has become extremely important for writing programs to generate Web pages. Converting data from one file format to another is crucial for getting various programs and databases to work together; this has been very important, for example, in the human genome project (3). Many of the most important Web resources in bioresearch have been developed, or at least linked together, using this language.

Perl has the reputation of being a fairly easy language for beginners to start using. Part of the reason for this is that there are simple ways to do many commonly performed tasks. This also has the effect of making many simple Perl programs very short. For example, the problem described in the introductory paragraph can be solved in fewer than ten lines of code; taking a restriction enzyme site list in GCG format (http://rebase.neb.com), the ten-line program skips the comments, extracts only the prototype enzymes and converts the single letter nucleotide ambiguity codes to the regular expression format used in Reference 2.

Perl Without Price

The ability to write simple but powerful data format conversion programs alone may make it worth the effort to download Perl and learn a bit about it. But that is just the beginning. Many Perl programs for a tremendous variety of purposes are freely available on the Web. Though the language was born on Unix (it ships with Linux and many other varieties), interpreters for running Perl programs on Windows®, Macintosh® and other operating systems are also available for free download.

BioComputing: Computer Tools for Biologists
Edited by Stuart M. Brown

Parsing with Perl

We now move to a more complicated example of a Perl program that writes Web pages; specifically, Web browser-based, multiple-choice quizzes. The example data sets are about molecular biology; they can be used as they are, or you can use the Perl program to write your own. The molecular biology practice quizzes, dictionary, associated data files and the Perl programs themselves, can be found at http://www.attotron.com/pub/quizmaker.htm.

If you only want to make one Web page, you are probably just as well off using a HyperText Markup Language (HTML) editor to write it. But if you want to automatically generate a large set of Web pages based on the content of some file or set of files, you may be well advised to seek programmatic assistance.

Our example aims to take "human-readable" data files and use them to create multiple-choice quizzes in a format similar to that described previously (1). Each of the multiple-choice answers has an associated feedback message to either instruct the user who has made an incorrect choice, or to praise the user who has answered correctly. Fundamentally, the program "parses" data from an input file, that is, it uses a set of rules to determine what type of information the various portions of data represent. For example, in the Quiz Data file, it knows that the first line is the title of the quiz. It then inserts the title into the appropriate positions in a new HTML document. Using other rules, it identifies questions, answer choices, feedback messages, etc., and outputs that information in the format of a JavaScript™-enabled Web page for an interactive quiz. Since the program knows the rules for translating a data file into a Web page, the quiz author can concentrate on questions, answers and feedback messages.

The names of the quiz data input files end in the extension *.quiz*. They can be edited with an ordinary text editor (like Windows NotePad), or with a word processor that can save as plain ASCII text without wrapping lines. Further details on the file format can be found on the QuizMaker Web site.

Computerized Cross Referencing

Merely "HTMLifying" data is rather straightforward, but this Perl program goes another step; it automatically creates a hypertext dictionary containing terms and definitions supplied by the author and links technical terms used in feedback messages from the quiz to the dictionary. A data file called *dictionary.txt* provides the definitions. The program *quizmaker.pl* first reads the dictionary into memory. Then it creates a cross-linked dictionary Web page file, in which each defined term used in a definition is hyperlinked to its own definition. This helps to address a common problem in technical literature; technical terms tend to be defined using other technical terms. One of the great attractions of the Web page format for data is the ability to follow links to find more information on a particular topic, and looking up definitions is a classic application (see the BioBit: Technical Dictionaries On-Line).

The *dictionary.txt* sample file that comes with the program contains definitions of molecular biology terms intended for student use. Using this as a template, aspiring hypertext dictionary authors can use *quizmaker.pl* to create automatically cross-linked dictionary pages for their own (sub)specialties.

Once the hypertext dictionary is created, *quizmaker.pl* opens each *.quiz* file in the current directory and creates a multiple-choice quiz. The data file format is straightforward; the first line is the title of the quiz, while subsequent lines contain questions followed by multiple-choice answers. Each answer has an associated feedback message to be presented to the quiz taker if that answer is chosen. If these explanatory messages contain image file names (in brief, any continuous word, no spaces allowed, which ends in .GIF, .JPG or common variants thereof, will be interpreted as the name of an image file), and they will be automatically converted to HTML image tags. As each quiz page is generated, any term used in a feedback message, which is defined in the dictionary, is automatically linked to its definition. The sample data files provided can serve as a starting point for creating your own quizzes.

As anyone who has tried to create an intricately inter-linked Web page can attest, creating large numbers of links can be monotonous, and it is easy to make mistakes. This task is well suited to programmatic solution. If additional terms are added to the dictionary in the future, the whole set of data files (dictionary.htm and all the various *.quiz* files) can simply be run through the Web page generating program again in a matter of minutes. This rewrites the entire set of Web pages, including the links to (and from) the new definition.

Quizmaker.pl is smart enough to infer that terms that differ only in a final "s" or "es" should link to the same definition (though it is not so clever as to recognize the relationship between irregular plurals like "terminus" and "termini"). It also tests to be sure not to try to link terms that are already part of another term. For example, if "DNA polymerase" links to a definition, it will not try to also link the word "DNA" inside that definition to a different place. It does this through the simple expedient of linking the terms containing the most words first, then checking to be sure that subsequent links are not added within existing ones.

All of this checking and cross-referencing takes quite a bit of processing; generating the dictionary and quizzes from the sample data set takes over a minute on a Windows 98 machine running at 450 MHz. Still, this is tremendously faster than cross-referencing by hand, and it only needs to be done when the dictionary and/or set of questions requires updating. Only quizzes whose data files are in the same directory as the script when the script is run will be processed, so it is simple to update a single quiz or a subset of quizzes.

ACKNOWLEDGMENT

Development of *quizmaker.pl* and the example data sets was supported in part by Grant number 1 R43 RR13645-01 from the National Center for Research Resources.

REFERENCES

1. **Horton, R.M.** 1997. Interactive internet instruction: Web quizzes and a new JavaScript™ QuizMaker. BioTechniques *23*:1048-1050.
2. **Horton, R.M.** 1999. Biological sequence analysis using regular expressions. BioTechniques *27*:76-78.
3. **Stein, L.** How Perl saved the human genome project. http://bio.perl.org/GetStarted/tpj_ls_bio.html (reproduced from The Perl Journal).

CHAPTER 21

BlastReport: A Perl Script to Facilitate the Use of Sequence Databases for Mapping and Clustering

Jeanette McClintick and Howard J. Edenberg

Indiana University School of Medicine, Indianapolis, IN, USA

Reprinted from BioTechniques 29:1272-1276 (December 2000)

There are many projects for which it is necessary to find or construct contigs or localized maps within a given chromosomal region. This can be very time consuming without some form of computer automation. We present here a method and a perl script that can be used to facilitate this using software freely available for most computers. The method uses blastcl3 from the National Center for Biotechnology Information (NCBI) to run Basic Local Alignment Search Tool (BLAST) searches of NCBI databases using text files containing multiple query sequences. The perl script BlastReport, available on our Web site, reads the BLAST output, reorganizes the results for ease of use and eliminates useless information.

There are several kinds of projects for which this is important. In genetic mapping experiments, chromosomal regions are initially defined by markers spaced at 5–10 cM intervals. It is then necessary to create detailed local maps of these regions to allow choice and ordering of additional markers for fine mapping and to select potential candidate genes for analysis. Markers for fine mapping are often chosen from different maps and have been placed with different levels of confidence. Ideally, marker maps and distances would be obtained from the complete genomic sequence, but that is not yet available for large regions of the human genome, and data for many other organisms is even more limited. One can, however, use currently available data to build local maps and identify genes and ESTs within the region. A second type of project involves assembling sequence data for the genomic region around a group of related genes and their associated regulatory elements.

The rapidly increasing availability of DNA sequence is an invaluable resource for these projects, but the very wealth of data makes searching through the databases and analyzing the results a time-consuming endeavor. The continual deposition of new sequences into the databases requires frequent repetition of searches, increasing the time required. Our suggested method and simple program can assist in these tasks.

To construct a map or contig starting from a marker, gene or EST, the sequence(s) can be searched against a database such as GenBank® (3) using the blastn program of the BLAST (1,2) tool available at NCBI (http://www.ncbi.nlm.nih.gov/blast/). Finished

BioComputing: Computer Tools for Biologists
Edited by Stuart M. Brown

sequences are found in the nonredundant (nr) database. Unfinished or draft sequences are found in the unfinished high throughput genomic sequences (htgs) database. The movement of sequences from the htgs database to the nr database upon completion of a BAC sequence requires one to search both databases. When a significant alignment is found, the ends of this new sequence can be used to search for adjoining sequences. Alternatively, if searching reveals that the sequence is on a still unfinished BAC (in unordered pieces), sequences from the BAC can be used as queries for further searching. A map of the region can be built by this iterative process. Other information about a BAC clone may be obtained from the clone registry at NCBI (http://www.ncbi.nlm.nih.gov/genome/clone/).

There are, however, limitations that can be both frustrating and time consuming. One is the very redundant nature of the nr database. Another is the retrieval of false positives generated when there is a repetitive sequence within the query. These two problems can lead to a very large number of irrelevant alignments that must be examined to find the identical matches. An identical match, if it is short, may not be found if there are too many long near-alignments. These problems can be minimized by appropriate choice of search strategies and by the use of the simple program we have developed. Because the searches must be repeated frequently until a region is completely sequenced, the time saved can be substantial.

The Web-based BLAST tool available at NCBI for this process is excellent when searching a single sequence. It can, however, be very time consuming to search a large number of query sequences (e.g., a set of markers in a region). We report an alternative method using software that is available for free. The process uses blastcl3 (4) (available for Mac OS®, Microsoft® Windows™ and several UNIX® environments; ftp://ncbi.nlm.nih.gov/blast/network/netblast/), and a Perl script, BlastReport, available from our Web site (http://bmbmac.biochemistry.iupui.edu/ Blastrep). [Perl is a scripting tool or language that is available on most UNIX machines, Mac OS and the Windows environment free of charge (http://www.Perl.com/ pub/language/info/software.html). Information about Perl is available at http://www.Perl.com and http://www. Perl.org.] Blastcl3 eliminates much of the manual and repetitive work of the BLAST searches by using a text file containing multiple query sequences in FastA format as input for BLAST searches. It runs locally and communicates to the BLAST server at NCBI. The results of these searches are written to a text file that has a format very similar to output of the browser-based BLAST search. Alternatively, the results can be returned in HTML format and viewed using a Web browser. These reports can, however, be very large and often contain alignments that are not useful. Judicious choice of parameters can reduce the clutter and increase the ability to detect meaningful alignments. Limiting the number of descriptions and alignments returned (parameters -v and -b, respectively) to about 10 each reduces the need to scan through many pages of alignments that are not likely to be meaningful. The output files can, nevertheless, be large and time consuming to review. BlastReport sorts through the BLAST output and formats it for rapid evaluation.

A great improvement in signal-to-noise ratio for the BLAST searches can be obtained by setting the EXPECTATION parameter -e to limit the alignments returned. The default value of 10 is much too large for this process. Choosing a value for EXPECTATION is a balancing act between eliminating poor alignments or repeat elements and losing significant alignments of small markers or exons. The BLAST programs calculate an EXPECTATION value for each alignment. This calculated EXPECTATION gives an indication of how likely an alignment of this quality will occur by chance for a database of this size. BLAST uses the value of the EXPECTATION parameter to limit the alignments returned. Any alignment whose calculated EXPECTATION is larger than the supplied value is not returned. [See Reference 2 or the BLAST Course available on the NCBI BLAST Web site for more information on EXPECTATION.]

Some microsatellite markers and ESTs contain repetitive elements such as Alu sequences, which have many alignments that are highly similar but do not represent real positives. For example, marker D16S748, which contains an Alu sequence, when used with the EXPECTATION parameter set to 1.0, yielded a text file that was 300 pages long but contained only three useful alignments. BLAST, by default, filters out the low-complexity regions such as the tandem repeats found in a microsatellite marker. This has the effect of producing two local alignments for one marker, one for each end. The EXPECTATION parameter must be set to return these shorter alignments without returning near-alignments to repetitive elements. With the default low-complexity filter on, an EXPECTATION of 1×10^{-20} (entered as 1e-20) works well. Turning off the low-complexity filter usually gives one alignment instead of two, but a smaller EXPECTATION (e.g., 1e-35) is needed to eliminate alignments to the repeated segment. A difference in length between the tandem repeat in the marker and in the genomic sequence frequently occurs in these highly polymorphic markers, so there are usually gaps in the alignments. If the difference in repeat length is large, two local alignments may be produced, one on each side of the gap. For cDNA sequences, an EXPECTATION parameter of 1×10^{-30} usually works well, although sequences representing short exons may be missed. If one suspects that an exon has been missed (see below), one can rerun the search using that exon as a query.

As a general guideline for modifying our suggested values, we note that perfect matches of 100 nucleotides give an EXPECTATION approximately 10^{-47}, 75 nucleotides about 10^{-33}, and 61 nucleotides about 10^{-25}. It is important to note that the quality of the sequence for most microsatellite markers is not as good as the genomic sequence currently being produced, so some mismatches are to be expected (Figure 1). A match with 98/110 identities gives an EXPECTATION of approximately 10^{-25}. Both the length of the query sequence and the size of the database affect the EXPECTATION calculated for a particular local alignment. These values are based on the database size in August 2000; the EXPECTATION will increase tenfold as the database grows tenfold. This parameter may need to be adjusted for future growth so that alignments are not lost. The recommended EXPECTATION settings have been developed for searches of the htgs and nr databases; both are currently on the order of a billion bases. When using smaller databases, the calculated expectations will be smaller for similar alignments.

A second example of the variability of calculated expectation values involves cDNAs. In searching htgs with the reference sequence NM_000807 (GABRA2; 2189 bases), the alignment of exon 4 to AC015523 had an EXPECTATION of 3e-30 (3×10^{-30}) for an exact 71/71 match. When a 90-base sequence from NM_000807 that included exon 4 was used as the query sequence, the EXPECTATION was 9e-32. The decrease in EXPECTATION is due to the decrease in query string length. This indicates that the EXPECTATION parameter may need to be adjusted to avoid missing small exons when using full cDNA sequences as search sequences (also see below).

BlastReport reformats the initial blastcl3 output to create a report that is much easier to evaluate. BlastReport eliminates many of the alignments to repetitive elements such as Alus by inspection of the alignments to find regions of the query sequence that have many alignments. A number of these regions will be listed, but some or all of the alignments themselves will be eliminated. This can remove between 10 and 100 pages from the report.

BlastReport produces three files. BRept_Full provides a summary for each query sequence, followed by the alignments that are most likely to be biologically significant

```
(A)              ----------------------=== D5S461 ===-------------------

Query= D5S461 Z50936 (CA) repeat AFM086xc1
         (252 letters)
(B)                                                                  Score      E
Sequences producing significant alignments:                          (bits)   Value

gb|AC008391.2|AC008391 Homo sapiens chromosome 5 clone CTC-236F1...    452   e-125
gb|AC021572.2|AC021572 Homo sapiens chromosome 18 clone RP11-143...    444   e-122
gb|AC016618.3|AC016618 Homo sapiens chromosome 5 clone CTD-2243O...    291   1e-76
gb|AC022431.2|AC022431 Homo sapiens chromosome 5 clone CTD-2583B...    246   5e-63
gb|AC011194.1|AC011194 Mus musculus chromosome 11 clone 196_F_5 ...    155   1e-35
gb|AC009180.7|AC009180 Homo sapiens chromosome 7, *** SEQUENCING...    147   4e-33

(C)--- D5S461 alignment analysis ---
  Number of alignments: 15
  Region(s) of alignment for Query: D5S461 (first 10 only)
     From:   1   To:  252   Expect: e-125    Match:   246/252     AC008391.2
     From:   1   To:  252   Expect: e-122    Match:   245/252     AC021572.2
     From:  79   To:  252   Expect: 1e-76    Match:   167/174     AC016618.3
     From:   1   To:  132   Expect: 2e-65    Match:   131/132     AC016618.3
     From:   1   To:  124   Expect: 5e-63    Match:   124/124     AC022431.2
     From:  67   To:  144   Expect: 1e-35    Match:     78/78     AC011194.1
     From:  67   To:  144   Expect: 4e-33    Match:     77/78     AC011194.1
     From:  67   To:  144   Expect: 4e-33    Match:     77/78     AC011194.1
     From:  67   To:  140   Expect: 4e-33    Match:     74/74     AC011194.1
     From:  67   To:  142   Expect: 6e-32    Match:     75/76     AC011194.1

        >>>>>>> ---------------------------------------- <<<<<<<<
                D5S461 may have a significant alignment.
(D)
D5S461             1 -    252
AC008391.2     23478 -  23725

D5S461             1 -    252
AC021572.2     30920 -  31167

D5S461            79 -    252         1 -    132
AC016618.3      2153 -   2326      2087 -   2218

D5S461             1 -    124
AC022431.2      8946 -   9069

(E) Only first 4 alignments will be printed.

>gb|AC008391.2|AC008391 Homo sapiens chromosome 5 clone CTC-236F12, WORKING DRAFT
SEQUENCE,
             37 unordered pieces Length = 140127

 Score =  452 bits (228), Expect = e-125
 Identities = 246/252 (97%), Gaps = 4/252 (1%)
 Strand = Plus / Plus

Query: 1     aagccatataaacattagcagaccacaataaaatcaatactgaaaatgcaacttaaggac 60
             ||||||||||||||||||||||||||||||||||||||||||||||||||||||||||||
Sbjct: 23478 aagccatataaacattagcagaccacaataaaatcaatactgaaaatgcaacttaaggac 23537

Query: 61    atggcactctctctctctctctctctctcacacacacacacacacacacacaca 120
             ||||||||||||||||||||||||||||    ||||||||||||||||||||||||
Sbjct: 23538 atggcactctctctctctctctctct----cacacacacacacacacacacaca 23593
                    ------ Deleted to conserve space ------
```

Figure 1. Example of output from BRept_Full. Query: microsatellite marker D5S461. DATABASE: htgs. EXPECTATION: 1×10^{-30}. Low complexity filter turned off. D5S461 had 1 alignment each to AC008391 and AC021572 and has two shorter alignments to AC016618. [Note that AC021572 is described as from chromosome 18; the exact match to this marker suggests that the chromosomal origin of AC021572 is misidentified. We used the BAC ends, AQ31730 and AQ37133, and five 500-700 nucleotide segments from this clone as search queries. Each of these aligned to one or more clone sequences annotated as chromosome 5, including AC008391, AC016618 and AC022431. Some of them also aligned to AC022944, labeled as chromosome 3.] The two short alignments with AC016618 occur because the query and the genomic segment differ in length of the dinucleotide repeat. There is also a repeat element in this marker between nucleotides 67 and 144 that is revealed in the alignment list by the presence of multiple, nonexact alignments to that region.

(Figure 1). BRept_Sum is a list of the regions of alignment only (Figure 2). BRept_Tab is a tab-delimited file containing query name, the name of the aligning sequence, the clone name if given, the first region of alignment and its expectation. BRept_Tab can be opened in a spreadsheet such as Microsoft Excel® to produce a table of alignments that can be sorted. Sorting by the aligning sequence (hit) makes it easy to see if that sequence contains multiple query sequences, such as two markers or a marker and a portion of a cDNA.

BlastReport prompts the user to specify whether markers or cDNAs were used for the search. Microsatellites and cDNA sequences differ in size and in the way that they align to genomic sequences. Microsatellite markers are 100–400 bp long and contain tandem repeats of 1–4 nucleotides that are 40–150 bp long. Since microsatellites are short, the entire sequence is usually found in one clone. On the other hand, cDNA sequences may have as many local alignments as they have exons, and the exons may be separated by large introns and be on different clones. The cDNA setting prints more data to adjust for the increased number of alignments expected for cDNAs. Individual exons or genomic segments should be analyzed using the marker setting. (see below for more on these limits.)

BRept_Full has an "eye-catcher" line for every query that has alignments (Figure 1A). This aids in scanning the report or searching using a text or word processor. The list of one-line descriptions with scores from the original BLAST output is next (Figure 1B), followed by a list of the top alignments (Figure 1C). For each alignment, the region of the alignment in the query sequence, the EXPECTATION, the ratio of matching residues to the total length of the local alignment and the alignment sequence identifier are given. This shows where alignments are found, how good they are and if there are repeat sequences in the query string (indicated by multiple matches to the same part of the query sequence). For each sequence producing a significant alignment, the regions of alignment are then shown (Figure 1D). For the cDNA setting, the alignments are presented in order of the query sequence rather than in order of the score (Figure 2). This helps the user to identify exons and to see if any are missing an alignment. If one region of a cDNA is missing, it might represent a small exon; we recommend that the search be repeated using the missing sequence with 5–10 flanking nucleotides. The most significant alignments are shown as they appear in the original BLAST report (Figure 1E). Within the BlastReport program there are easily modified parameters for marker and cDNA settings that limit the number of alignment regions, sequences with alignments and full alignments printed in

```
Query= SLC6A2 NM_001043 solute carrier family 6 (neurotransmitter)

SLC6A2            1 -   326      323 -   456      454 -   694      694 -   834
AC007602.3    93813 - 93488    80921 - 80788    78549 - 78309    65343 - 65203

SLC6A2          831 -   969      967 -  1072     1069 -  1197     1195 -  1309
AC007602.3    58470 - 58332    56377 - 56272    55111 - 54983    54163 - 54049

SLC6A2         1310 -  1439     1437 -  1539     1538 -  1639     1639 -  1807
AC007602.3    52489 - 52360    51919 - 51817    50833 - 50732    50248 - 50080

SLC6A2         1877 -  2136
AC007602.3    48066 - 47807
```

Figure 2. Example of output from BRept_Sum. NM_001043 is the reference sequence for cDNA *SLC6A2*, a noradrenalin transporter. Regions that align are sorted in sequence order along the query rather than in order of scores. This aids in the identification of exons in the cDNA. AC007602.3 is a working draft sequence containing six unordered pieces (deposited 6 April 2000). In this case, all 13 exons (spanning 2136 bp in the reference sequence) match sequences spanning 46006 bp within a 71312-bp contig.

BRept_Full (Figure 1C, 1D, and 1E). The number of one-line descriptions (Figure 1B) is limited by the value of parameter -v used in the original blastcl3 run. See the program documentation at our Web site for full details on changing these values. Since the original blastcl3 output is not deleted by BlastReport, changes in parameters can be made and BlastReport rerun as needed. At the end of BRept_Full is a list of the query sequences that had no alignment in the original BLAST report, eliminating 40 lines of useless information for each of these.

The strategy presented for use of blastcl3 followed by BlastReport, with appropriate parameters, can greatly improve the efficiency with which multiple markers or other query sequences can be used to find matches in the databases. This strategy decreases the workload and increases the ability to evaluate the matches returned. The software, available free, will run on a small workstation or desktop computer; we use an iMac, but the software is also available for Windows and UNIX. This method does not require the downloading of NCBI data to a local computer along with the problem of retrieving new or updated sequences on a regular basis. These features make it an inexpensive alternative for researchers with limited computer resources and/or budgets to build local maps or find contiguous genomic regions.

ACKNOWLEDGMENTS

This research was supported by U.S. Public Health Service grant nos. AA06460 (to H.J.E.) and U10 AA08403 (The Collaborative Study on the Genetics of Alcoholism; H. Begleiter, PI) from the National Institute of Alcohol Abuse and Alcoholism.

REFERENCES

1. **Altschul, S.F., W. Gish, W. Miller, E.W. Myers and D.J. Lipman.** 1990. Basic local alignment search tool. J. Mol. Biol. *215*:403-410.
2. **Altschul, S.F., T.L. Madden, A.A. Schäffer, J. Zhang, Z. Zhang, W. Miller and D.J. Lipman.** 1997. Gapped BLAST and PSI-BLAST: a new generation of protein database search programs. Nucleic Acids Res. *25*:3389-3402.
3. **Benson, D.A., I. Karsch-Mizrachi, D.J. Lipman, J. Ostell, B.A. Rapp and D.L. Wheeler.** 2000. GenBank. Nucleic Acids Res. *28*:15-18.
4. **Madden, T.L., R.L. Tatusov and J. Zhang.** 1996. Applications of network BLAST server. Methods Enzymol. *266*:131-141.

Update to: BlastReport: A Perl Script to Facilitate the Use of Sequence Databases for Mapping and Clustering

Jeanette McClintick and Howard J. Edenberg

Indiana University School of Medicine, Indianapolis, IN, USA (edenberg@iupui.edu)

BlastReport remains useful as a simple way to collect Basic Local Alignment Search Tool (BLAST)-aligned sequences from National Center for Biotechnology Information (NCBI) databases for use in mapping and contig construction. There were two aspects to our initial report: (*i*) suggestions for parameters that return reasonable numbers of hits that are likely to be interesting; and (*ii*) the Perl script itself, BlastReport, which trims and organizes the often voluminous output of such searches.

We have updated BlastReport to support format changes in newer versions of NCBI's Blastcl3. BlastReport has been tested with the latest version of Blastcl3 (network blast from NCBI; modules are dated April 2001 to July 2001 depending on the operating system for which the module was compiled). Changes were also made to accommodate sequence position numbers of 7 digits, necessary for positions in genomic contigs of 1 million bases or more.

The program and a readme file explaining the parameters that must be changed for the program to run on your computer (e.g., file/directory locations) can be downloaded from (http://bmbmac.biochemistry.iupui.edu/blastrep/Home.html).

NCBI has added functions to its Blastcl3 that can be used to refine a blast search. Searches can be restricted to results of an Entrez2 lookup in the same manner as on NCBI's interactive Blast Web page, by using the -u parameter. For example, to limit a search of the nonredundant (nr) or high-throughput genomic sequence (HTGS) database to human sequences use: -u human[ORGN]. NCBI has made available genome databases that may be searched by Blastcl3. Table1 gives the parameters for three sample queries. All three examples use the program blastn to search one of NCBI's nucleotide databases with a nucleotide sequence as the query. Contact NCBI (blast-help@ncbi.nlm.nih.org) for the names of other genomic databases, which are being added.

The first example searches the human genome database. This is the NCBI draft assembly of the public human genome sequence. The aim of this database is to have the entire genome represented once, so multiple hits usually reflect a repetitive sequence. This database is excellent for determining position of markers on the genome within assembled contigs.

The second example searches the nr database, but restricts the search to mouse sequences. Searches can be restricted to any organism by similar restrictions in the Entrez2 lookup parameter. This is very useful if your question is targeted to a single organism (e.g., Rattus norvegicus[ORGN]) or group of organisms (e.g., Primate[ORGN] or Eukaryota[ORGN]). You may also limit the search by modification date: 2002[Modification Date] will limit the search to sequences that have been added to the database or modified in 2002. Any of the restrictions found on the standard BLAST Web page should work.

Table 1. Parameters for Restriction Searches

Parameter	Keyword	Human Genome[a] Value	Mouse nr[b] Value	Mouse Contigs[c] Value
Program	-p	blastn	blastn	blastn
Database	-d	hs_genome/genome	nr	mouse_contig/contigs
Expectation	-e	1e-35	1e-35	1e-35
Number of descriptions	-v	10	10	10
Number of alignments	-b	10	10	10
Entrez2 lookup	-u		mouse[ORGN]	

[a]Search human genome database.
[b]Search the nr database, restricted to mouse sequences.
[c]Search the mouse genome contigs database.

The third example blasts the mouse genome contigs database, which contains curated NT contigs (nonredundant finished sequence for the mouse). At this time (March 2002), this database is still quite small, less than 5% of the genome. For phase 0, 1, 2, and 3 bacterial artificial chromosomes (BACs), continue to search the HTGS database, using the -u parameter to limit the search to mouse sequence.

In addition to these parameters, you must supply the name of the input file (-i) and the name for the output file (-o). (Note: for Mac/OS users, use the parameter values in the Blastcl3 form omitting the parameter keyword.)

CHAPTER 22

The Sequence Manipulation Suite: JavaScript Programs for Analyzing and Formatting Protein and DNA Sequences

Paul Stothard

Department of Biological Sciences, University of Alberta, Edmonton, Alberta, Canada

Reprinted from BioTechniques 26:1102-1104 (June 2000)

JavaScript is an object-based scripting language that can be interpreted by most commonly used Web browsers, including Netscape® Navigator® and Internet Explorer®. In conjunction with HTML form elements, JavaScript can be used to make flexible and easy-to-use applications that can be accessed by anyone connected to the Internet (3). The Sequence Manipulation Suite (http://www.ualberta.ca/~stothard/javascript/) is a collection of freely available JavaScript applications for molecular biologists. It consists of over 30 utilities for analyzing and manipulating sequence data, including the following:

- **Codon Plot** accepts a DNA sequence and generates a graphical plot consisting of a horizontal bar for each codon. The length of the bar is proportional to the frequency of the codon in the codon frequency table you enter. Codon frequency tables for numerous organisms are available online (4) (http://www.kazusa.or.jp/codon/). Use Codon Plot to find portions of DNA sequence that may be poorly expressed, or to view a graphic representation of a codon usage table (by using a DNA sequence consisting of one of each codon type).
- **Codon Usage** accepts a DNA sequence and returns the number and frequency of each codon type. Since the program also compares the frequencies of codons that code for the same amino acid (synonymous codons), you can use it to assess whether a sequence shows a preference for particular synonymous codons.
- **CpG Islands** searches for CpG islands in a DNA sequence using the method of Gardiner-Garden and Frommer (2). CpG islands are often found in the 5′ regions of vertebrate genes, and therefore this program can be used to highlight potential genes in genomic sequences.
- **DNA/Protein Pattern Find** accepts a sequence along with a set of search patterns and returns the number and positions of sites that match the patterns. The search patterns can contain “wild cards”, which allow you to detect a variety of similar sequences using a single pattern.
- **DNA/Protein Stats** returns the number of occurrences of each residue in the sequence you enter. Percentage totals are also given for each residue, and for certain

BioComputing: Computer Tools for Biologists
Edited by Stuart M. Brown

groups of residues, allowing you to quickly compare the results obtained for different sequences.

- **Filter DNA/Protein** removes non-protein or non-DNA characters from text. Use this program when you want to remove digits and blank spaces from a sequence to make it suitable for other applications. All programs in the Sequence Manipulation Suite automatically filter the sequences you enter before proceeding.
- **GenBank® Feature Extractor** accepts a GenBank file as input and reads the sequence feature information described in the feature table, according to the rules outlined in the GenBank release notes (ftp://ncbi.nlm.nih.gov/genbank/gbrel.txt). The program concatenates the relevant sequence segments and returns each sequence feature in FASTA format. GenBank Feature Extractor is particularly helpful when you want to derive the sequence of a cDNA from a genomic sequence that contains many introns.
- **GenBank to FASTA** accepts a GenBank file as input and returns the entire DNA sequence in FASTA format. Use this program when you want to quickly remove all of the non-DNA sequence information from a GenBank file.
- **GenBank Trans Extractor** accepts a GenBank file as input and returns each of the protein translations described in the file in FASTA format. GenBank Trans Extractor should be used when you are more interested in the predicted protein translations of a DNA sequence than the DNA sequence itself.
- **Group DNA/Protein** adjusts the spacing of DNA or protein sequences and adds numbering. You can specify the group size (the number of residues per group) and the number of residues per line. The output of this program can serve as a convenient reference for use with some of the other programs in the Sequence Manipulation Suite (such as DNA/Protein Pattern Find) since the numbering and spacing allow you to quickly locate specific residues.
- **Ident and Sim** accepts a pair of aligned sequences (in FASTA format) and calculates their identity and similarity. Identity and similarity values are often used to assess whether or not two sequences share a common ancestor or function. Since many alignment programs do not calculate these values, you may find Ident and Sim quite useful when making pairwise comparisons.
- **Multiple Align Show** accepts a group of aligned sequences (in FASTA format) and formats the alignment to your specifications. You can specify the number of residues per line, and whether to use colored text or colored backgrounds to highlight matching residues. You can also set a consensus level, which specifies the percentage of residues that need to be identical in a column of the alignment for highlighting to be added. Use Multiple Align Show to enhance the output of sequence alignment programs.
- **ORF Finder** searches for open reading frames (ORFs) in the DNA sequence you enter. The program returns the range of each ORF along with its protein translation. Use ORF Finder to search newly sequenced DNA for potential protein-encoding segments.
- **Primer Show** accepts a DNA sequence along with a set of primer sequences and returns a textual map showing the annealing positions of the primers. The translation of the DNA sequence can be shown in the reading frames you specify. You can also choose the number of bases per line of the map, and whether to show the DNA in its single-stranded or double-stranded form. The primer sequences you enter can contain "wild cards", a feature that allows Primer Show to handle degenerate primers. Use this program to produce a useful reference figure, particularly when you have

designed a large number of primers for a particular template.

- **Protein Molecular Weight** accepts a protein sequence and calculates the molecular weight. You can append copies of commonly used epitopes and fusion proteins using the supplied list. Use Protein Molecular Weight when you want to predict the location of a protein of interest on a gel in relation to a set of protein standards.
- **Random DNA/Protein Sequence** generates a random sequence of the length you specify. Random sequences can be used to evaluate the significance of sequence analysis results.
- **Rest and Trans Map** accepts a DNA sequence and returns a textual map showing the positions of restriction endonuclease cut sites. The translation of the DNA sequence is also given, in the reading frame you specify. You can choose the number of bases per line of the map, and whether to show the DNA in its single-stranded or double-stranded form. Use the output of this program as a reference when planning cloning strategies, particularly when you need to consider the reading frames of the molecules you are ligating.
- **Restriction Summary** accepts a DNA sequence and returns the number and positions of restriction endonuclease cut sites. Use this program when you want to quickly determine whether or not an enzyme cuts a particular segment of DNA, and to produce a table to complement the output of Rest and Trans Map.
- **Reverse Complement** converts a DNA sequence into its reverse, complement or reverse-complement counterpart. You may want to work with the reverse complement of a sequence if it contains an ORF on the reverse strand.
- **Reverse Translate** accepts a protein sequence and uses a codon usage table to generate a graph that can be used to find regions of minimal degeneracy at the nucleotide level. Use Reverse Translate to design PCR primers for amplifying genes that encode similar proteins.
- **Shuffle DNA/Protein** randomly shuffles a sequence. Shuffled sequences can be used to evaluate the significance of sequence analysis results, particularly when sequence composition is an important consideration.
- **Simple Plot** calculates DNA sequence composition using a sliding window. The results are returned as a set of x and y values. The x-value is the position of the first base in the window of bases used in the calculation, and the y-value is the result of the calculation. Use Simple Plot when you are looking for segments of a DNA sequence with a particular base composition, rather than a specific sequence of bases.
- **TestCode** accepts a DNA sequence and calculates the TestCode value as described by Fickett (1). The reading frame of the DNA sequence does not need to be considered when performing this analysis. Use TestCode to predict whether or not a sequence encodes a protein.

In addition to the features described above, the interface to each application displays (where applicable) the genetic code, restriction enzyme set and codon usage table it uses so that you can make changes before performing an operation. Each program also comes with a complete set of default inputs to help illustrate the desired data formats.

To access a program in the Sequence Manipulation Suite, simply point your Web browser to (http://www.ualberta.ca/~stothard/javascript/) and click on the name of the program you want to use. The time needed for the applications to complete their tasks is dependent on how much input you give them. Try short sequences first to gauge how they perform on your computer.

Each program in the Sequence Manipulation Suite writes its results to the same output window, which appears when the first analysis is performed. Whenever you perform an additional analysis, the results are simply appended to the existing output data. By scrolling up and down in the window, you can easily compare the results of a particular operation performed on a series of sequences, or you can compare the results of many different operations performed on a single sequence. You can also copy a sequence segment of interest from the output window and paste it back into a Sequence Manipulation Suite program for further analysis or formatting.

To save or print the contents of the output window, copy the contents and paste them into your favorite text editor. If the spacing of the text looks incorrect after it has been pasted, switch to a fixed-width font such as Courier. The color and font of the text shown in the output window may not be conserved after copying and pasting, depending on which Web browser and text editor you are using.

The Sequence Manipulation Suite can be used online, or a copy of the programs can be downloaded and saved on your own computer for use offline (http://www.ualberta.ca/~stothard/javascript/download.html). One advantage of using a local copy of a program is that its default settings can be modified and saved to suit your preferences. For example, you might want to replace the codon usage tables with a table generated for the organism you study. You might also want to replace the restriction enzyme sets with a set consisting of the enzymes you have access to in your lab. Anyone with HTML experience is encouraged to make these changes. Readers familiar with JavaScript may also want to build new programs by combining and modifying existing portions of the Sequence Manipulation Suite.

Additional questions or comments regarding the Sequence Manipulation Suite can be sent by e-mail to the author at stothard@ualberta.ca.

REFERENCES

1. **Fickett, J.W.** 1982. Recognition of protein coding regions in DNA sequences. Nucleic Acids Res. *10*:5303-5318.
2. **Gardiner-Garden, M. and M. Frommer.** 1987. CpG islands in vertebrate genomes. J. Mol. Biol. *196*:261-282.
3. **Horton, R.M.** 1999. Biological sequence analysis using regular expressions. BioTechniques *27*:76-78.
4. **Nakamura, Y., T. Gojobori and T. Ikemura.** 2000. Codon usage tabulated from the international DNA sequence databases: status for the year 2000. Nucleic Acids Res. *28*:292.

Update to: The Sequence Manipulation Suite: JavaScript Programs for Analyzing and Formatting Protein and DNA Sequences

Paul Stothard

Department of Biological Sciences, University of Alberta, Edmonton, Alberta, Canada (stothard@ualberta.ca)

The Sequence Manipulation Suite is now hosted by Bioinformatics.Org, an organization that promotes the development of Open Source software for the biological sciences. It can be accessed at (http://bioinformatics.org/sms), as well as from numerous mirror sites, most of which can be found using Web search engines.

Several minor enhancements have been made to the Sequence Manipulation Suite, including the addition of new features and options to many of the programs. For example, Filter DNA/Protein can now remove specific types of non-DNA and non-protein characters, or it can replace them with alternative characters of your choice. You could use this added versatility, for example, to quickly replace all the International Union of Pure and Applied Chemistry (IUPAC) degenerate characters in a DNA sequence with "N" to make the sequence suitable for programs that do not recognize 2-fold or 3-fold degenerate sites. GenBank® Feature Extractor now has an option that allows sequence feature segments to be converted to uppercase, making it possible to view features in the context of the entire DNA sequence described in the GenBank record. Furthermore, coding sequences that are converted to uppercase can be pasted into other programs in Sequence Manipulation Suite that generate protein translations (Primer Show, Rest and Trans Map, Show Translation, and Translate). These all have a new translation option (called "uppercase"), which causes the uppercase segments to be treated as one continuous reading frame. Intervening lowercase segments are ignored during the translation, allowing genomic sequences to be viewed with their biologically relevant protein product. Multiple Align Show now includes more options for controlling the appearance of sequence alignments. The precise font type, size, and color of text that is used in the output can be adjusted. Also, different highlight colors can be assigned to different groups of residues. This feature can be used to reveal protein regions that have conserved biochemical properties. Finally, a new numbering option allows alignments made using partial sequences to contain the correct numbering relative to the complete sequences.

One limitation of the Sequence Manipulation Suite is that the programs, which are written in JavaScript, execute slowly in certain Web browsers. To address this problem, and to provide features difficult to implement in JavaScript, some of the more popular utilities have been rewritten in Perl. These have been combined into a new program called Sequence Extractor, which can be accessed through its Web interface at (http://bioinformatics.org/seqext). The primary purpose of Sequence Extractor is to generate useful reference figures for molecular biologists working with protein and DNA sequences. It accepts sequence files in raw, FASTA, GenBank, or European Molecular Biology Laboratory (EMBL) format. It parses the sequence record and then generates a hypertext markup language (HTML) Web page that displays the DNA sequence with protein translations, restriction sites, and primer annealing sites (primer sequences can be entered with the sequence record). What sets Sequence Extractor apart from other restriction map and feature viewing programs is that JavaScript code is embedded throughout the resulting figure, to make it much more dynamic than static text output. For example, pointing to any restriction site label in the figure displays the precise position of the site and the recognition sequence of the enzyme in the Web browser's status bar. Similarly, pointing to primer sites or protein translations displays additional information not shown in the figure text. Clicking on two restriction site or two appropriate primer site labels causes the sequence between them to appear in a separate window, essentially performing a virtual restriction digest and band isolation or polymerase chain reaction (PCR). When primer sites are clicked, the sequence of the primers replaces the underlying "template" sequence, allowing engineered restriction sites to be incorporated into the product. These features greatly assist in the isolation of sequence fragments of interest, and in the construction of DNA constructs in silico. Finally, the protein translations and DNA sequences corresponding to coding regions can be obtained in a separate window using links that Sequence Extractor provides at the top of the figure. In summary, Sequence Extractor can be used in place of many of the utilities in the Sequence Manipulation Suite and provides enhanced output not available from other existing programs.

Section III

Similarity Searching

This section contains a number of short "Internet OnRamp" articles and a few longer review articles that all focus on the same basic concept—to educate the reader about the use of Web-based similarity search tools. This is an extremely worthwhile subject; most biologists use these tools frequently, yet few are truely expert in their use. More than half of the articles deal directly with the use of NCBI's BLAST search engine—offering tips and tricks for setting up queries and in the interpretation of results. This is especially important for difficult searches such as when using short sequences as a query or when searching for distantly related genes. A few Web addresses have changed since these articles were originally published, but otherwise most of the advice they give is still accurate and relevant.

CHAPTER 23

Sequence Similarity Searches on the World Wide Web

Stuart M. Brown

NYU Medical Center, New York, NY, USA

Reprinted from BioTechniques 24:248-250 (February 1998)

If you have just determined the sequence of an interesting bit of DNA, one of the first questions you are likely to ask yourself is "has anybody else seen anything like this before?" Fortunately, there has been a very successful international effort to collect all known sequences (both DNA and protein) into databases so that they can be searched. These databases (and searching tools) are freely available on the Web. The problem is that these databases are HUGE and, as a result, you must compare your sequence with the vast number of other sequences.

Sequence comparison is the most powerful and most reliable method of answering biological questions about the evolutionary relationships between genes. Recent improvements in the statistical methods used by similarity searching computer programs allow the biologist to make conclusions with a high level of certainty. Now that the entire genomes of some organisms have been completely sequenced, it is possible to use similarity-search methods to provide definitive answers to questions such as *"Is there a copy of this interesting gene in this organism?"*

A database search is a similarity search, but it is frequently—although incorrectly—referred to as homology searching. The term "homology" implies a common evolutionary relationship between two traits. Just because two sequences share a stretch of nearly identical nucleotides (or amino acids [aa]) does not mean that they are directly descended from a common ancestor. Homologous proteins share a common 3-dimensional structure, while proteins with a chance similarity of aa sequence do not. Over evolutionary time, two homologous proteins may diverge to the point that sequence similarity cannot be detected, yet they may still retain their common structure. Homologous proteins generally have similar biological functions, but this is not a rigorous requirement; conversely, discovery of homology is not proof of function.

Of course, a high level of similarity is a strong indication of homology. As a rule of thumb, 25% identity over a stretch of 100 aa can be considered to be good evidence of common ancestry for two sequences. Homologous sequences are usually similar over an entire sequence, or sometimes just over one functional domain, but it is never correct to state that two sequences are "50% homologous." Homology is an all or nothing decision. Be careful in asserting homology between short regions of two sequences; matches that

BioComputing: Computer Tools for Biologists
Edited by Stuart M. Brown

are more than 50% identical in a 20–40 aa region can occur by chance.

At present, the two most popular programs for similarity searching are Basic Local Alignment Search Tool (BLAST) (1) and FASTA (3). They use similar approaches to reach similar answers; although, there are some subtle differences. The two programs will generally give slightly different results: in terms of which sequences are found to be similar, their relative rankings by similarity score and the statistical significance that is assigned to those similarity scores. It is generally thought that FASTA is more sensitive for DNA-DNA comparisons, but BLAST is usually faster. The choice of which program to use is often governed by factors of convenience rather than mathematical rigor; but to be thorough, it is best to perform searches with both programs.

For most researchers, the Web has become the best method to make similarity searches. Web servers are available to everyone, they have the most current data, they are usually fast and your computer is not tied up while your search is computing. However, the addresses and the services offered by various Web servers are subject to change without warning, so it is wise to have some alternatives lined up when your computing is dependent on the kindness of strangers.

Protein similarity searches can find much more distant similarities than comparisons of DNA sequences (ca. 2.5 billion vs. 100 million years of evolutionary divergence). This is true for several reasons. (*i*) The DNA "alphabet" is restricted to only 4 letters, while the aa alphabet has 20 letters, so the probability of chance matches is much greater with DNA-DNA comparisons. (*ii*) Two differing DNA bases can only be scored as a mismatch, while two aa can share varying degrees of similarity based on their physical and chemical properties, etc. (*iii*) The protein databanks are much smaller than the DNA databanks, so searches can be more sensitive without incurring too many false positives.

Generally, if your query sequence is protein, you will search protein databanks, and with DNA sequence, you will search nucleotide data. However, it is possible to automatically translate your DNA sequence into aa in all six reading frames (BLASTX and FASTX) and compare it to protein databases, or to compare your protein sequence to the six reading frame translations of all DNA database sequences (TFASTA and TBLASTN). Surprisingly, a protein-protein search using a translated DNA sequence as a query against a protein databank, or a protein query against a translated DNA databank, is much more sensitive than a DNA-DNA search. It is generally better to perform all searches as protein-protein comparisons unless you are studying DNA sequences that do not code for protein.

For some searches, it is important to find every possible database match to a given sequence; for others, it may be advantageous to limit the search to just humans, to mammals or to the animal kingdom. The measurement of statistical significance in similarity searches is dependent on the size of the database. As the database gets bigger, weak homologies become less significant (i.e., the chance of finding equally strong matches due to random similarities increases). The more you can restrict your search, the faster it will run, the better your significance scores will be and the fewer false-positive hits you will have to sort through.

The scores reported by similarity programs are called "expectation values", but they should be treated just like the "P values" from traditional types of statistical tests. These "expectation values" represent the number of times that a match with the observed similarity score would occur by chance if the database actually contained no sequences that were truly similar to your query sequence (i.e., in a database of random sequences). Generally, expectation values less than 0.05 or 0.02 are considered significant. The results of a similarity search can be divided into three categories: (*i*) highly significant matches between nearly identical proteins, (*ii*) insignificant matches that are clearly due to chance similarities and (*iii*) an intermediate range of matches with scores at the borderline of statistical significance.

These intermediate values have been described as the "twilight zone" (2). Each "twilight zone" protein can be used as the query for another similarity search. True homologues will have significant matches, not only with your original query sequence but also to other members of that protein family (i.e., other sequences that received highly significant scores in the original search). Unrelated sequences will not match any other proteins in the family of your original query sequence (4).

BLAST and FASTA produce similar output files. First there is a short description of the program and a list of the databases and program options chosen. Then there is a list of all of the database sequences that matched your query sequence. A number is assigned to each of these sequences that represents the quality of the match. The list is presented in descending order, so that the best matches are at the top of the list. However, the most biologically significant matches are not always the ones ranked highest in the list.

A quick look at this list of hits can provide a lot of information. *Are there any hits with expectation values of zero or smaller than* e^{-100}*?* These would be nearly exact matches—most likely the same gene in related species or a recently duplicated gene in a single species. *Are there lots of hits with highly significant scores and a very gradual decline from the* e^{-100} *down to 0.01?* This would indicate that your query sequence is a member of a large protein family and would suggest further research on that family in protein domain databases and in the literature. The databases are now so large that it is extremely unlikely to find that a query sequence has no significant matches. If you do have such a result, to be sure it is correct, re-examine your experimental protocols and the computational steps that were followed to generate that sequence.

REFERENCES

1. **Altschul, S.F., W. Gish, W. Miller, E.W. Myers and D.J. Lipman.** 1990. Basic local alignment search tool. J. Mol. Biol. *215*:403-410.
2. **Doolittle, R.F.** 1986. Of ORFs and URFs: A Primer on How to Analyze Derived Amino Acid Sequences. University Science Books, Mill Valley, CA.
3. **Pearson, W.R.** 1990. Rapid and sensitive sequence comparison with FASTP and FASTA, p. 63-98. *In* R.F. Doolittle (Ed.), Methods in Enzymology, Vol. 183. Academic Press, San Diego.
4. **Pearson, W.R.** 1997. Identifying distantly related protein sequences. CABIOS *13*:325-332.

CHAPTER 24

Searching ESTs

Stuart M. Brown

NYU Medical Center, New York, NY, USA

Reprinted from BioTechniques 24:604-606 (April 1998)

A top priority in the multidisciplinary effort to sequence the human genome is the identification of all of the genes. Since only 2%-3% of human genomic DNA actually encodes proteins, it makes sense to focus sequencing efforts on these "expressed" regions first. This is the basis for a number of expressed sequence tag (EST) projects.

ESTs are short (200–500 bp) "single-pass" sequencing reads taken from both the 3′ and 5′ ends of cDNA clones. The sequence data are then fed directly from automated sequencers into the data banks with no error correction and only automatic annotation. The EST sequences are not complete genes nor complete cDNAs, but are merely gene "Tags." GenBank® contains over 1.4 million of these EST sequences. The EST section (dbEST) now constitutes more than half of GenBank and is growing much faster than the other sections.

It is likely that the vast amount of EST data contains at least part of virtually all expressed human genes. Exceptions most likely include genes expressed in tiny quantities, in rare specialized cell types or at restricted developmental periods. Various EST sequencing projects have addressed this issue by sequencing ESTs from a variety of tissues and developmental stages, but it is impossible to be absolutely comprehensive. Comparisons of EST data from various tissues can be used as a measure of tissue-specific expression patterns. There are some genes such as histones and non-coding RNA genes that are underrepresented in the EST databanks because their RNA transcripts lack poly(A) tails.

EST sequences are short and contain many errors, particularly frameshifts, which make the identification of open reading frames (protein coding regions) very difficult. The dataset is huge and highly redundant, containing thousands of copies of many common genes. Yet the elusive value of ESTs is clear: every gene is probably in there somewhere. Searches of EST databanks are an essential component of the search for new disease genes.

The Basic Local Alignment Search Tool (BLAST) similarity search engine, at the Web site of the National Center for Biotechnology Information (NCBI; http://www.ncbi.nlm.nih.gov/BLAST/), can be used to search for EST sequences that are similar to a specific query sequence. Since EST sequences are likely to contain many incorrect bases and insertion/deletions, it is best to use the new Gapped BLAST (BLAST Version 2.0) tool (http://www.ncbi.nlm.nih.gov/cgi-bin/BLAST/nph-newblast?Jform=0). Also, since protein-protein comparisons are much more sensitive than DNA-DNA comparisons, use

BioComputing: Computer Tools for Biologists
Edited by Stuart M. Brown

TBLASTN for protein query sequences (compares your protein to 6-frame translated ESTs) and TBLASTX for DNA query sequences (translates your query in six frames and compares it to 6-frame translated ESTs). In addition to BLAST searches of the entire dbEST database, NCBI offers subsets of mouse ESTs, human ESTs and "other ESTs" for searching. Do not be confused by other database options such as sequence tagged sites (dbSTS; short marker sequences for genome mapping projects), genome survey sequences (gss) or high-throughput genomic sequences (htgs); the data in these sections are not derived from cDNAs and are not directly comparable to ESTs.

When might a typical molecular biologist benefit from a search of EST data?

(*i*) If similarity searches of other databases have not yielded any significant matches, then the ESTs can be used as a databank of last resort. If a sequence matches nothing in the EST databank, then the chances are very good that it is not an expressed sequence.

(*ii*) The EST databanks contain new members of known protein families. Researchers who suspect that they have identified a new protein by function may discover candidates among ESTs that are similar but not identical to other known proteins.

(*iii*) EST sequences include many mutations and allelic variations of known genes. Investigators interested in gene diversity or seeking polymorphisms as genetic markers can obtain useful data from the EST databanks.

If an interesting EST sequence is identified in the databanks, the original cDNA clone from which that sequence was derived can be obtained from the IMAGE consortium for $45 for the first clone and $24 for each additional clone (http://www-bio.llnl.gov/bbrp/image/).

The redundancy of the EST databanks can be used to partially compensate for the high error rate and the fragmentary nature of the individual sequences. The reverse transcription reactions that are used to create cDNA libraries do not all run to completion. The pool of cDNA clones contains many partial sequences of varying lengths; therefore, ESTs databases will contain partially overlapping sequences from the same gene. Several groups are working on procedures to cluster all ESTs that are derived from a single gene. In this way, partial and fragmentary sequences can be assembled into contigs that correspond to whole gene transcripts. Thus, the vast, low-quality EST databanks can be collapsed into a single consensus set of nonredundant transcripts from all expressed human genes.

Investigators can conduct routine similarity searches against the EST clusters, which require only a tiny fraction of the disk storage space used by the complete EST databank; these searches can be computed much more quickly. Similarity matches to the EST clusters will be much more informative than matches to the current set of minimally annotated ESTs. Those wishing to examine tissue-specific gene frequencies or allelic sequence variations can extract the original EST sequences used to create the consensus from a central repository (i.e., dbEST at the NCBI Web site).

There are three different sets of clustered human ESTs available on the Web. Each set has been created with different software using different assumptions as to data quality, minimum overlap etc. The three sets of consensus sequences are available for download, so investigators can use their own searching software against these databases.

UniGene

The NCBI, home of GenBank and BLAST, has created UniGene, a collection of human EST sequences from GenBank that have been grouped into clusters by sequence similarity. Some of these clusters correspond to genes that have been identified by other means (i.e., functional cloning), while others are genes of unknown function. As of January 1998, UniGene contains 39377 clusters. This is a much more manageable set of sequences to search than the 1409359 sequences in the complete EST section of GenBank (January 1998). The UniGene collection is also indexed by tissue of origin, so it is possible to search for genes that are expressed in a particular tissue. For more information, see the http://www.ncbi.nlm.nih.gov/UniGene/index.html address.

TIGR Human Gene Index

The Institute for Genomic Research (TIGR) has created a Human Gene Index (HGI) for their own collection of ESTs (http://www.tigr.org/tdb/hgi/hgi.html). As of July 1997 (HGI Version 3.2), TIGR has identified 60388 "tentative human consensus" (THC) sequences by assembling over 600000 ESTs into complete "virtual transcripts". THCs are consensus sequences based on two or more ESTs that overlap for at least 40 bases with at least 95% sequence identity. THCs retain information about the source library and abundance of the ESTs from which they are composed. TIGR also provides a free Web-based similarity search service that can be used to compare query sequences against the HGI. Similarity results are returned by e-mail in less than 5 min. Clones corresponding to sequences in the TIGR EST database can be obtained from the American Type Culture Collection (ATCC, Rockville, MD, USA).

STACK

The South African National Bioinformatics Institute (SANBI) at the University of the Western Cape has created an EST consensus database known as Sequence Tag Alignment and Consensus Knowledgebase (STACK; http://ziggy.sanbi.ac.za/stack/). It consists of all available human EST sequences in GenBank, clustered, aligned and made into joined consensus records. STACK consensus sequences are often longer than TIGR or UniGene clusters. STACK creates separate consensus clusters for ESTs from each of 17 different tissue types.

SANBI provides a free Web-based BLAST search utility that can be used to compare query sequences against the STACK clusters. Similarity searches can be calculated as DNA-DNA (BLASTN), protein query against 6-frame translations of the STACK database (TBLASTN) or 6-frame translation of a DNA query against 6-frame translations of the STACK database (TBLASTX). There is also the option of returning search results by e-mail.

The EST databanks give researchers access to basic data from the genome project almost as soon as it is collected in the laboratory. This allows those investigators who are willing to sift through mountains of low-quality data to discover some hidden jewels. For most others, the pre-analyzed clustered EST databases will be much more useful. The promise of EST data continues to be "your gene is out there." You just have to have the patience and persistence to look for it and to be smart enough to know when you have found it.

Update to: Searching ESTs

Stuart M. Brown

NYU Medical Center, New York, NY, USA

The public EST database continues to grow rapidly, although the number of new human ESTs being submitted is pateauing while the numbers from the mouse, *Arabidopsis*, and others continue to increase. As of Novemver 2001, GenBank contained over 9.5 million ESTs, including 3.9 million human and 2.3 million from mouse. Overall, the majority of new sequences currently being contributed to the EST database are from organisms other than human and from various subtraction and screening experiments designed to identify differentially expressed genes.

The EST databases have become invaluable resources in the current effort to annotate complete genome sequences. Currently, the most reliable method to find genes in the assembled genomic sequences is by comparing genomic DNA to ESTs using BLAST or a similar tool. All of the major "Genome Browsers" (UCSC Golden Path, Ensembl, GenBank Genome Viewer, and Celera Genomics) now incorporarte similarity to EST sequences in the annotation of each gene on the assembled genome sequence. Incyte Genomics (Palo Alto, CA, USA) uses data from clusters of ESTs to predict alternate splice forms of genes.

However, some of the optimism about the completeness of EST databases has been tempered by the efforts to align clusters of ESTs to the genome. In many cases, more than one cluster aligns to a single genomic locus, but it is not possible to unambiguously determine alternative splicing patterns that could create these non-overlapping clusters. On the other hand, many EST clusters align with multiple locations on the genome—presumably due to gene duplications and the presence of pseudogenes. Many genes predicted by computer from the genomic sequence seem to be missing from the EST collections and the overall gene count for humans seems to be too low. A large number of genes may be expressed at such low levels or under such rare conditions of development, tissue specificity, and/or environmental response that they have never been detected by bulk EST sequencing projects.

CHAPTER 25

Advanced Similarity Searches on the Web: Gapped BLAST, PSI-BLAST, FASTA 3.0 and INCA

Stuart M. Brown

NYU Medical Center, New York, NY, USA

Reprinted from BioTechniques 25:214-216 (August 1998)

In the ever changing world of bioinformatics, the two most popular programs for sequence similarity searching, Basic Local Alignment Search Tool (BLAST) and FASTA, have both recently been improved. BLAST Version 2.0 is now available at the National Center for Biotechnology Information (NCBI) Web site, and FASTA 3.0 is available both as free software for most computer systems and on several Web servers. Iterative Neighborhood Cluster Analysis (INCA) is a Java™ applet that runs a series of BLAST searches to collect all sequences that are similar to a query sequence.

BLAST Gets Better

The NCBI, home of GenBank®, ENTREZ and BLAST, has introduced some major improvements in the BLAST program used to search protein and DNA databases for sequence similarities. The BLAST Web site (http://www.ncbi.nlm.nih.gov/BLAST/) now offers two new options, Gapped BLAST and Position-Specific Iterated (PSI)-BLAST (1). These new options represent significant changes in the BLAST algorithm, but the upshot for the researcher is more sensitivity to weak, but biologically meaningful, sequence similarities.

The original BLAST program does not allow for gaps (insertions or deletions) within pairs of aligned sequences. As a result, if two sequences share extensive regions of similarity, but have several gaps, BLAST only identifies the longest ungapped region of the match or produces separate alignments and similarity scores for segments of the alignment that are separated by gaps. The longest ungapped region might not provide a statistically significant similarity score. Gapped BLAST allows the introduction of gaps (insertions and deletions) into alignments. Gapped BLAST produces longer continuous alignments rather than multiple short segments. Also, the scoring of gapped results tends to be more biologically meaningful than ungapped results. The Web interface for BLAST 2.0 has also been improved. At the top of the *BLAST Results* output, there is now a bar graph showing the regions of alignment of each matching sequence with the query sequence and with color coding for the quality of the match.

BioComputing: Computer Tools for Biologists
Edited by Stuart M. Brown

The basic BLAST algorithm compares two sequences by breaking them into sets of short words and then looks for close matches between pairs of words from the two sequences. Wherever statistically significant matches are found between word pairs, BLAST tries to create an alignment by extending the match in both directions. The new gapped BLAST algorithm requires matches between two different pairs of words located near each other on the two sequences before it tries to create an alignment. Because the chance of randomly finding two different matching words between two unrelated sequences is much lower than the chance for a single random match, the gapped BLAST algorithm is able to use a less stringent cutoff score for considering two similar words to be a match. The net result is a more sensitive search with less computer time wasted trying to extend random word matches, as well as better, longer and more meaningful alignments.

In addition to improving sensitivity, generating longer aligned regions and improving the predictive power of the statistical scores, gapped BLAST is about three times faster than the traditional BLAST algorithm. This feat of mathematical wizardry is very important because the rate of growth in the GenBank database currently exceeds the rate of improvement in computer processor speeds (1).

There is another new tool available on the BLAST Web site called PSI-BLAST. PSI-BLAST uses groups of aligned sequences to improve the sensitivity of similarity searching. PSI-BLAST first performs an initial gapped BLAST search of the database. Then, it makes a multiple alignment from the significant hits. The multiple alignment is then used to construct a position-specific scoring matrix (a table of amino acid frequencies at each position in the sequence). This matrix, also known as a profile, is then used instead of a query sequence for another BLAST search. New alignments found in the second search are incorporated into the scoring matrix, and the process is repeated (iterated) until no more significant hits are found. Remarkably, this entire process uses the same amount of computer time as a series of BLAST searches equal to the number of times the matrix search is repeated.

PSI-BLAST is able to find highly diverged members of protein families. This is much faster and easier than the procedure that many investigators follow for collecting meaningful hits from a BLAST (or FASTA) search, building a multiple alignment and then creating profiles to search for additional similar protein sequences. This remarkable PSI-BLAST program is very new and still in its evaluation phase, but it could have a major impact on the process of identifying the function of new genes as they are sequenced by the Genome Projects. You can use the new BLAST programs on the NCBI's Web server (http://www.ncbi.nlm.nih.gov/BLAST/).

FASTA 3.0 Copes with Frameshifts

It is almost always advantageous to perform similarity searches with protein rather than DNA sequences. This is true for three reasons: (*i*) the information content per residue of protein sequence is greater than DNA because the amino acid alphabet is much larger than the DNA alphabet (20 vs. 4 letters); (*ii*) nonidentical amino acids can be scored for similarity using a mutation frequency matrix such as PAM or BLOSUM; and (*iii*) the protein databanks are much smaller than the DNA databanks, so the likelihood of chance matches is reduced.

Both the BLAST and FASTA programs have an automatic translation feature, so that a DNA sequence can be translated, and then the resulting amino acid sequence can be used as a query sequence against protein databanks. However, certain types of DNA sequence data, such as expressed sequence tags (ESTs) and raw reads from automated sequencers, contain frameshifts that interfere with the automatic translation into protein.

The result is that a similarity search is only performed for the largest piece of the sequence that does not contain a frameshift.

The problem of similarity searching with DNA sequences that contain frameshifts has been addressed by Dr. William Pearson in his new version of the FASTA program (3). FASTA 3.0 now contains the program TFASTX, which compares a DNA sequence to a protein database, translating in three frames and allowing for frameshifts. FASTA 3.0 uses the same methods and statistics as FASTA 2.0, but it is designed in a modular fashion and runs not only on conventional UNIX® workstations but also on multiprocessors in parallel.

FASTA 3.0 is available for searching GenBank and SwissProt at several Web servers. The DNA DataBase of Japan (DDBJ) Web site offers the complete FASTA 3.0 package (http://watson.genes.nig.ac.jp:8080/fasta_blast/welcome-e.shtml) including the programs FASTA, TFASTA, FASTX and TFASTX as well as all of the command line options including choice of amino acid scoring matrixes, ktup and the number of alignments reported. The European Bioinformatics Institute (EBI) also has a FASTA 3.0 Web server (http://www2.ebi.ac.uk/fasta3/) or (http://croma.ebi.ac.uk/fasta3/).

Users who wish to run FASTA 3.0 on their own computers can obtain the program code from Dr. Pearson's anonymous File Transfer Protocol (FTP) server (ftp://ftp.virginia.edu/pub/fasta/).

Users of the Wisconsin Sequence Analysis Package (GCG) should take note that the current version of GCG (Version 9.1) uses FASTA 2.0, but it is possible to install FASTA 3.0 on the same computer that runs GCG so that the two programs can share the same databases.

INCA

A single BLAST or FASTA search with one query sequence will often fail to find all of the members of a protein family (or some other group of related sequences) because not all of the sequences in the group are significantly similar to the initial query. However, if each of the sequences that do have significant similarity to the initial query are used as queries for additional searches, more members of the family will be found. This process can be repeated (iterated) using sequences identified in the second round of searching and repeated again until no new similar sequences are found. Since homology is transitive, if sequence A is homolgous to sequence B, and B is homolgous to C, then A is homolgous to C.

INCA is an automated way of doing such a series of BLAST searches (2). INCA runs a BLAST search on a given query sequence. Then INCA runs BLAST on each of the significantly similar sequences found in the first iteration. All of the significant matches from the searches are saved in a composite cluster. INCA is a Java applet that runs inside a Web browser (a Java-enabled Web browser such as Netscape 4.0 or Internet Explorer 4.0 is required). Access INCA on the Web site (http://itsa.ucsf.edu/~gram/home/inca/).

REFERENCES

1. **Altschul, S.F., T.L. Madden, A.A. Schäffer, J. Zhang, Z. Zhang, W. Miller and D.J. Lipman.** 1997. Gapped BLAST and PSI-BLAST: a new generation of protein database search programs. Nucleic Acids Res. *25*:3389-3402.
2. **Graul, R.C. and W. Sadée.** 1997. Evolutionary relationships among proteins probed by an iterative neighborhood cluster analysis (INCA). Alignment of bacteriorhodopsins with the yeast sequence YRO2. Pharm. Res. *14*:1533-1541.
3. **Zhang, Z., W.R. Pearson and W. Miller.** 1997. Aligning a DNA sequence with a protein sequence. Comput. Biol. *4*:339-349.

Update to: Advanced Similarity Searches on the Web: Gapped BLAST, PSI-BLAST, FASTA 3.0 and INCA

Stuart M. Brown

NYU Medical Center, New York, NY, USA (browns02@med.nyu.edu)

The vast expansion of genomic DNA sequences in the public databanks has made it more important than ever to use protein similarity searching whenever possible. In a casual search of GenBank with a cDNA sequence generated in the lab using the NCBI BLASTN server using the default **nr** (non-redundant) DNA database, a dozen or more of the top-scoring hits may be genomic sequence contigs with no other useful annotation. This is particularly true for human sequences, but *Drosophila*, *Caenorhabditis elegans*, and mouse genomic sequences are also filling up the DNA databanks. Genome derived "hypothetical" sequences can also be found in the NCBI's **nr** protein database, but fewer matches will be found and many of these hypothetical sequences have at least some annotation as to their function or similarity to other proteins (or domains) with known function.

In addition to searching protein databases whenever possible, it is also useful to search smaller databases, such as those restricted to a single species. The NCBI BLAST server allows you to choose yeast, *Escherichia coli*, and *Drosophila* genomes instead of the **nr** database from a "Choose Database" dropdown menu located right above the "**BLAST!**" button that you click to start the search. A bit farther down the page is another dropdown menu that allows you to choose any one of 27 different species or major taxonomic groups such as viruses, bacteria, eukaryotes, mammals, etc. You can also specify your own database subset using any valid Entrez query terms including AND\OR\NOT Boolean operations. All of these advanced options are now combined into a single BLAST Web page rather than dividing them up on "Basic" and "Advanced" pages.

The NCBI has also set up specialized BLAST Web pages for "genome" sequences. There is a page dedicated to human genome searches (http://www.ncbi.nlm.nih.gov/genome/seq/page.cgi?F=HsBlast.html&&ORG=Hs) and a similar page for the *Arabidopsis thaliana* genome (http://www.ncbi.nlm.nih.gov:80/PMGifs/Genomes/ara.html). There is another page where you may choose any (or any group) of 37 different eukaryotic genome data sets, in their current unfinished status (http://www.ncbi.nlm.nih.gov/cgi-bin/Entrez/genom_table_cgi?organism=euk) and a similar page for searching 129 bacterial and 14 archaeal genomes (http://www.ncbi.nlm.nih.gov/cgi-bin/Entrez/genom_table_cgi).

Elsewhere on the Web, many single species genome projects have set up BLAST servers that allow a variety of specialized searches on their own data. Many of these contain data that has not been submitted to GenBank such as preliminary sequencing reads, incomplete contigs, preliminary protein predictions, etc.

Drosophila melanogaster genome project: Flybase
 http://flybase.bio.indiana.edu/

C. elegans Genome Database: Wormbase
http://www.wormbase.org/
Mouse Genome Informatics (Jackson Laboratory)
http://www.informatics.jax.org/mgihome/
Arabidopsis Genome Initiative, MIPS Arabidopsis Database (MATDB)
http://mips.gsf.de/proj/thal/db/index.html
The official database of the *Plasmodium falciparum* Genome Project: PlasmoDB
http://plasmodb.org/
The Institute for Genome Research (TIGR) maintains searchable genome databases for hundreds of organisms
http://www.tigr.org/tdb/
The Genome Web at the UK Human Genome Mapping Project Resource Centre maintains a set of links to hundreds of genome databases
http://www.hgmp.mrc.ac.uk/GenomeWeb

The improvements in BLAST software over the past 2 to 3 years have led to a dramatic increase in its popularity over FASTA. Free FASTA servers on the Web are becoming harder to find.

Fasta 3.0 servers on the Web:

DNA DataBase of Japan (DDBJ)
http://spiral.genes.nig.ac.jp/homology/fasta-e.shtml
National Institute of Agrobiological Sciences, Tsukuba, Japan
http://www.dna.affrc.go.jp/htdocs/Blast/FastA.html
GenomeNet, Kyoto University, Japan
http://www.fasta.genome.ad.jp
GENESTREAM II network server, Institut de Génétique Humaine, Montpellier, France
http://www2.igh.cnrs.fr/bin/fasta-guess.cgi
Institut Pasteur, France
http://bioweb.pasteur.fr/seqanal/interfaces/fasta.html
Infobiogen, France (in French)
http://www.infobiogen.fr/services/analyseq/cgi-bin/fasta_in.pl
Indiana University Biology Department, USA
(a FASTA search of subsets of GenBank defined by species or keyword)
http://iubio.bio.indiana.edu/srsfasta/

CHAPTER 26

BLAST on the Web

Bruno A. Gaëta

eBioinformatics, Sydney, Australia

Reprinted from BioTechniques 28:436-440 (March 2000)

The last few decades have seen an explosion in the amount of DNA and protein sequence available through public databases. The major nucleotide sequence databases double in size approximately every 14 months, and the number of completely sequenced organism genomes is constantly increasing. It is now impractical for scientists to maintain local copies of the sequence databases and, as a result, the Internet is now the main avenue of access to this information.

Sequence databases are typically searched using keywords or a sequence. Keyword search engines such as the Entrez system provided by the National Center for Biotechnology Information (http://www.ncbi.nlm.nih.gov) search the annotation section of the sequence database record. The major application of these databases involves sequence similarity searching, where sequences similar to a query sequence supplied by the investigator are identified in a sequence database. In most cases, this operation identifies potential homologues of the query sequence, that is, database sequences sharing a common evolutionary history with the query sequence. Homologous sequences can provide a clue as to the function of the query sequence, its evolutionary history and its structure. For example, if an unknown protein sequence is found to be markedly similar over its whole length to the sequence of a protein of known structure in the database, the two proteins are likely to have similar structures and related functions. Other applications include the identification of coding regions in uncharacterized genomic DNA by looking for DNA regions that can be translated into an amino acid sequence similar to that of a known protein.

BLAST (Basic Local Alignment Search Tool) (1) is the most commonly used suite of programs for sequence database similarity searching. It allows rapid searching for sequences similar to a query sequence. The BLAST programs provide access to a large number of biological sequence databases all over the Internet, including central repositories of public sequence data such as NCBI, the Japanese GenomeNet server (http://www.blast.genome.ad.jp/) and EBI, the European Bioinformatics Institute (http://www.ebi.ac.uk). BLAST is also used to provide access to genome sequence data provided by large-scale sequencing centers such as the Institute for Genome Research (http://www.tigr.org) and the Sanger Centre (http://www.sanger.ac.uk). The package is also used as a front-end to smaller collections of sequence data such as organism-specific genome databases, for example, the HIV sequence database at http://hiv-web.lanl.gov/index.html. Web-based sequence analysis service providers such as Bio-

BioComputing: Computer Tools for Biologists
Edited by Stuart M. Brown

Navigator (http://www.bionavigator.com) also use the BLAST programs as a central method for accessing their collections of sequence data. A partial list of publicly available BLAST servers is available at http://gifts.univ-mrs.fr/BLASTULA/Blastula_ links.html.

All these sites make use of BLAST, but the implementation of the package can vary greatly between different sites. Differences include the version of the program used, the choice of programs available and the optional program parameters that can be modified by the user.

BLAST searches for similar sequences in the database and creates alignments between the query sequence and the matching sequences. Version 1 of the BLAST package produces ungapped alignments, but later versions create alignments that can contain gaps and provide a better comparison of sequences that diverged through residue substitution and also insertion and deletion mutations. There are two main implementations of this method: the gapped BLAST (2) developed at NCBI, and WU-BLAST (http://blast.wustl.edu/), which provides access, for example, to the databases maintained at the EBI. Both implementations produce gapped alignments but use a slightly different approach and set of parameters. The database to be searched will usually dictate the version of BLAST that can be used since most Web sites maintain only one version of the package.

The database to be searched will also dictate which BLAST programs can be used. There are three BLAST programs for searching nucleotide sequence databases (**blastn**, **tblastn** and **tblastx**) and two programs for searching protein sequence databases (**blastp** and **blastx**; the more recently developed PSI-BLAST and PHI-BLAST programs that are variations on **blastp** are not discussed here).

The first program, **blastn**, searches a nucleotide sequence database using a nucleotide sequence. It is the only BLAST program to produce nucleotide sequence alignments. Typical uses of **blastn** include checking whether a newly obtained DNA sequence has already been published, identifying coding regions in genomic DNA by searching for encoded (or related) mRNA sequences and extending sequence fragments by searching collections of partially characterized sequences such as EST (Expressed Sequence Tags) databases. **Blastp** compares a protein sequence to a protein sequence database. This program is typically used to identify new protein sequences, retrieve homologues of a given protein, identify the domain structure of a protein or determine the structure of a protein by identifying related proteins of known structure. **Blastx** translates a nucleotide sequence into protein and searches a protein sequence database. It is particularly useful for identifying potential coding regions in newly sequenced DNA. **Tblastn** uses a protein query sequence to search a translated nucleotide sequence database and is useful for detecting unidentified proteins encoded in a nucleotide sequence database. Finally, **tblastx** compares a nucleotide sequence with a nucleotide sequence database, after translating both sequences into proteins. This approach greatly increases the sensitivity of the search when dealing with protein-coding regions but is substantially slower than the direct comparison performed by **blastn**.

Internet BLAST Access

Different sites on the Internet provide access to different databases and to different sets of BLAST programs. For example, the EBI provides access to **blastp**, **blastn** and **blastx**, but not to **tblastn** and **tblastx**. The Sanger Centre only offers access to nucleotide databases and only to **blastn**, **tblastn** and **tblastx**. NCBI maintains both nucleotide and protein sequence databases and so provides access to all five programs. However, to limit the load on the server computer, NCBI does not provide Web access to its large nonredundant nucleotide sequence database using the computer-intensive **tblastx**. An alternative is provided by private sequence analysis service providers such as BioNavigator, which allows **tblastx** searching of its nonredundant nucleotide database.

When planning a database search, first decide which database to search. For most common applications such as checking whether a sequence is new, identifying the function of an unknown sequence or collecting homologous sequences for further analysis, the best starting point is a nonredundant database that brings together the majority of the available data. These databases are maintained at a number of Web sites around the world including NCBI, EBI, GenomeNet and BioNavigator. Which site to choose depends on personal preferences regarding the user interface and the available output options. For example, the NCBI BLAST server includes a useful graphical overview of the results in the output page, and the BioNavigator BLAST output viewer makes it easy to save the database sequences identified by the BLAST search for further analysis.

When searching specialized databases and database subsets, the choice of BLAST servers is frequently more limited. Users will often have only one option available on the Web if they want to access a specific set of data (e.g., the TIGR gene indices that are available for searching only on the TIGR Web site). The situation is especially complicated when searching for ESTs because there are many proprietary EST collections available only at specific sites on the Web, in addition to the major public domain collection maintained at NCBI.

BLAST Parameter Configuration

The popularity of BLAST has made it one of the most reliable sequence analysis software packages currently available, and one that will perform adequately in most cases without modification of its default parameter configuration. This is why many BLAST Web sites do not allow users to change the program parameters, with the exception of the database to be searched or the BLAST program to be used. In some cases, however, it may be useful to change some of the parameter values to improve the results when the option is available.

The default BLAST configuration works well in general with query sequences of medium length (100–1000 nucleotides or 50–200 amino acids). Searching with longer query sequences can, on occasion, retrieve preferentially longer sequences from the database, which may de-emphasize shorter significant matches. This can present a problem when the aim of the search is to identify significant short regions of similarity in a longer sequence (e.g., to determine the domain structure of a new protein or identify short exons in a comparatively long genomic DNA fragment). For these types of searches, try splitting long query sequences into shorter overlapping fragments (1000 nucleotides or 200 amino acids each) and use each as a separate BLAST query to see if interesting additional hits are identified in the database.

BLAST searches using short query sequences may commonly not report any matches in the database. This is because BLAST reports only significant matches that are unlikely to be the result of random sequence similarity. Since matches with a short sequence are likely to have occurred by chance, the program does not report them. While this is useful when searching for homologous sequences, it prevents using BLAST for searching for potential candidates matching a short protein sequence (such as one obtained by amino-terminal sequencing of an unknown spot on a two dimensional protein gel). A notable application of BLAST with short sequences involves checking whether PCR primers are likely to match many sequences in a genomic DNA preparation and give rise to the wrong PCR products. Using the primer sequence as a query in a **blastn** search of a large nucleotide sequence database could do this, but since the matches would be happening by chance, they would be unlikely to be reported by the program. When a search with a short sequence does not retrieve any (or enough) database hits, a workaround is to raise the value of the Statistical Expectation Threshold parameter (**Expect** on the NCBI page, **MAX.EXP** on the EBI WU-BLAST server). This controls the level of statistical significance that will be reported by the program. The default value for this parameter is 10, corresponding to a cutoff level of similarity where 10 reported database matches are expected to have been found by chance alone. Raising the threshold will cause the program to report more chance matches and may provide an answer in this case. Note that on many BLAST servers, the maximum expectation threshold value that can be selected from the Web form is 1000, which is not enough for some applications involving short query sequences (some servers allow this limit to be bypassed by typing a line command). Raising the statistical expectation is the first thing to do when a search does not report any matches when some were expected. However, remember that most matches with a high statistical expectation score are more likely to be due to chance sequence similarity than to biological relationships.

Another parameter available on most BLAST servers lets the user control whether the query sequence should be filtered before the search. Filtering removes from the query sequence regions of limited amino acid composition or simple repeats. These "low complexity" regions are of concern because they can cause a match to be identified as more significant than it really is or to retrieve many false positives. As a consequence, most BLAST servers filter query sequences by default to mask these regions before starting the search. A low complexity region that has been masked can appear as a string of Ns or Xs (protein BLAST programs) in the query sequence within the sequence alignment region of the BLAST output. In some cases (if the low complexity region is at the beginning or the end of the query sequence), a filtered region may not be shown at all in the BLAST output.

Since filtering the query sequence effectively removes some data from the search, it is often useful to repeat a search with the filtering turned off to obtain better sequence alignments or to check whether the low complexity region adds essential information to the search query. This is particularly important if a search retrieves no matches with filtering turned on.

Most BLAST servers also allow users to change the scoring matrix, or substitution matrix, for protein-protein comparison. This matrix is a table assigning a score to all possible pairings of amino acids in an alignment and controls how the program handles the likelihood that one amino acid can substitute for another between homologous proteins. Most of the modern substitution matrices have been derived empirically, and while the default matrix, BLOSUM62, works best with the majority of sequences, other matrices may perform better when trying to detect distant homologies in particular cases. Basically, the lower the number associated with a BLOSUM matrix, the more distant homologies it can detect, but the more nonrelated sequences it is likely to report. For example, BLOSUM45 is thought to perform better than BLOSUM62 for detecting distant homologies. This rule is reversed for the PAM family of matrices: the PAM matrices associated with low numbers, such as PAM30, are more suited to detecting closely related sequences, whereas PAM250 is better at detecting distant homologies (again, with a corresponding increase in false positives). Overall, the BLOSUM matrices that are derived from the

BLOCKS database of conserved protein regions (3) are considered better performers than the older PAM matrices, which are derived from a smaller set of aligned globular proteins (4). However, many servers still provide access to PAM matrices because they may perform better in some cases, especially with short query sequences.

While BLAST contains more parameters that can be configured to improve its sensitivity or performance in specific cases, keep in mind that this method was designed first and foremost for speed. It may be useful to tweak some parameters in specific cases and experiment with different values, but for critical searches it is also a good idea to try a slower, more rigorous search method such as the Smith-Waterman algorithm. Implementations of this method include the program **ssearch** in the fasta package (5) available on the GeneStream server (http://vega.crbm.cnrs-mop.fr/) and BioNavigator, and programs running on specialized hardware (http://www.irisa.fr/SAMBA, http://www.ch.embnet.org/software/GMFDF_form.html, http://www2.ebi.ac.uk/Bic/ and http://decypher.stanford.edu/expsw.htm).

REFERENCES

1. **Altschul, S.F., W. Gish, W. Miller, E.W. Myers and D.J. Lipman.** 1990. Basic local alignment search tool. J. Mol. Biol. *215*:403-410.
2. **Altschul, S.F., T.L. Madden, A.A. Schäffer, J. Zhang, Z. Zhang, W. Miller and D.J. Lipman.** 1997. Gapped BLAST and PSI-BLAST: a new generation of protein database search programs. Nucleic Acids Res. *25*:3389-3402.
3. **Henikoff, S. and J.G. Henikoff.** 1992. Amino acid substitution matrices from protein blocks. Proc. Natl. Acad. Sci. USA *89*:10915-10919.
4. **Schwartz, R.M. and M.O. Dayhoff.** 1978. Matrices for detecting distant relationships, p. 353-358. *In* M.O. Dayhoff (Ed.) Atlas of Protein Sequence and Structure, vol. 5, suppl. 3, Natl. Biomed. Res. Found., Washington, DC.
5. **Pearson, W.R. and D.J. Lipman.** 1988. Improved tools for biological sequence analysis. Proc. Natl. Acad. Sci. USA *85*:2444-2448.

Update to: BLAST on the Web

Bruno A. Gaëta

University of New South Wales, Sydney, Australia (bagaeta@fastmail.fm)

Since the publication of this article, a number of the World Wide Web (WWW) sites cited have moved, changed, or are no longer available: National Center for Biotechnology Information (NCBI) (http://www.ncbi.nlm.nih.gov/) has updated its WWW site to offer additional variations on Basic Local Alignment Search Tool (BLAST), including MEGABLAST, which searches for highly similar nucleotide sequences and can handle very long query sequences, and specially configured versions of blastn and blastp, for searching for exact matches to short query sequences (e.g., oligonucleotide primer sequences and short peptide fragment sequences). NCBI now also allows tblastx searches of its nonredundant nucleotide database.

The Sanger Centre (http://www.sanger.ac.uk) is now called the Wellcome Trust Sanger Institute and offers access to both protein and nucleotide databases through blastn, blastp, blastx, tblastn, and tblastx. The Sanger Institute also hosts the Ensembl genome resource that provides blastn, tblastn, and tblastx searches of the human genome and other genomes.

The BLASTULA listing of BLAST servers no longer appears to be available. Alternative listings of BLAST servers are available at numerous other sites, e.g., (http://mbcr.bcm.tmc.edu/homology.html).

The BioNavigator service is no longer generally available. An alternative integrated site for sequence analysis (including BLAST) is PISE at the Pasteur Institute (http://bioweb.pasteur.fr/seqanal/blast/intro-uk.html). BioNavigator (renamed BioManager) is available in Australia and limited to international subscribers through ANGIS (http://www.angis.org.au).

The Genestream server (http://vega.crbm.cnrs-mop.fr/) and the Stanford Decypher service (http://decypher.stanford.edu/expsw.htm) for searching using the Smith-Waterman algorithm no longer appear available. Accelerated Smith-Waterman database searches are still available at (http://www.irisa.fr/SAMBA/), (http://www.ch.embnet.org/software/GMFDF_form.html), and European Bioinformatics Institute (EBI) (new uniform resource locator [URL]) (http://www.ebi.ac.uk/MPsrch/).

CHAPTER 27

BLAST...What Next?

Bruno A. Gaëta

eBioinformatics, Sydney, Australia

Reprinted from BioTechniques 28:874-876 (May 2000)

The vast amount of sequence information available through public and private sources is changing the face of biology. To access this information, researchers must become familiar with the computer tools available to analyze sequences, many of which are distributed on the Web. BLAST (1,2) allows investigators to identify sequences in a sequence database that are similar to a supplied sequence. However, this search is often only the first step in a more complex series of analyses in which the matches identified by the search are processed to yield additional information on the query sequence and its biological relationships.

Result Visualization

The first BLAST output processing programs added no information to the BLAST results but made them easier to view. The standard output consists of an often long list of matching database sequences, followed by pairwise alignments between the query sequence and each of the best matching database sequences. Browsing this long text file is made easier by the hypertext links between the database hits in the list and their corresponding pairwise alignments. The description of each hit can also be linked with the original database entry; both features are now standard for BLAST.

Some implementations of BLAST provide alternative views of the results, besides the standard output view. For example, the BLAST server at the National Center for Biotechnology Information (NCBI) (http://www.ncbi.nlm.nih.gov) offers the option to replace the series of pairwise alignments with one large multiple sequence alignment of all matching database sequences and the query sequence. The BLAST server provided by the National Center for Genome Resources (http://seqsim.ncgr.org/newBlast.html) saves the BLAST output as a tab- or comma-delimited text file that can be imported directly into a spreadsheet for display, filtering and further processing.

MSPcrunch (6) is another output processor designed for high-throughput analysis of large genomic sequences. Unlike programs that simply reformat the BLAST output, MSPcrunch filters the results and recalculates some of the scores to emphasize hits that may be biologically significant. It also filters out alignments of poor biological significance caused by regions of biased composition and combines small but significant

BioComputing: Computer Tools for Biologists
Edited by Stuart M. Brown

ungapped alignments into a larger, more biologically relevant gapped alignment. The program also provides a "big picture" output option to represent the alignments graphically, together with a range of output formats suitable for entering into semiautomatic genome annotation tools and databases. MSPcrunch is available on the Web at the Pasteur Institute (http://bioweb.pasteur.fr/seqanal/blast/intro-uk.html) and the Telethon Institute of Genetics and Medicine (http://www.tigem.it/LOCAL/BLAST/blast.html). The program is also available at http://www.embnet.se/new/mspup.html, where users can enter their own BLAST outputs generated with other BLAST servers. Note that the current version of MSPcrunch is not compatible with some versions of BLAST, including the gapped BLAST version 2 available at NCBI, which already provides gapped alignments and a graphical display.

Multiple Sequence Alignments

The NCBI gapped BLAST interface makes it possible to reformat the standard output to produce a multiple sequence alignment of all matched sequences—either as full sequences or showing only the differences between the database sequences and the query sequence. However, this alignment is not in a format that can be input easily into other programs for further analysis and is therefore suitable mainly for display and visualization. On the other hand, MView (3), a multiple sequence alignment conversion, viewing and reformatting tool, can take a BLAST output and convert it into a multiple sequence alignment in a range of formats for display on a Web page or for input into other analysis programs. MView is available on the Web through the Pasteur Institute Web server (http://bioweb.pasteur.fr/seqanal/interfaces/mview_blast.html) and has also been integrated into the PIR Integrated Sequence Analysis Environment (http://www-nbrf.georgetown.edu/pirwww/search/piriesa.html) for reformatting the output of BLAST searches of the PIR database. Integrated Web-based bioinformatics analysis systems such as BioNavigator (http://www.BioNavigator.com) allow the researcher to save the database hits so they can be entered directly into separate multiple sequence alignment programs such as ClustalW (7). Instead of using the alignments in the BLAST output, the alignment program recalculates the alignments and aligns the database sequences rather than just the matching sequence segments found by BLAST.

Phylogeny

Multiple sequence alignments are commonly used as data sets for the construction of phylogenetic trees. Using an integrated platform such as BioNavigator or the Pasteur Institute's PISE interface (http://bioweb.pasteur.fr/intro-uk.html) makes it easy to run a BLAST search and extract the matching database sequences, then align them to produce phylogenetic trees. (A number of programs and methods are available on different Web sites for constructing the phylogenetic trees from the alignment.) Note that all tree-building programs make certain assumptions and that using these programs without being aware of their underlying model can result in misleading phylogenetic trees. A BLAST server has also been set up at http://www.Bork.EMBL-Heidelberg.DE/Blast2e/ for differentiating orthologues (homologous genes or proteins in different species) and paralogues (homologues in the same species that are the result of gene duplication) using a phylogenetic approach. This server performs a BLAST search and offers the option of creating a multiple sequence alignment from the results and deriving a phylogenetic tree from the alignment. The tree can be compared to the known phylogeny of the species from which the sequences were obtained. In principle, a phylogenetic tree constructed from orthologous sequences should be identical to the species tree.

Profiles and Hidden Markov Models

A multiple sequence alignment can also be used to construct a mathematical model such as a sequence profile (also called position-specific scoring matrix or PSSM) or a hidden Markov model (HMM) of the group of sequences in the alignment. If the sequences all belong to the same protein family, the model can be used as a representation of this family and can be used to search a sequence database for new members of the family or to test if a new sequence belongs to the family. Searching for new members of a sequence family with a profile or an HMM is more sensitive than searching with an individual member of this family because the model brings together information from many sequences and is less affected by individual variation in sequences.

The easiest way to use profiles for database searching is to make use of the program PSI-BLAST (position-specific iterated BLAST) (2) developed at NCBI. PSI-BLAST starts with a standard blastp search of a protein database, and sequences with sufficient similarity to the query are then aligned together and used to build a profile to search the database again for new matching sequences. The new matching sequences can be added to the profile and the search repeated until no new sequences are identified. PSI-BLAST is available from an increasing number of Web sites including NCBI, BioNavigator and the DNA database of Japan (http://crick.genes.nig.ac.jp/homology/welcome-e.shtml). PSI-BLAST at NCBI also allows the program to be run interactively, with users selecting sequences to be added to the profile after each round of searching. It is also possible to perform the same type of analysis manually, using a different type of model for the sequence family. An integrated platform makes it much easier to go from a BLAST output to a multiple sequence alignment. It then creates the model from the multiple sequence alignment and finally searches a database using the model to identify new members of the protein family. For example, BioNavigator provides a Web front end to the HMMer package (4), which builds HMMs from multiple sequence alignments and then uses them to search sequence databases.

Sequence Motifs

Many sequence families, domains and functions can be identified using specific combinations of conserved residues that form recognizable patterns or motifs. Motif databases, such as PROSITE (available at http://www.expasy.ch), collect protein motifs to facilitate sequence identification. Searches for sequence patterns can yield information complementary to that provided by a BLAST search, and the two approaches can be combined when identifying the function of new proteins.

BEAUTY (8) is a program based on BLAST that allows the researcher to identify protein sequences using both sequence similarity information and information about conserved patterns and motifs present in the query sequence and the matching database sequences. The program performs a BLAST search of databases of conserved protein regions and motifs and reports which regions of the query sequence matched known motifs and conserved domains. Investigators can then decide whether or not the query sequence contains all the conserved sequence patterns required for performing a particular biological function and can provide a much stronger indication of the function of a protein or a domain than simple sequence similarity. BEAUTY is available through the Baylor College of Medicine Search Launcher (http://dot.imgen.bcm.tmc.edu:9331/seq-search/protein-search.html).

Another BLAST variant combining sequence similarity and searches for specific motifs is PHI-BLAST (pattern hit initiated BLAST) (9) available through the NCBI Web site. PHI-BLAST requires as input both a query sequence and a particular sequence pattern present in this sequence. PHI-BLAST searches for sequences containing the input sequence pattern that are similar to the query sequence around this pattern. This approach is more specific than just searching for the conserved motif, which may match many false positives in the database.

Protein Structure

A common approach to predict the structure of a protein from its sequence is to use the sequence as query in a search of a database of proteins of known structure. The structure of matching sequences can then be used as a guide for creating a model of the structure of the unknown protein. This approach works well when the query protein has close homologues of known structure; it is the basis for the SWISS-MODEL (5) automated structure prediction system (http://www.expasy.ch/swissmod/SWISS-MODEL.html). The Prodom database of protein domains (http://www.toulouse.inra.fr/prodom.html) provides a BLAST server to search for domains contained in a query sequence and allows this information to be passed directly to the SWISS-MODEL server for subsequent protein structure prediction. BioNavigator takes a more manual approach to structure prediction using the same principle: the researcher can run a BLAST search of the PDB protein structure database, then enter the query sequence and its matching structure into other programs for building a structural model of the query protein.

BLAST is most commonly used as a stand-alone program but can also be the basis for more involved sequence analysis. This analysis can be carried out using dedicated programs such as BEAUTY and MSPcrunch or by extracting the database hits identified by the BLAST search and entering them into other sequence analysis programs. Integrated bioinformatics packages such as BioNavigator greatly facilitate the latter task compared with standard BLAST interfaces available on the Web, which require users to copy and paste or download individual hits one by one. These integrated services also offer a large collection of different methods for analyzing the sequences identified by BLAST.

REFERENCES

1. **Altschul, S.F., W. Gish, W. Miller, E.W. Myers and D.J. Lipman.** 1990. Basic local alignment search tool. J. Mol. Biol. *215*:403-410.
2. **Altschul, S.F., T.L. Madden, A.A. Schäffer, J. Zhang, Z. Zhang, W. Miller and D.J. Lipman.** 1997. Gapped BLAST and PSI-BLAST: a new generation of protein database search programs. Nucleic Acids Res. *25*:3389-3402.
3. **Brown, N.P., C. Leroy and C. Sander.** 1998. MView: a Web compatible database search or multiple alignment viewer. Bioinformatics *14*:380-381.
4. **Eddy, S.** 1998. HMMER User's Guide. Biological sequence analysis using profile hidden Markov models.

http://hmmer.wustl.edu/

5. **Guex, N., A. Diemand and M.C. Peitsch.** 1999. Protein modelling for all. Trends Biochem. Sci. *24*:364-367.
6. **Sonnhammer, E.L. and R. Durbin.** 1994. A workbench for large-scale sequence homology analysis. Comput. Appl. Biosci. *10*:301-307.
7. **Thompson, J.D., D.G. Higgins and T.J. Gibson.** 1994. CLUSTAL W: improving the sensitivity of progressive multiple sequence alignment through sequence weighting, position-specific gap penalties and weight matrix choice. Nucleic Acids Res. *22*:4673-4680.
8. **Worley, K.C., B.A. Wiese and R.F. Smith.** 1995. BEAUTY: an enhanced BLAST-based search tool that integrates multiple biological information resources into sequence similarity search results Genome Res. *5*:173-184.
9. **Zhang, Z., A.A. Schäffer, W. Miller, T.L. Madden, D.J. Lipman, E.V. Koonin and S.F. Altschul.** 1998. Protein sequence similarity searches using patterns as seeds. Nucleic Acids Res. *26*:3986-3990.

Update to: BLAST...What Next?

Bruno A. Gaëta

University of New South Wales, Sydney, Australia (bagaeta@fastmail.fm)

Since the publication of this article, the following World Wide Web (WWW) sites have moved, changed, or are no longer available:

The National Center for Genome Resources no longer offers its public search site (http://seqsim.ncgr.org/newBlast.html).

The Telethon Institute of Genetics and Medicine's Basic Local Alignment Search Tool (BLAST) server (http://www.tigem.it/LOCAL/BLAST/blast.html) is no longer available at this writing.

The BioNavigator service is available only on a limited basis through subscription to the Australian National Genomic Information Service (http://www.angis.org.au). An alternative integrated site for sequence analysis (including BLAST) is PISE at the Pasteur Institute (http://bioweb.pasteur.fr/seqanal/blast/intro-uk.html). PISE includes the program seqsblast (http://bioweb.pasteur.fr/seqanal/interfaces/seqsblast.html), which can extract hit sequences from BLAST outputs for subsequent processing, e.g., for building a multiple sequence alignment.

The European Molecular Biology Laboratory (EMBL) BLAST server for differentiating orthologs and paralogs (http://www.Bork.EMBL-Heidelberg.DE/Blast2e/) has a new uniform resource location (URL) (http://dove.embl-heidelberg.de/Blast2e/).

CHAPTER 28

Strategies for Searching Sequence Databases

Hugh B. Nicholas, Jr., David W. Deerfield II, and Alexander J. Ropelewski

Pittsburgh Supercomputing Center, Pittsburgh, PA, USA

Reprinted from BioTechniques 28:1174-1191 (June 2000)

ABSTRACT

We provide a detailed overview of the choices inherent in performing a sequence database search, including the choice of algorithm, substitution matrix and gap model. Each of these choices has implications that can be described as restrictions on the underlying model of sequence evolution, the expected degree of divergence between the query sequence and the database sequences (if one uses an evolutionary based matrix), as well as the sensitivity and selectivity of the search. We conclude with a series of recommendations for researchers performing these searches based on our experience and literature studies.

INTRODUCTION

Database searching is the application of knowledge gained from previous examinations of well-characterized sequences to the problem of discovering the biochemistry and physiology of a newly discovered gene or its protein product. The objective of a database search is to distinguish sequences related to the query sequence by some model (e.g., evolution) from unrelated sequences. This objective is different from the objective of aligning sequences, which is to discover the most likely history of changes between sequences already inferred to be related. This difference in objectives implies a very different set of choices for parameters and strategies. A database search is a computational investigation and, like a laboratory investigation, must be performed thoughtfully. A database search performed with injudicious parameters will lead to wrong answers and missed discoveries. While the default parameters on most database search servers are appropriate to a wide range of the most common circumstances, no single set of parameters will be the best for all searches. Thus there are at least three distinct situations where other parameters are likely to give better results.

The first is when sequences in the database that are homologous to the query sequence

BioComputing: Computer Tools for Biologists
Edited by Stuart M. Brown

are evolutionarily highly diverged. The second is when either the query sequence or its homologues in the database are too short to achieve a statistically significant match using the default parameters. The third is when the researcher is interested in finding homologues in only a limited range of species whose evolutionary divergence from the species in which the query sequence was found is different from that implied by the default parameters. Examples of all of these situations are reported in the literature cited in this paper. This discussion is intended to help researchers recognize and adjust their database searches to these circumstances.

Database searching helps to evaluate whether a newly determined sequence is related to a previously determined and characterized gene or protein through a common evolutionary ancestor. Homology, relatedness through a common evolutionary ancestor, is not directly observable. In sequence database searching, we observe sequence likeness or similarity. If the likeness is great enough, we may infer that the two sequences are homologous. Thus, much of the previously determined knowledge we apply in database searching involves how to best measure sequence likeness and how to assess whether the observed degree of sequence likeness is sufficient to allow us to infer that the sequences are homologous; that is, related through divergent evolution.

The similarity scores for pairs of sequence residues, either amino acids or nucleotides, used to assess sequence likeness are the major sources of previous biological knowledge for database searching. Similarity matrices provide a quantifiable measure of the ability of one residue to replace another, instead of assuming that all residues are equally conserved and that all mismatches are equally bad, as does the unitary scoring matrix. Similarity matrices have been developed reflecting different degrees of evolutionary divergence. The point accepted mutation (PAM) (11) and block sum (BLOSUM) (14) similarity matrices widely used in protein database searches are two good examples. These matrices incorporate the information of which residues have successfully replaced each other during the course of evolution. Similar matrices are available, if not widely used, for nucleic acid sequences (30). The nucleic acid matrices can incorporate knowledge about differential rates of transitions and transversions in the same way that some amino acid substitutions are judged more favorable than others in protein similarity matrices.

The algorithm used in searching the database incorporates the knowledge of how to find the maximum degree of sequence likeness and is the second most important source of previous knowledge. The three most widely used programs today, Smith-Waterman (28), FASTA (25), and BLAST (3,4), place different restrictions on the simple evolutionary model on which database searching is based. Smith-Waterman is the most rigorous algorithm and does not place any heuristic restrictions on the evolutionary model; it is both the most sensitive and the least selective algorithm. The actual pattern of evolutionary changes between a newly determined sequence and any homologue in the database can be incompatible with the heuristic restrictions imposed by either BLAST or FASTA. Alternatively, the additional selectivity that results from these restrictions can sometimes be an advantage. There is no single program that is always best at finding distantly related sequences for all gene and protein families, although the Smith-Waterman algorithm is most often the best (23,24).

Virtually all sequence database search programs now return an estimate of the statistical significance of the matches observed between the query sequence and sequences in the database. The statistical significance is an attempt to determine how often a score of a given value would be observed from the comparison of a random sequence to the length and composition of the query sequence and the sequences in the database.

Finally, the sequence database itself represents a large store of previously acquired knowledge. Making the best use of this knowledge can save many months of expensive laboratory experimentation and allow limited resources to be used to acquire truly new knowledge. The size of this potential gain is the determining factor in deciding how much effort to devote to any particular database search.

SEARCH ALGORITHM

In the next section, we will describe the operation of the three most widely used database searching algorithms: Dynamic Programming Algorithm [Smith-Waterman (28,33), Needleman-Wunsch (21) and Sellers (26,27)], Word Search [FASTA (25)] and Maximal Segments Pairs [BLAST (3)].

Dynamic Programming Algorithm

The dynamic programming algorithm is mathematically rigorous and, given a specific substitution matrix and gap model, is guaranteed to find the optimum score and alignment (21,26,28,33). Several variants of the dynamic programming algorithm are useful in different situations. The first use of the dynamic programming algorithm for sequence analysis was by Needleman and Wunsch (21) and independently by Sellers (26). These equivalent variants of dynamic programming provide a global alignment between two sequences, which forces the alignment to include the entire length of both sequences. The Smith-Waterman algorithm (28,33) yields a local alignment, which provides the regions within the pair of sequences that are the most similar given the choice of scoring matrix and gap penalties. A third variant of the dynamic programming algorithm first proposed by Sellers (27), often called the quasi-global alignment algorithm, in a well-behaved case, aligns an entire shorter sequence within the best matching region of a longer sequence. It has been shown that the Smith-Waterman (local alignment) algorithm is the most effective of the database searching dynamic programming algorithms for finding similar sequences, although the Needleman-Wunsch (global alignment) or Sellers (quasi-global) may be more appropriate for problem-specific final alignment.

The Smith-Waterman algorithm is a recursive mathematical equation (28,33). Recursive means that the results are computed in steps with any subsequent step depending on the answers to previous steps. The Smith-Waterman algorithm can be expressed as:

$$SW_{i,j} = \max\{SW_{i-1,j-1} + s(a_i,b_j);\ SW_{i-k,j} + gap_j;\ SW_{i,j-k} + gap_i;\ 0\}$$

where, for an affine gap penalty:

$$gap_j = open\_gap + (i-k+1)*extend\_gap$$

$$gap_i = open\_gap + (j-k+1)*extend\_gap \qquad \text{[Eq. 1]}$$

$SW_{i,j}$ is the Smith-Waterman score for the partial alignment ending at residue i of sequence a and residue j of sequence b. It can be visualized in a table form (Figure 1). The value of $SW_{i,j}$ is the maximum of four different terms. The first term, $SW_{i-1,j-1} + s(a_i,b_j)$, corresponds to extending the alignment by one residue from each sequence in which $s(a_i,b_j)$ is the similarity score for aligning the i residue in sequence a with the j residue in sequence b. The second term, $SW_{i-k,j} + gap_j$, describes extending the alignment by including residue j from sequence b and inserting a gap of k residues in length in sequence a. The third term, $SW_{i,j-k} + gap_i$, is the equivalent term for inserting a gap into sequence b. The fourth term, 0, does not allow the computed score ($SW_{i,j}$) to become negative. Thus, the best scoring regions do not have to overcome the effects of surrounding

Table 1. Smith-Waterman Alignment Table

		g	c	t	g	g	a	a	g	g	c	a	t
	0	0	0	0	0	0	0	0	0	0	0	0	0
g	0	5	0	0	5	**<u>5</u>**	0	0	5	5	0	0	0
c	0	0	10	3	0	1	**<u>1</u>**	0	0	1	10	3	0
a	0	0	3	6	0	0	6	**<u>6</u>**	0	0	3	15	8
g	0	5	0	0	11	5	0	2	**<u>11</u>**	5	0	8	11
a	0	0	1	0	4	7	10	5	**<u>4</u>**	7	1	5	4
g	0	5	0	0	5	9	3	6	10	**<u>9</u>**	3	0	1
c	0	0	10	3	0	2	5	0	3	6	**<u>14</u>**	7	0
a	0	0	3	6	0	0	7	10	3	0	7	**<u>19</u>**	12
c	0	0	5	0	2	0	0	3	6	0	5	12	15
t	0	0	0	10	3	0	0	0	0	2	0	5	17
						g	a	a	g	-	g	c	a
						g	c	a	g	a	g	c	a

Similarity Scores: DNA PAM 47, Match = 5, Mismatch = -4; Open Gap = 0, Extend Gap = -7

regions of low similarity to achieve a high score. For a global alignment (Needleman-Wunsch), the fourth term is eliminated (21,26). That is, the partial scores within the table are allowed to become negative and the boundaries of the alignment are defined by the ends of the sequences.

The alignment for a local search corresponds to the path that was used to compute the maximum score (19 in Table 1). This is accomplished in the program by starting at the maximum score and backtracking along the path until the first zero is reached. The bold and underlined numbers in Table 1 indicate this path. The MaxSegs algorithm developed by Waterman and Eggert (33) allows retrieval from the same table of additional, lower-scoring, nonintersecting alignments, which frequently provide valuable additional information about the relationship between the aligned sequences.

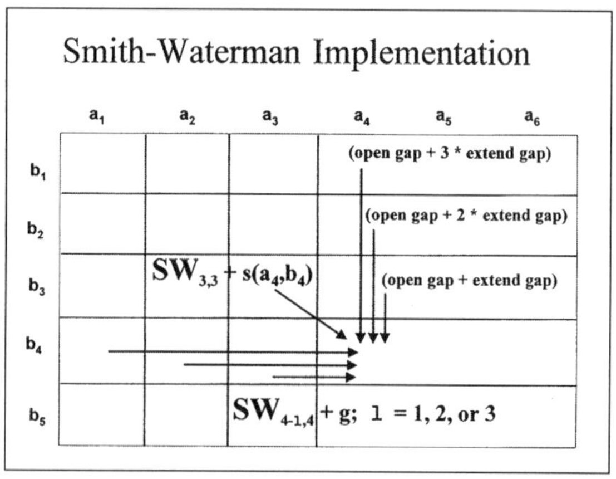

Figure 1. The two-way table layout of the calculations performed by the Smith-Waterman algorithm. Each column in the table corresponds to one residue of sequence A, a_i, and each row corresponds to one residue of sequence B, b_j. Each cell in the table, designated $SW_{i,j}$, corresponds to a partial alignment of sequence A with sequence B that ends at the sequence residues a_i and b_j. Each arrow in the table represents a term that is computed and evaluated during the calculation of table cell $SW_{4,4}$.

The dynamic programming algorithm is computationally demanding—all terms in Equation 1 for each cell in Table 1

Table 2. Schematic Representation of the Heuristic Algorithm

Step 1:Initial Word Search • Identity for FASTA, expanded list for BLAST • Initial filtering for both (BLAST requires two words on the same diagonal) Step 2:Initial Alignment • FASTA links strong diagonals • BLAST expands good extended-word matches along the diagonal in both directions (final BLAST I alignment) • Significance test Step 3:Final Alignment • FASTA and prior versions of BLAST perform a bounded (by window size) restricted Smith-Waterman • Most recent BLAST performs a centered (on dipeptide in highest scoring 11-amino acid peptide) restricted Smith-Waterman • Final significance test—BLAST uses the statistics underlying the MSP, while FASTA uses the database as a reference distribution.

must be computed to get the alignment score. Recovering the actual alignment requires either the entire table in memory, which for long sequences may require more memory than is available, or doubling the number of computations, which makes the required memory proportional to the length of the longest sequence (20). These computational requirements place database searches with the Smith-Waterman algorithm beyond the capabilities of personal computers.

HEURISTIC ALGORITHMS

To reduce the computational requirements of routine database searches, a number of heuristic algorithms have been developed. The most popular heuristics are FASTA and BLAST. The heuristics can, in general, be described as a three-step process (Table 2). The first step is the initial heuristic-specific screening to identify which region (diagonal) of both the query and database sequence contains sufficient similarity for further consideration. The second step is the creation of an initial alignment in the identified region, often followed by a check to see if this alignment is statistically significant. The third step is the final alignment, usually a restricted version of the Smith-Waterman algorithm, in the region identified in Table 2, Steps 1 and 2. Substantial computational resources are saved because of the heuristic-specific reduction of the region (Table 2, Step 1) that must be investigated followed by the savings in the restricted Smith-Waterman alignment (Table 2, Step 3). The following sections will discuss the heuristic-specific limitations in FASTA and the various implementations of BLAST.

Word Search

The FASTA heuristic-specific screening is a word search. In a word (or k-tuple) search, both the query and library sequence are divided into overlapping "words" of a specific length (Figure 2). The lists of words from the query and library sequence are compared, and the diagonal with the most matching words is taken as the region most likely to contain the best alignment between the two sequences. The results from the word search are used to determine if the two sequences have a region of sufficient similarity to merit further examination, and combinations of the words are joined to create an initial alignment. If it contains sufficient sequence similarity, then a restricted Smith-Waterman alignment is performed on the region of the alignment table centered on the diagonal with the highest score and bounded by the window size. The results of this restricted Smith-Waterman alignment are reported as the optimal score. The computational savings can be seen for the comparison of two 300-residue proteins: a total of 90 000 cells would need to be computed for an unrestricted Smith-Waterman, while a restricted Smith-Waterman would require, at most, 9100 cells (assuming a window-size of 16 on the main diagonal).

The FASTA region-defining heuristic can be interpreted as a series of restrictions on the sequence evolution model used in comparing the sequences. The word-size parameter, usually two for proteins and six for nucleic acids, defines the first restriction. Using this restriction, FASTA constrains the evolution between a pair of sequences to preserve a number of unchanged dipeptides or hexanucleotides. The second restriction is the window size that limits the total sum of insertions or deletions one sequence can accumulate with respect to the other sequence in the course of evolution. The limitations in the FASTA approach can be shown in two pathological examples. The first example would have two proteins that share 50% identity, but the proper alignment consists of alternating match and mismatches. With a word size of two, there would be no word matches along the main diagonal of the dot plot for the sequences (although there will potentially be random or spurious word matches on the off-diagonals), and the proper alignment would probably not be found. The second case consists of two proteins that are almost identical, except the second protein has a 20-residue insertion into the middle of the sequence. If the window size is 16, then the restricted Smith-Waterman alignment will have insufficient alignment space to jump the 20-residue insertion. Thus, the resulting alignment score will either not be significant enough to be identified or the sequence before or after the insertion (whichever had the higher diagonal scores) will be reported, while the observation that the proteins were basically identical (with only one long insertion) will be missed. In practice, these pathological cases are very unlikely, but the word size restriction has been shown to affect database search results (23,24).

```
SEQUENCE
q l n f s a g w
     FASTA                BLAST
q l                 ql,qm,hl,zl
  l n               ln,lb
    n f             nf,af,ny,df,qf,ef,gf,hf,kf,sf,tf,bf,zf
      f s           fs,fa,fn,fd,fg,fp,ft,fb,ys
        s a         no words scored 8 or more, including sa
          a g       ag
            g w     gw,aw,rw,nw,dw,qw,ew,hw,iw,kw,mw,pw,
                    sw,tw,vw,bw,zw,xw
```

Figure 2. Word Lists as they are created for FASTA and BLAST. The word size is 2 in both examples shown here. For the BLAST example, we used the PAM120 matrix, and the threshold was arbitrarily assigned a score of 8. The sequence in this example is the adipokinetic hormone II from migratory locust.

Maximal Segment Pairs

Maximal segment pair (MSP) alignments are defined by the first and fourth terms in the Smith-Waterman algorithm (Equation 1). Thus, the MSP alignment consists of long stretches of matches and mismatches without gaps. The advantage of this method is that the statistical significance of each found segment is easily determined. The disadvantage is that computational requirements can be comparable to Smith-Waterman, which effectively precludes this approach for routine database searching.

The BLAST algorithm contains a heuristic for reducing the computational requirements of the MSP algorithm. Like FASTA, BLAST divides the sequence into a list of overlapping words (Figure 2). BLAST extends the list to include all words that score above a specific matrix-defined threshold for the specified matrix. The value of the threshold limits the number of matches that will survive the first step (screening). The score for common, highly mutable peptides aligned with themselves may not be above the threshold required for inclusion in the list, and hence they would not be included. However, BLAST does provide the option of forcing the word into the list as long as the exact word is present in the sequence. The expansion of the word list to include all words that match with a suitably high score provides a large increase in sensitivity over using only words identically present in the sequence.

The alignment for the original BLAST algorithm was created by taking each identified matching word and then extending this match in both directions along the diagonal (without gaps) until the alignment score went below a cutoff point. The original algorithm reported the best segment if it was significant (see below). The two components of the BLAST heuristics can, as with those in FASTA, be interpreted as restrictions to the evolutionary model. The word size, usually three for proteins and 11 for nucleic acids, requires that there be at least one word-size segment within the conserved region that scores above the threshold. The second restriction is that a statistically significant ungapped segment must be conserved between the two sequences. For an average protein, the minimum length of this ungapped segment is around 35 residues.

The BLAST algorithm has undergone several refinements and improvements while attempting to maintain selectivity, increase sensitivity, decrease computational requirements and provide a better resulting alignment. The sensitivity was increased while the computational requirements were decreased by first lowering the threshold for the words to be identified and then requiring that there be at least two identified, nonoverlapping words on a diagonal before extending the alignment. Thus, more words are identified, yet the number of initial alignments examined (by extending matches along the diagonal in both directions) is decreased. The quality of the resulting alignment has been improved by first providing high-scoring segments that were near the maximal scoring segment. Later, using an approach similar to FASTA, the maximal scoring segment is used to define a band that uses the Smith-Waterman algorithm to find a gapped alignment within the band.

The new gapped BLAST circumvents the problem of being restrained within an alignment region bounded by the window size while avoiding the high computational cost of an unrestricted Smith-Waterman alignment by extending the alignment out from a central high-scoring pair of aligned amino acids in a way analogous to how BLAST extends the initial maximal segment pair alignment. The initial pair of aligned amino acids is chosen as the middle pair of the highest scoring, 11-residue window in the high-scoring segment pair alignments. The Smith-Waterman algorithm is then used to extend the alignment in both directions until the score falls below a fixed percentage of the highest score computed in the Smith-Waterman phase. This method will find the highest-scoring Smith-Waterman alignment if two conditions are met. First, the calculation is extended until a score of zero is reached. Using a higher threshold for stopping the calculation the way BLAST does may risk not finding the complete alignment in return for a large savings in computer resources. The second criterion that must be met is that the initial pair of amino acids selected as the midpoint from which to extend the alignment must actually be part of the one that would be reported as the best by a complete Smith-Waterman alignment of the pair of sequences. The selection criteria appear to be effective in selecting an appropriate pair of amino acids from which to extend the alignment. Nonetheless, a full Smith-Waterman or Needleman-Wunsch alignment is recommended for publication.

Comparison of Search Algorithms

Both heuristics, FASTA and BLAST, approximate the Smith-Waterman algorithm and were developed to reduce the computational requirements to make routine database searches feasible. The heuristics will compute the same score for each alignment in a database search as the full Smith-Waterman if (*i*) the heuristic correctly identifies the proper region (Table 2, Step 1); (*ii*) the initial screening decision to continue analysis is correct (Table 2, Step 2); and (*iii*) the final alignment is correctly described by the restricted Smith-Waterman (Table 2, Step 3). The refinements in identifying the proper region and screening the result have established these heuristics as effective database search tools. Note that the computation of the significance for each alignment is program specific.

The two parameters used to describe the performance of a search algorithm are sensitivity and selectivity. Sensitivity is the number of related sequences found in a database search, while selectivity is the number of unrelated sequences identified in a database search. Ideally, one would want a program that is highly sensitive (recovers all related sequences) while being selective (no false positives). The Smith-Waterman algorithm often shows better sensitivity than either heuristic, although the heuristics have continued to be refined and optimized so that the Smith-Waterman's advantage is now relatively small (4,5). All algorithms demonstrate similar selectivity, with BLAST usually the most selective. The decreased sensitivity and increased selectivity of the heuristics have been attributed to the initial word search and screening (Table 2, Steps 1 and 2), which tend to eliminate more diverged sequences and false positives (24). With the advances in algorithms, effective search parameters and statistical significance models, the most common problem encountered by the algorithms is sufficient sensitivity to find all related database sequences (4,5,24). It is not uncommon for only half of the known related sequences to be identified in a trial database search (4,5,24).

For proteins, the BLAST word-based heuristic is usually more sensitive than FASTA; however, the combination of a long word size and modifications of the heuristic to improve protein sensitivity decreases the BLAST performance for nucleic acids relative to FASTA.

The failure to find all of the sequences alluded to above is the limitation that gives rise to the first of the circumstances mentioned previously where the default search parameters can be inappropriate for a specific sequence database search. A sequence database search need not find all of the homologues but only find one or more characterized homologues that provide leads for further experimentation. However, for a particular newly determined query sequence there may be only a few homologues in the sequence database, all of which are highly diverged from the query sequence. None of these few homologues may be discovered by a search using the default settings. This is because the default settings have been appropriately chosen to find the largest number of sequences for query sequences that are representative of the known members of the family for a large number of families that are present in large numbers in the sequence databases. Thus the default parameters are generally set for a moderate level of sequence divergence rather than a very high level. Much of the rest of this review addresses what to do in this circumstance.

SIMILARITY MATRICES

To use any of these search algorithms, one must quantify whether the substitution of one residue for another is likely to conserve the physical and chemical properties necessary to the structure and function of the protein or is more likely to disrupt essential structural and functional features of the protein. This choice of similarity matrix is the most critical determinant of which sequences will be reported as similar to the query sequence in the database search report. Numerous approaches have been used to create such quantifications, referred to as similarity matrices. Similarity matrices have been based on explicit or implicit (empirical) evolutionary models, structural properties such as Chou-Fasman propensities, chemical properties such as charge, polarity, and shape, as well as combinations such as those used in the structure-genetics matrix. Regardless of the underlying approach, all similarity matrices are attempts to quantify whether a mutation preserves or disrupts the function of a protein. Note that the underlying approach of the substitution matrix defines the basis for the similar sequences that will be found and reported in a database search. Thus, if you wish to infer an evolutionary relationship between similar sequences, you should use an evolution-based matrix. The following discussion describes the approaches for several early similarity matrices, followed by a presentation of the theory and methodology for the creation of the two most commonly used evolution-based matrices.

Early Matrices

Early sequence alignment programs used the unitary scoring matrix, in which all matches were equally good and all mismatches were equally penalized. This scoring matrix was sometimes appropriate for DNA and RNA comparisons (in which transitions equal transversions and the sequences are moderately diverged), but not for protein comparisons.

Many alternatives to the unitary scoring matrix have been suggested. One of the earliest suggestions was a scoring matrix based on the minimum number of bases that must be mutated to convert a codon for one amino acid into a codon for a second amino acid. This matrix, known as the minimum mutation distance matrix, has succeeded in identifying more distant relationships among protein sequences than the unitary matrix approach. The minimal mutation distance matrix is an improvement over the unitary matrix because it incorporates knowledge about the process of mutating one amino acid into another. However, it is limited because the minimum mutation distance matrix does not include codon usage or selection processes.

Another improvement over the unitary matrix is a scoring matrix based on selected physical, chemical or structural properties shared and not shared by the 400 pairs of amino acids. Specific instances of this approach work well for some sequences but not for others. The approach works best if the matrix is based on properties that have been strongly conserved during the evolution of the sequence family. This is because the properties matrix attempts to specify the criteria that determine whether or not a mutation can survive and be fixed in a population. However, this approach suffers from problems of balancing the contributions of the different properties to the positive selection of mutations and from ignoring the different rates at which different mutations are generated.

Empirical Evolutionary Matrices and Log-Odds Scores

The biggest improvement achieved over the unitary matrix (and other theoretically based matrices) was based on the empirical study of the evolutionary replacements of amino acids. Margaret Dayhoff pioneered this approach in the 1970s when she made an extensive study of the frequencies in which amino acids substituted for each other during evolution (11). The studies involved carefully aligning all of the proteins in several families and then constructing phylogenetic trees for each family. This approach incorporates both the generation and selection of mutations and has been successful in sequence alignment applications. We present below more details of two widely used families of these empirically based matrices.

Computing Similarity Matrices

A similarity score (S_{ij}) is computed using information theory to provide a measure for the probability of residue i replacing residue j in an alignment (Equation 2).

$$S_{ij} = \log_2(R_{ij}) = \log_2(q_{ij}/p_i p_j) \qquad \text{[Eq. 2]}$$

where q_{ij} is the relative frequency with which residues i and j are observed to replace each other in related sequences.

In Dayhoff's original research, a separate q_{ij} and q_{ji} were tabulated. Dayhoff and others have since made the simplifying assumption that $q_{ij} = q_{ji}$ and averaged these values. The terms, p_i and p_j, are the expected probabilities for these residues and are usually the frequencies at which residues i and j occur in the database. The product, $p_i p_j$, is the frequency at which these residues would be expected to replace each other if the pattern of mutations were random. The similarity score (S_{ij}) is the base 2 logarithm of the ratio (R_{ij}) of the observed and expected frequencies of mutation. The base 2 logarithm allows the direct computation of entropy (see below) and a straightforward calculation of statistical significance from the score. The similarity matrices in common use have generally been scaled either by using logarithms to the base ten (the Dayhoff PAM matrices) or by multiplying by a factor of two or three (the BLOSUM matrices). This scaling makes no differences in which sequences are reported as similar by a database search (1,3) but is done to allow the similarity matrices to be represented as integer numbers with a lesser loss of accuracy. Scores above zero ($S_{ij} > 0.0$) indicate that two residues replace each other more often than would be expected if the replacement rate were random. Likewise, scores below zero indicate that residues replace each other less often than would be expected if the replacement rate were random. Thus, a positive alignment score means that the pattern of identities and substitutions described by an alignment is more likely to result from previously observed evolutionary processes than to result from random replacements.

The average amount of information available per position in an alignment computed using a specific matrix is referred to as either the information content or relative entropy, which is reported in bits. A bit is the amount of information required to distinguish between two possibilities, such as distinguishing between an alignment of homologous

Table 3. BLOSUM and PAM Similarity Matrices Equating the Matrices Based on the Entropy

BLOSUM		PAM		
Matrix	Entropy	Matrix	Entropy	% Identity
BLOSUM90	1.18	PAM100	1.18	43
BLOSUM80	0.99	PAM120	0.98	38
BLOSUM60	0.66	PAM160	0.70	30
BLOSUM52	0.52	PAM200	0.51	25
BLOSUM45	0.38	PAM250	0.36	20

sequences from an alignment of random sequences. The equation for calculating entropy for a similarity matrix is shown in Equation 3 and given for selected matrices in Table 3.

$$\text{Entropy} = \sum_{i=1,20} \sum_{j=1,20} q_{ij} \cdot S_{ij} \qquad \text{[Eq. 3]}$$

Counting the Replacements

The primary difference between the two major evolution-based similarity matrices is in the computations of the replacement frequency (q_{ij}). The PAM matrices use counts derived from an explicitly tree-like, branching evolutionary model. The BLOSUM matrices use counts derived from highly conserved blocks within an alignment.

The PAM or Dayhoff Family of Matrices

Margaret Dayhoff and co-workers performed a careful global alignment (including both highly conserved and variable regions) on a number of closely related sequences, which allowed them to create the appropriate evolutionary trees (11). The ancestral sequences are computed for each of the internal nodes in the tree using the principle of minimum replacements or maximum parsimony. They then tabulated which residues were conserved and which ones were mutated for each branch in the evolutionary tree (Figure 3). Not all mutations were observed (e.g.,W→C) because of the small amount of data available. The tabulated counts were then converted into the appropriate replacement frequencies for a single point mutation. Using these underlying replacement frequencies for a single point mutation, Dayhoff and co-workers asked the question: what would the replacement frequencies be after "*n*" PAM were observed within a stretch of 100 residues? Matrix multiplication of the single point mutational frequencies results in replacement frequencies that can then be converted (Equation 2) into similarity scores representing the expected substitution pattern after a selected number of mutations. An estimated percent identity for each matrix can be computed, assuming a database-consistent distribution of residues. This approach has been applied successfully to both nucleic acids (30) and proteins (11).

Since each matrix is developed for a specific PAM distance, how well do the matrices work if the sequences that are to be identified are at a different PAM distance? That is, if two sequences are diverged by 240 PAM units, would a PAM40 matrix identify them in a database search? The efficiency of a matrix at the various PAM distances is computed as the ratio of the sequence pairwise score for the matrix in question divided by the sequence pairwise score with the properly diverged matrix (1) (Figure 4). A more complete graph can be seen at http://www.psc.edu/biomed/data/graphs.html. A database search with a different matrix is indicated (1) if the new matrix is able to increase the alignment score by 2 bits (or a significance factor of four). For proteins, this level is approximately (34-2)/(34) (where 34 bits is required for a significant database result) or 0.94 efficiency. Figure 4 shows the efficiency of four PAM matrices. Note that no matrix effectively covers the entire range of percent identity above an efficiency of 0.94. The PAM160 matrix provides the best coverage of the range of interesting sequences typically found in a database search (which is similar to the BLOSUM62). One would need two or three different matrices for an effective search of all divergence distances in a database search. We provide recommendations for matrix selection from the literature in Table 5.

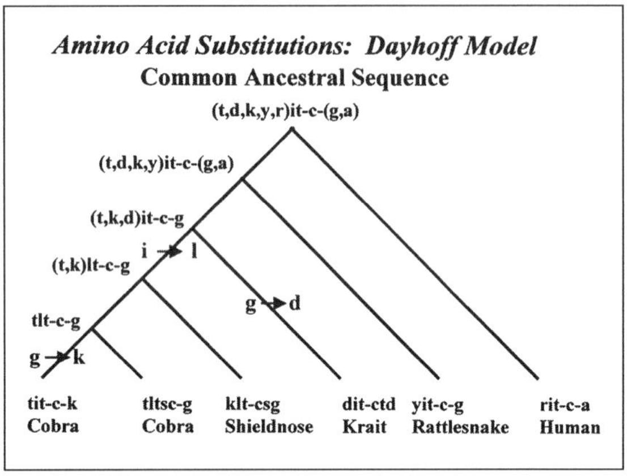

Figure 3. An evolutionary tree with the recreated ancestral sequences at the internal nodes and three amino acid substitutions indicated on branches of the tree. The observed amino acid substitutions tabulated in computing the Dayhoff PAM similarity matrices were counted by using such recreated ancestral sequences and substitutions.

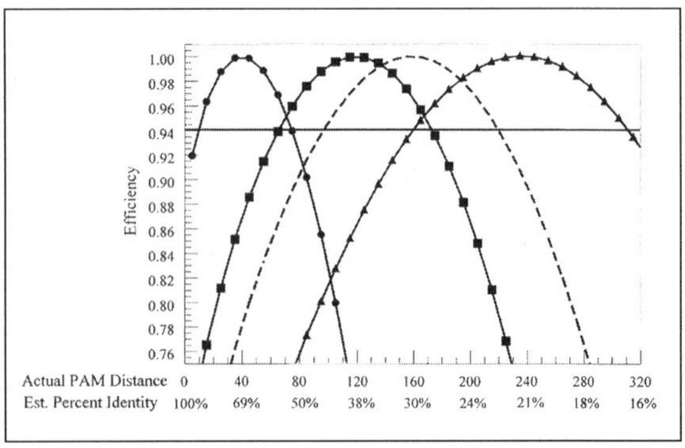

Figure 4. Efficiency plots for the protein scoring by the PAM40 (closed circles), PAM120 (closed boxes), PAM160 (dashed line) and PAM240 (closed triangles) of sequences that are at an "actual PAM distance" apart. The estimated (assuming a database-consistent distribution of residues) percent identity is also provided for reference. The solid line (efficiency = 0.94) represents the level at which using a second matrix would be beneficial.

The BLOSUM Family of Matrices

Henikoff and Henikoff created multiple sequence alignments of related proteins and identified conserved regions without gaps, the BLOCKS database, which serves as the source of the data for the BLOSUM matrices. These BLOCKS constitute the regions that would most likely be found in a database search. To create a specific BLOSUM matrix, in each BLOCK, all sequences that share at least *n* percent identity (where *n* = 50% for the BLOSUM50) are grouped together to create weighted set representations (Figure 5). The counts are then tabulated between the weighted set representations within each BLOCK (Figure 5). The counting procedure is the sum-of-pairs evolutionary model that allows direct evolution between any pair of sequences in the family rather than through a branching process involving intermediate sequences (Figure 3). The counts are then added across all BLOCKS and used to compute the q_{ij} values (Equation 2).

The efficiency of the BLOSUM series of matrices can also be estimated but not directly computed because these matrices are not determined based on a single set of single point mutation frequencies that are then matrix-multiplied to a given set of distances. The BLOSUM62 is often the default matrix in a number of search programs. The BLOSUM62 matrix, which is similar in efficiency to the PAM160 matrix (Figure 4), has the widest span across the range of interesting sequences found in a database search. Thus, the BLOSUM62 is often the default matrix in database searching programs. But, once again, for complete coverage of all divergence distances, multiple matrices should be used.

Differences between the PAM and BLOSUM Models

The BLOSUM and PAM matrices are the most widely used amino acid similarity matrices for database searching and sequence alignment. There are three primary differences between the PAM and BLOSUM matrices: (*i*) all PAM matrices are derived using matrix multiplication from a single set of single point mutation frequencies derived from an explicit evolutionary model, while BLOSUM matrices are derived with a sum-of-pairs evolutionary model using the BLOCKS database with different weight set representations (sequence groups); (*ii*) the PAM matrices are based on mutations observed in both highly conserved and variable regions in a global alignment; while the BLOSUM matrices are based exclusively on local, highly conserved regions without gaps; and (*iii*) the PAM matrices are based on a limited number of observations (e.g., C→W was not observed), while the BLOSUM matrices are based on a large number of observations.

In empirical tests of the effectiveness of the two series of matrices, both generally perform well (11,14). However, when comparing the results of a single database search, the BLOSUM matrices will usually perform slightly better (15,17,24). This likely reflects the fact that the BLOSUM matrices are based on the replacement patterns found in more highly conserved regions of the sequences. These patterns discovered in data-

Multiple sequence alignment:

```
Bpi Bovine      npGivaRItqkgLdyacqqgvltlQkele
 Bpi Human      npGvvvRIsqkgLdyasqqgtaalQkelk
Cept Human      eaGivcRItkpaLlvlnhetakviQtafq
 Lbp Human      npGlvaRItdkgLqyaaqegllalQsell
Lbp Rabbit      npGlitRItdkgLeyaaregllalQrkll
```

Weighted set representation:

Group 1: [n][p][G][½i,½v][v][½a,½v][R][I][½s,½t][q][k][g][L][d][y][a][½s,½c][q][q][g][½t,½a][½a,½l][½a,½t][l][Q][k][e][l][½k,½e]

Group 2: [e][a][G][i][v][c][R][I][t][k][p][a][L][l][v][l][n][h][e][t][a][k][v][i][Q][t][a][f][q]

Group 3: [n][p][G][l][½i,½v][½a,½t][R][I][t][d][k][g][L][½q,½e][y][a][a][½q,½r][e][g][l][l][a][l][Q][½s,½r][½e,½k][l][l]

Replacement Counts:

```
v -> t = 0.0 (within the same group)
v -> a = 0.5 * 1.0           = 0.5
v -> l = 0.5 * (0.5 + 0.5) = 0.5
t -> a = 0.5 * 1.0           = 0.5
t -> l = 0.5 * (0.5 + 0.5) = 0.5
a -> l = 1.0 * (0.5 + 0.5) = 1.0
l -> l = 0.0 (within the same group)
```

Figure 5. Counting replacements for a BLOSUM70 similarity matrix. Three different groups were defined: group 1 = Bpi (bovine and human), group 2 = Cept (human) and group 3 = Lbp (human and rabbit). Replacements within a group were not counted.

Table 4. Significance Levels Using a Protein or a Nucleic Acid Query

Protein Percent Identity	Protein PAM Distance	Protein Sig. Len. (34 bits)	Protein Info. Per. Residue	N.A. to Protein Efficiency	N.A. Info. Per. Codon	N.A. Sig. Len. (41 bits)	N.A. PAM Distance	N.A. Percent Identity
100	0	8	4.17	**1.44**	**6.0**	**21**	**0**	**100**
83	20	12	2.95	**1.23**	**3.63**	**34**	**16**	**86**
63	**50**	**17**	**2.00**	1.01	**2.02**	**61**	**40**	**69**
43	**100**	**29**	**1.18**	**0.73**	0.86	143	80	51
32	**150**	**45**	**0.76**	**0.51**	0.39	315	120	40

Data derived from Dayhoff et al. (5) and States et al. (6).
N.A. = Nucleic Acid; Info. Per. = Information Per; Sig. Len. = Significant Length

base searches serve as anchor points in alignments involving complete sequences. It is reasonable to expect that the replacements that occur in highly conserved regions will be more restricted than those that occur in highly variable regions of the sequence.

The PAM matrices still perform relatively well despite the small amount of data underlying them. The most likely reasons for this are the care used in constructing the alignments and phylogenetic trees used in counting replacements and the fact that they are explicitly based on a simple model of evolution. They still perform better than many of the more modern matrices that have been less carefully constructed. Both the PAM and BLOSUM matrices generally perform better than matrices based explicitly on criteria other than observed replacement frequencies (17) (e.g., properties, structure-genetic and minimum mutation distance).

Scores for Nucleic Acids

For coding regions of genes, it is generally better to use the translated amino acid sequence for a database search (2). Many factors can interfere with nucleic acid-based searches. One of these is the compositional bias found in many organisms and organelles. Another is that some DNA sequences are derived from mRNA while others are genomic DNA with introns, and the exons may be too short to give a significant alignment with an mRNA-derived sequence. A third reason is that for distant relationships, the protein sequence alignment will generally contain more information (Table 4).

For noncoding regions, there are additional considerations in the choice of search algorithm. BLAST uses a very long word size for nucleic acid sequences. Also, the modifications to the BLAST heuristic that improve its sensitivity for protein sequences do not work well for nucleic acid sequences because nucleic acids have only a four-letter alphabet and the similarity scores are usually calculated with equal frequencies of replacement for all of the nucleotides (transitions equal transversions). Thus, FASTA is more sensitive than BLAST for even moderately diverged nucleic acid sequences and should be used instead of BLAST if a heuristic is desired.

Transition versus Transversion

It is possible to use different sets of evolutionary matrices for nucleic acids just as have been recommended for proteins. Two theoretical PAM models have been proposed, one

Table 5. Recommended Matrices to Be Used for a Database Search

PAM Matrices		BLOSUM Matrices	
3 Matrices	2 Matrices	3 Matrices	2 Matrices
PAM40	PAM120	BLOSUM60	BLOSUM60
PAM120	PAM250	BLOSUM40	BLOSUM35
PAM240		BLOSUM30	

using a uniform evolution model assuming equal rates of transitions and transversions and another using a biased model with a 3 to 1 transition to transversion ratio (29). Note that the uniform model at PAM 47 provides substitution values similar to the ones historically recommended by Smith et al. (29).

There have also been a few empirical studies on gene-pseudogene evolution (7,16). These studies provide replacement frequencies that can be easily converted into empirical log-odds scoring matrices such as the BLOSUM protein matrices already described. These derived log-odds matrices may prove useful for researchers interested in searching for pseudogenes.

SCORING INSERTIONS AND DELETIONS

The gap penalty represents the probability that an insertion will start or continue relative to either a match or mismatch. Thus, the value of the gap penalty must be consistent with and coupled to the values contained within the substitution matrix. Furthermore, the value of the gap penalty must be consistent with the goal of the alignment—finding similar sequences in a database search or aligning sequences to provide the best estimate of the evolutionary history. There have been many approaches to quantify the gap penalty (12,32).

The most commonly used gap penalty consists of two terms, an open gap penalty and extended gap penalty. These terms allow the gap penalty to separate the cost of opening an insertion versus extending the preexisting gap. This two-term gap penalty performs better both in database searches (24) and in providing the best alignment (12). In practice, the open gap penalty is usually relatively large, while the extended gap penalty is more modest.

Database searching scores for related sequences are usually dominated by relative long ungapped segments. Thus, the gap penalty value has only a modest effect on computing the score for related sequences, but the expected score for a random sequence alignment increases with a decrease in the gap penalty. Decreasing the gap penalty increases the chances for identifying spurious sequences (decreasing selectivity). For database searches, a general rule of thumb is that the open gap penalty should be larger than the largest match score and twice the largest mismatch penalty, while the extend gap penalty is usually set at a minimal value of -1.

The final alignment represents the best estimate of the evolutionary history of changes that occurred during the divergence of the two sequences from their most recent ancestor. Studies of protein structural alignments (22) and pseudogenes (13) indicate that insertions and deletions are usually short (one or two residues in length) and become more frequent as the sequences become more highly diverged. Producing the best alignment usually requires reducing the open gap penalty and increasing the extended gap penalty, which results in the formation of small gaps in variable regions. Thus, the contrast between database searching and final alignment is easy to see—database searching identifies long, highly conserved regions, while the final alignment contains shorter conserved regions separated by small gaps.

STATISTICAL SIGNIFICANCE

A database search will yield a number of high-scoring alignments, which leads to the question: is the score or the length of the alignment of the query and database sequence greater than what one would expect from a random sequence? BLAST uses an explicit approach to assess significance, while FASTA and most implementations of the dynamic programming algorithm use a semi-empirical approach.

Explicit Approach: Using Information Theory and the Length of the Alignment

Each alignment contains a certain amount of information per position, which is approximately the product of the entropy of the matrix used (Table 3) and the length of the alignment (1,18,19) or, more exactly, the alignment score in bits. Current statistical models have not developed the ability to compute entropy for a gap. Thus, the information content of only ungapped alignments can be explicitly computed (18,19).

The amount of information in a database search required for an alignment to be significant is the base 2 logarithm of the product of the length of query sequence and the total number of residues in the database. From these two relationships, we can determine the minimum required length of an ungapped alignment (Equation 4).

$$\text{Minimum Significant Alignment Length} = \frac{\log_2(\text{query length} \cdot \text{residues in db})}{\textit{entropy of matrix}}$$

[Eq. 4]

Thus, using a 500-residue query sequence to search a 40 million-residue database using a BLOSUM62 matrix (Table 3) requires a minimum alignment of 36 residues to be significant. BLAST (maximal segment pairs) uses this approach, along with a similar sum statistic, to determine significance in the second step (Table 2). If the query sequence is shorter than the required minimum length, then no match from a database search can be judged to be significant, and one needs to select a higher entropy matrix.

Semi-Empirical Approach: Using the Database Search to Create a Reference Distribution

FASTA and various implementations of the dynamic programming algorithm (e.g., Smith-Waterman) allow the alignment to contain gaps. Thus, the explicit approach described above cannot be used to determine the significance of the alignment resulting from the query and a database sequence. A number of models has been developed to allow the estimation of the maximum score (cutoff score) for a random sequence to be aligned with the query sequence. The majority of these models are based on the simple observation that, with respect to any newly determined query sequence, the database contains primarily random sequences. That is, the pertinent random process is evolution from independent, unrelated ancestral sequences rather than from a common ancestral sequence.

The results of the database search are usually presented as a histogram of score versus frequency of occurrence. There is a large peak in the histogram representing the bulk of the database, the random sequence portion and then a number of isolated peaks representing the high-scoring hits. It is the large peak representing the random sequence portion of the database that needs to be fit. Two approaches will be described in detail. First, the histogram of scores for the random sequence portion of the database is fitted using an extreme value distribution. This allows the determination of the mean for the histogram, the size of a standard deviation and the computation of the maximum random score. Second, the decay after the peak of the histogram of scores is empirically observed to be exponential (9,10,32). When the log of the frequency is plotted, this decay is a straight line. Extrapolating this line to a frequency of one sequence (essentially the axis intercept) computes a score larger than might be reasonably expected from a random sequence in the database (9,10). Thus, any appreciably higher score is statistically rare. This is the approach used in many implementations of the dynamic programming algorithm.

Both approaches to determine the cutoff score are based on a well-defined random sequence portion of the database search. The shape of this region depends on the gap penalty being large enough (32). If the gap penalty is too low, the scores are observed to decay much more slowly, broadening the distribution of scores, which may confuse either approach to fit the histogram and obscure the significant hits.

All significance computations are based on a given probability of finding and aligning specific residues, usually the database composition. Thus, a region formed almost exclusively of one or two residues (e.g., a stretch as small as 20–30 residues of poly-C or poly-G) can skew the significance computation and lead to unrealistic results. Methods exist to remove these low-complexity regions before database searches (2).

FUTURE CONSIDERATIONS

It is not uncommon for only half of the related sequences to be identified in a trial database search with a well-characterized, large family of sequences (4,5,24). What happens when a database search reveals no related sequences? Are there no related sequences or are the already sequenced members of the family too diverse from the query to be identified in a database search? Assuming the latter, one needs to improve the sensitivity of the search. Radical improvement in sensitivity will require a major shift in the underlying model of evolution implied by the database search. There is a maximum sensitivity for the combination of search algorithm and substitution matrix that cannot be improved. The heuristics, which are approximations of the Smith-Waterman algorithm, are approaching the quality of the Smith-Waterman and probably have limited room for improvement. The evolution model expressed in the general database searching programs depends on residue-specific mutation rates regardless of the position of the residue in either the three-dimensional structure or the function of the residue in the protein. With this simplistic model, there is little overall difference between a highly diverged sequence and a random sequence.

In our opinion, the best hope for improving the sensitivity of the database searching algorithm is to improve the underlying evolutionary model, probably through the use of more sophisticated statistical models, such as the maximum likelihood alignments (6,31) or related methods using hidden Markov representations (8). These algorithms evaluate all possible alignments rather than exclusively consider the best scoring alignment. In general, highly diverged homologous sequences have many alignments that are almost as good as the best ones, while alignment scores for nonhomologous sequences decrease rapidly from the best score. Thus, addition over the alternative alignments allows more sensitive discrimination between highly diverged homologous sequences and unrelated, nonhomologous sequences. These algorithms are at least as computationally demanding as dynamic programming algorithms, and there are still no effective heuristic approximations available because the algorithms themselves are still at the research stage.

One would assume that finding related sequences should become easier in the future because the sequence databases are doubling in size almost yearly, and several complete genomes have been sequenced. Thus, the potential number of hits should increase. While this number will increase, the nonhomologous portion of the database will increase at least as fast. Actually, this doubling of the sequence databases may increase the difficulty in identifying related sequences. This can be clearly seen when using the explicit approach to compute significance. Each year, an additional bit of information will be required for an alignment to be significant because the database doubles in size. Over the next five years, this will result in lengthening the minimum required ungapped alignment from 36 to almost 45 residues. One approach to ease the problems of growth in the database is to reduce its size by creating a nonredundant database. That is, for each cluster of highly identical sequences, one sequence is kept, while the others are removed. Alternatively, species- or genus-specific databases are also being created. While these approaches do limit the effects of the growth of the database, neither approach is ideal and can result in lost data. For example, sequentially searching a number of species-specific databases will provide a greater measure of significance than is actually valid.

Alternatively, one can use an abstraction to represent groups of sequences, creating what is in essence a nonredundant database. The group of sequences is usually a sequence family, while the abstraction can be a consensus sequence, weighted set representations, hidden Markov model (HMM) or position-specific scoring matrix (PSSM). This approach is now being used by the classification libraries ProSite, BLOCKS (PSSM), PRINTS (PSSM) and Pfam (HMM). Many of the abstractions allow a more accurate model for evolution, that is, defining some positions as rigorously conserved, while others allow random mutations.

CONCLUSION

In this paper, we have reviewed the four major aspects of sequence database searching: algorithm, substitution matrix, gap penalty and significance calculation. We have described how these components relate and how the choices affect the results of a database search. Based on this discussion, we recommend the following procedure for carrying out an effective database search: (*i*) use a local favorite program (Smith-Waterman, BLAST or FASTA) or the Web server of your choice; (*ii*) use at least two and preferably three similarity tables (Table 5); and (*iii*) if Smith-Waterman or FASTA is used, be sure the open gap penalty is large enough.

If the initial runs do not uncover any similar sequences and a heuristic has been used, repeat the search using a highly diverged matrix and the Smith-Waterman algorithm. Alternatively, assuming an evolution-based matrix was used, use a highly diverged matrix from the other series (e.g., if three BLOSUM matrices were used, then try the PAM320).

To produce the final alignment between two sequences, choose a matrix that reflects the appropriate divergence for the two sequences (Table 2), reduce the gap penalty relative to the values used in the database search and use the MaxSegs (Waterman-Eggert) version of the dynamic programming algorithm to provide the best local alignment and also search for repeats in the sequences.

If there are questions about which matrix or open gap penalty should be used, try several choices to see which parts of the alignment are invariant and which parts are more suspect.

ACKNOWLEDGMENTS

This overview is based on a tutorial developed under a grant from the National Institutes of Health National Center for Research Resources grant no. P41 RR06009. This material has been presented in workshops supported by the Research Resource grant and as a part of a workshop on nucleic acid and protein sequence analysis presented annually at the PSC. The latter workshop is funded by grant no. T-15 HG00015 from the NIH's National Human Genome Research Institute. The authors acknowledge the helpful comments and discussions with the other Pittsburgh instructors: Dr. Michael Gribskov from the San Diego Supercomputer Center, Dr. Gary Churchill from Jackson Labs, Dr. Stephen Altschul from the NCBI, Dr. Michael Waterman from University of Southern California and Dr. Gary Stormo from Washington University in St. Louis. We would like to express our appreciation to the referees whose comments helped us more clearly express several important points in this overview.

REFERENCES

1. **Altschul, S.F.** 1991. Amino acid substitution matrices from an information theoretic perspective. J. Mol. Biol. *219*:555-665.
2. **Altschul, S.F., M.S. Boguski, W.Gish and J.C. Wooten.** 1994. Issues in searching molecular sequence databases. Nat. Genet. *6*:119-129.
3. **Altschul, S.F., W. Gish, W. Miller, E.W. Myers and D.J. Lipman.** 1990. Basic local alignment search tool. J. Mol. Biol. *215*:403-410.
4. **Altschul, S.F., T.L. Madden, A.A. Schaffer, J. Zhang, Z. Zhang, W. Miller and D.J. Lipman.** 1997. Gapped BLAST and PSI-BLAST: a new generation of protein database search programs. Nucleic Acids Res. *25*:3389-3402.
5. **Agarwal, P. and D.J. States.** 1998. Comparative accuracy of methods for protein sequence similarity search. Bioinformatics *14*:40-47.
6. **Bishop, M. and E. Thompson.** 1986. Maximum liklihood alignment of DNA sequences. J. Mol. Biol. *190*:159-165.
7. **Blake, R.D., S.T. Hess and J. Nicholson-Tuell.** 1992. The influence of nearest neighbor on the rate and pattern of spontaneous point mutations. J. Mol. Evol. *34*:189-200.
8. **Bucher, P. and K. Hoffman.** 1996. A sequence similarity algorithm based on a probabilistic interpretation of an alignment scoring system, p. 44-51. *In* D. States, P. Agarwal, T. Gaasterland, L. Hunter and R. Smith (Eds.), ISMB-4. AAAI Press, Menlo Park, CA.
9. **Collins, J.F. and A.F.W. Coulson.** 1988. The significance of protein sequence similarities, p. 474-487. *In* R.F. Doolittle (Ed.), Methods in Enzymology, Vol. 183, Molecular Evolution: Computer Analysis of Protein and Nucleic Acid Sequences. Academic Press, San Diego, CA.
10. **Collins, J.F., A.F.W. Coulson and A. Lyall.** 1988. The significance of protein sequence similarities. Comput. Appl. Biosci. *4*:67-71.
11. **Dayhoff, M.O., R.M. Schwartz and B.C. Orcutt.** 1978. A model of evolutionary change in proteins, p. 345-352. *In* M.O. Dayhoff (Ed.), Atlas of Protein Sequence and Structure. National Biomedical Research Foundation, Washington DC.
12. **Fitch, W.M. and T.F. Smith.** 1983. Optimal sequence alignments. Proc. Natl. Acad. Sci. USA *80*:1382-1386.
13. **Gu, X. and W.H. Li.** 1995. The size distribution of insertions and deletions in human and rodent pseudogenes suggests logarithmic gap penalties for sequence alignment. J. Mol. Evol. *4*:464-473.
14. **Henikoff, S. and J.G. Henikoff.** 1992 Amino acid substitution matrices from protein blocks. Proc. Natl. Acad. Sci. USA. *89*:10915-10919.
15. **Henikoff, S. and J.G. Henikoff.** 1993. Performance evaluation of amino acid substitution matrices. Proteins: Structure, Function Genetics. *17*:49-61.
16. **Hess, S.T., J.D. Blake and R.D. Blake.** 1994. Wide variations in neighbor-dependent substitution rates. J. Mol. Biol. *236*:1022-1033.
17. **Johnson, M.S. and J.P. Overington.** 1993. A structural basis of sequence comparisons: an evaluation of scoring methodologies. J. Mol. Biol. *233*:716-738.

18.**Karlin, S. and S.F. Altschul.** 1990. Methods for assessing the statistical significance of molecular sequence features by using general scoring schemes. Proc. Natl. Acad. Sci. USA *87*:2264-2268.
19.**Karlin, S. and S.F. Altschul.** 1993. Applications and statistics for multiple high-scoring segments in molecular sequences. Proc. Natl. Acad. Sci. USA *90*:5873-5877.
20.**Myers, E.W. and W. Miller.** 1988. Optimal alignments in linear space. Comput. Appl. Biosci. *4*:11-17.
21.**Needleman, S.B. and C.D. Wunsch.** 1970. A general method applicable to the search for similarities in the amino acid sequences of two proteins. J. Mol. Biol. *48*:443-453.
22.**Pascarella, S. and P. Argos.** 1992. Analysis of insertions/deletions in protein structures. J. Mol. Biol. *20*:461-471.
23.**Pearson, W.R.** 1991. Searching protein sequence libraries: comparison of the sensitivity and selectivity of the Smith- Waterman and FASTA algorithms. Genomics *11*:635-650.
24.**Pearson, W.R.** 1995. Comparison of methods for searching protein sequences databases. Protein Science *4*:1145-1160.
25.**Pearson, W.R. and D.J. Lipman.** 1988. Improved tools for biological sequence comparison. Proc. Natl. Acad. Sci. USA. *85*:2444-2448.
26.**Sellers, P.H.** 1974. On the theory and computation of evolutionary distances. SIAM J. Appl. Math. *26*:787-793.
27.**Sellers, P.H.** 1979. Pattern recognition in genetic sequences. Proc. Natl. Acad. Sci. USA *76*:3041-3041.
28.**Smith, T.F. and M.S. Waterman.** 1981. Identification of common molecular subsequences. J. Mol. Biol. *147*:195-197.
29.**Smith, T.F., M.C. Waterman and C. Burkes.** 1985. The statistical distribution of nucleic acid similarities. Nucleic Acids Res. *13*:645-656.
30.**States, D.J., W. Gish and S.F. Altschul.** 1991. Improved sensitivity of nucleic acid database search using application-specific scoring matrices methods. Methods Enzymol. *3*:66-77.
31.**Thorne, J.L., H. Kishino and J. Felsenstein.** 1991. An evolutionary model for maximum likelihood alignment of DNA sequences. J. Mol Evol. *33*:114-124.
32.**Vingron, M. and M.S. Waterman.** 1994. Sequence alignment and penalty choice. J. Mol. Biol. *235*:1-12.
33.**Waterman, M.S. and M. Eggert.** 1987. A new algorithm for subsequence alignments with application to tRNA-rRNA comparisons. J. Mol. Biol. *197*:723-728.

Update to: Strategies for Searching Sequence Databases

Hugh B. Nicholas, Jr., David W. Deerfield II, and Alexander J. Ropelewski

Pittsburgh Supercomputing Center, Pittsburgh, PA, USA (nicholas@psc.edu)

INTRODUCTION

Sequence database searching is a complex activity that is built upon the areas evolution, mathematics, and computer sciences. Since the initial publication of our review, there has been noteworthy progress in several aspects of sequence database searching. We believe that one of the most significant is the development of new methods for computing empirical evolutionary substitution frequencies for different residues and similarity scores derived from these observed substitution frequencies (9,10). A recent suggestion (4) for making use of the properties of evolutionary rate matrices in studying protein families may also provide guidance in distinguishing the low scoring homologous sequences reported from a sequence database search from the high scoring nonhomologous sequences reported in the same search. One of the most critical areas underlying sequence database searching is the statistics of sequence alignments and the statistical significance of sequence database search results. There have been noteworthy results in understanding the factors influencing the accuracy of database searching statistics (1), for computing significance probabilities based on state-of-the-art approximations for gapped alignments (8), and in new approaches to the problem of alignment statistics (6). Lastly, the important practical problem of organizing and managing the shear volume of reported hits from a sequence database search, so that all of unique significant matches with different regions of the query sequence are emphasized, is being addressed (2).

SIMILARITY MATRICES

Muller and Vingron (10) present a procedure, the resolvent approach, for computing similarity matrices based on the same simple Markov evolutionary model used by M.O. Dayhoff et al. (3) without having to go through the very labor-intensive steps of creating phylogenetic trees and ancestral sequences at each internal node in these trees. This allowed them to create a series of variable time (VT) similarity matrices that they label VT(t) (e.g., VT160), which are matrices where (t) is the evolutionary time measured in Dayhoff's point accept mutation (PAM) units. These VT similarity matrices have the same desirable theoretical properties that have strongly contributed to the continued effectiveness and usefulness of the original PAM matrices, but are based on almost 3 million aligned amino acid pairs drawn from the Systematic Researching (SYSTERS) (7) database of protein families. This new method makes use of data from moderately and highly divergent alignments, as well as the highly similar alignments required by the Dayhoff procedure. Thus, it reduces the bias toward highly mutable positions inherent in the Dayhoff procedure. The VT160 similarity matrix is very similar to the Blosum62 similarity matrix and less similar to the PAM160 similarity matrix. These matrices are available on the web from the uniform resource locator (URL): (http://www.dkfz.de/tbi/people/tmueller/model/index.html).

Later, Muller et al. (9) presented a maximum likelihood approach to estimating substitution matrices based on the simple Markov evolutionary model, which is more rigorous than the resolvent approach discussed above. Although the computational requirements for maximum likelihood make it unlikely that this method will be applied to very large datasets, such as that used in the resolvent method, the maximum likelihood method produced better results than the resolvent method on small and medium sized datasets. The only tests performed on either the matrices produced by the resolvent or maximum likelihood methods are simulation tests, to see how well they reproduce the parameters used to generate artificial datasets. In these tests, both methods performed better than the Dayhoff estimation procedure. Neither the resolvent nor maximum likelihood methods have been tested in a realistic sequence database searching application. However, given the theoretical rigor of the estimation procedures and the large amount of data used in their derivation, it seems reasonable to expect that they will be found to perform well when such tests are made and, thus, will be valuable in practical sequence database searching applications. The maximum likelihood estimation method might also be particularly valuable in conjunction with the Markov matrices derived from the small datasets proposed by Devauchelle et al. (4).

Devauchelle et al. (4) observed that if the Dayhoff Markovian model of protein evolution is an accurate model for a homologous family of proteins, then a single rate matrix, which reflects the unique evolutionary environment of that protein family, can model the multiple sequence alignment of that family of homologous proteins. They tested whether or not a simple Markov model accurately describes an alignment of 12 proteins in the mitochondrial DNA of 26 species and found the simple Markov model an adequate, if imperfect representation. Further, each pair of sequences in the alignment of the protein family will yield a rate matrix for a specific degree of evolutionary divergence that, within sampling error, is equal to the underlying rate matrix for the family adjusted to the specific degree of evolutionary divergence by raising the underlying matrix to the power given by the divergence. This prediction was also borne out by their tests. Since the homologous sequences found by a sequence database search form such an homologous family, these results could, in the future, provide the basis for a sensitive test for which sequences reported from a sequence database search with scores in the "twilight zone" (5) are actually homologous and which are high scoring random matches of nonhomologous sequences.

ALIGNMENT AND DATABASE SEARCH STATISTICS

Altschul et al. (1) have examined the recently proposed island technique for computing Basic Local Alignment Search Tool (BLAST) database searching statistics. The island technique involves looking at the score for each island, a positive region of scores completely surrounded by zero scores, in a path graph computed by the Smith-Waterman algorithm (see Table 1 in the original paper). In addition to a significant speed up in the computations, the use of the island technique allows a more accurate estimate of the tail behavior of the score distribution for comparison of short sequences. It also makes explicit the tradeoff between parameter estimate bias due to finite length of the sequences compared and the stochastic errors involved in estimating the parameters. The effect of unusual sequence composition on the estimated statistics is also discussed.

Richard Mott (8) has derived an empirical formula that can be used to evaluate the statistical significance of database searches performed using the Smith-Waterman algorithm or its equivalent for profile searches. The formula takes into account the length and composition of the sequences, the average entropy of the scoring matrix, and an affine gap penalty. The formula was derived from a large number of simulations and the results were shown to be accurate over a wide range of conditions. The paper also discusses factors, such as very short sequences or large differences in sequence lengths, that cause difficulties in estimating the statistical significance of sequence alignments. Software for sequence database searching incorporating the formula is available on-line through the URL (http://www.well.ox.ac.uk/~rmott/ariadne.html).

EXAMINING THE RESULTS OF SEQUENCE DATABASE SEARCHES

Sequence databases have gotten so large that one can have several hundred positive hits to a single family of proteins or to a single domain that is present in several different families of proteins. This volume of hits can obscure other positive hits to protein domains that are distantly related to the query or are only present in one or a few of the sequence database proteins. This is particularly true when the score from a matching rare domain is low relative to the more common matches. Berman et al. (2) have developed computer algorithms to help deal with this problem and describe how these algorithms implement several different winnowing criteria in different circumstances, including the type of winnowing currently available in BLAST (11).

REFERENCES

1. **Altschul, S.F., R. Bundschuh, R. Olsen, and T. Hwa.** 2001. The estimation of statistical parameters for local alignment score distributions. Nucleic Acids Res. *29*:351-361.
2. **Berman, P., Z. Zhang, Y.I. Wolf, E.V. Koonin, and W. Miller.** 2000. Winnowing sequences from a database search. J. Comp. Biol. *8*:293-302.
3. **Dayhoff, M.O., R.M. Schwartz, and B.C. Orcutt.** 1978. A model of evolutionary change in proteins, p. 345-352. *In* M.O. Dayhoff (Ed.), Atlas of Protein Sequence and Structure 5(3). National Biomedical Research Foundation, Washington.
4. **Devauchelle, C., A. Grossmann, A. Henaut, M. Holschneider, M. Monnerot, J.L. Risler, and B. Torresani.** 2000. Rate matrices for analyzing large families of protein sequences. J. Comp. Biol. *8*:381-399.
5. **Doolittle, R.F.** 1986. Of URFs and ORFs: A Primer on How to Analyze Derived Amino Acid Sequences. University Science Books, Mill Valley California.
6. **Drasdo, D., T. Hwa, and M. Lassig.** 2000. Scaling laws and similarity detection in sequence alignment with gaps. J. Comp. Biol. *8*:115-141.
7. **Krause, A. and M. Vingron.** 1998. A set-theoretic approach to database searching and clustering. Bioinformatics *14*:430-438.
8. **Mott, R.** 2000. Accurate formula for P-values of gapped local sequence and profile alignments J. Mol. Biol. *300*:649-659.
9. **Muller, T., R. Spang, and M. Vingron.** 2002. Estimating amino acid substitution models: a comparison of Dayhoff's estimator, the resolvent approach and a maximum likelihood method. Mol. Biol. Evol. *19*:8-13.
10. **Muller, T. and M. Vingron.** 2000. Modeling amino acid replacement. J. Comp. Biol. *8*:761-776.
11. **Zhang, Z., P. Berman, and W. Miller.** 1998. Alignments without low-scoring regions. J. Comp. Biol. *5*:197-210.

CHAPTER 29

Approaches to Detection of Distantly Related Proteins by Database Searches

K. Cattell, B. Koop, R.S. Olafson, M. Fellows, I. Bailey[1], R.W. Olafson and C. Upton

University of Victoria, and [1]BC Systems Corporation, Victoria, BC, Canada

Reprinted from BioTechniques 21:1118-1125 (December 1996)

ABSTRACT

The searching of protein databases as a method of identifying newly sequenced genes is commonplace in molecular biology laboratories. However, it is a procedure that is not usually formally taught to students, and method cookbooks discuss it only briefly. This article uses a single family of highly diverged uracil-DNA glycosylases, which fall into two distinct groups, to highlight some of the difficulties associated with identification of such proteins by database searching.

INTRODUCTION

The process of using a protein primary sequence to search protein databases for similar sequences is common to most molecular biology laboratories. The goal is usually to infer the function of a protein by its similarity to a previously characterized protein, thereby saving significant amounts of time and effort. Thus, database searching has become a very important tool in experimental protocols, and the process of database searching should be regarded as a complex experiment in itself.

There are numerous computer programs that are capable of searching protein databases for similar sequences, but results can differ dramatically. It is difficult and time-consuming to maintain multiple programs and current versions of the protein sequence databases in appropriate formats. Additionally, search programs are not always available in formats for all computers, or they may run slowly on older machines. These problems have been solved for many researchers with the introduction of database searching by remote computing. This involves sending data by e-mail to a powerful distant computer that is capable of performing the searches very rapidly, and which also uses e-mail to return the results. More recently, in some cases, it has become possible to access the remote computer by using the World Wide Web (WWW; Reference 3). Pedro Coutinho

BioComputing: Computer Tools for Biologists
Edited by Stuart M. Brown

maintains an excellent catalogue of such sites, which can be accessed through the WWW (http://www.public.iastate.edu/~pedro/research_tools.html). Additional database searches may look for the presence of amino acid motifs or conserved domains, which are characteristic of many protein families. Such analyses may use the PROSITE (4) and BLOCKS (5) databases and rely on the previous characterization of a group of related proteins. Other approaches include the BEAUTY program (14), which attempts to integrate similarity and motif-based searching, and Blast3, which performs multiple alignments on sequences in the "hit list" (2).

In similarity searches, the time taken to search a database is dependent upon two invariant factors: the size of the database and size of the query protein; and upon two variables: the processing capacity of the computer and the particular program used. The use of "massively parallel" computer architectures addresses the first of these variables, although the choice of computer program and its associated parameter settings remains the most influential factor. Since the primary goal is the recognition of significantly related sequences that may have very low similarity, database searching protocols must attempt to optimize the speed of the search while providing maximum sensitivity.

Unfortunately, the output of database searches are not always simple to interpret. First, one should remember that there may not be any biologically significant similar proteins in the database, and that in such an example, the sequences at the top of the results list are there by chance. Second, also note that if two proteins are indeed related, several additional factors may still influence the success of database searching programs: (*i*) the overall percentage identity and percentage similarity between the proteins, (*ii*) the number of gaps required to align the regions of similarity and (*iii*) the distribution of the identical and similar residues within the protein sequences. As an example of the last point, it may be easier to recognize similarity between two proteins with 20% identity overall (global) if there is at least one small contiguous region (local, without gaps) with significantly higher similarity.

In this article, we wish to present some observations that significantly influence the success of detecting similarities between distantly related proteins and especially those without hot spots of high local similarity. These are in part based upon the identification of a poxviral uracil DNA glycosylase (12). While, with this example, we demonstrate the success of the programs NW_Align and FASTA, it is not our intention to rigorously compare the various search programs, but to use the results from searches with this protein family to illustrate a number of key points.

MATERIALS AND METHODS

Uracil DNA glycosylases from eight organisms were used in the database searches: Shope fibroma virus (SFV), vaccinia virus (VV), fowlpox virus (FPV), herpes simplex virus type-1 (HSV), Epstein Barr virus (EBV), *Saccharamyces cerevisiae* (yeast), *Streptococcus pneumoniae* (Strep.) and *Escherichia coli* (*E. coli*) with the following SWISS-PROT Database Accession numbers: P32941, P20536, P21968, P10186, P12888, P12887, P23379 and P12295, respectively. There are also a number of other uracil DNA glycosylases in the protein database.

Database searches (SWISS-PROT No. 32, containing 49 340 protein sequences) were performed at the following locations: (*i*) Blastp, blast@ncbi.nlm.nih.gov (BLOSUM62 matrix); (*ii*) Blitz, Blitz@embl-heidelberg.de (PAM120 matrix); (*iii*) FASTA v1.5, fasta@genome.ad.jp (PAM250 matrix; KTUP=1); (*iv*) DFlash, dflash@watson.ibm.com

Table 1. Percentage Amino Acid Identity Among the Eight Uracil-DNA Glycosylases

	SFV	VV	FPV	HSV	EBV	Yeast	Strep.
VV	70						
FPV	59	55					
HSV	21	23	24				
EBV	17	18	20	42			
Yeast	18	21	21	44	40		
Strep.	22	21	24	34	39	41	
E. coli	19	21	23	45	47	51	47

(PAM250 matrix); (*v*) FASTA v2.0 (BLOSUM50 matrix; KTUP=1) and (*vi*) NW_Align (RBO matrix) and were run locally on an Indy workstation (Silicon Graphics Inc., Mountain View, CA, USA). Help for the e-mail servers can be obtained by sending a message with "help" (no quotes) as the subject or as the message. Unless otherwise indicated, searches were performed using default parameters.

The NW_Align program is a module of the SEQSEE package (13), which runs on UNIX machines. It uses the Needleman-Wunsch algorithm (7), which scores global alignments and permits the introduction of gaps into the alignment. It also uses a nonstandard weight matrix (RBO), which was derived initially to function in similarity-based secondary structure prediction (13).

E-mail servers performed searches within highly variable times; often, results were not returned until the next day. FASTA searches, using v2.0 locally, took 3–4 min using KTUP=1. Blastp searches performed remotely on the WWW take less than 10 s of central processing unit (cpu) time, and results are usually returned within a minute, unless the load on the server is high (http://www.ncbi.nlm.nih.gov/BLAST). NW_Align searches, which were run with alignment scores sorted by raw score divided by sequence length, took approximately 1 h (e-mail: seqsee@procyon.biochem.ualberta.ca for further information).

RESULTS

It is instructive to review how this project began. Initial BLAST (1) and FASTA (9) database searches using an unidentified open reading frame (ORF) from SFV revealed similar ORFs in two other poxviruses: vaccinia virus and fowlpox virus. These three poxvirus proteins were clearly related, possessing amino acid identity ranging from 55% to 70% (Table 1), but no other significant matches were observed in the database searches.

Subsequent more rigorous searches using the newly available NW_Align module of SEQSEE as the search engine (13) placed several uracil-DNA glycosylases in the "hit list". While none of these uracil-DNA glycosylases produced the best score (after the poxvirus family of proteins), they were from very different organisms (HSV-1, yeast, *E. coli* and *S. pneumoniae*) and therefore unlikely to be highly similar. In fact, these enzymes share a low of 34% and a maximum of 51% identical amino acid residues (Table 1). In this situation, the significance of the matches was increased because the query detected homologous proteins that were themselves distantly related. Alignment of the poxvirus proteins with the uracil-DNA glycosylases (Figure 1) confirmed the matches detected by the NW_Align program were correct. Thus, two elements were critical in recognizing the match between the two sets of proteins: first, the search matched several members of a protein family, and second, the person running the search understood the significance of a single protein matching several members of protein family that were themselves distantly related.

The similarity between the original family of uracil-DNA glycosylases is limited to discrete regions of the protein sequences, and only a subset of these are also conserved in the poxviral proteins. Note that a number of small gaps must be inserted into the two groups of sequences to produce an optimal alignment (Figure 1). Subsequently, expressed protein from the SFV ORF demonstrated that this protein does possess the predicted uracil-DNA glycosylase activity (12). This represented the first DNA repair enzyme to be characterized in poxviruses and is especially interesting since it is essential for the replication of at least one poxvirus (10), whereas this enzyme is nonessential in other organisms. The PROSITE motif for uracil DNA glycosylases was subsequently modified to permit identification of the poxvirus enzymes.

To discover why the link between these two groups of homologous proteins was only detected by the NW_Align program, we have performed database searches using three poxviral (group A) and five other (group B) uracil-DNA glycosylases as the query

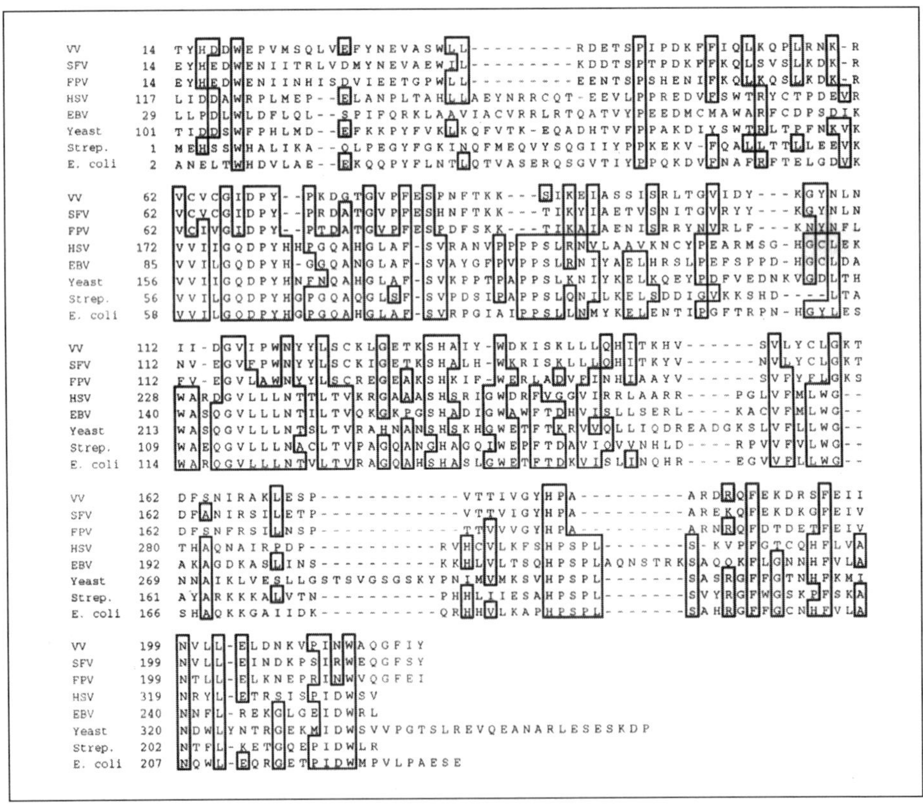

Figure 1. Alignment of the eight uracil-DNA glycosylases. The proteins have been truncated at the N-termini to match the start of the shortest sequence. Residues conserved in at least 50% of the sequences have been boxed. Protein sequences were manually aligned using SeqVu (Garvan Institute for Medical Research, Sydney, NSW, Australia).

Table 2. Detection of Uracil-DNA Glycosylases by Several Database Search Programs

	SFV	VV	FPV	HSV	EBV	Yeast	Strep.	*E. coli*
Blastp			HSV Strep. *E. coli*				FPV	FPV
FASTA v1.6	HSV	HSV	HSV *E. coli*	FPV VV				FPV
Blitz			HSV Strep. *E. coli*	FPV			FPV	FPV
DFlash			Strep.				FPV	
NW_Align	HSV Strep. *E. coli* Yeast	HSV Strep. *E. coli*	HSV Strep. *E. coli* Yeast	FPV VV SFV	FPV	FPV	FPV VV	FPV VV SFV

Each protein detected members within the same group. Therefore, only hits from the non-query group are reported. Default parameters were used.

sequences with several different search programs available through the Internet. In these searches, we have simply scored the presence or absence of matches anywhere in the hit list returned with default parameters. Although this scoring system includes matches that may not normally be considered significant, it mimics the process by which we originally identified the poxvirus uracil-DNA glycosylases. As expected, all programs successfully matched each query sequence with the other members of the query group. However, when scored for the ability to match a uracil-DNA glycosylase from one group with members of the other group, the programs performed differently (Table 2). It is also clear that all of the sequences in a group did not produce similar search results with a particular program. The best example of this can be seen with the FPV sequence, which consistently performed better than the other two poxvirus sequences (Table 2). The Blastp search with this query detected three other uracil-DNA glycosylases from the other group (HSV, Strep. and *E. coli*), which would have been sufficient to call attention to these matches even if they were not very high in the hit list. Equivalent searches with the other two poxvirus proteins (SFV and VV) failed to match any of the other group. This result can probably be ascribed to the slightly higher similarity of the FPV protein to the members of the other group (Table 1). The FPV protein shares 0%–4% greater identity with members of the other group than do the SFV or VV proteins, and high scoring amino acid substitutions may also influence the results. Similarly, searches with the other query group frequently matched only the FPV protein in the poxvirus group. Such matches were usually low in the hit list, surrounded by insignificant matches, and therefore this single match would not be sufficient to allow a researcher to recognize the existence of a relationship between the two protein groups. For this, it is necessary to produce hits with each of the poxvirus proteins and also be aware that these viruses are quite dissimilar. In this series of searches, the NW_Align program performed the best, since each of the poxvirus proteins were detected with the HSV and *E. coli* queries. NW_Align performs global alignments using the Needleman-Wunsch algorithm (7) and is well suited to this particu-

Table 3. Detection of Uracil-DNA Glycosylases by BLITZ with a PAM240 Matrix

	SFV	VV	FPV	HSV	EBV	Yeast	Strep.	*E. coli*
Blitz	HSV *E. coli*	HSV *E. coli*	HSV *E. coli*	FPV VV			FPV VV	FPV VV
PAM240	Strep.	Strep.	Strep.	SFV			SFV	SFV

With all searches, each protein detected members within the same group. Therefore, only hits from the non-query group are reported.

Table 4. Detection of Uracil-DNA Glycosylases by FASTA v2.0

	SFV	VV	FPV	HSV	EBV	Yeast	Strep.	*E. coli*
FASTA	HSV *E. coli*	HSV *E. coli*	HSV *E. coli*	FPV VV	FPV VV	VV	FPV VV	FPV VV
v2.0	EBV Strep.	EBV Strep. Yeast	EBV Strep.	SFV			SFV	SFV
Position in Hit list	1 2 3 5	1 2 3 5 12	1 2 3 4	1 9 20	3 26	14	1 2 3	1 2 3

With all searches, each protein detected members within the same group. Therefore, only hits from the non-query group are reported. Position in the hit list is given after the removal of members of the query group.

lar search situation. However, we have observed that Blastp and FASTA perform better than NW_Align when searching with short sequences (data not shown); therefore, one must tailor the search program to the task at hand.

Although it was not our intention to make an exhaustive comparison of the search programs as performed by others (8), we did change from the default comparison matrix in several instances. This had little effect on the search results with the notable exception of BLITZ, where changing to a PAM240 matrix significantly increased the sensitivity of the searches with both groups as the query sequences (Table 3). This was not unexpected since high-numbered PAM matrices are usually better for remotely related sequences, but this point emphasizes the need for users to be aware of a program's default settings. However, for most programs there was little to be gained by switching from the default matrices when searching for distantly related proteins.

The e-mail server used to perform FASTA searches used v1.5 of the program (currently runs v2.0); therefore, in an attempt to provide a current evaluation of FASTA, we repeated the searches using FASTA v2.0 on a local machine. This version of FASTA uses the BLOSUM50 matrix and includes statistical analysis of the search. The results shown in Table 4 clearly demonstrate an improvement in the detection of these distantly related proteins. However, again this shows the importance of which protein is used as the query sequence. For example, searching with the yeast protein only detected one of the poxvirus proteins, whereas the others in the group detected two or all of the poxvirus proteins. A further difference between the query sequences can be observed by examining where the hits are ranked in the hit list (after removal of hits from search query group of proteins). Searching with the *E. coli* or HSV protein resulted in the detection of all three poxvirus uracil-DNA glycosylases (Table 4). However, the *E. coli* query placed the proteins 1, 2 and 3 in the hit list, while HSV placed them 1, 9 and 20. This is important since the actual alignments are weak, and the significance of the results must be analyzed by the researcher scanning the hit list for related sequences. It is therefore essential that the search program returns a sufficient number of hits to allow this visual scanning. One could envision that a chance hit against a protein with many very close homologues in the database could push other significant hits so far down the hit list that they are not returned by the search program or are disregarded by the researcher.

DISCUSSION

The identification of distantly related protein sequences is important from many viewpoints. We have used an extreme example to provide insights into how best one can go about detecting such relationships. However, it will always be a difficult and frustrating task because it is hard to know when a negative result is a true negative, and one cannot afford to waste time chasing false positives. Uracil-DNA glycosylases are present in very diverse organisms and have probably evolved from a single ancestral gene. While alignment of these proteins demonstrates that there are a number of highly conserved amino acids, they are distributed throughout the protein sequence, and there are no large blocks that are conserved amongst all members of the family. A further complication results from the need to introduce a number of short gaps into the groups of sequences for optimal alignment (Figure 1). Indeed, these gaps are most likely responsible for the relatively poor results using Blastp with this particular family of proteins. A crystallographic study of one of the enzymes demonstrated that the active site is formed from noncontiguous regions (6), and this may account for the lack of any highly conserved block. This family of proteins may prove to be a useful control group in other tests of database search programs.

Our example of the identification of the poxvirus uracil-DNA glycosylases is not a unique situation. A poxvirus interferon-gamma binding protein was identified in an identical manner (11). These poxviral proteins are believed to have evolved from a eukaryotic interferon-gamma receptor, but the similarity is very low. Currently, Blastp does not detect these proteins, but NW_Align and FASTA match the viral proteins with both the human and mouse interferon-gamma receptors. Again, the significance of these matches is enhanced by an appreciation of the relatively low similarity between the human and mouse receptors.

It is not our intention to suggest that any one search program is always better than another. While the various algorithms, matrices and implementations clearly produce different results, our aim was to highlight the less obvious differences and suggest approaches to database searching that might increase the chances of detecting distantly related sequences. These are equally pertinent when searching the databases with a single unknown protein or with a group of proteins looking for new family members.

Our observations are summarized in Table 5.

In conclusion, since large amounts of resources are used in the determination of

Table 5. Summary of Observations

1) The person performing the search can provide valuable insight into the biological significance of sequences in the hit list.
2) The significant matches are not always at the top of the hit list.
3) If working with a group of unidentified proteins, then searches should be performed with all members of the group. Not all proteins within a family may be able to match with all other members of the family in a standard database search.
4) When using a group of known proteins to look for new members of a family, the new candidate members themselves should be used to search the database since they are likely to match several members of the family if they are true members.
5) Searches should return at least the top 50 matches to permit manual scanning for similar proteins that are not closely grouped. Large families of unrelated proteins may force weak matches so far down the hit list that it appears unlikely that a protein with such a low ranking could be significantly related to the query.
6) Searches should be repeated using different programs.
7) Searches should be repeated at regular intervals because of the rapid growth of the databases. By the time one reads "no matches were observed in the database" in a journal, this result is out of date.
8) Current versions of the search programs and databases should be used. This information is not always obvious from the search results or help files.
9) Beware of false positives.

DNA/protein sequences, one should be prepared to expend a significant amount of time on the analysis of this data.

ACKNOWLEDGMENTS

This work was funded by a Natural Sciences and Engineering Research Council of Canada (NSERC) Operating Grant to C.U. and Canadian Genome Analysis and Technology and NSERC grants to B.K. The help of the BC Systems Corporation is gratefully acknowledged. We would like to thank David Wishart and Robert Boyko for help with NW_Align.

REFERENCES

1. **Altschul, S.F., W. Gish, W. Miller, E.W. Myers and D.J. Lipman.** 1990. Basic local alignment search tool. J. Mol. Biol. *215*:403-410.
2. **Altschul, S.F. and D.J. Lipman.** 1990. Protein database searches for multiple alignments. Proc. Natl. Acad. Sci. USA *87*:5509-5513.
3. **Atwell, R., F. Gibbins and C. Upton.** 1995. Using a World Wide Web server as a local organizer for protein and DNA sequences. BioTechniques *19*:966-970.
4. **Bairoch, A.** 1991. PROSITE: a dictionary of sites and patterns in proteins. Nucleic Acids Res. *19*:2241-2245.
5. **Henikoff, S. and J.G. Henikoff.** 1994. Protein family classification based on searching a database of blocks. Genomics *19*:97-107.
6. **Mol, C.D., A.S. Arvai, G. Slupphaug, B. Kavli, I. Alseth, H.E. Krokan and J.A. Tainer.** 1995. Crystal structure and mutational analysis of human uracil-DNA glycosylase: structural basis for specificity and catalysis. Cell *80*:869-878.

7. **Needleman, S.B. and C.D. Wunsch.** 1970. A general method applicable to the search for similarities in the amino acid sequence of two proteins. J. Mol. Biol. *48*:443-453.
8. **Pearson, W.R.** 1995. Comparison of methods for searching protein sequence databases. Protein Sci. *4*:1145-1160.
9. **Pearson, W.R. and D.J. Lipman.** 1988. Improved tools for biological sequence comparison. Proc. Natl. Acad. Sci. USA *85*:2444-2448.
10. **Stuart, D.T., C. Upton, M.A. Hilman, E.G. Niles and G. McFadden.** 1993. A poxvirus-encoded uracil DNA glycosylase is essential for virus viability. J. Virol. *67*:2503-2512.
11. **Upton, C., K. Mossman and G. McFadden.** 1992. Encoding of a homolog of the IFN-gamma receptor by myxoma virus. Science *258*:1369-1373.
12. **Upton, C., D.T. Stuart and G. McFadden.** 1993. Identification of a poxvirus gene encoding a uracil DNA glycosylase. Proc. Natl. Acad. Sci. USA *90*:4518-4522.
13. **Wishart, D.S., R.F. Boyko, L. Willard, F.M. Richards and B.D. Sykes.** 1994. SEQSEE: a comprehensive program suite for protein sequence analysis. CABIOS *10*:121-132.
14. **Worley, K.C., B.A. Wiese and R.F. Smith.** 1995. BEAUTY: an enhanced BLAST-based search tool that integrates multiple biological information resources into sequence similarity search results. Genome Res. *5*:173-184.

Update to: Approaches to Detection of Distantly Related Proteins by Database Searches

Chris Upton

University of Victoria, Victoria, BC, Canada (cupton@uvic.ca)

In our original manuscript, we described some of the difficulties of detecting distantly related proteins by database searches. Although bioinformatics has grown rapidly (3), and database search algorithms, such as BLASTP and FASTA, have undergone improvements in the last few years, one still cannot be certain of finding all biologically significant matches in protein database searches using these programs. To illustrate this point, we have again used the family of uracil DNA glycosylases (UNG). We searched a protein database with 15 UNG protein sequences and scored whether they were found by each other and noted the expected value (Figure 1). We used UNG proteins from 5 chordopoxviruses (FPV-1, FPV, MYX, VV, VAR), 2 entomopoxviruses, and 7 other organisms, which included eukaryotic, prokaryotic, and herpes sequences. From the results in Figure 1, it is clear that the entomopoxvirus proteins are more similar to the eukaryotic, prokaryotic, and herpesvirus sequences than other poxvirus UNGs and that all members of each group match the other members of the group in both BLASTP (Figure 1A) and FASTA (Figure 1B). It is also apparent that with these search programs, proteins of one group do not reliably "hit" members of the other group. For example, the yeast UNG does not match any of the orthopoxvirus proteins nor is matched by them in either BLASP or FASTA searches. Other proteins have intermediate results. Thus, if one was working with the vaccinia virus protein because of its interesting role in DNA replication and only searched the protein database with this particular poxvirus orthologs and only used BLASTP, it would not be discovered that these poxvirus proteins are members of the UNG family.

Although the points listed in Table 5 (of the original manuscript) should still help researchers detect low similarities, such as those that exist between these two groups of UNGs, a new interative searching strategy has been described in recent years together with several computer programs that make it relatively easy to perform.

The Position-Specific Iterative Basic Local Alignment Search Tool (PSI-BLAST) program (1) is a modification of BLAST, such that after the first BLASTP search, a position-specific scoring matrix (PSSM) is automatically constructed from a multiple alignment of the best hits. The PSSM is used in a second BLASTP search, and the process is repeated until convergence is achieved. Users can control which sequences are used to build the PSSM. An advantage of PSI-BLAST is that it can be run remotely from the same Web site that makes BLASTP available (http://ww.ncbi.nlm.nih.gov). Figure 1C illustrates the significant improvement of PSI-BLAST at detecting the relationship between the two groups of UNGs. Almost every sequence was detected by 3 search iterations. The failure of herpes simplex virus-1 (HSV-1) to detect 2 of the poxvirus sequences is probably due to the large number of herpesvirus UNGs that are in the database. Thus, the PSSM becomes heavily weighted in favor of herpes UNGs and is less useful for detecting the distantly related sequences. This also highlights the fact that PSI-BLAST has the potential to produce misleading hits with apparently significant E values after detecting a large unrelated family of proteins.

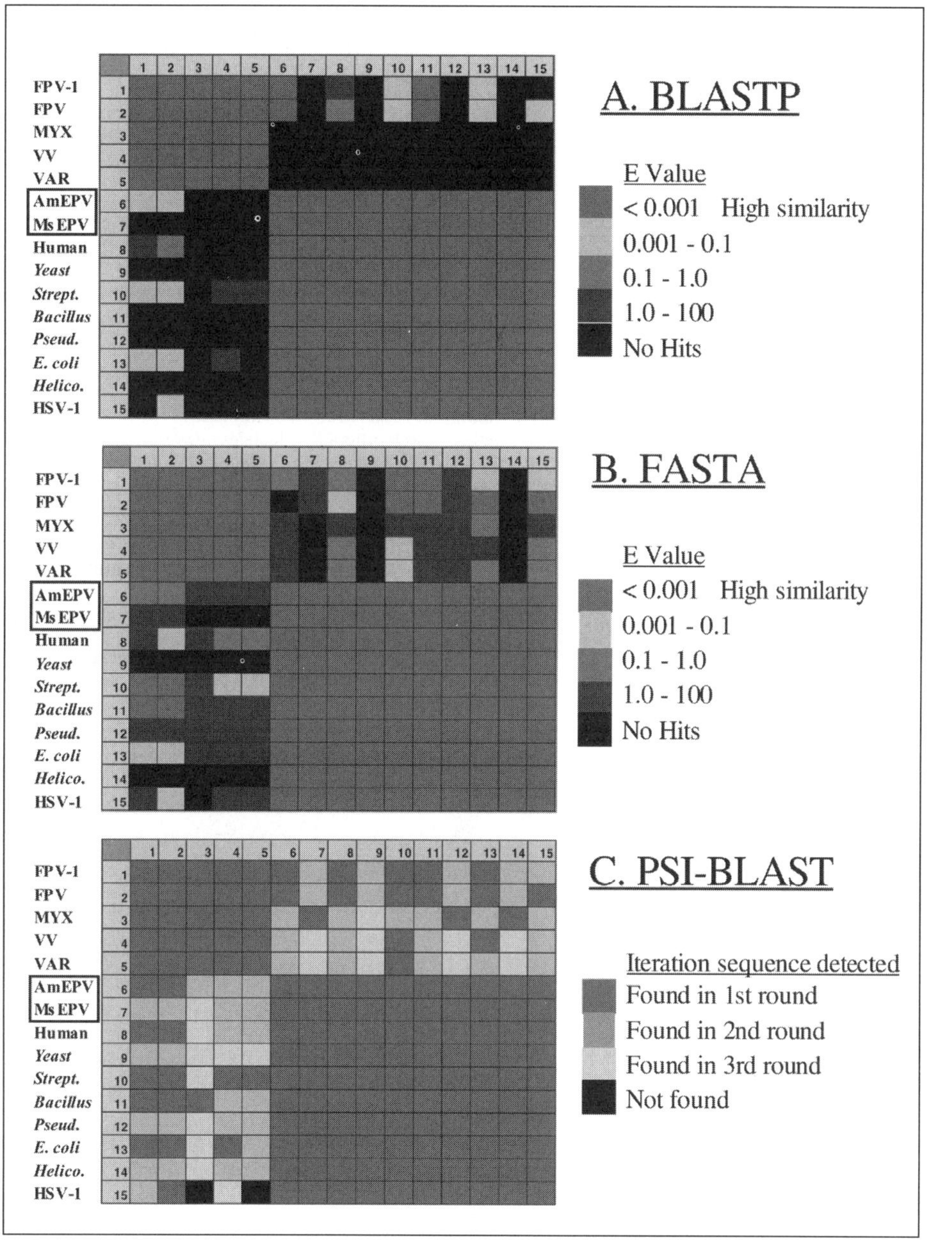

Figure 1. Database searches with uracil DNA glycosylase. (A) BLASTP, (B) FASTA ktup (wordsize) = 2, and (C) PSI-BLAST. Query sequences are shown along the vertical axis and matches are indicated on the horizontal axis. FPV-1, fowlpox (A35216); FPV, fowlpox (NP_039025); MYX, myxoma virus (NP_001611); VV, vaccinia virus (NC_001559); VAR, variola virus (NC_001611); AmEPV, Amsacta moorei entomopoxvirus (NC_002520); MsEPV, Melanoplus sanguinipes entomopoxvirus (T28369); Yeast, *Saccharomyces cerevisiae* (NC_001145); Strept., *Streptomyces coelicolor* (CAB95800); Bacillus, *Bacillus subtilis* (S39712); Pseud., *Pseudomonas denitrificans* (I36145); E. coli, *Escherichia coli* K12 (DGECU); Helico., *Helicobacter pylori* (NP_223984); HSV-1, herpes simplex virus-1 (BAA00274). **(See color plate A4.)**

Saturated BLAST (2) uses an Intermediate Sequence Search (ISS) method. It is based on the idea that a sequence that is intermediate to two remotely homologous proteins may have significant alignment scores to both and can be used to establish a relationship between the original two proteins. The program runs multiple BLASTP searches with ISS via a graphical user interface, automatically identifying new sequences to be used as query sequences in the next round of BLAST searches. Saturated BLAST runs under LINUX and is available from (http://bioinformatics.ljcrf.edu/xblast).

These new iterative programs have the potential to detect very distantly related proteins, but they can generate misleading results, and the user must interpret the output carefully.

REFERENCES

1. **Altschul, S.F., T.L. Madden, A.A. Schäffer, J. Zhang, W. Miller, and D.J. Lipman.** 1997. Gapped BLAST and PSI-BLAST: a new generation of protein database search programs. Nucleic Acids Res. *25*:3389-3402.
2. **Li, L., F. Pio, K. Pawlowski, and A. Godzik.** 2000. Saturated BLAST: an automated multiple intermediate sequence search used to detect distant homology. Bioinformatics *16*:1105-1110.
3. **Rehm, B.H.A.** 2001. Bioinformatic tools for DNA/protein sequence analysis, functional assignment of genes and protein classification. Appl. Microbiol. Biotechnol. *57*:579-592.

CHAPTER 30

Database Searches with Multiple Oligopeptides Containing Ambiguous Residues

Michael H. Vodkin, Robert J. Novak, and Gerald L. McLaughlin[1]

Illinois Natural History Survey, Champaign and University of Illinois, Urbana, IL;[1]Indiana University School of Medicine, Indianapolis, IN, USA

Reprinted from BioTechniques 21:1116-1117 (December 1996)

Several techniques in molecular biology frequently yield partial and ambiguous data on genes and gene products. For instance, N-terminal sequence analysis of oligopeptide cleavage products generates this type of sequence data. Typically, data generated from blotted or HPLC-resolved peptides consist of disconnected and unordered oligopeptides derived from N-terminal analysis of fragments resulting from complete or partial trypsin, chymotrypsin or CNBr digestion; such sequences are also "linked" if they were derived from the same isolated polypeptide, e.g., a band identified after sodium dodecyl sulfate polyacrylamide gel electrophoresis (SDS-PAGE) and blotting. To empirically identify the protein represented by such data, labor-intensive sequencing, with or without cloning, is frequently required. We became interested in defining a strategy to more reliably identify the source protein from existing sequence databases without an investment of additional laboratory experiments.

Several algorithms are readily available to rapidly search the databases for proteins or nucleic acids that are identical to or related to a specified query sequence. One popular program is the basic local alignment search tool (BLAST) (Reference 1; see Availability). However, BLAST can search databases (e.g., SWISS-PROT) with only moderate sensitivity. At the National Institutes of Health (NIH) address (see Availability), BLAST can search the updated, nonredundant protein or nucleic acid databases.

A disadvantage of BLAST is that very limited ambiguity is allowed at each position. For amino acids, "X" designates an unknown, "B" designates aspartate or asparagine, "Z" designates glutamate or glutamine and "-" designates a gap of indeterminate length. Table 1 shows an actual example of such data. When the individual oligopeptides listed in Table 1 were used to search the SWISS-PROT database, multiple related and unrelated sequences with similar or identical scores were retrieved. Even by comparing the individual lists for common, multiple hits, it was not possible to determine a unique candidate protein that was related to all or most of the oligopeptides.

BioComputing: Computer Tools for Biologists
Edited by Stuart M. Brown

Table 1. N-terminal Sequence of Tryptic Peptides of a Protein from Rat-Derived *P. carinii*

Oligopeptide No.	Residues
1	(K,R)X(E,L)TNLEGLP
2	I(C,R)DEYQ(N,K)A
3	(Q,A,Y)(E,L)(D,S)(L,I)QIATYR
4	(T,S,L)(A,K)(A,F)(D,S)(D,T)FQALR
5	QFP
6	(G,N,S)(V,L)(Q,E,Y)(D,S)L(Q,E)(E,I)

A protein was isolated using detergent extraction and column chromatography from the lungs of rats heavily infected with *P. carinii*. The protein was resolved on a SDS-polyacrylamide gel and found to have a mobility corresponding to 40–60 kDa. It was blotted to an Immobilon® filter (Millipore, Bedford, MA, USA), eluted and partially cleaved with trypsin. Each fragment was resolved by HPLC, and N-terminal analysis was performed on 6 fragments. Each amino acid residue is designated by a single letter. Multiple letters within parentheses indicate ambiguity at that site in the biochemical analysis. "X" is an unknown residue.

Another approach tested was to search with BLAST in pairwise or N-wise combinations of the oligopeptides, either as a continuous string of residues or as a broken string with hyphens designated as a discontinuity. (BLAST at the NIH supports the latter syntax; however, BLAST at some other addresses does not.) The first method created strings of characters that were not originally juxtaposed and thus did not allow the correct identification. The second method, when used in various pairwise combinations, still did not detect homologous proteins in the database.

We therefore utilized an alternative search algorithm and show its utility for identifying a protein in the database when query sequences include several linked oligopeptide fragments with some ambiguous amino acid residues. FindPatterns, or Find (a subset of the GCG package; see Availability), was used for the peptide data set in Table 1. Find has more versatility for managing ambiguous residues and multiple, discontinuous oligopeptides. Each individual residue of the query oligopeptide can be specified as either unknown (X) or up to a 20-fold ambiguity (every amino acid candidate at a position is encoded, as in Table 1). The gap size between the unordered fragments can also be specified with a minimum and maximum number of residues. The maximum, in the case of chemical or proteolytic digestion products of an unknown but purified peptide or protein, was set as slightly greater than the molecular weight of the protein as determined by relative mobility on SDS-PAGE (in this example of a 40–60-kDa polypeptide set at 500 residues).

Three of the 10 tested pairwise combinations, with all ambiguities inserted, jointly detected the family of eukaryotic vimentins (intermediate filament proteins) in one of the two possible orientations of the oligopeptides. Indeed, all of the oligopeptides listed on Table 1, except for the tripeptide 5, were successfully aligned and oriented with the vimentin sequences of vertebrates when searched with the identified vimentin proteins; Figure 1 shows the example of rat vimentin. Nineteen of the 28 unambiguous residues (scored as one of 3 for oligopeptide 5), or 68%, were identical to the aligned rat vimentin. For 3 of the 5 (60%) aligned oligopeptides, a potential R or K cleavage site was also located 1 or 2 residues upstream of the N-terminal sequence, as predicted by the speci-

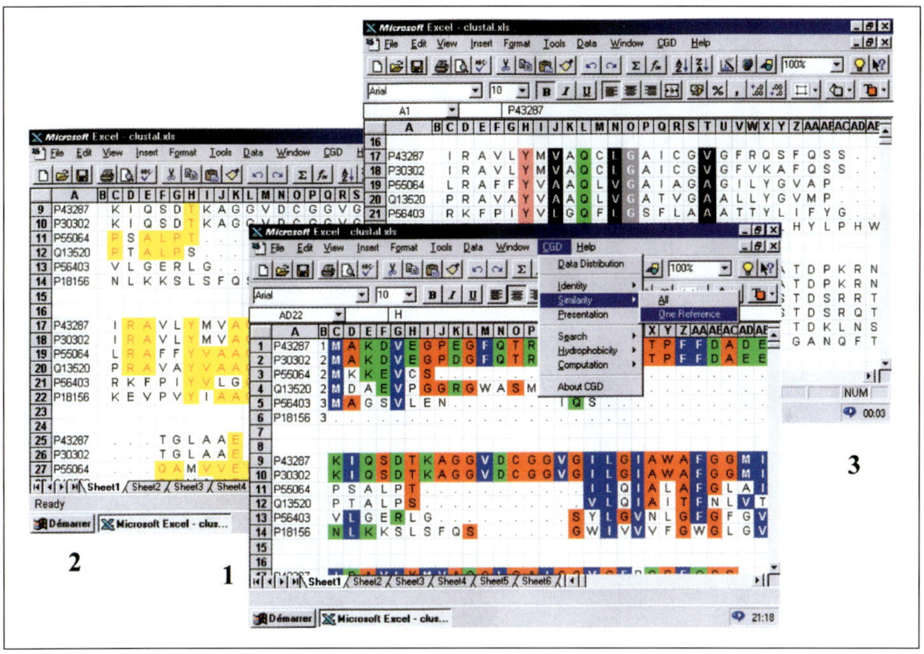

Chapter 7, Figure 1. Three examples of multiple alignment presentations with CGD. (1) Similarities with sequence P43287 according to structural properties of the amino acids, (A, C, G, P, S, T, W, Y), (R, N, D, Q, E, H, K), (I, L, M, F, V). Numbers typed in the first block of column B permit the manual indexing of each sequence for computing the residue frequencies (see the weighting method). (2) Identities with sequence P55064. (3) Similarities between all sequences according to chemical properties of the amino acids, (D, E), (H, R, K), (C, S, T), (N, Q), (M), (A, I, L, V), (F, W, Y), (P), (G).

Chapter 7, Figure 3. Logo obtained from the 42 MIP sequence alignment in region of loop B. Colors are assigned according to the Blockmaker criteria (17): red for negative (D, E), blue for positive (H, K, R), light grey for polar uncharged (C, S, T), green for amide (N, Q), yellow for methionine (M), black for hydrophobic (A, I, L, V), orange for aromatic (F, W, Y), purple for proline (P) and grey for glycine (G).

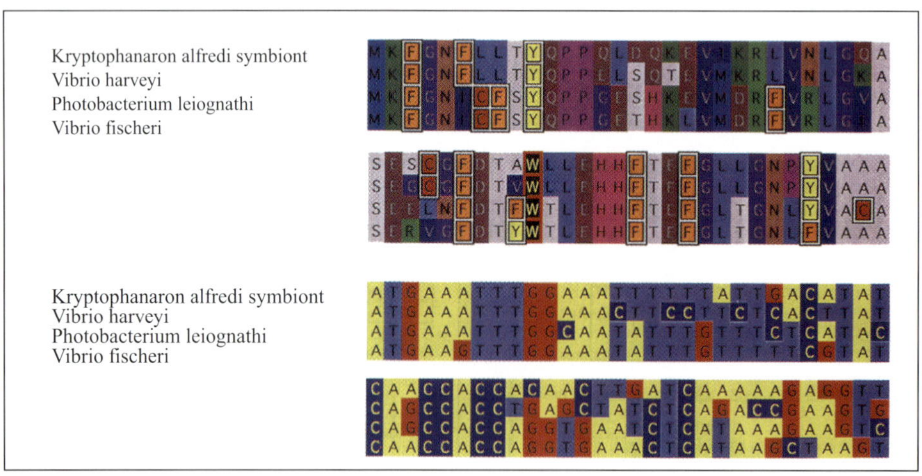

Chapter 8, Figure 1. Examples of colored sequences. Upper panel: Partial alignment of *luxA* (bacterial luciferase alpha subunit) amino acid sequences. Lower panel: Partial alignment of *luxA* nucleotide sequences. References for sequences: (1, 2, 5, 6).

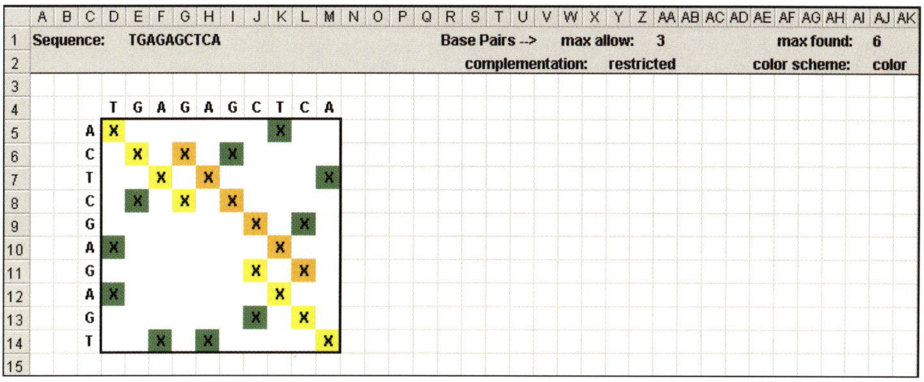

Chapter 9, Figure 2. Dimer analysis worksheet. The primer sequence 5′-TGAGAGCTCA-3′ (sequence no. 7 in Figure 1) is shown following the analysis for intramolecular/intermolecular complementarity. ROCK constructs a square matrix with the sequence across the top (5′ to 3′) and down the left side (3′ to 5′). At each row-column intersection, an "X" indicates complementation. In this example, three contiguous complementary base pairs was the maximum allowed (cell AA1). Two regions (yellow) composed of four contiguous pairs were detected near the termini of the sequence, and a single region (red) of six contiguous pairs was found near the middle of the sequence.

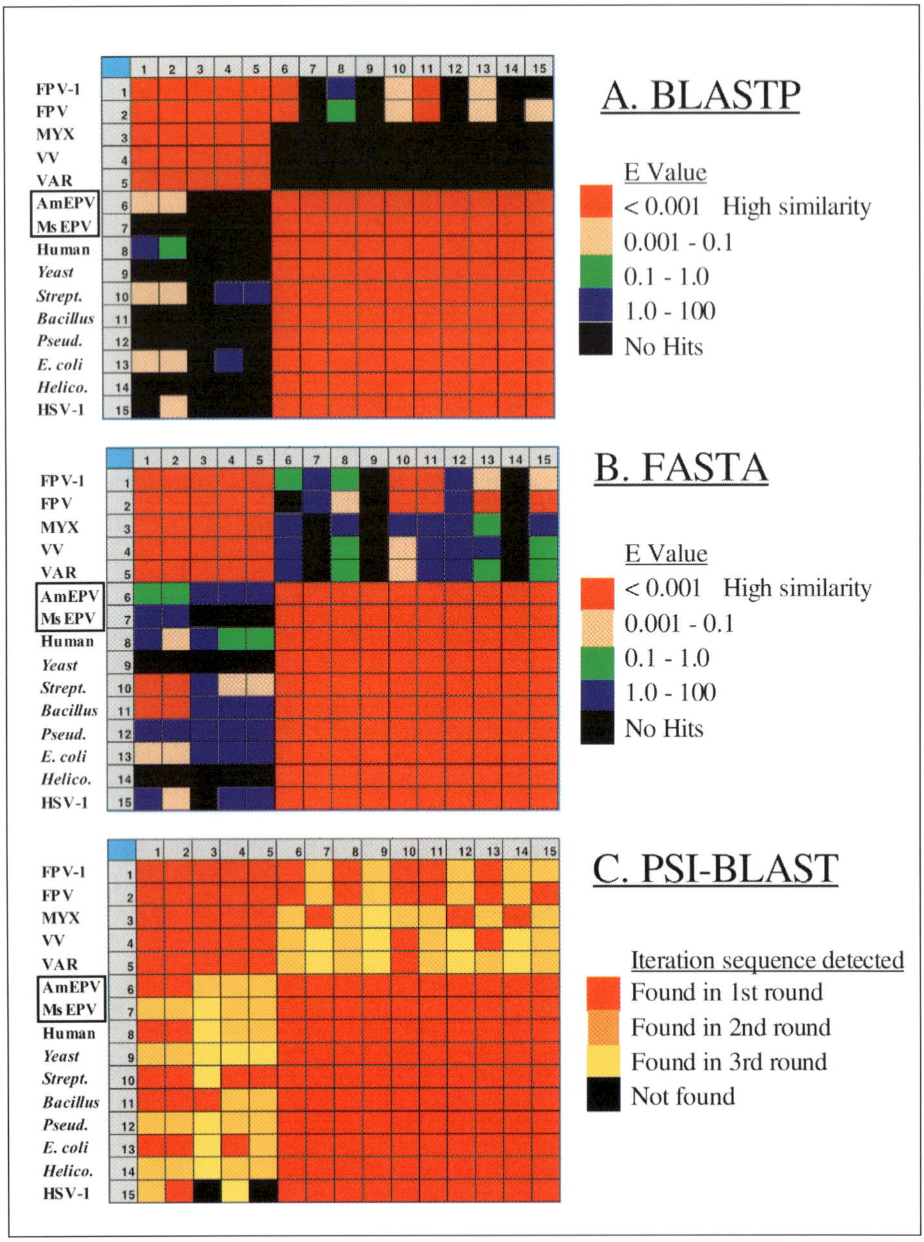

Chapter 29, Figure 1. Database searches with uracil DNA glycosylase. (A) BLASTP, (B) FASTA ktup (wordsize) = 2, and (C) PSI-BLAST. Query sequences are shown along the vertical axis and matches are indicated on the horizontal axis. FPV-1, fowlpox (A35216); FPV, fowlpox (NP_039025); MYX, myxoma virus (NP_001611); VV, vaccinia virus (NC_001559); VAR, variola virus (NC_001611); AmEPV, Amsacta moorei entomopoxvirus (NC_002520); MsEPV, Melanoplus sanguinipes entomopoxvirus (T28369); Yeast, *Saccharomyces cerevisiae* (NC_001145); Strept., *Streptomyces coelicolor* (CAB95800); Bacillus, *Bacillus subtilis* (S39712); Pseud., *Pseudomonas denitrificans* (I36145); E. coli, *Escherichia coli* K12 (DGECU); Helico., *Helicobacter pylori* (NP_223984); HSV-1, herpes simplex virus-1 (BAA00274).

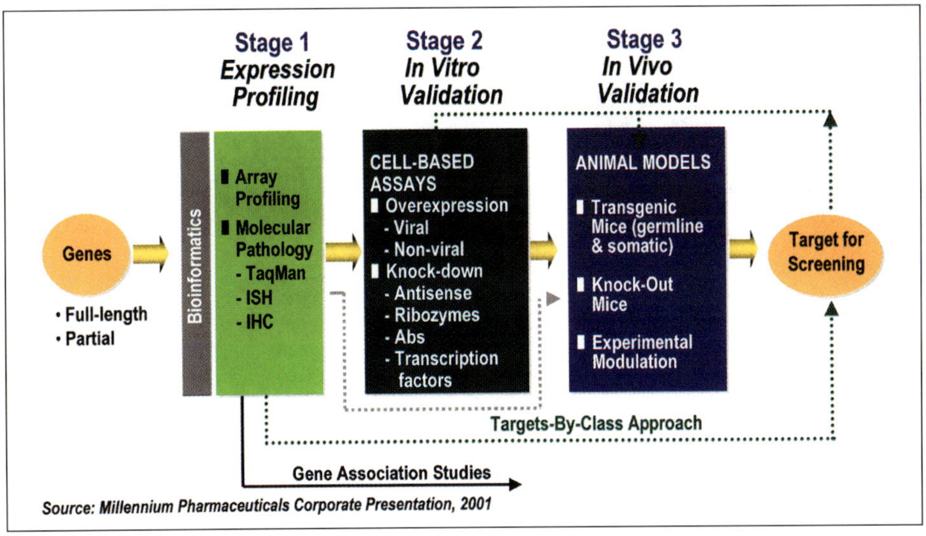

Chapter 31, Figure 2. Efficient processes for finding more targets.

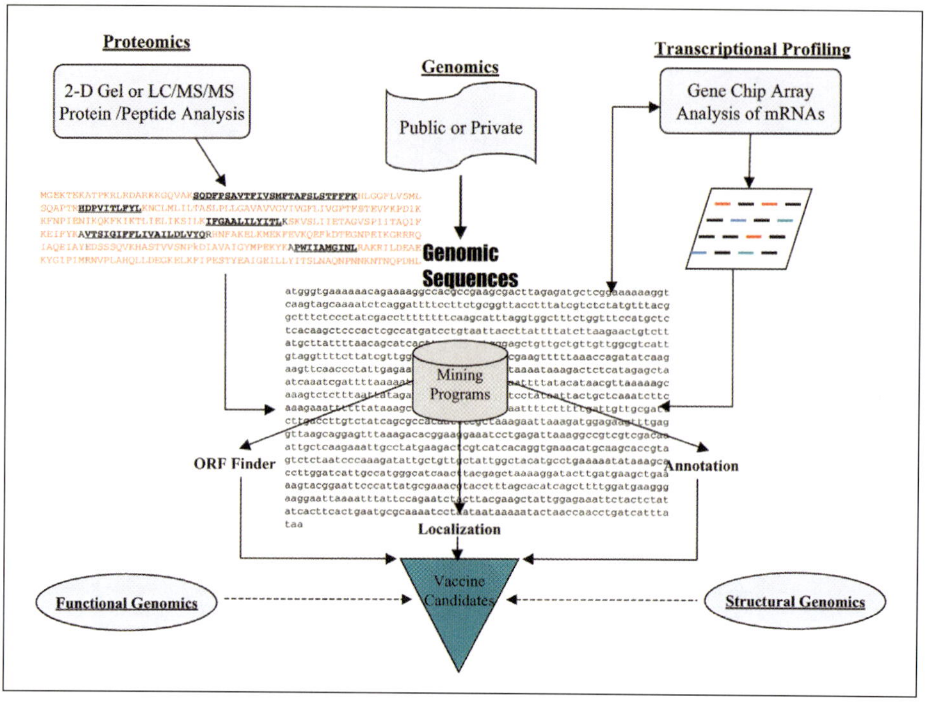

Chapter 32, Figure 1. Various research disciplines of bioinformatics. The center of bioinformatics is genomic DNA sequence that feeds into various research disciplines. Genomics mines the sequence and predicts various features such as genes and their putative function. Limited protein sequence generated by proteomics is searched through the translated genomic sequence for gene identification. Transcriptional profiling is dependent on genomics for gene identification and subsequent chip design. Structural genomics predictions are based on protein sequence and comparison with various structural databases. Functional genomics helps to correctly define gene function using both computational and biological information.

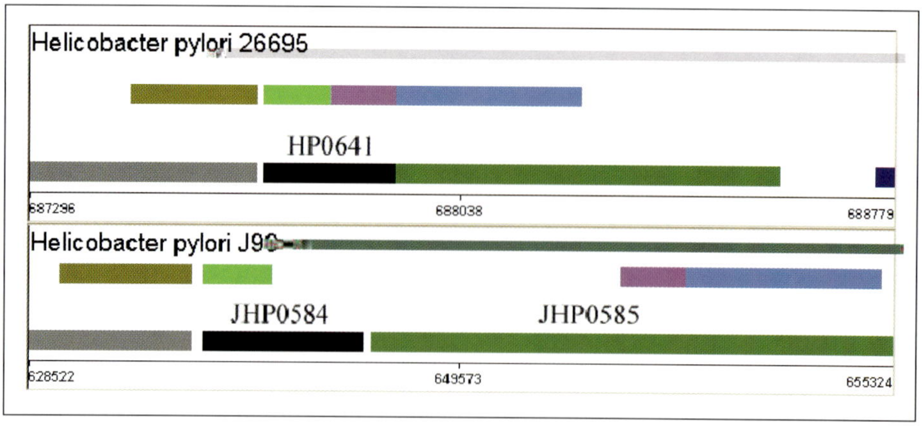

Chapter 34, Figure 1. The rearrangement between HP0641, JH0584, and JHP0585. The first half of HP0641 matches the first 120 bp of JHP0584, while the second half matches the middle 110 bp of JHP0585. This reveals that deletion or insertion took place in one of the genomes at this locus.

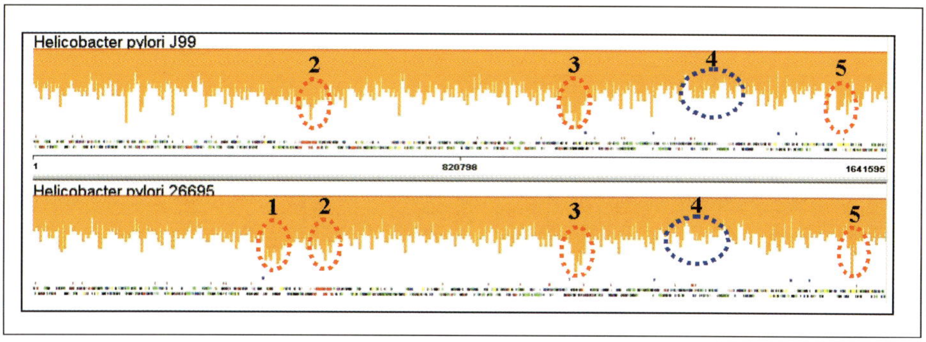

Chapter 34, Figure 2. The distinct GC regions in both stains. The picture was taken from a local view of the VGE*b*, displaying the GC content of both genomes. The top of the orange region represents 50% of GC content, the longer the orange bars, the lower the GC content. Here we see the two genomes generally have GC content < 50%. The red circles indicated the regions with particularly low GC content, while the blue circle displays the regions with exceptionally high GC content.

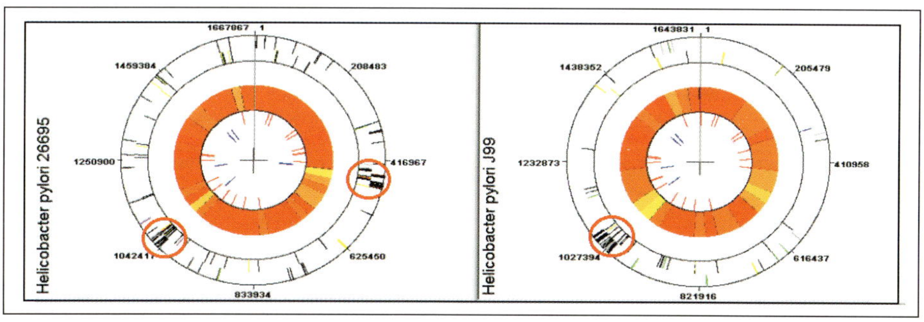

Chapter 34, Figure 3. The distribution of strain-specific genes in both genomes. The red circles indicate that the strain-specific gene clusters called plasticity zones. These clusters are always relatively low in GC content.

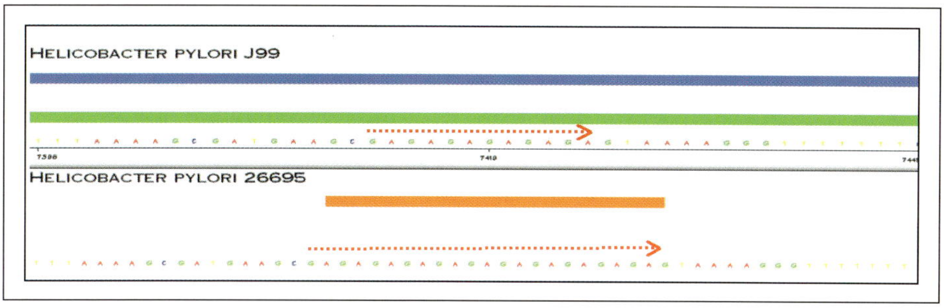

Chapter 34, Figure 4. Both are in "on" status. In the beginning of HP0007 and JHP0009, there are different numbers of GA repeats, indicating an "on" expression status.

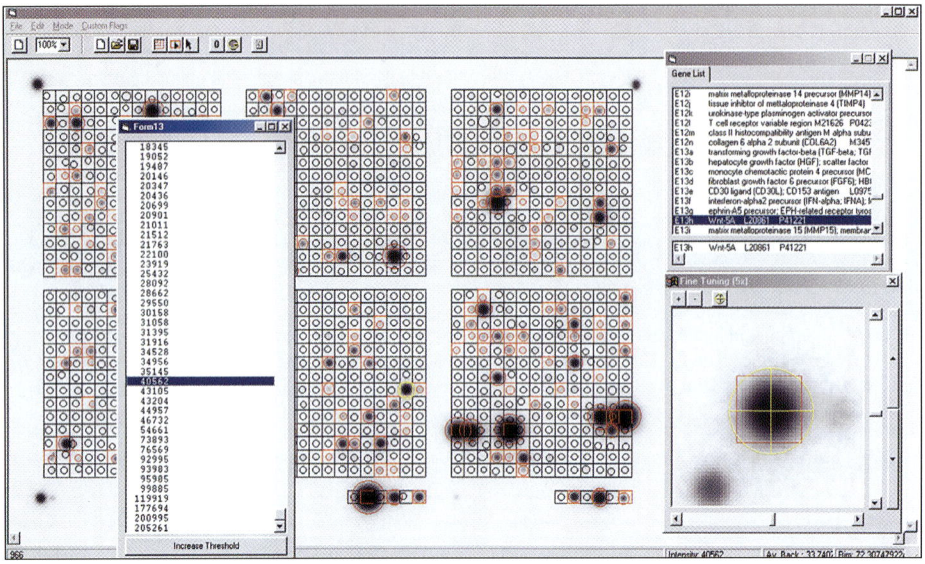

Chapter 36, Figure 1. ArrayExplorer screen display. ArrayExplorer has four linked operational windows. Each window can be resized and repositioned on the screen. The main window displays the entire array image with the overlaid reference grid. Red boxes indicate signals above the background threshold; black boxes are below this threshold. The *Fine Tuning*, *Gene List*, and *Threshold* windows allow easy navigation and manual adjustment of the grid. Operations are facilitated through the menu items, the toolbar buttons, and the designated shortcut keys.

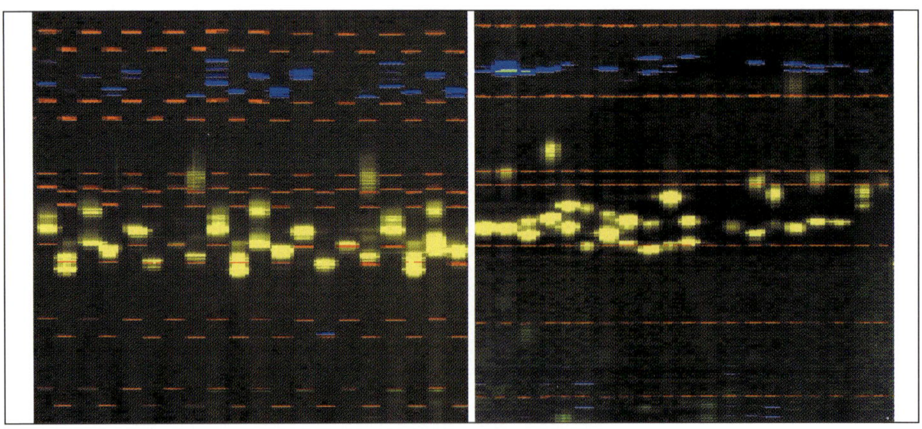

Chapter 46, Figure 1. Magnified sections of a typical 64-lane shark's-tooth comb (left) and 64+4 Rapid-Load membrane comb (right). Each lane is a four-locus multiplex reaction amplified from individual anomuran crabs, *Petrolisthes cinctipes*, and uses the GeneScan-500 ROX size standard from Applied Biosystems. We have zoomed in on the 200–450 bp portion of the center of each gel (lanes 22–42) for comparison.

ficity of trypsin (2). Taken together, these data indicate that the protein isolated from the infected rat lung is definitely related to vimentin. However, the amount of divergence for the unambiguous residues suggests that it is not identical to bona fide rat or to any other published vertebrate vimentin. No sequences from invertebrates are currently available for comparison; we hypothesize that it represents the Pneumocystis carinii vimentin analog, but another possibility is a more distant member of the rat vimentin family.

The probability of encountering multiple oligopeptides in an individual database protein is the product of the oligopeptides' individual probabilities. One reason for our success of matching the query peptides to the database with Find was the ability to incorporate more of the data, i.e., to also include the ambiguous residues in the search. Another reason is that the pairwise searching algorithm practically eliminates the chance of encountering a random single, very similar short peptide sequence in the same database. In our experience, this type of problem has increased significantly as the sequence databases have grown in size, and new approaches such as those we have used are necessary to address such random false homologies.

The strategy that we have defined to address the problem with discontinuous oligopeptides can be refined and extended to oligonucleotide sequences. For instance, similar searches can be designed to determine whether short oligonucleotides (e.g., random-amplified polymorphic DNA [RAPD] primers) are linked in the same nucleotide database entry and to help identify origins of linked but discontinuous DNA sequences. In summary, especially when multiple, short, ambiguous and linked sequences are available, the use of more versatile algorithms, such as Find with pairs of sequences, can provide a powerful advantage for database searching.

```
              .         .         .         .         .
   1 .GVESLQEXXXXXXXXXXXXXXXXXXXXXXXXXXXXXXXSKADDFQALRX 49
       ||||||                               ||:|  .|||
 221 RKVESLQEEIAFLKKLHDEEIQELQAQIQEQHVQIDVDVSKPDLTAALRD 270
              .         .         .         .         .
  50 XXXXXXXXXXXXXXXXXXXXXXXXXXXXXXXXXXXXXXXXXXXXXXXXXX 99

 271 VRQQYESVAAKNLQEAEEWYKSKFADLSEAANRNNDALRQAKQESNEYRR 320
              .         .         .         .         .
 100 XXXXXXXXXXXXXXXXXXXXXXXXXXXXXXXXXXXXXXXXXXXICDEYQN 149
                                                : ||.||
 321 QVQSLTCEVDALKGTNESLERQMREMEENFALEAANYQDTIGRLQDEIQN 370
              .         .         .         .         .
 150 AXXXXXXXXXXXXXXXXXXXALDIQIATYRXXXXXXXXXXXXXXXXXXXX 199
                         ||||:|||||
 371 MKEEMARHLREYQDLLNVKMALDIEIATYRKLLEGEESRISLPLPNFSSL 420
              .
 200 XKXETNLEGLP 210
        |||||:||
 421 NLRETNLESLP 431
```

Figure 1. Alignment of ordered oligopeptides to rat vimentin. The oligopeptides listed in Table 1 were searched in various pairwise combinations against the SWISS-PROT database with FindPatterns, Mismatch=4, minimal spacing between oligopeptides set at two and maximal set at 500 residue (GCG package). The following is an actual example of the syntax, using oligopeptides 1 and 2, in both orientations as the query (after $ or other prompt): find/batch/mis=4/pat=(K,R)X(E,L)TNLEGLP(X){2,500}I(C,R)DEYQ(N,K)A find/batch/mis=4/pat=I(C,R)DEYQ(N,K)A(X){2,500}(K,R)X(E,L)TNLEGLP. After it was determined that all the oligopeptides except for the tripeptide number 5 could be aligned with the available vimentin sequences, a protein was artificially created to maintain the appropriate spacing as determined by each oligopeptide's matched position, using X's for unknown amino acids, and to resolve ambiguous amino acid residues maximizing identity between the query and rat vimentin sequence (SW:VIME_RAT). This artificial protein (top line) was then aligned to rat vimentin (bottom) with GAP (GCG package).

AVAILABILITY

BLAST is accessible from Genetics Computer Group (GCG), Program Manual for the Wisconsin Package, Version 8.1, July 1995, 575 Science Dr., Madison, WI 53711, USA, E-mail (blast@ncbi.nlm.nih.gov) or at several World Wide Web sites (the primary one for this study is at the BCM Search Launcher, Human Genome Center, Houston, TX: http://gc.bcm.tmc.edu:8088/search-launcher/launcher.html).

ACKNOWLEDGMENTS

The primary peptide sequence data came from Dr. S.F. Queener and Dr. Lori Bolyard, Department of Pharmacology, Indiana University School of Medicine, Indianapolis, IN, and we thank them for permission to use the data to demonstrate the search strategy. We also wish to thank Dr. J. Siegel, Illinois Natural History Survey and Dr. R. Weigel, University of Illinois, Department of Veterinary Pathobiology for reading earlier drafts of the manuscript. This work was supported in part by a grant to R.J.N. from the Illinois Department of Natural Resources (SENR TM34). Computer support was provided to M.H.V. from the Pittsburgh Supercomputer Center (PSCB DMB890077P).

REFERENCES

1. **Altschul, S.F., W. Gish, W. Miller, E.W. Myers and D.J. Lipman.** 1990. Basic local alignment search tool. J. Mol. Biol. *215*:403-410.

2. **Voet, D. and J. Voet.** 1995. Biochemistry, 2nd ed. John Wiley & Sons, New York.

Update to: Database Searches with Multiple Oligopeptides Containing Ambiguous Residues

Michael H. Vodkin[1], Robert J. Novak[2], and Gerald L. McLaughlin[3]

[1]University of Illinois, Division of Environmental Health and Safety, Urbana, IL; [2]Illinois Natural History Survey, Champaign and University of Illinois, Urbana, IL; [3]National Institute of Health, NIAID, Bethesda, MD, USA (r-novak1@uinc.edu)

All the software mentioned in the original article is still extant, although the Web site addresses have changed. The new addresses are as follows: Genetics Computer Group (GCG) is located at (http://www.accelrys.com/), Entrez for Basic Local Alignment Search Tool (BLAST) searches is located at (http://www.ncbi.nlm.nih.gov/blast/Blast.cgi), and the Baylor College of Medicine (BCM) suite is located at (http://searchlauncher.bcm.tmc.edu/). GCG has added an additional database that nominally could be used with FindPatterns. However, this database is not particularly useful, since the information represents a subset of the other available databases. There are now several alternative versions of the BLAST search, but none is designed to use input data with ambiguous amino acid residues at multiple sites or gaps of unknown residue length. Thus, FindPatterns remains a valid tool to search databases with the above type of input. Retrospective studies could also be initiated if there was a method to systematically access such archived data. As we look ahead, the principle and concept of this paper may be applicable to the expanding fields of proteomics and genomics.

Section IV

Gene Expression and Genomics

Microarray technology has rapidly become a popular tool that allows biologists to measure levels of expression of thousands of genes in a single experiment. Consequently, many biologists are suddenly facing the challenge of quantifying images containing thousands of spots that correspond to expression levels of thousands of genes, and then analyzing data files containing thousands of gene expression measurements. This is an obvious challenge and an opportunity for software developers. While a number of commercial software packages are now available, the free software described in these articles offers many innovative solutions to these problems.

The availabilty of complete genome sequences for many organisms allows many new approaches to the challenges of drug development and vaccine disovery. It is now possible to work backwards from a list of all of the genes and all of the proteins in a microorganism to narrow the seach for drug and vaccine targets.

Impact of Genomics in Drug Discovery

Susan J. Ward

Millennium Pharmaceuticals, Inc., Cambridge, MA, USA

Reprinted from BioTechniques 31:626-634 (September 2001)

For the past decade, the promise of genomics has been eagerly anticipated. Among many aspirations, the genome was felt by many to be the key to revolutionizing innovation and productivity in drug discovery. Now that the sequence of the genome is known, will the promise be realized?

The Promise of Genomics in Drug Discovery

Drug discovery had its origins late in the 19th century with the manufacture of natural products and semi-synthetic products such as aspirin. By the turn of the 21st century, the pharmaceutical industry had grown to a $385 billion business. Current drugs are known to modulate only approximately 480 gene products. Early in the 1990s, it was widely thought that the human genome would contain in the region of 150 000 genes—thus, the existing drugs modulated only a small fraction of gene function. Therefore, the pharmaceutical industry, spearheaded by a landmark $145 million deal between SmithKline Beecham and Human Genome Sciences in 1993, invested heavily in genomics research to be at the front of the race to identify new genes as targets for important new therapies.

The Pharmaceutical Innovation and Productivity Gap

Over the past decade, the pharmaceutical industry has undertaken a number of big moves—investment in genomics research, outsourcing of critical clinical development, major mergers and acquisitions—all in an attempt to bridge a major deficit in innovation and productivity. To achieve even a 10% annual growth rate, Top Tier Pharma companies today need to bring to market 3–5 major new products each year—but none do (Figure 1). Average annual new drug approvals in the U.S. doubled from the 1970s to the 1990s, while R&D costs more than quintupled (2) (i.e., the industry is spending incrementally more than double for what it delivers). Currently, the cost to discover and develop a successful new therapeutic product reflects as much as a 75% expenditure on research and products that ultimately fail (5).

Genomics to the Rescue?

The primary goal of genomics research in the pharmaceutical industry in the 1990s was

BioComputing: Computer Tools for Biologists
Edited by Stuart M. Brown

to identify not only new molecular targets but also more of them, and to be first to gain proprietary rights to use those targets. This goal has been successful—many, many more targets are known, intellectual property abounds, and some drugs emanating from these targets will indeed provide great benefits to patients (1). Moreover, the efficiency of identifying targets from the genome has increased markedly. The process of identifying genes, profiling them, and establishing a moderate level of validation of their potential utility in addressing disease was once a lifetime's work—now, many elements of this process have been industrialized. Millennium Pharmaceuticals, for example, is noted for its industrialization of the gene-to-target process (Figure 2) and striking yearly advances in making the process more efficient—in 2000, Millennium more than doubled its output from this process with no change in costs over 1999 expenditure, purely from process mapping and process enhancement.

To date, however, relatively few targets from genomics research have produced clinical candidates—the forerunners being Human Genome Sciences and GlaxoSmithKline. Despite great advances in the efficiency of identifying new targets, it is even less clear that the fundamental issue of productivity has been addressed. To understand why finding more new targets from genomics does not in and of itself solve the productivity and innovation gap in the industry, we need to look at the sources of lost productivity in the R&D process.

Cost of Failure in the Pharmaceutical R&D Process

For every dollar spent in the R&D process, an average of 25 cents is spent on the product that is eventually brought successfully to the patient (Figure 3). Historically, drugs have failed about equally for the following reasons: poor efficacy (the wrong target in the first place), poor drug disposition, unacceptable toxicities or side effects, and "other" (including erroneous commercial projections, changes in management, fiscal constraints, etc.). Of course, the later in development that a product fails, the more will have been spent to reach the no-go point. Thus, some of the biggest gains in productivity will be made by eliminating future failures sooner and avoiding as much as feasible of the 75% of total R&D that is

2000 Pharma Sales (Example Companies Near Size)	Anticipated Sales From Current Products in 2010	Annual Real Growth Target (2000-2010)	Sales Gap for New Products to Fill in 2010	Estimated Number of NMEs Required to Fill Gap (Over 10 Years)	Year 2010 Required NME Output
$15 Billion (e.g., GlaxoSmithKline, Aventis, Bristol-Myers Squibb)	**$10 B**	**8%**	**$22 B**	**40-45**	**5.5-6.0**
		6%	**$17 B**	**30-35**	**4.0-4.5**
		4%	**$12 B**	**20-25**	**2.5-3.0**
$8 Billion (e.g., Eli Lilly, Bayer, Schering Plough)	**$5.5B**	**8%**	**$12 B**	**20-25**	**3.0-3.5**
		6%	**$9 B**	**15-20**	**2.0-2.5**
		4%	**$6 B**	**10-15**	**1.0-1.5**

- **Top-tier players must average 3 significant NME launches per year to continue rapid growth**
- **Medium-sized companies must produce at least 1 NME per year to sustain average growth**

Source: Executive Briefing, High Performance Drug Discovery: An Operating Model for a New Era, Accenture, 2000

Figure 1. Number of new NMEs required to meet 10-year growth objectives.

spent on ultimate failures. Clearly, drug discovery that has a higher-quality output—a clinical candidate that is less likely to fail—can make a big dent in the productivity gap. A change of only 5% in the attrition rate creates as much as a $100 million increase in net present value (NPV) of the pipeline in a company previously delivering one new molecular entity (NME) a year at industry-average productivity.

Potential Value of Genomics on Productivity in Drug Discovery

Increasingly throughout the 1990s, genomics has been used broadly to identify additional gene family members and to create screens that chemists can use to identify drug candidates that are much more selective for their target protein than is typical historically. Identification of orthologs in species used for in vivo efficacy models or toxicological and drug disposition studies has also proven valuable to understanding drug behavior, especially in cases when compounds have differential affinity or activity at the target protein across species. Though there are many anecdotes to suggest their value, data on the extent to which these advances have translated into improved success rates in the clinic are not yet available.

The genome has provided us with many new "targets" for pharmaceutical drug research and development. The value of having more targets is that it should provide the basis for making *better choices* of targets to work on. The key to *making* these better choices is using the knowledge and tools of genomics to better understand the molecular pathways underlying disease. It stands to reason that the more one understands the complete linkage of genes, transcription factors, enzymes, receptors, modulators, etc., and their aberrations in disease, the more one can make rational choices of the best point in a pathway to intervene—choices driven by the sensitivity of the desired output from a pathway to different points of intervention, accessibility, and tractability to modulation by drugs, etc. A variety of genomically driven tools and technical capabilities will be important to make the "better choices" revolution real.

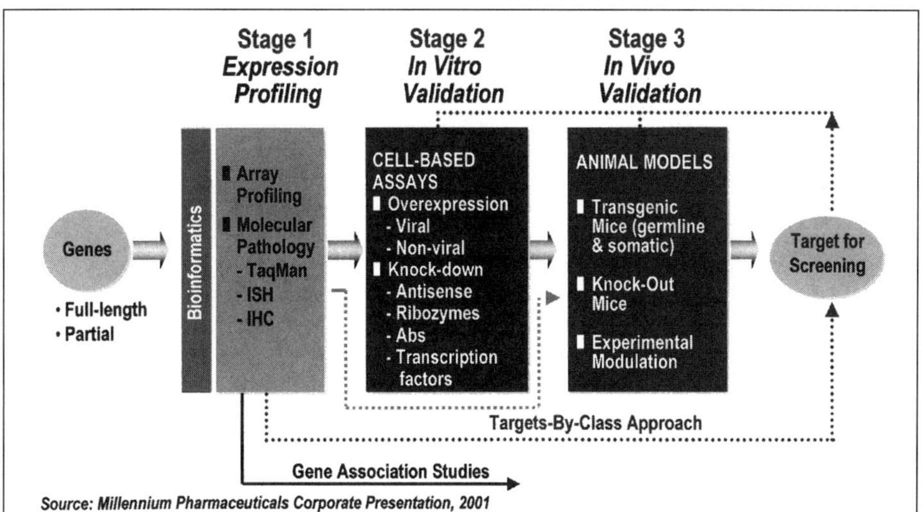

Figure 2. Efficient processes for finding more targets. (See color plate A5.)

Transcriptional Profiling

Transcriptional profiling is a key technology used to profile genes and identify their association with disease. Over- or under-expression of genes in a given situation provides a valuable window into the underlying molecular state. Comparisons that reflect perturbation of a system in multiple ways—for example, between diseased and normal cells, between drug-treated and placebo-treated cells, or between diseased cells responsive to therapy and those resistant to it—are particularly valuable. While regulated genes themselves may or may not be targets for drug intervention, they represent excellent clues that, with the help of software algorithms linking genes with proteins and known functions, can point researchers toward known pathways, new elements, or new pathways in disease.

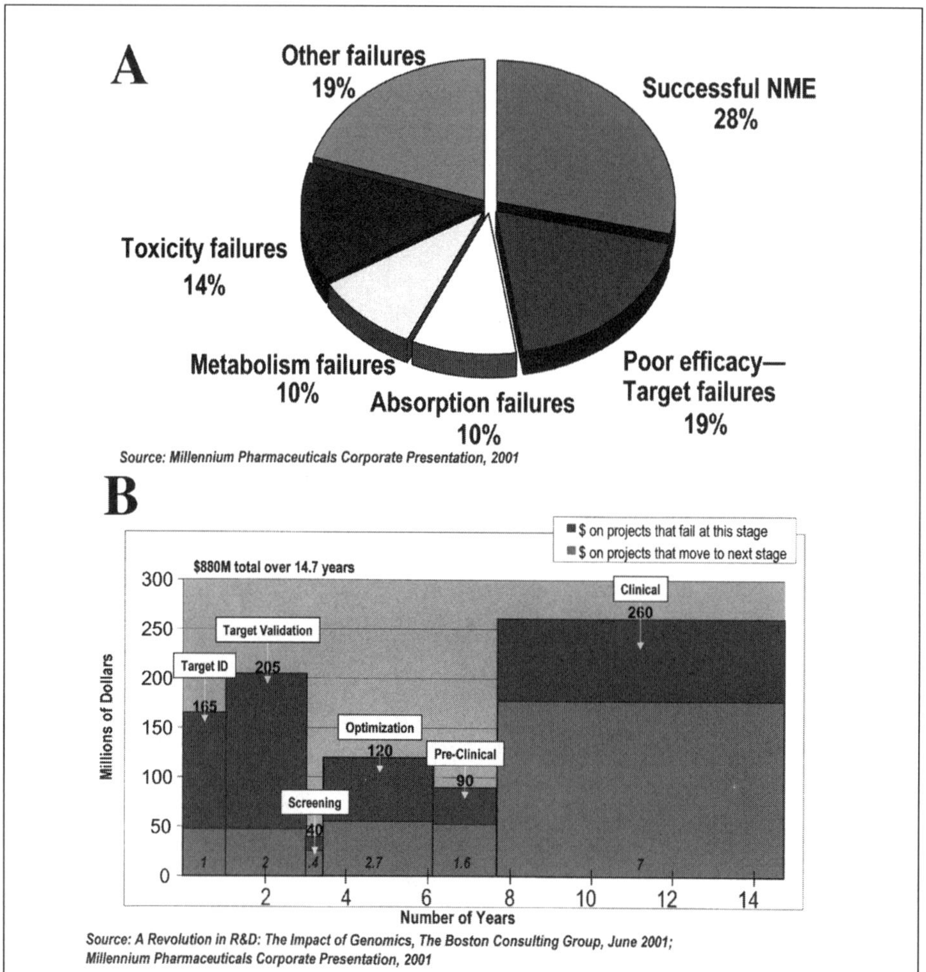

Figure 3. Causes of failure—where do the wasted dollars go in the industry-standard process? (A) Causes of project termination. (B) Investment by stage and outcome.

Model Systems

Most research is conducted on model systems of disease, which have historically had varying degrees of success in accurately predicting drugs that will have efficacy in Phase 2 clinical trials. Clearly, the power of model systems to more efficiently identify good drug targets and drug candidates is predicated almost entirely on the extent to which the model system truly reflects the human disease state. The capabilities of genomics, particularly transcriptional profiling and proteomic analysis, provide a powerful assessment at the molecular level of how well a model system reflects disease and, thus, the likelihood that it will adequately predict the relevance of therapeutic targets and drug action.

Genomics tools can have comparable value in assessing, or even creating, new systems that model drug disposition or toxicity. The recent introduction of an isozyme of human cytochrome P450 into mice in which the murine version had been knocked out is likely just the beginning of a new paradigm of model systems to assess drug disposition. Finally, as genomics technologies accelerate the rate at which model systems become amenable to formatting on a micro-scale, the medicinal chemistry cycle of design-make-test will continue to become much more efficient, taking additional costs out of the drug discovery process.

Chemical Genomics

The convergence of combinatorial chemistry, transcriptional profiling, proteomics, informatics, and miniaturization technologies has fostered the emergence of "chemical genomics" as an approach to target and drug discovery. In this paradigm, a desired phenotype or phenotypic change in a model system is sought by screening large libraries of diverse compounds. "Hits" are not only starting points for medicinal chemistry but also represent tools to dissect the molecular events and pathways underlying the observed phenotypic change. Pioneers of this approach include Stuart L. Schreiber (4) (Howard Hughes Medical Institute, Harvard University), Peter G. Schultz (3) (Scripps Research Institute), and Vertex Pharmaceuticals Incorporated.

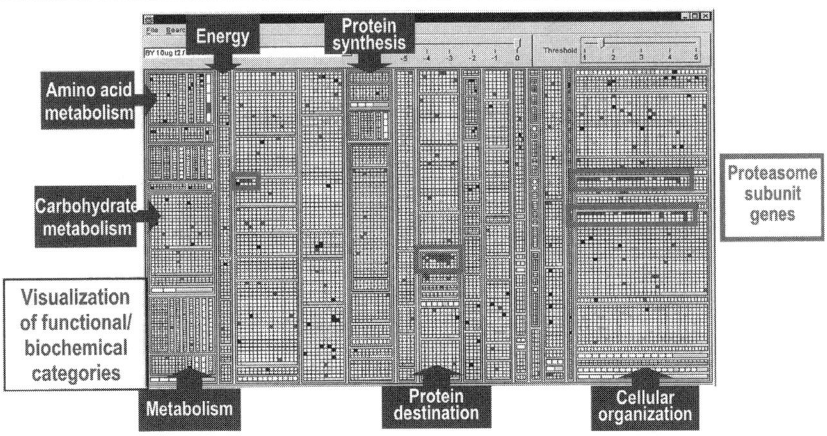

Source: Ronald K. Blackman, Jim Robinson, James Fleming, and Vala Thoroddsen, Millennium Technology at a Glance Presentation, 2001

Figure 4. A better way to visualize the data: linking biology with transcriptional profiling.

In one sense, this paradigm is a return to the empirical drug discovery paradigm of 20–50 years ago, when screening of potential drugs/drug starting points was often conducted in functional systems to identify a desired phenotypic change (e.g., drop in blood pressure, inhibition of seizures, change in electrical responsiveness of guinea-pig ileum, growth of cells in culture, etc.), without knowledge of the molecular entity for which compounds tested may have had affinity. Though largely a process "blind" to the molecular events in disease, this paradigm was in fact quite successful and brought many important drugs to patients. Conducted with today's powerful tools, this concept can be conducted at a massively higher scale (orders of magnitude), at a greatly reduced cost, and by using model systems and phenotypic read-outs that can be validated on a molecular level. Coupled with the potential of chemical leads to also illuminate important pathways and the molecular targets at which they interact, this approach has the potential to meld the best benefits and attributes of the two fundamentally different drug discovery paradigms of the last century.

Knowledge Management

The explosion of genomics information over the past decade is likely to dwarf the upcoming explosion of genomics, proteomics, and chemical genomics information of the post-genomic era. Ironically, the sheer volume of information can reduce, rather than enhance, understanding and productivity, unless the information is ordered logically. A simple example of this principle is illustrated in Figure 4, which shows the transcriptional response of yeast cells to PS-341, an agent under investigation by Millennium Pharmaceuticals for the treatment of cancer. Viewing the array of 5500 expressed genes is uninformative; viewing the same information through a filter that organizes genes by their function shows very clearly that the compound affects the proteasome.

The key drivers of turning the existing and upcoming masses of data and information into knowledge will be software algorithms and computer systems that can sensibly filter information into usable knowledge. Technologies that turn information into computable form, broad ontological frameworks adopted by the community at large, and simulation technologies will all be critical.

The Promise and the Reality

In this author's opinion, the promise of genomics is intact. This article has explored the value of genomics in drug discovery, illustrating a few benefits already realized and the potential for many more. The value of genomics will likely extend far beyond drug discovery, helping to identify patients whose disease, based on its molecular fingerprint, is better suited to one particular therapy over another—for example, as already happens in the prescribing of Herceptin and Tamoxifen preferentially in patients with HER-2-expressing or estrogen-dependent cancers. As a community, we are still on the early part of the learning curve, and it will take time for the full promise of genomics to be realized. But it is already evident that genomics can be a key driver of meeting the innovation and productivity gap in the pharmaceutical industry.

REFERENCES

1. **Amato, J.V. (Ed.).** 2001. The Fruits of Genomics: Drug Pipelines Face Indigestion Until the New Biology Ripens. Lehman Brothers and McKinsey & Company, New York.

2. **Kaitin, K.I. (Ed.).** Impact Report, Volume 2. Tufts Center for the Study of Drug Development, Tufts University, Boston, MA.
3. **Martin, A. and P.G. Schulz.** 1999. Opportunities at the interface of chemistry and biology. Trends Biochem. Sci. *24*:M24-M28.
4. **Schreiber, S.L.** 2000. Target-oriented and diversity-oriented organic synthesis in drug discovery. Science *287*:1964-1969.
5. **Tollman, P., P. Guy, J. Altschuler, A. Flanagan, and M. Steiner.** 2001. A Revolution in R&D: The Impact of Genomics. BCG Focus. The Boston Consulting Group, Boston, MA.

CHAPTER 32

Bioinformatics: Use in Bacterial Vaccine Discovery

Robert J. Zagursky and David Russell[1]

Wyeth Lederle Vaccines, West Henrietta, NY, and
[1]Idea Integration, Fairport, NY, USA

Reprinted from BioTechniques 31:636-659 (September 2001)

ABSTRACT

Bioinformatics has now become a common laboratory name for groups studying genomic sequences. It is composed of many different, yet interrelated scientific fields such as genomics, proteomics, and transcriptional profiling. The availability of complete genomic sequences, especially prokaryotic organisms, allows one to rapidly identify, analyze, and clone genes of interest. For bacterial vaccine discovery, one can "mine" the genomic sequence for potential surface targets using various algorithms, characterize these gene targets, and produce primers for cloning, all before one enters the wet laboratory. This review will focus on various genomic mining tools/algorithms available for predicting open reading frames and their associated annotation (if known), physical and functional characterization, and cellular localization. Finally, examples are given of how all of this is being used for the identification of potential bacterial vaccine candidates.

INTRODUCTION

Bioinformatics is a name coined in the 1990s in response to the increase in biological information and the ease of sharing this information using the Internet. Bioinformatics is defined as "a scientific discipline that encompasses all aspects of biological information acquisition, processing, storage, distribution, analysis, and interpretation that combines the tools and techniques of mathematics, computer science, and biology with the aim of understanding the biological significance of a variety of data" (14). Hence, bioinformatics covers a multiple of scientific disciplines and encompasses various areas of genomic analysis, including: *(i)* computational genomics (the storage and analysis of DNA sequence using various software), *(ii)* proteomics (laboratory methods used to determine what proteins are actually expressed and possible cellular location), *(iii)* transcriptional profiling (the analysis of mRNAs at different stages of cell growth and conditions), *(iv)* functional genomics (combined function prediction and experimental biological analysis of gene function), and

BioComputing: Computer Tools for Biologists
Edited by Stuart M. Brown

(v) structure-function determinations (predicted as well as determined 3-D structures and comparison to known protein structures and functions). Figure 1 shows the integration of these various scientific disciplines used by pharmaceutical companies in their attempt to help identify drug and vaccine targets. This review focuses on how bioinformatics, specifically computational genomics, is being applied in the search for protein subunit vaccines against bacterial diseases. (For a general review on bioinformatics, see Reference 100.)

Even today, with the best of medicines available, the leading cause of death in young children, worldwide, is infectious disease, as shown in Figure 2. Preventing infection or limiting the effects of the infection by vaccination dates back to the first scientific study and development of a safe vaccine by Jenner in 1796. Jenner utilized a live cowpox virus to vaccinate humans against smallpox virus. Today, there are many safe human vaccines directed toward preventing disease caused by various infectious agents (61). Present-day man is fortunate to live in a world free of smallpox and nearly free of polio thanks to the World Health Organization's (WHO) efforts of trying to eradicate diseases by vaccination. Advances in vaccine development have often been the result of applying basic research discoveries. For example, multivalent vaccines comprised of different capsular polysaccharides from *Streptococcus pneumoniae*, one of the major causes of bacterial meningitis, have been in use since 1946 and prevent bacteremic disease in older children and adults. Yet, these same vaccines were found to be largely ineffective in children less than two years of age and of variable effectiveness in children less than five years of age because of the immaturity of their immune systems (40). The capsular polysaccharides, which are also non-immunogenic in infants, became highly effective vaccine components when conjugated with protein antigens (37). Using glycoconjugate technology, an effective vaccine for prevention of invasive pneumococcal disease in infants has recently been developed (15). This vaccine, licensed in early 2000, provides approximately 100% protection against the seven most prevalent serotypes of *S. pneumoniae*.

A multivalent vaccine to prevent bacterial meningitis and sepsis caused by *Neisseria meningitidis* is also needed. In the case of this pathogen, there are at least 13 different serogroups, based on the bacterial capsular polysaccharide composition, with the majority of the disease caused by groups A, B, C, Y, and W-135 (61). Polysaccharide-based vaccines are available (for older children and adults) against serogroups A, C, Y, and W-135, and a serogroup C conjugate vaccine is available in several European countries. Unfortunately, the serogroup B capsular polysaccharide is immunobiologically similar to polysialic acid expressed on human cells and is poorly immunogenic in humans. Thus, trying to generate antibodies to this polysaccharide could potentially induce autoimmune disease and harm the vaccinee. Therefore, a protein antigen-based vaccine is being pursued (102).

As indicated above, several existing vaccines rely on generating an effective immune response against the bacterial polysaccharide capsule. Other effective vaccines are directed toward bacterial toxins (e.g., diphtheria, tetanus). For organisms that make no capsular polysaccharide or do not produce toxin-mediated diseases, other targets have to be identified. The most promising candidates are surface-exposed or secreted proteins. In the "pre-genomic sequencing era", a common approach used in early vaccine discovery programs was to generate polyclonal or monoclonal antisera in animals by administering either live or inactivated bacteria. The sera were analyzed in an immuno-assay format against either whole bacteria or a bacterial protein preparation. The major reactive proteins, usually less than a dozen, were further characterized by purifying the native proteins and analyzing the subsequent sera, generated from vaccinating animals with the same proteins. The sera were tested in functional immunobiological assays and then in an animal model. The proteins that elicited functional immunobiological activity were taken into the next phase of basic research. This second phase consisted of N-terminal amino acid sequencing, cloning and expression of the genes into a safe bacterial host, such as the well-known laboratory strain, K12 of *E. coli*, and analyzing the purified engineered proteins in various functional immunobiological assays. The total time it took to reach the end of the second phase could be as short as six months or as long as several years. Today, if one has the genomic sequence of an organism, one can rapidly identify literally hundreds of potential vaccine candidates during the first phase of the study and rapidly clone the genes by PCR technology into various expression vectors. This approach has recently been termed "reverse vaccinology" (103). The bottleneck now seems to be in the testing of the large number of gene candidates. Newer technologies, such as the linear expressed elements DNA vaccination method, are being developed that should reduce this time (127).

COMPUTATIONAL GENOMICS

Today, in what one might consider being the "post-genomic sequencing era", there is a tremendous amount of raw sequence data and related information. The total amount of information available to a researcher is becoming "mind boggling" and may give one a feel-

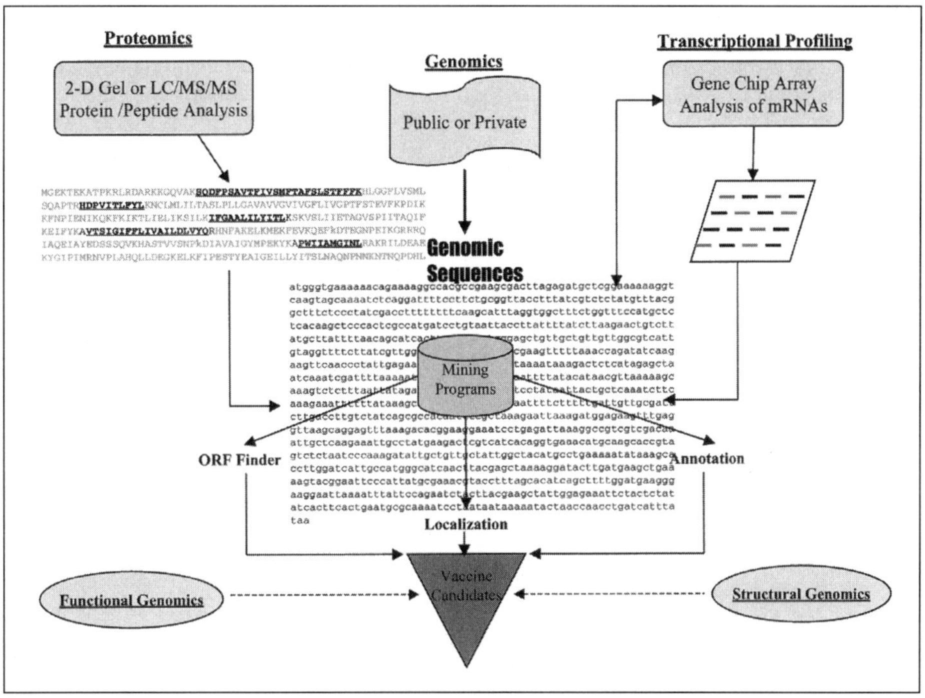

Figure 1. Various research disciplines of bioinformatics. The center of bioinformatics is genomic DNA sequence that feeds into various research disciplines. Genomics mines the sequence and predicts various features such as genes and their putative function. Limited protein sequence generated by proteomics is searched through the translated genomic sequence for gene identification. Transcriptional profiling is dependent on genomics for gene identification and subsequent chip design. Structural genomics predictions are based on protein sequence and comparison with various structural databases. Functional genomics helps to correctly define gene function using both computational and biological information. **(See color plate A6.)**

ing of data overload. In just six years since the first complete bacterial genome was sequenced (42), about 200 microbial genomes (about 40 complete) are being sequenced (see http://wit.integratedgenomics.com/GOLD/ and Reference 71). Having the genomic sequence of a bacterial organism allows one to search (mine) the genome and predict genes (protein and RNA), promoters and other regulatory sites, as well as the origin of replication and other features. To know the total genetic sequence of an organism was once thought to be the "Holy Grail" in genetics. However, understanding and deciphering all that information and assigning functions to the genes is turning out to be a long and exciting learning process for all genomes being sequenced. For vaccine discovery, it is most important to identify those genes, whether their function is known or not, that encode surface-exposed antigens. This is in contrast to drug discovery research in which knowledge of the function of a gene is paramount. Thus, of the predicted genes of a bacterial genome (gene density is roughly one gene per 1000 nucleotides—smallest genome, *Mycoplasma genitalium*, about 5.8 kbp; largest genome, *Nostoc punctiforme* about 7.5 Mbp), a screen of the genetic information for predicted surface or secreted proteins is done well before a researcher enters a wet laboratory. A schematic flowchart of the work progression toward having a vaccine candidate is shown in Figure 3. The first step in any mining operation is to identify the genes or open reading frames (ORFs) contained within the genomic sequence.

ORF (Gene) Finding

In the scientific literature today, an ORF and gene are being used interchangeably to define the region between two in-frame stop codons that may contain a putative start

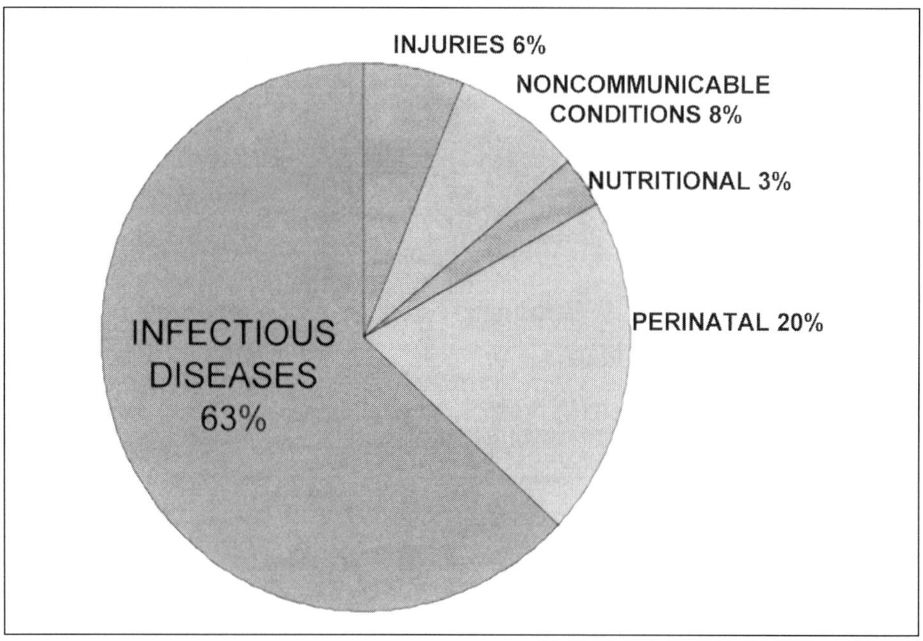

Figure 2. Causes of death in children worldwide. Distribution of main causes of death among children, ages 0–4 years—estimates are for 1998, worldwide. Data from the 1999 World Health Organization Report on Infectious Disease can be accessed at http://www.who.int/infectious-disease-report/index-rpt99.html.

codon regardless of whether there is biological proof that the encoded protein even exists. A number of ORF finder programs exist today to identify genes in a prokaryotic genome. The main feature of many of these programs is the use of some type of "learning" software that is trained on a dataset. Based on the training dataset, a model is built that incorporates information about codon frequency and probability of a certain nucleotide or state following another nucleotide or state. (For theoretical overview on sequence models, see Reference 36.) This is done in both sequence directions of the DNA looking in all six reading frames. Some of the ORF finders invoke other features for predicting genes such as allowed amount of nucleotide overlap between two genes, the start and stop codons to be used, and whether a ribosomal binding site or Shine-Dalgarno box is present (116). One of the best-known programs is GeneMark (17), which uses a Markov chain model based on a training set of known genes and noncoding DNA regions extracted from an external database for learning. The training set and the sequence to be analyzed should be of a similar statistical composition for best results. An *E. coli* training set would not be effective in predicting genes from a completely different species. A recent enhancement in this program uses a hidden Markov model (HMM) framework and RBS pattern to improve gene prediction (78). Another well-known and widely used program is GLIMMER, which was developed in part at The Institute for Genomic Research (TIGR) and uses a technique called interpolated Markov models (109) that trains itself on the genome being analyzed. GLIMMER first extracts large ORFs from the genome being studied as

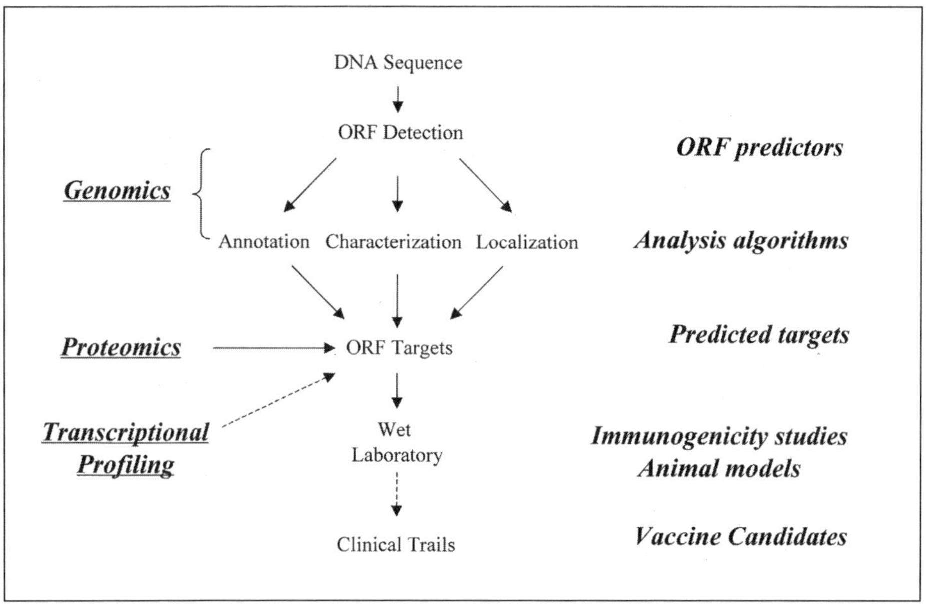

Figure 3. Flow diagram of vaccine discovery research involving bioinformatics. Starting at the top, ORFs are predicted from genomic DNA and further analyzed using various algorithms to help define function and location. Predicted surface-localized candidates are combined with experimentally determined surface proteins identified using proteomics, which results in a preliminary vaccine target list. From this list, the genes are then fed into the next phase of laboratory cloning, expression, and antibody generation. Successful gene candidates that make it through this phase are then tested in various pre-clinical immunological assays and animal model studies. Only those candidates that show positive results are further analyzed before going into clinical studies. Other research disciplines, such as transcriptional profiling, may contribute to this flow pathway.

Table 1. Listing of Various Bioinformatic Resources and Description

Sequence Databases		
NCBI	http://www.ncbi.nlm.nih.gov/	NCBI GenBank, PubMed, Entrez, and other databases of published sequence data
EBI	http://www.ebi.ac.uk/	European Bioinformatics Institute
DDBJ	http://www.ddbj.nig.ac.jp/	DNA Data Bank of Japan
PIR	http://www.mips.biochem.mpg.de/	Protein Information Resource
Domain or Pattern/Motif Databases and Tools		
PROSITE	http://www.expasy.ch/prosite/	Database of protein families based on patterns and weight matrices
Blocks	http://blocks.fhcrc.org/	Multiply aligned, ungapped segments corresponding to conserved regions of proteins
Pfam	http://www.sanger.ac.uk/ Software/Pfam/	Pfam-A contains a human-curated protein domain family. Pfam-B contains automatically clustered families.
ProDom	http://www.toulouse.inra.fr/ prodom.html	Automatically generated domains using a novel procedure based on recursive PSI-BLAST searches
PRINTS	http://www.bioinf.man.ac.uk/ dbbrowser/PRINTS/	Database of protein fingerprints consisting of multiple usually non-overlapping conserved motifs for a protein family
Subcellular Location		
PSORT	http://psort.nibb.ac.jp/	Expert system for protein localization
ProtLock	Reference 21	Statistical discrimination of amino acid content
Chou and Elrod	Reference 24	Discriminant function algorithm
LocateProtein	Reference 113	NN prediction
Signal Sequence Determination		
SignalP	http://www.cbs.dtu.dk/services/ SignalP/	NN and/or HMM signal sequence prediction
Sigcleave	EMBOSS	Weight matrix
SPScan	GCG	Weight matrix

Table 1. Listing of Various Bioinformatic Resources and Description (Continued)

Transmembrane Prediction		
DAS	http://www.sbc.su.se/~miklos/DAS/	Dot-plot analysis against a protein dataset
SOSUI	http://sosui.proteome.bio.tuat.ac.jp/sosuiframe0E.html	Average hydrophobicity with average amphiphilicity index and positive-inside rule
TopPred II	http://www.sbc.su.se/~erikw/toppred2/	Sliding window average of hydrophobicity and positive-inside rule
PHDhtm	http://cubic.bioc.columbia.edu/predictprotein/	Trained NN and multiply aligned sequences
TMHMM	http://www.cbs.dtu.dk/services/TMHMM/	HMM trained on protein dataset
PRED-TMR	http://o2.db.uoa.gr/PRED-TMR/	Statistical analysis of SwissProt proteins
Lio and Vannucci	Reference 74	Wavelet change-point
Software		
EMBOSS	http://www.hgmp.mrc.ac.uk/Software/EMBOSS/	Command line sequence analysis package
GCG	http://www.accelrys.com/	Command line sequence analysis package
SeqWeb	http://www.accelrys.com/	Web-based front end for GCG analyses
Lasergene99	http://www.dnastar.com/default.html	Windows sequence manipulation and analysis package
MacVector	http://www.accelrys.com/	Macintosh sequence manipulation and analysis package
Vector NTI	http://www.informaxinc.com/	Windows sequence manipulation and analysis package

the training set and then uses an HMM trained on the extracted dataset to find genes. Other programs such as ORPHEUS use a DNA to protein database similarity search to derive statistical characteristics of ORFs (44), and SELFID, a program similar to GeneMark, that uses an *ab initio* iterative Markov modeling procedure and can process unassembled sequences (8). The accuracy of ORF finding can be as high as 99% based on current knowledge and the latest ORF finders (33). Ultimately, predictions need to be validated by laboratory experiments, and new ORFs will continually be found that were not predicted by software. For instance, a database of new gene products characterized since the completion of the *E. coli* genome (16) is available at http://web.bham.ac.uk/bcm4ght6/res.html and is maintained by Gavin Thomas (131).

A number of factors can affect the true determination of an ORF such as the start and stop codons used. The major start codon is AUG, while GUG and UUG are minor codons. Rare start codons AUU, CUG, AAA, GGA, and UGC have also been found (2,16). There are even cases when multiple proteins are encoded within the same gene that use the same stop codon but different start codons such as the *clpA* and *clpB* genes of *E. coli* (95,114). Another example is the *nolA* gene from *Bradyrhizpbium japonicum* that produces three functionally distinct proteins (77). As in the case of start codons, there are a number of different stop codons (UAG, UAA, and UGA), but exceptions have been found both in mycoplasma and in some mitochondria. In these, UGA and UGG encode for tryptophan (65). The UGA codon can also direct the incorporation of the modified amino acid selenocysteine, but, for UGA to be decoded (or recoded), additional elements must be present (75). Ribosomal -1 frameshifting can also account for extended protein sequences at the 3′ end of an ORF, such as in the case of the *Bacillus subtilis cdd* gene (81). There are known cases of genes totally embedded within another gene but out-of-frame such as the Holin-like protein of *Staphylcoccus aureus* bacteriophage 187 (76) and the *Rz1* gene of *E. coli* bacteriophage λ (55). The phenomenon of joining two ORFs by translational bypassing has been demonstrated for T4 gene *60* (58). Identification of an RBS in front of a potential start codon is helpful in defining an ORF (115). However, because the complementarity to the 3′ end of the 16S rRNA does not have to be exact, computational methods for RBS identification are still imperfect. Computer analysis for finding RBS among various prokaryotic genomes does reveal a correlation between the RBS properties of a given genome and the potential gene start prediction accuracy (43,54). However, there are cases when there is no detectable base pairing between the region upstream of start codons and the 3′ end of the 16S rRNA (43). There are also cases of zero to just a few nucleotides of leader sequence upstream of the translation start codon. For example, there are zero nucleotides of *cI* repressor gene of bacteriophage λ transcribed from the Prm promoter (52) or the *cre* recombinase gene of bacteriophage P1 transcribed from the p*R*3 promoter (124), and there are two nucleotides from the bop gene in archaebacterium *Halobacterium halobium* (32). Finally, the size of the ORF has a direct correlation with the accuracy of the prediction method, with large ORFs being the most accurate [one of the largest known prokaryotic ORFs identified to date is *syrE*, 28231 nucleotides (51)]. At the opposite end of the size spectrum, minigenes (or peptides) are very difficult to predict and can be as small as two amino acids, λ*bar* (57) and *E. coli* MR (129).

Homology Search

After predicting ORFs, the next phase is to apply a battery of algorithms designed to extract as much information about the ORF as possible. One of the most widely used methods to define function, and hence gene annotation, is to perform a search of existing sequence databases to determine if a similar protein sequence has an associated function. The two main algorithms used for homology searching are basic local alignment search tool (BLAST) (3) and FASTA (99). BLAST uses a heuristic algorithm to detect relationships among sequences that share only isolated regions of similarity. The BLAST family of programs allows one to search protein or DNA databases regardless of whether the query sequence is amino acid or nucleic acid. FASTA is a heuristic algorithm that uses a modification of the optimization method described by Needleman and Wunsch (87) for global alignment and Smith and Waterman (118) for local alignment and may be more sensitive than BLAST. Like BLAST, FASTA also allows one to search protein or DNA databases regardless of whether the query sequence is amino acid or nucleic acid. (For an overview of searching sequence databases, see Reference 89.) Two more advanced versions of the BLAST program that run faster due to requiring a "two-hit" method are Gapped BLAST and position-specific iterated BLAST (PSI-BLAST) (4). The new gapped alignment algorithm uses dynamic programming to extend a central pair of aligned residues in both directions. The PSI-BLAST program uses a position-specific score matrix (also called profiles or motifs) to detect weak but biologically relevant sequence similarities. Although this program may identify weak homologies to annotated entries in a database, it does require manual inspection of the output data between each iteration. Another search program for finding weak homology is HMMER (http://hmmer.wustl.edu) (36). This allows one to use a multiple sequence alignment of known protein family members to build an HMM to search for new family members within a database or use a single protein sequence to search the Pfam HMM database as discussed below.

It is interesting that even today a significant fraction (30%–50% of predicted ORFs) of both previously and newly sequenced genomes encode uncharacterized genes, including a significant number of strictly "orphan" genes, "ORFans" (ORFs with no detectable sequence similarity to proteins of other genomes) (1,41).

Sequence Databases

At present, there are three major publicly available sequence database facilities (Table 1). The National Center for Biotechnology Information (NCBI) at the National Institutes of Health (NIH) was established in 1988 and has assumed responsibility for maintaining the sequence database GenBank® since 1992. GenBank consists of nucleotide and protein sequences including those from EBI and DDBJ (see below) as well as molecular modeling database (MMDB) and other sequence databases (12). The NCBI site contains other information such as clusters of orthologous groups (COGs) of proteins—phylogenetic classification), links to literature databases (such as PubMed), tools for data mining (such as BLAST and PSI-BLAST), access to Entrez (search and retrieval system), and much more.

The EMBL Nucleotide Sequence Database was established in 1980 at the European Molecular Biology Laboratory in Heidelberg, Germany. Since 1994, the EMBL Database has been maintained at the European Bioinformatics Institute (EBI), an EMBL outstation, located at the Wellcome Trust Genome Campus in Hinxton near Cambridge, UK. The EBI operates a number of databases, namely the EMBL Nucleotide Sequence Database including sequences from GenBank and DDBJ, translated protein sequences from EMBL (TrEMBL), SWISS-PROT (curated non-redundant protein sequences), Ensembl (automatic annotation on eukaryotic genomes), Macromolecular Structure Database (MSD), and many other databases (9). The EBI site contains additional information such as links to protein domain and motif database (InterPro) tools for data mining, database searching (SRS), homology searching (such as FASTA), analysis tools (such as ClustalW), and much more.

The DNA Data Bank of Japan (DDBJ) was established in 1986 and is located at the Center for Information Biology, National Institute of Genetics (NIG), Mishima, Japan (128). DDBJ is the sole DNA data bank in Japan and also contains sequences from GenBank and the EMBL Data Library. It provides similar tools for data retrieval (such as SRS), homology searching (such as PSI-BLAST, SSEARCH, and FASTA), analysis programs for multiple sequence alignment (such as ClustalW), secondary protein structure prediction (such as SSThread), and much more. All three data banks exchange data nightly and therefore contain the same data.

The Protein Information Resource (PIR), in collaboration with the Munich Information Center for Protein Sequences (MIPS) and the Japan International Protein Information Database (JIPID), produces the most comprehensive, well-annotated, and non-redundant protein sequence database in the public domain, called the PIR-International Proteins Sequence Database (Table 1) (10). A listing of other sequence and types of databases can be found in the annual database issue of Nucleic Acids Research (see Volume 29, Number 1, January 1, 2001).

Word of caution. With the fast pace of genomic sequences being deposited into the databases, there is also much pressure to provide annotation to the sequences. This comes with a risk of the data not being highly accurate (45,119). Hence, the error rate of annotations for various genomes can be at least 8%, as in the case for *M. genitalium* (18). Other genomes such as *Chlamydia trachomatis* and *Methanococcus jannaschii* have been reported to contain annotation errors in the database (72,83). The probable main cause for errors is the sole reliance on automated computer methods to assign some type of a function based on similarity to other sequences that may contain annotation errors or using too low of a minimal comparison threshold value. Another major problem with the current automated annotation approaches is inconsistency (119). This puts pressure on the hand-curated databases (such as Amos Bairoch's SWISS-PROT database at http://www.expasy.ch/sprot/sprot/) to check the literature, reanalyze the sequence using other resources such as PROSITE or Pfam, and weed out those annotations without supportive information (68). The recent publication of the complete genomic sequence of *Pseudomonas aeruginosa* avoided the perpetuation of annotation errors by enlisting 61 researchers for the Pseudomonas Community Annotation Project to define ORFs and assign function (125). There is also a need to re-annotate completed genomes on an ongoing basis, as what was recently done for the genome of *Mycoplasma pneumoniae* (31), as newer algorithms and information become available.

Domain or Pattern/Motif Databases and Tools

As we have seen above, additional analysis other than just a BLAST search alone is needed for gene annotation. There exists a number of different domain/pattern or motif databases, each based on a slightly different philosophy with regard to sequence features that make up a subclass of proteins with known or presumed function (Table 1). PROSITE is a database of protein families and domains based on patterns and profiles/weight matrices (60). The profiles are constructed by various different techniques such as HMMs (38). Each protein family or domain listed in PROSITE comes with documentation providing information on the structure and function of these proteins. Blocks are multiply aligned, ungapped segments corresponding to the most conserved regions of proteins. The list of protein families used to construct the Blocks database were from Prosite, but, in the recent version, Blocks+, additional families are obtained from ProDom, Prints, Pfam, and DOMO (56). The user may search the Blocks database using a comparison BLIMPS search program or, as an alternative, the IMPLALA program, which searches PSI-BLAST-generated subsets of the Blocks database (56). Pfam is a database of protein domain families (11). Pfam-A contains curated multiple sequence alignments for each protein domain or conserved region generated from a non-redundant protein set of sequences from the SWISS-PROT and TrEMBL databases (68). With the latest 6.1 release of Pfam, supplemented domains, Pfam-B, have been constructed using the ProDom database. Profile HMMs have been built from the Pfam alignments and are a useful tool for finding these domains in a new sequence even if the homology is weak. ProDom is another database of protein domain families generated from the SWISS-PROT and TrEMBL sequence databases including the Pfam-A families (28). A set of complete genomes (archaea, bacteria, and eukaryote) is used to generate ProDom-CG. Both databases are available from the Web. PRINTS is a database of protein family "fingerprints" that occurred because of the limitations of single-motif pattern-matching methods that yield binary diagnoses with no biological context with which to assess the significance of the result (7). In the alignment of proteins within a family, more than one motif may exist. Hence, PRINTS yields information as to the number of fingerprint matches a new sequence has to a family, as well as researched information about the family and, where possible, links to known structural or functional data. The PRINTS database now includes a composite of SWISS-PROT and TrEMBL and is called PRINTS-S (6). A listing of other types of domain and motif databases can be found in the annual database issue of Nucleic Acids Research (see Volume 29, Number 1, January 1, 2001).

There are a number of standalone tools that allow one to find motifs within a dataset. A software tool called TEIRESIAS is capable of detecting and reporting all existing patterns in an input set of sequences without pairwise alignments (107). TEIRESIAS scans the input sequences and locates elementary patterns that are then used in a convolution phase to construct the maximal pattern(s) in the dataset. The Gibbs motif sampling has been used to define motifs within immunoglobulin fold proteins, helix-turn-helix DNA binding proteins, and β-barrel strands in porins (88). Finally, a study was done to evaluate seven other different computational motif-finding methods (BLOCKMAKER, ITERALIGN, MATCHBOX, MEME, PIMA, PROBE, and SAM) using a set of 20 reverse transcriptase protein sequences (63).

Subcellular Location

In bacterial genomes, 30%–50% of predicted ORFs are genes with no known associated function. In these cases when the standard search algorithms have failed to produce a function, methods based on amino acid composition alone may at least help to define the localization of the gene product (Table 1). One of the best known methods is an expert system called PSORT (84,85) that uses various kinds of knowledge rules, based on 106 proteins with known subcellular locations, for predicting the location of proteins in bacteria. This program predicts the location of a protein in one of four subcellular regions, cytoplasm, inner membrane, periplasm, or outer membrane, with an accuracy of 83%. PSORT gives additional information such as determination of a signal sequence using McGeoch's method (80), cleavable signal sequence using von Heijne's method (136), lipoprotein based on a combination of McGeoch's method and the consensus motif by von Heijne (138), and transmembrane segments using ALOM software (69). A program called ProtLock does a statistical discrimination analysis of the amino acid content of a protein to predict the cellular location of the protein (21). ProtLock defines five categories of subcellular location (integral membrane, anchored membrane, extracellular, intracellular, and nuclear) that are used for identifying proteins with about 76% accuracy based on a total training set of 1000 proteins. Using correlation analysis of the amino acid composition and cellular location of a protein, Reinhardt and Hubbard (105) have developed a program to define the three subcellular locations in prokaryotic organisms (cytoplasmic, extracellular, and periplasmic) based on neural networks (NNs) trained on sequences whose subcellular location was annotated. The reported accuracy of Reinhardt and Hubbard's (105) NN was 81% based on jackknife analysis using a dataset of 993 proteins extracted from the SWISS-PROT database. A discriminant function algorithm developed by Chou and

Elrod (24) gave an accuracy prediction range of 87%–91% for the same set of three subcellular locations. Chou and Elrod (24) used the dataset produced by Reinhardt and Hubbard (105) (with an additional four proteins) for training and evaluation; however, their method of performing the jackknife test was slightly different than Reinhardt and Hubbard. Using an improved neural network implementation, Schneider (113) developed a program called *LocateProtein* with a cross-validated average prediction accuracy of 93% based on the same dataset as Chou and Elrod (24). *LocateProtein* only seeks to classify a protein as either cytoplasmic or non-cytoplasmic.

Signal Sequence Determination

Most of the bacterial proteins that are translocated to the cell membrane contain a signal sequence (also called signal peptide or SP) at the N-terminal region of the protein. This sequence consists of three distinct regions: an N-terminal positively charged n-region, a hydrophobic h-region, and a neutral polar c-region (92). For cleavage, the residues at positions -3 and -1 must be small and neutral (137). A program called SignalP uses a combined NN approach to predict the recognition of SPs and their cleavage sites (92). A more recent HMM version of SignalP has been developed and is also available on the Web (90). The performance values of SignalP between the NN and HMM versions show the NN method to be slightly better at predicting the correct bacterial cleavage site, while the HMM method gives minor improvement in discriminating true SP from non-secretory peptide sequences. Another program for predicting SPs is Sigcleave program that uses the weight matrix method of von Heijne (136). SPScan from Genetics Computer Group (Madison, WI, USA) is similar to Sigcleave but also uses the weight matrices given in Nielsen et al. (91) in concert with McGeoch's description of a minimum acceptable signal peptide (80). Since these programs are based on different prediction methods, it may be wise to use multiple programs in concert. There are also various programs available for predicting N-terminal sorting signals (see Table 1 and Reference 25). Recently, Chou (23) has shown that the discriminate function techniques applied to subcellular localization of proteins can also be used to identify signal sequences and predict their cleavage sites. The discriminate function algorithm detects common features of signal sequences without regard for differing lengths, sequences, and composition.

Transmembrane Segments

One of the main differences between Gram-negative and Gram-positive bacteria is the cell wall structure. Gram-negative bacteria have two lipid bilayers—an inner membrane layer and an outer membrane layer. Between these two membrane layers lies the periplasmic space that contains the peptidoglycan matrix. Gram-positive bacteria have only a single membrane layer that is surrounded on the outside by the peptidoglycan matrix.

Inner Membrane Proteins

Typically in the inner membrane of Gram-negative bacteria and in the single membrane of Gram-positive bacteria, integral membrane proteins are held in place by hydrophobic interactions with the lipid bilayer. Aside from lipoproteins that are anchored to the membrane via a N-terminal lipid moiety, membrane proteins consist of at least one strongly predicted α-helical transmembrane segment (67). These segments typically consist of a stretch of hydrophobic amino acids, 18–24 residues long. One of the early prediction methods for being able to distinguish transmembrane segments from hydrophobic segments of soluble proteins used a sliding window (39). This method allowed both the residue hydrophobicity value and the α-helix hydrophobic moments to be computed and displayed in a hydrophobic moment plot of the entire protein. Examples of transmembrane prediction methods currently available over the Web are the discriminant analysis method (69) employed in the PSORT program mentioned above, the dense alignment surface (DAS) method (29), and the system SOSUI (Table 1) (59). The accuracy in predicting transmembrane helices in the correct location can be approximately 95% (120). However, predicting the correct topology of the transmembrane segments within a protein (i.e., how it transverses the membrane, inward or outward) is at best about 75% accurate for a single sequence-based hydrophobicity/discriminant analysis and as high as 86% accurate using multiple protein alignments. The amino acid composition and the observation that there is a greater abundance of positively charged amino acids in the cytoplasmic segment versus the periplasmic segment of integral membrane proteins ("positive-inside" rule) play an important role in determining the topology (67,86,135). There are various programs that predict the transmembrane helices and topology in a protein. TopPred II (26) incorporates the positive-inside rule and compiles other existing knowledge about the topology. PHD-htm performs a local alignment to generate a profile of a protein and then two levels of neural network systems along with a refinement of topology using the positive residue content of the non-transmembrane loop regions to predict transmembrane segments (108). TMHMM (120) uses an HMM for predicting transmembrane segments and has recently been tested on complete genomes predicting that 20%–30% of all genes encode membrane proteins (70). PRED-TMR uses statistical analysis of transmembrane proteins in the SwissProt database for defining potential termini of transmembrane regions (97). Nilsson et al. (94) concluded that by using the consensus of five different membrane protein topology prediction methods, a high level of topology prediction can be achieved by a simple majority vote approach. A recent paper by Lio and Vannucci (74) uses a non-parametric method based on a wavelet change-point analysis and new propensity scale to predict transmembrane segments. A comparison of this wavelet change-point analysis system with a number of other transmembrane prediction programs was demonstrated in the paper. Both the source code and database are available upon request of the authors.

Outer Membrane Proteins

The outer membrane proteins (OMPs) form a distinct class of integral membrane proteins (for overview of OMPs, see References 19 and 122). Porins are a major class of OMPs. These proteins typically contain regions of amphipathic β-pleated sheet structure (alternating hydrophobic/hydrophilic amino acids) that transverse the membrane 8-22 times and form β-barrel structures. Schirmer and Cowan (112) used a modified β-side-hydrophobicity and a sliding window width of four residues for every second residue to correctly predict 16 out of the 18 β-strands of LamB. Neuwald et al. (88) used a new Gibbs sampling algorithm to detect motif-encoding regions or patterns. When applied to a set of 32 bacterial OMPs, the sampler detected repeats that corresponded to β-strands (34). Gromiha et al. (50) used a rule-based approach for identifying transmembrane β-strands for a limited set of three different 16 β-strands porins. A neural network algorithm was trained on a set of seven atomically determined porins to predict outer membrane β-strands (34). This program can be accessed at: http://loop8.biologie.uni-konstanz.de/~kay/om_topo_predict.html. Topology analysis of OmpA predicted at least 16 transmembrane β-strands. However, high-resolution structure analysis of OmpA reveals only eight transmembrane β-strands (98). Recently, Gnanasekaran et al. (47) have developed a method using structurally conserved profiles from structurally conserved porin regions to identify or discriminate β-strands. However, it is recommended that other algorithms be used in combination with this method for searching other protein databases. In the genomic analysis of *Helicobacter pylori*, strain 26695, a family of 32 putative OMPs was identified based on the "characteristic alternating hydrophobic residues at their carboxy termini" (132). Four of these OMPs were identified as porins based on identity to published sequences. Unfortunately, all of these methods have shown only limited success in defining transmembrane β-strands, and better methods are needed.

Short Nucleotide Repeat Regions

Prokaryotic genomes contain regions of repetitive DNA in the form of homopolymeric tracts [i.e., poly(dA), poly(dG), etc.] and homogeneous or heterogeneous tracts. For a review on short DNA repeat sequences in prokaryotic genomes, see van Belkum et al. (133). Gain or loss of the repeat regions through slipped-strand mispairing is one mechanism of genotypic variation (82). Since the publication of the first bacterial genome, many types of short repeat sequences have been identified in pathogenic organisms. These repeats localize within the coding or promoter region of genes involved in virulence. One of the mechanisms that bacteria have in trying to evade the host immune system is to change its appearance by either phase variation (the turning "on" or "off" of a gene) or antigenic variation (modification of the gene production). In the genomic sequence of *N. meningitidis* serogroup A strain Z2491, there are about 26 tandem repeats ranging from homopolymeric tracts to di-, tetra-, and pentanucleotide repeats found mostly within the coding sequences. The repeats were found mainly in surface-exposed genes or genes involved in the biosynthesis or modification of surface structures (96). In another strain of *N. meningitidis*, serogroup B strain MC58, 65 potential phase variable genes were identified, and, of these, only 13 were previously recognized (110). This shows the fruitfulness of genomic analysis for the identification of novel genes presumed to be responsible for immunogenic variation. Repeat sequences have also been found in other human pathogens such as *Haemophilus influenzae* and *H. pylori* (62,111).

The ability to identify these potential immunogenic variable genes is important in the decision of whether the gene(s) might be a good vaccine candidate. Certainly, if the gene product were not expressed or present during pathogenesis, it most likely would not be considered a worthwhile vaccine target. A DNA tandem repeat finder program has been developed by Benson (13) and is available over the Web at http://c3.biomath.mssm.edu/trf.html.

Software and Computers

There are a number of commercially available programs that can perform many of the analysis functions described above (Table 1). Probably the most widely used and known UNIX® software is the Wisconsin Package® (Accelrys, San Diego, CA, USA). SeqWeb® is a Web interface of the GCG software. Both of these software packages are also available from Accelrys. Other desktop operating sequence analysis software are Lasergene99 (Microsoft® Windows™ or Macintosh®) (DNASTAR, Madison, WI, USA); MacVector® (Macintosh) and OMIGA (Windows), both from GCG or Oxford Molecular Group; and Vector NTI® Suite (Windows) from InforMax (North Bethesda, MD, USA). The European Molecular Biology Open Software Suite (EMBOSS) is an extensive, free, open-source UNIX software analysis package developed for the needs of the molecular biology community. Most research facilities customize the software needed for data analysis and how the output data is presented based on their specific needs. This process is constantly evolving as new software and methods are being developed.

Dealing with the vast quantity of data produced through computational genomics provides a relatively daunting task in itself. Many different efforts exist to help standardize the input and output formats that various new analysis methods handle or provide. In general, however, the standards that are most widely used are *de facto* standards that have grown out of a need for simplicity.

The quantity of data generated in analyzing an entire genome can be enormous. For a relatively small bacterial genome like *C. trachomatis*, about 1 Mbp, as much as 500 MB data can easily be generated in the process of storing all of the different analysis results for the ORFs. While that quantity of data is not terribly significant given the size of today's mass storage devices, managing that quantity of data to allow easy storage and retrieval requires some thought and planning. In any significant genomic mining program, time and effort must be put into developing a scheme for storing the data on top of the mechanism for generating that data.

As with every field related directly to computers, all of the major hardware platforms are represented in one way or another with different available software (MacVector runs on Macintosh, Lasergene 99 runs under Windows, and GCG and EMBOSS run under most flavors of UNIX). Many of the tools described in earlier sections have been developed in C or Perl and could potentially be ported to any platform desired. In the simplest case, a new computational genomics laboratory would do well with a UNIX machine with 512 MB RAM and 32 GB disk space. This would provide the computational power and data storage capabilities necessary to begin analyzing whole bacterial genomes.

PUTTING IT ALL TOGETHER

A nice example of using genomics for bacterial vaccine development is the group from Chiron S.p.A. (Siena, Italy), who, in collaboration with TIGR, mined the genome of *N. meningitidis*, serogroup B (NmB) to identify possible vaccine candidates (101,130). The approach Pizza et al. (101) took was to search the genome for potentially encoded novel surface-exposed or exported proteins in the development of a protein-based vaccine. The first step was to determine the complete genomic sequence of a virulent strain of NmB and prediction of ORFs (130). ORFs that were determined to code for cytoplasmic proteins were not investigated further. All other ORFs were further analyzed to identify proteins that were exposed to the outside of the cell using various programs such as PSI-BLAST, FASTA, MOTIFS, FINDPATTERNS, and PSORT. Databases such as ProDom, Pfam, and Blocks were used to predict surface-associated proteins such as transmembrane domains, leader peptides, homologies to known surface proteins, lipoprotein signature, outer membrane anchoring motives, and host cell binding domains such as the tripeptide, arginine-glycine-aspartic acid (RGD) (66). Out of a total of 2158 predicted ORFs, 570 were predicted to be surface exposed or exported. These ORFs were cloned and expressed as fusion proteins in *E. coli*. Three hundred fifty of the proteins were successfully expressed, purified, and used to immunize mice. The sera obtained were tested in a whole cell ELISA, fluorescence-activated cell sorter (FACS) system, and for bactericidal activity (a property known to correlate with vaccine efficacy in humans) to verify cell surface presence of the proteins. Eighty-five proteins were found to be strongly positive in at least one of the assays above. Of these, seven representative proteins that were positive in all three assays and whose genes were not predicted to be phase variable were selected for further study. Surprisingly, these seven surface exposed proteins are conserved among all 31 *N. meningitidis* strains tested. Presumably, any or all of these seven vaccine ORF candidates would be moved into the next preclinical phase of vaccine development.

In a recent report, Wizemann et al. (140) used similar genomic algorithms, as mentioned above, for the prediction of surface-localized proteins from the genomic sequence of *S. pneumoniae*. A major difference in their approach than what was done with NmB was to test their vaccine candidates directly in an animal model to screen the proteins for vaccine efficacy. In this way, they were able to identify two vaccine candidates that conferred protection in mice but were not detected on laboratory-grown bacteria by FACS analysis. Either the amount or accessibility of these antigens on the bacteria is limited or these proteins represent a novel class of proteins that are only expressed during in vivo infection.

In another example, Gomez et al. (48) searched the genome of *Mycobacterium tuberculosis* for potential secreted proteins. They used various mining tools such as SignalP, SPScan, TMpred, and lipoprotein PrositeScan to identify various classes of secreted proteins. A subclass of the predicted secreted proteins was confirmed in the laboratory by constructing *phoA′* fusions.

These few examples highlight current approaches whereby genomic analysis and screening is done in silico before any wet laboratory research. Although we have become smarter, faster, and better at predicting vaccine candidates, the need and the time required to do experimental laboratory and clinical studies still remains unchanged in the development of safe and efficacious vaccines.

WHAT NEXT?

So where do we go from here? Much work is still needed, as even with the best algorithms available today, proteins are being found on the cell surface of the bacterium that were not predicted to be surface exposed. Examples of such findings are the ribosomal protein L12 of *Neisseria gonorrhoeae* as being membrane associated and surface exposed (121); a histone-like protein and S4 and S5 ribosomal proteins were isolated from the cell wall of *Mycobacterium leprae* (79); and, in the analysis of OMPs of *H. pylori*, translational elongation factor EF-Tu, urease α and β subunits, catalase and others were found (93). More and more examples of this type will be forthcoming as we continue the analysis of bacterial surface proteins. What does this mean? The answer is simple: researchers should not limit their analysis to just genomic mining.

Transcription profiling using oligonucleotide or DNA microarrays is a technology geared for high-throughput genomic mRNA expression analysis. This method uses genomic gene predictions for the design of the array and subsequent hybridization of the mRNA, either directly or indirectly (cDNA), to the array for detection and has shed tremendous light on the gene expression profile of the entire bacterial genome grown under various conditions (5,106) or as a means to monitor bacteria-host cell interactions (27,30). However, there are some technical difficulties using this method for identifying vaccine candidates: *(i)* the level of mRNA transcripts may not correlate with the actual amount of protein expressed (5,53); *(ii)* the stability of most bacterial mRNAs are in the order of minutes (104); and *(iii)* patterns of gene expression are not the ideal method for discovery of vaccine targets (20).

An area of concern using just genomics and/or transcriptional analysis for protein expression is the difficulty in defining post-transcriptional modifications of the expressed protein and in predicting genes due to frameshifts or alternative RNA splice sites (117), hence the need to identify all the proteins made by the genome, termed proteome, through proteomics. Proteomics is now at the level of sensitivity and high throughput to be able to complement genomics and give a more realistic analysis of the actual proteins expressed by the organism. In a search for potential vaccine candidates in *H. pylori*, Nilsson et al. (93) used a combination of liquid-phase isoelectric focusing and gel separation for the isolation of *H. pylori* proteins. These proteins were then digested with trypsin and analyzed on a matrix-assisted laser desorption/ionization time-of-flight (MALDI-TOF) mass spectrometer with subsequent database searching for the identification of the protein. Chakravarti et al. (22) also showed how the combined use of genomics and proteomics can help identify potential vaccine candidates for *H. influenzae* as well as *H. pylori*. For a recent general review on how vaccinology will benefit from genomic and proteomic technologies, see Grandi (49).

Knowing that no one method alone can identify all of the proteins of interest, researchers are combining all three disciplines of genomics, transcription profiling, and proteomics (142). Whether this will be the future for vaccine discovery programs remains to be seen. The next key level of importance for bacterial vaccine discovery is to be able to determine what proteins are made during infection, in vivo, as it is already known that type III effector proteins and virulence factors are only expressed during eukaryotic cell contact (64). The questions now being asked are: "Are we missing good vaccine candidates by focusing our study of laboratory-grown bacteria?" and "How do we identify these vaccine targets induced in vivo?"

NOW AND BEYOND

The recent release of the draft sequence of the human genome (73,134) has opened the door for newer methods to be developed both in the analysis and detection of genes and their proteins. Many of these methods can and will be applied to the discovery of new pharmaceutical targets and hopefully will lead to the development of new treatments. On the bacterial vaccine discovery side, better approaches are needed to provide insight into pathogenesis that may lead to new interventions to prevent and treat infection (126). However, until that time, as more and more genomic sequences are being completed, improved laboratory methods and algorithms are needed to help sort and understand all of the present data, especially the substantial proportion of ORFs that have no known function (139). In this latter case, new computational approaches and genomic context are being developed to help define putative gene function (46,141). Comparisons between genomes, especially of the same genus, may help to define vaccine candidates (123). In a way, this marks the end of the early genome mining era and the start of a relatively new area of scientific discipline called functional genomics (35,46).

ACKNOWLEDGMENTS

The authors wish to thank Deb Chakravarti, Leah Fletcher, George Maalouf, John McMichael, Stephen Olmsted, Burkhard Rost, Steven Salzberg, Guenter Stoesser, Yoshio Tateno, George Weinstock, and Rose Marie Woodsmall for providing valuable editorial comments on various sections of the manuscript. We also appreciate the help of Susan Hoiseth and Laura York with part of the Introduction, as well as Terri Ward and Eileen Shirley for their endless librarian help.

REFERENCES

1. **Alimi, J., O. Poirot, F. Lopez, and J. Claverie.** 2000. Reverse transcriptase-polymerase chain reaction validation of 25 "orphan" genes from *Escherichia coli* K-12 MG1655. Genome Res. *10*:959-966.

2. **Alm, R.A., L.S. Ling, D.T. Moir, B.L. King, E.D. Brown, P.C. Doig, D.R. Smith, B. Noonan et al.** 1999. Genomic-sequence comparison of two unrelated isolates of the human gastric pathogen *Helicobacter pylori*. Nature *397*:176-180.

3. **Altschul, S.F., W. Gish, W. Miller, E.W. Myers, and D.J. Lipman.** 1990. Basic local alignment search tool. J. Mol. Biol. *215*:403-410.

4. **Altschul, S.F., T.L. Madden, A.A. Schaffer, J. Zhang, Z. Zhang, W. Miller, and D.J. Lipman.** 1997. Gapped BLAST and PSI-BLAST: a new generation of protein database search programs. Nucleic Acids Res. *25*:3389-3402.

5. **Ang, S., C.Z. Lee, K. Peck, M. Sindici, U. Matrubutham, M.A. Gleeson, and J.T. Wang.** 1679. Acid-induced gene expression in *Helicobacter pylori*: study in genomic scale by microarray. Infect. Immun. *69*:1679-1686.

6. **Attwood, T.K., M.D. Croning, D.R. Flower, A.P. Lewis, J.E. Mabey, P. Scordis, J.N. Selley, and W. Wright.** 2000. PRINTS-S: the database formerly known as PRINTS. Nucleic Acids Res. *28*:225-227.

7. **Attwood, T.K., D.R. Flower, A.P. Lewis, J.E. Mabey, S.R. Morgan, P. Scordis, J.N. Selley, and W. Wright.** 1999. PRINTS prepares for the new millennium. Nucleic Acids Res. *27*:220-225.

8. **Audic, S. and J.M. Claverie.** 1998. Self-identification of protein-coding regions in microbial genomes. Proc. Natl. Acad. Sci. USA *95*:10026-10031.

9. **Baker, W., A. van den Broek, E. Camon, P. Hingamp, P. Sterk, G. Stoesser, and M.A. Tuli.** 2000. The EMBL nucleotide sequence database. Nucleic Acids Res. *28*:19-23.

10. **Barker, W.C., J.S. Garavelli, Z.L. Hou, H.Z. Huang, R.S. Ledley, P.B. McGarvey, H.W. Mewes, B.C. Orcutt et al.** 2001. Protein Information Resource: a community resource for expert annotation of protein data. Nucleic Acids Res. *29*:29-32.

11. **Bateman, A., E. Birney, R. Durbin, S.R. Eddy, K.L. Howe, and E.L. Sonnhammer.** 2000. The Pfam protein families database. Nucleic Acids Res. *28*:263-266.

12. **Benson, D.A., I. Karsch-Mizrachi, D.J. Lipman, J. Ostell, B.A. Rapp, and D.L. Wheeler.** 2000. GenBank. Nucleic Acids Res. *28*:15-18.

13. **Benson, G.** 1999. Tandem repeats finder: a program to analyze DNA sequences. Nucleic Acids Res. *27*:573-580.

14. **Benton, D.** 1996. Bioinformatics—principles and potential of a new multidisciplinary tool. Trends Biotechnol. *14*:261-272.

15. **Black, S., H. Shinefield, B. Fireman, E. Lewis, P. Ray, J.R. Hansen, L. Elvin, K.M. Ensor et al.** 2000. Efficacy, safety and immunogenicity of heptavalent pneumococcal conjugate vaccine in children. Northern California Kaiser Permanente Vaccine Study Center Group. Ped. Infect. Dis. J. *19*:187-195.

16. **Blattner, F.R., G. Plunkett, III, C.A. Bloch, N.T. Perna, V. Burland, M. Riley, J. Collado-Vides, J.D. Glasner et al.** 1997. The complete genome sequence of *Escherichia coli* K-12. Science *277*:1453-1474.

17. **Borodovsky, M. and J. McIninch.** 1993. Prediction of gene locations using DNA Markov chain models, p. 231-248. *In* H.A. Lim, J.W. Fickett, C.R. Cantor, and R.J. Robbins (Eds.), The Second International Conference on Bioinformatics, Supercomputing, and Complex Genome Analysis. World Scientific Publishing, River Edge, N.J.

18. **Brenner, S.E.** 1999. Errors in genome annotation. Trends Genet. *15*:132-133.

19. **Buchanan, S.K.** 1999. Beta-barrel proteins from bacterial outer membranes: structure, function and refolding. Curr. Opin. Struct. Biol. *9*:455-461.

20. **Carucci, D.** 2001. Genomic tools for gene and protein discovery in malaria: toward new vaccines. Vaccine *19*:2315-2318.

21. **Cedano, J., P. Aloy, J.A. Perez-Pons, and E. Querol.** 1997. Relation between amino acid composition and cellular location of proteins. J. Mol. Biol. *266*:594-600.

22. **Chakravarti, D., M. Fiske, L. Fletcher, and R. Zagursky.** 2000. Application of genomics and proteomics for identification of bacterial gene products as potential vaccine candidates. Vaccine *19*:601-612.

23. **Chou, K.-C.** 2001. Prediction of protein signal sequences and their cleavage sites. Proteins Struct. Funct. Genet. *42*:136-139.

24. **Chou, K.C. and D.W. Elrod.** 1998. Using discriminant function for prediction of subcellular location of prokaryotic proteins. Biochem. Biophys. Res. Commun. *252*:63-68.

25. **Claros, M.G., S. Brunak, and G. von Heijne.** 1997. Prediction of N-terminal protein sorting signals. Curr. Opin. Struct. Biol. *7*:394-398.

26. **Claros, M.G. and G. von Heijne.** 1994. TopPred II: an improved software for membrane protein structure

predictions. Comput. Appl. Biosci. *10*:685-686.
27.**Coombes, B.K. and J.B. Mahony.** 2001. cDNA array analysis of altered gene expression in human endothelial cells in response to *Chlamydia pneumoniae* infection. Infect. Immun. *69*:1420-1427.
28.**Corpet, F., F. Servant, J. Gouzy, and D. Kahn.** 2000. ProDom and ProDom-CG: tools for protein domain analysis and whole genome comparisons. Nucleic Acids Res. *28*:267-269.
29.**Cserzo, M., E. Wallin, I. Simon, G. von Heijne, and A. Elofsson.** 1997. Prediction of transmembrane α-helices in prokaryotic membrane proteins: the dense alignment surface method. Protein Eng. *10*:673-676.
30.**Cummings, C.A. and D.A. Relman.** 2000. Using DNA microarrays to study host-microbe interactions. Emerg. Infect. Dis. *6*:513-525.
31.**Dandekar, T., M. Huynen, J.T. Ragula, B. Ueberle, C.U. Zimmermann, M.A. Andrade, T. Doerks, L. Sanchez-Pulido et al.** 2000. Re-annotating the *Mycoplasma pneumoniae* genome sequence: adding value, function and reading frames. Nucleic Acids Res. *28*:3278-3288.
32.**DasSarma, S., U.L. RajBhandary, and H.G. Khorana.** 1984. Bacterio-opsin mRNA in wild-type and bacterio-opsin deficient *Halobacterium halobium* strains. Proc. Natl. Acad. Sci. USA *81*:125-130.
33.**Delcher, A.L., D. Harmon, S. Kasif, O. White, and S.L. Salzberg.** 1999. Improved microbial gene identification with GLIMMER. Nucleic Acids Res. *27*:4636-4641.
34.**Diederichs, K., J. Freigang, S. Umhau, K. Zeth, and J. Breed.** 1998. Prediction by a neural network of outer membrane β-strand protein topology. Protein Sci. *7*:2413-2420.
35.**Dunham, I.** 2000. Genomics—the new rock and roll? Trends Genet. *16*:456-461.
36.**Durbin, R., S.R. Eddy, A. Krogh, and G. Mitchison.** 1998. Biological Sequence Analysis: Robabilistic Models of Proteins and Nucleic Acids. Cambridge University Press, Cambridge.
37.**Eby, R.** 1995. Pneumococcal conjugate vaccines, p. 695-718. *In* M.F. Powell and M.J. Newman (Eds.), Pharmceutical Biotechnology, vol. 6. Plenum Press, New York.
38.**Eddy, S.R.** 1996. Hidden Markov models. Curr. Opin. Struct. Biol. *6*:361-365.
39.**Eisenberg, D., E. Schwarz, M. Komaromy, and R. Wall.** 1984. Analysis of membrane and surface protein sequences with the hydrophobic moment plot. J. Mol. Biol. *179*:125-142.
40.**Fedson, D.S., D.M. Musher, and J. Eskola.** 1988. Pneumococcal vaccine, p. 553-607. *In* S.A. Plotkin and W.A. Orenstein (Eds.), Vaccines, 3rd ed. Saunders Co., Philadelphia.
41.**Fischer, D. and D. Eisenberg.** 1999. Finding families for genomic ORFans. Bioinformatics *15*:759-762.
42.**Fleischmann, R.D., M.D. Adams, O. White, R.A. Clayton, E.F. Kirkness, A.R. Kerlavage, C.J. Bult, J.F. Tomb et al.** 1995. Whole-genome random sequencing and assembly of *Haemophilus influenzae* Rd. Science *269*:496-512.
43.**Frishman, D., A. Mironov, and M. Gelfand.** 1999. Starts of bacterial genes: estimating the reliability of computer predictions. Gene *234*:257-265.
44.**Frishman, D., A. Mironov, H.W. Mewes, and M. Gelfand.** 1998. Combining diverse evidence for gene recognition in completely sequenced bacterial genomes. Nucleic Acids Res. *26*:2941-2947.
45.**Galperin, M.Y. and E.V. Koonin.** 1999. Functional genomics and enzyme evolution. Homologous and analogous enzymes encoded in microbial genomes. Genetica *106*:159-170.
46.**Galperin, M.Y. and E.V. Koonin.** 2000. Who's your neighbor? New computational approaches for functional genomics. Nat. Biotechnol. *18*:609-613.
47.**Gnanasekaran, T., S. Peri, A. Arockiasamy, and S. Krishnaswamy.** 2000. Profiles from structure based sequence alignment of porins can identify β stranded integral membrane proteins. Bioinformatics *16*:839-842.
48.**Gomez, M., S. Johnson, and M.L. Gennaro.** 2000. Identification of secreted proteins of *Mycobacterium tuberculosis* by a bioinformatic approach. Infect. Immun. *68*:2323-2327.
49.**Grandi, G.** 2001. Antibacterial vaccine design using genomics and proteomics. Trends Biotechnol. *19*:181-188.
50.**Gromiha, M.M., R. Majumdar, and P.K. Ponnuswamy.** 1997. Identification of membrane spanning β strands in bacterial porins. Protein Eng. *10*:497-500.
51.**Guenzi, E., G. Galli, I. Grgurina, D.C. Gross, and G. Grandi.** 1998. Characterization of the syringomycin synthetase gene cluster. A link between prokaryotic and eukaryotic peptide synthetases. J. Biol. Chem. *273*:32857-32863.
52.**Gussin, G.A., A.D. Johnson, C.O. Pabo, and R.J. Sauer.** 1983. Repressor and cro protein: structure, function, and role in lysogenization, p. 93-121. *In* R.W. Hendrix, J.W. Roberts, F.W. Stahl, and R.A. Weisberg (Eds.), Lambda II. CSH Laboratory Press, Cold Spring Harbor, NY.
53.**Gygi, S.P., Y. Rochon, B.R. Franza, and R. Aebersold.** 1999. Correlation between protein and mRNA abundance in yeast. Mol. Cell. Biol. *19*:1720-1730.
54.**Hannenhalli, S.S., W.S. Hayes, A.G. Hatzigeorgiou, and J.W. Fickett.** 1999. Bacterial start site prediction. Nucleic Acids Res. *27*:3577-3582.

55.**Hanych, B., S. Kedzierska, B. Walderich, B. Uznanski, and A. Taylor.** 1993. Expression of the Rz gene and the overlapping Rz1 reading frame present at the right end of the bacteriophage lambda genome. Gene *129*:1-8.

56.**Henikoff, J.G., E.A. Greene, S. Pietrokovski, and S. Henikoff.** 2000. Increased coverage of protein families with the blocks database servers. Nucleic Acids Res. *28*:228-230.

57.**Hernandez-Sanchez, J., J.G. Valadez, J.V. Herrera, C. Ontiveros, and G. Guarneros.** 1998. lambda bar minigene-mediated inhibition of protein synthesis involves accumulation of peptidyl-tRNA and starvation for tRNA. EMBO J. *17*:3758-3765.

58.**Herr, A.J., J.F. Atkins, and R.R. Gesteland.** 2000. Coupling of open reading frames by translational bypassing. Annu. Rev. Biochem. *69*:343-372.

59.**Hirokawa, T., S. Boon-Chieng, and S. Mitaku.** 1998. SOSUI: classification and secondary structure prediction system for membrane proteins. Bioinformatics *14*:378-379.

60.**Hofmann, K., P. Bucher, L. Falquet, and A. Bairoch.** 1999. The PROSITE database, its status in 1999. Nucleic Acids Res. *27*:215-219.

61.**Hoiseth, S.** 2000. Vaccines, Bacterial, p. 767-778. *In* J. Lederberg (Ed.), Encyclopedia of Microbiology, Second ed., vol. 4. Academic Press.

62.**Hood, D.W., M.E. Deadman, M.P. Jennings, M. Bisercic, R.D. Fleischmann, J.C. Venter, and E.R. Moxon.** 1996. DNA repeats identify novel virulence genes in *Haemophilus influenzae*. Proc. Natl. Acad. Sci. USA *93*:11121-11125.

63.**Hudak, J. and M.A. McClure.** 1999. A comparative analysis of computational motif-detection methods, p. 138-149. *In* R.B. Altman, A.K. Dunker, L. Hunter, T.E. Klein and K. Lauderdale (Eds.), Pacific Synmposium on Biocomputing '99, World Scientific, Singapore.

64.**Hueck, C.J.** 1998. Type III protein secretion systems in bacterial pathogens of animals and plants. Microbiol. Mol. Biol. Rev. *62*:379-433.

65.**Inamine, J.M., K.C. Ho, S. Loechel, and P.C. Hu.** 1990. Evidence that UGA is read as a tryptophan codon rather than as a stop codon by *Mycoplasma pneumoniae*, *Mycoplasma genitalium*, and *Mycoplasma gallisepticum*. J. Bacteriol. *172*:504-506.

66.**Isberg, R.R. and G. Tran Van Nhieu.** 1994. Binding and internalization of microorganisms by integrin receptors. Trends Microbiol. *2*:10-14.

67.**Jones, D.T., W.R. Taylor, and J.M. Thornton.** 1994. A model recognition approach to the prediction of all-helical membrane protein structure and topology. Biochemistry *33*:3038-3049.

68.**Junker, V., S. Contrino, W. Fleischmann, H. Hermjakob, F. Lang, M. Magrane, M.J. Martin, N. Mitaritonna, C. O'Donovan, and R. Apweiler.** 2000. The role SWISS-PROT and TrEMBL play in the genome research environment. J. Biotechnol. *78*:221-234.

69.**Klein, P., M. Kanehisa, and C. DeLisi.** 1985. The detection and classification of membrane-spanning proteins. Biochim. Biophys. Acta. *815*:468-476.

70.**Krogh, A., B. Larsson, G. von Heijne, and E. Sonnhammer.** 2001. Predicting transmembrane protein topology with a hidden Markov model: application to complete genomes. J. Mol. Biol. *305*:567-580.

71.**Kyrpides, N.C.** 1999. Genomes OnLine Database (GOLD 1.0): a monitor of complete and ongoing genome projects world-wide. Bioinformatics *15*:773-774.

72.**Kyrpides, N.C. and C.A. Ouzounis.** 1999. Whole-genome sequence annotation: "going wrong with confidence". Mol. Microbiol. *32*:886-887.

73.**Lander, E.S., L.M. Linton, B. Birren, C. Nusbaum, M.C. Zody, J. Baldwin, K. Devon, K. Dewar et al.** 2001. Initial sequencing and analysis of the human genome. Nature *409*:860-921.

74.**Lio, P. and M. Vannucci.** 2000. Wavelet change-point prediction of transmembrane proteins. Bioinformatics *16*:376-382.

75.**Liu, Z., M. Reches, I. Groisman, and H. Engelberg-Kulka.** 1998. The nature of the minimal "selenocysteine insertion sequence" (SECIS) in *Escherichia coli*. Nucleic Acids Res. *26*:896-902.

76.**Loessner, M.J., S. Gaeng, and S. Scherer.** 1999. Evidence for a holin-like protein gene fully embedded out of frame in the endolysin gene of *Staphylococcus aureus* bacteriophage 187. J. Bacteriol. *181*:4452-4460.

77.**Loh, J., M.G. Stacey, M.J. Sadowsky, and G. Stacey.** 1999. The *Bradyrhizobium japonicum* nolA gene encodes three functionally distinct proteins. J. Bacteriol. *181*:1544-1554.

78.**Lukashin, A.V. and M. Borodovsky.** 1998. GeneMark.hmm: new solutions for gene finding. Nucleic Acids Res. *26*:1107-1115.

79.**Marques, M., S. Mahapatra, D. Nandan, T. Dick, E.N. Sarno, P. Brennan, and M. Pessolani.** 2000. Bacterial and host-derived cationic proteins bind a2-laminins and enhance *Mycobacterium leprae* attachment to human Schwann cells. Microbes Infect. *2*:1407-1417.

80.**McGeoch, D.J.** 1985. On the predictive recognition of signal peptide sequences. Virus Res. *3*:271-286.

81.**Mejlhede, N., J.F. Atkins, and J. Neuhard.** 1999. Ribosomal -1 frameshifting during decoding of *Bacil-*

lus subtilis cdd occurs at the sequence CGA AAG. J. Bacteriol. *181*:2930-2937.

82.**Moxon, E.R., P.B. Rainey, M.A. Nowak, and R.E. Lenski.** 1994. Adaptive evolution of highly mutable loci in pathogenic bacteria. Curr. Biol. *4*:24-33.

83.**Mushegian, A.R.** 2000. Reply to Kyrpides and Ouzounis. Mol. Microbiol. *35*:697-698.

84.**Nakai, K.** 2000. Protein sorting signals and prediction of subcellular localization. Adv. Protein Chem. *54*:277-344.

85.**Nakai, K. and M. Kanehisa.** 1991. Expert system for predicting protein localization sites in gram-negative bacteria. Proteins *11*:95-110.

86.**Nakashima, H. and K. Nishikawa.** 1992. The amino acid composition is different between the cytoplasmic and extracellular sides in membrane proteins. FEBS Lett. *303*:141-146.

87.**Needleman, S.B. and C.D. Wunsch.** 1970. A general method applicable to the search for similarities in the amino acid sequence of two proteins. J. Mol. Biol. *48*:443-453.

88.**Neuwald, A.F., J.S. Liu, and C.E. Lawrence.** 1995. Gibbs motif sampling: detection of bacterial outer membrane protein repeats. Protein Sci. *4*:1618-1632.

89.**Nicholas, H.B., Jr., D.W. Deerfield, II, and A.J. Ropelewski.** 2000. Strategies for searching sequence databases. BioTechniques *28*:1174-1191.

90.**Nielsen, H., S. Brunak, and G. von Heijne.** 1999. Machine learning approaches for the prediction of signal peptides and other protein sorting signals. Protein Eng. *12*:3-9.

91.**Nielsen, H., J. Engelbrecht, S. Brunak, and G. von Heijne.** 1997. Identification of prokaryotic and eukaryotic signal peptides and prediction of their cleavage sites. Protein Eng. *10*:1-6.

92.**Nielsen, H., J. Engelbrecht, S. Brunak, and G. von Heijne.** 1997. A neural network method for identification of prokaryotic and eukaryotic signal peptides and prediction of their cleavage sites. Int. J. Neur. Syst. *8*:581-599.

93.**Nilsson, C.L., T. Larsson, E. Gustafsson, K.A. Karlsson, and P. Davidsson.** 2000. Identification of protein vaccine candidates from *Helicobacter pylori* using a preparative two-dimensional electrophoretic procedure and mass spectrometry. Anal. Chem. *72*:2148-2153.

94.**Nilsson, J., B. Persson, and G. von Heijne.** 2000. Consensus predictions of membrane protein topology. FEBS Lett. *486*:267-269.

95.**Park, S.K., K.I. Kim, K.M. Woo, J.H. Seol, K. Tanaka, A. Ichihara, D.B. Ha, and C.H. Chung.** 1993. Site-directed mutagenesis of the dual translational initiation sites of the clpB gene of *Escherichia coli* and characterization of its gene products. J. Biol. Chem. *268*:20170-20174.

96.**Parkhill, J., M. Achtman, K.D. James, S.D. Bentley, C. Churcher, S.R. Klee, G. Morelli, D. Basham et al.** 2000. Complete DNA sequence of a serogroup A strain of *Neisseria meningitidis* Z2491. Nature *404*:502-506.

97.**Pasquier, C., V.J. Promponas, G.A. Palaios, J.S. Hamodrakas, and S.J. Hamodrakas.** 1999. A novel method for predicting transmembrane segments in proteins based on a statistical analysis of the SwissProt database: the PRED-TMR algorithm. Protein Eng. *12*:381-385.

98.**Pautsch, A. and G.E. Schulz.** 2000. High-resolution structure of the OmpA membrane domain. J. Mol. Biol. *298*:273-282.

99.**Pearson, W.R. and D.J. Lipman.** 1988. Improved tools for biological sequence comparison. Proc. Natl. Acad. Sci. USA *85*:2444-2448.

100.**Persson, B.** 2000. Bioinformatics in protein analysis. Proteomics Funct. Genomics Protein Struct. Anal. *88*:215-231.

101.**Pizza, M., V. Scarlato, V. Masignani, M.M. Giuliani, B. Arico, M. Comanducci, G.T. Jennings, L. Baldi et al.** 2000. Identification of vaccine candidates against serogroup B meningococcus by whole-genome sequencing. Science *287*:1816-1820.

102.**Rappuoli, R.** 2001. Conjugates and reverse vaccinology to eliminate bacterial meningitis. Vaccine *19*:2319-2322.

103.**Rappuoli, R.** 2001. Reverse vaccinology, a genome-based approach to vaccine development. Vaccine *19*:2688-2691.

104.**Rauhut, R. and G. Klug.** 1999. mRNA degradation in bacteria. FEMS Microbiol. Rev. *23*:353-370.

105.**Reinhardt, A. and T. Hubbard.** 1998. Using neural networks for prediction of the subcellular location of proteins. Nucleic Acids Res. *26*:2230-2236.

106.**Richmond, C.S., J.D. Glasner, R. Mau, H. Jin, and F.R. Blattner.** 1999. Genome-wide expression profiling in *Escherichia coli* K-12. Nucleic Acids Res. *27*:3821-3835.

107.**Rigoutsos, I. and A. Floratos.** 1998. Combinatorial pattern discovery in biological sequences: The TEIRESIAS algorithm. Bioinformatics *14*:55-67.

108.**Rost, B., R. Casadio, and P. Fariselli.** 1996. Refining neural network predictions for helical transmembrane proteins by dynamic programming. Ismb *4*:192-200.

109.**Salzberg, S.L., A.L. Delcher, S. Kasif, and O. White.** 1998. Microbial gene identification using interpolated Markov models. Nucleic Acids Res. *26*:544-548.
110.**Saunders, N.J., A.C. Jeffries, J.F. Peden, D.W. Hood, H. Tettelin, R. Rappuoli, and E.R. Moxon.** 2000. Repeat-associated phase variable genes in the complete genome sequence of *Neisseria meningitidis* strain MC58. Mol. Microbiol. *37*:207-215.
111.**Saunders, N.J., J.F. Peden, D.W. Hood, and E.R. Moxon.** 1998. Simple sequence repeats in the *Helicobacter pylori* genome. Mol. Microbiol. *27*:1091-1098.
112.**Schirmer, T. and S.W. Cowan.** 1993. Prediction of membrane-spanning β-strands and its application to maltoporin. Protein Sci. *2*:1361-1363.
113.**Schneider, G.** 1999. How many potentially secreted proteins are contained in a bacterial genome? Gene *237*:113-121.
114.**Seol, J.H., S.J. Yoo, M.S. Kang, D.B. Ha, and C.H. Chung.** 1995. The 65-kDa protein derived from the internal translational start site of the clpA gene blocks autodegradation of ClpA by the ATP-dependent protease Ti in *Escherichia coli*. FEBS Lett. *377*:41-43.
115.**Shine, J. and L. Dalgarno.** 1974. The 3′-terminal sequence of *Escherichia coli* 16S ribosomal RNA: complementarity to nonsense triplets and ribosome binding sites. Proc. Natl. Acad. Sci. USA *71*:1342-1346.
116.**Shmatkov, A.M., A.A. Melikyan, F.L. Chernousko, and M. Borodovsky.** 1999. Finding prokaryotic genes by the "frame-by-frame" algorithm: targeting gene starts and overlapping genes. Bioinformatics *15*:874-886.
117.**Smith, R.** 2000. Probing proteomes—seeing the whole picture? Nat. Biotechnol. *18*:1041-1042.
118.**Smith, T.F. and M.S. Waterman.** 1981. Identification of common molecular subsequences. J. Mol. Biol. *147*:195-197.
119.**Smith, T.F. and X. Zhang.** 1997. The challenges of genome sequence annotation or "the devil is in the details". Nat. Biotechnol. *15*:1222-1223.
120.**Sonnhammer, E.L., G. von Heijne, and A. Krogh.** 1998. A hidden Markov model for predicting transmembrane helices in protein sequences. Ismb *6*:175-182.
121.**Spence, J.M. and V.L. Clark.** 2000. Role of ribosomal protein L12 in gonococcal invasion of Hec1B cells. Infect. Immun. *68*:5002-5010.
122.**Stathopoulos, C.** 1999. Bacterial outer membrane proteins: topological analyses and biotechnological perspectives. Membr. Cell Biol. *13*:3-21.
123.**Stephens, R.S.** 2000. Chlamydial genomics and vaccine antigen discovery. J. Infect. Dis. *181*:S521-S523.
124.**Sternberg, N., B. Sauer, R. Hoess, and K. Abremski.** 1986. Bacteriophage P1 cre gene and its regulatory region. Evidence for multiple promoters and for regulation by DNA methylation. J. Mol Biol. *187*:197-212.
125.**Stover, C.K., X.Q. Pham, A.L. Erwin, S.D. Mizoguchi, P. Warrener, M.J. Hickey, F.S. Brinkman, W.O. Hufnagle et al.** 2000. Complete genome sequence of *Pseudomonas aeruginosa* PA01, an opportunistic pathogen. Nature *406*:959-964.
126.**Sun, Y.-H., S. Bakshi, R. Chalmers, and C.M. Tang.** 2000. Functional genomics of *Neisseria meningitidis* pathogenesis. Nat. Med. *6*:1269-1273.
127.**Sykes, K.F. and S.A. Johnston.** 1999. Linear expression elements: a rapid, in vivo, method to screen for gene functions. Nat. Biotechnol. *17*:355-359.
128.**Tateno, Y., S. Miyazaki, M. Ota, H. Sugawara, and T. Gojobori.** 2000. DNA data bank of Japan (DDBJ) in collaboration with mass sequencing teams. Nucleic Acids Res. *28*:24-26.
129.**Tenson, T., J.V. Herrera, P. Kloss, G. Guarneros, and A.S. Mankin.** 1999. Inhibition of translation and cell growth by minigene expression. J. Bacteriol. *181*:1617-1622.
130.**Tettelin, H., N.J. Saunders, J. Heidelberg, A.C. Jeffries, K.E. Nelson, J.A. Eisen, K.A. Ketchum, D.W. Hood et al.** 2000. Complete genome sequence of *Neisseria meningitidis* serogroup B strain MC58. Science *287*:1809-1815.
131.**Thomas, G.H.** 1999. Completing the *E. coli* proteome: a database of gene products characterised since the completion of the genome sequence. Bioinformatics *15*:860-861.
132.**Tomb, J.F., O. White, A.R. Kerlavage, R.A. Clayton, G.G. Sutton, R.D. Fleischmann, K.A. Ketchum, H.P. Klenk et al.** 1997. The complete genome sequence of the gastric pathogen *Helicobacter pylori*. Nature *388*:539-547.
133.**van Belkum, A., S. Scherer, L. van Alphen, and H. Verbrugh.** 1998. Short-sequence DNA repeats in prokaryotic genomes. Microbiol. Mol. Biol. Rev. *62*:275-293.
134.**Venter, J.C., M.D. Adams, E.W. Myers, P.W. Li, R.J. Mural, G.G. Sutton, H.O. Smith, M. Yandell et al.** 1304. The sequence of the human genome. Science *291*:16.
135.**von Heijne, G.** 1992. Membrane protein structure prediction. Hydrophobicity analysis and the positive-inside rule. J. Mol. Biol. *225*:487-494.

136.**von Heijne, G.** 1986. A new method for predicting signal sequence cleavage sites. Nucleic Acids Res. *14*:4683-4690.
137.**von Heijne, G.** 1985. Signal sequences. The limits of variation. J. Mol. Biol. *184*:99-105.
138.**von Heijne, G.** 1989. The structure of signal peptides from bacterial lipoproteins. Protein Eng. *2*:531-534.
139.**Weinstock, G.M., D. Smajs, J. Hardham, and S.J. Norris.** 2000. From microbial genome sequence to applications. Res. Microbiol. *151*:151-158.
140.**Wizemann, T.M., J.H. Heinrichs, J.E. Adamou, A.L. Erwin, C. Kunsch, G.H. Choi, S.C. Barash, C.A. Rosen et al.** 2001. Use of a whole genome approach to identify vaccine molecules affording protection against *Streptococcus pneumoniae* infection. Infect. Immun. *69*:1593-1598.
141.**Wolf, Y.I., I.B. Rogozin, A.S. Kondrashov, and E.V. Koonin.** 2001. Genome alignment, evolution of prokaryotic genome organization, and prediction of gene function using genomic context. Genome Res. *11*:356-372.
142.**Yoshida, K., K. Kobayashi, Y. Miwa, C.M. Kang, M. Matsunaga, H. Yamaguchi, S. Tojo, M. Yamamoto et al.** 2001. Combined transcriptome and proteome analysis as a powerful approach to study genes under glucose repression in *Bacillus subtilis*. Nucleic Acids Res. *29*:683-692.

Update to: Bioinformatics: Use in Bacterial Vaccine Discovery

Robert J. Zagursky and David Russell[1]

Wyeth Vaccines, West Henrietta, NY, and [1]*LPA System, Fairport, NY, USA (zagursrj@wyeth.com)*

Progress continues to be made in bacterial vaccine discovery. Since the original publication of this article, an even larger bacterial sequence has been completed, new software algorithms have been produced, proteomic techniques continue to provide new insight, and research continues towards more precise identification of potential vaccine targets.

According to Integrated Genomics (http://wit.integratedgenomics.com/GOLD/gold.html), the sequence of *Bradyrhizobium japonicum* has been completed. At 10231 kbp, this sequence exceeds the originally reported largest sequenced bacterial genome by almost 3 million bp.

Development and refinement of tools and techniques continues on a swift pace. A new program called BETAWRAP (http://theory.lcs.mit.edu/betawrap) has been developed to detect right-handed parallel β-helix motifs in proteins (1). Detection of these β-helix motifs has identified several vaccine class proteins in those sequences that have been analyzed. Building on past development, researchers in the laboratory of Fiona Brinkman (http://www.pathogenomics.bc.ca/brinkman/research.html) are updating and enhancing the PSORT algorithm to improve accuracy.

The push towards proteomic analysis has lead to the development of new laboratory tools and techniques as well. The development of high-density protein–peptide microarrays (8) shows real promise for proteomics. These arrays can be used as a diagnostic tool, to help determine protein function, and for antibody screening. In the latter case, screening of a proteome chip can be used to identify in vivo-induced antigens as was first demonstrated by Suk et al. using a genomic expression library (7). New techniques continue to appear in the literature as more researchers achieve success using proteomics in vaccine discovery programs. For example, in the analysis of 2-dimensional gel protein spots, a technique using in vitro transcription–translation reactions (TTR) helped identify several *Neisseria meningitidis* membrane proteins (3).

In two recent papers, genomic or genomic–proteomic methods for identifying bacterial vaccine candidates have been described. A genomic analysis of *Porphyromonas gingivalis*, a Gram-negative bacterium and causative agent for periodontitis, was done using FASTA and PSORT algorithms to identify potential vaccine candidates (4). Even more recently, a combined genomic–proteomic analysis of *Chlamydia pneumoniae*, a Gram-negative obligate intracellular pathogen responsible for respiratory infections, and its association with atherosclerosis was described (2). This analysis of *C. pneumoniae* identified 28 elementary-body surface-exposed antigens.

The promise of bioinformatics continues to grow as old goals are reached and new goals are set. Where once knowledge of the entire genomic sequence was considered the "Holy Grail" that would answer all questions, new goals must be refined and set to continue forward with development of new vaccines. It is not enough simply to know what proteins may be present; protein expression under specific conditions must be understood to more precisely target vaccine development. Although it is possible to mimic some of the in vivo conditions in vitro (6), the current push for genomic technology is to determine bacterial gene expression during actual pathogenesis (5). Technically, this is very difficult, due to the low amount of bacterial RNA isolated from infected tissue or recovery of enough bacteria for proteome analysis. However, given time, these problems will eventually be overcome through continued research.

REFERENCES

1. **Bradley, P., L. Cowen, M. Menke, J. King, and B. Berger.** 2001. BETAWRAP: successful prediction of parallel beta-helices from primary sequence reveals an association with many microbial pathogens. Proc. Natl. Acad. Sci. USA *98*:14819-14824.
2. **Montigiani, S., F. Falugi, M. Scarselli, O. Finco, R. Petracca, G. Galli, M. Mariani, R. Manetti, et al.** 2002. Genomic approach for analysis of surface proteins in *Chlamydia pneumoniae*. Infect. Immun. *70*:368-379.
3. **Norais, N., R. Nogarotto, E.T. Iacobini, I. Garaguso, R. Grifantini, G. Galli, and G. Grandi.** 2001. Combined automated PCR cloning, in vitro transcription/translation and two-dimensional electrophoresis for bacterial proteome analysis. Proteomics *1*:1378-1389.
4. **Ross, B.C., L. Czajkowski, D. Hocking, M. Margetts, E. Webb, L. Rothel, M. Patterson, C. Agius, et al.** 2001. Identification of vaccine candidate antigens from a genomic analysis of *Porphyromonas gingivalis*. Vaccine *19*:4135-4142.
5. **Sassetti, C. and E.J. Rubin.** 2002. Genomic analyses of microbial virulence [review]. Curr. Opin. Microbiol. *5*:27-32.
6. **Schoolnik, G.K.** 2002. Functional and comparative genomics of pathogenic bacteria [review]. Curr. Opin. Microbiol. *5*:20-26.
7. **Suk, K., S. Das, W. Sun, B. Jwang, S.W. Barthold, R.A. Flavell, and E. Fikrig.** 1995. Borrelia burgdorferi genes selectively expressed in the infected host. Proc. Natl. Acad. Sci. USA *92*:4269-4273.
8. **Templin, M.F., D. Stoll, M. Schrenk, P.C. Traub, C.F. Vohringer, and T.O. Joos.** 2002. Protein microarray technology. Trends Biotechnol. *20*:160-166.

CHAPTER 33

Retriever and *CompareTable*, Two Informatics Tools for Data Analysis of High-Density Oligonucleotide Arrays

Allan R. Brasier

The University of Texas Medical Branch,
Galveston, TX, USA

Reprinted from BioTechniques 32:100-109 (January 2002)

ABSTRACT

High-density oligonucleotide arrays are widely employed for detecting global changes in gene expression profiles of cells or tissues exposed to specific stimuli. Presented with large amounts of data, investigators can spend significant amounts of time analyzing and interpreting this array data. In our application of GeneChip® arrays to analyze changes in gene expression in viral-infected epithelium, we have needed to develop additional computational tools that may be of utility to other investigators using this methodology. Here, I describe two executable programs to facilitate data extraction and multiple data point analysis. These programs run in a virtual DOS environment on Microsoft® Windows® 95/98/2K operating systems on a desktop PC. Both programs can be freely downloaded from the BioTechniques *Software Library (www.BioTechniques.com). The first program,* Retriever, *extracts primary data from an array experiment contained in an Affymetrix® text file using user-inputted individual identification strings (e.g., the probe set identification numbers). With specific data retrieved for individual genes, hybridization profiles can be examined and data normalized. The second program,* CompareTable, *is used to facilitate comparison analysis of two experimental replicates.* CompareTable *compares two lists of genes, identifies common entries, extracts their data, and writes an output text file containing only those genes present in both of the experiments. The output files generated by these two programs can be opened and manipulated by any software application recognizing tab-delimited text files (e.g., Microsoft NotePad® or Excel®).*

INTRODUCTION

High-density DNA microarrays are a rapidly emerging technology that allows for the measurement of gene expression profiles in cells or tissues. Microarrays have provided important insights into cellular genetic responses to environmental manipulations (3), stimulation of growth and signal transduction pathways (4,7,12), alterations in malignant transformation (9), and responses to viral infection (8,21,22), to name but a few. This

BioComputing: Computer Tools for Biologists
Edited by Stuart M. Brown

methodology now enables investigators to measure a significant fraction of the cellular transcriptosome as a single experimental point. As a consequence, the analysis, quantitation, and interpretation of this data is now becoming a major challenge for the scientific community.

High-density (micro- and macro-) arrays are ordered representations of complementary nucleic acid sequences ("targets") immobilized on a nylon or glass support; the various methods for producing arrays have been comprehensively reviewed (14–16,18). After the selected genetic targets have been selected and immobilized on a solid support, fluorescent or radioactive complementary DNA probes are synthesized from complex biological samples and hybridized to the array. The amount of hybridization to each target on the array is then quantitated and used to determine relative changes in mRNA abundance in the original sample.

Of the many array technologies available, high-density oligonucleotide arrays are a highly popular methodology that employs photolithographic masking and modified phosphoramidite-based oligonucleotide synthesis chemistry to synthesize short (approximately 25-mer) oligonucleotide probe sequences in situ (GeneChip®; Affymetrix, Santa Clara, CA, USA; described in References 11 and 15). Each gene is represented by 15–20 different query sequences distributed throughout the length of the cDNA; each query sequence is paired with a mismatch sequence for the determination of signal specificity. Biotinylated cDNA probes are made from complex RNA samples and, after hybridization, are detected by staining with streptavidin-phycoerythrin complexes.

The hybridization intensity for each probe cell is determined and a series of algorithms

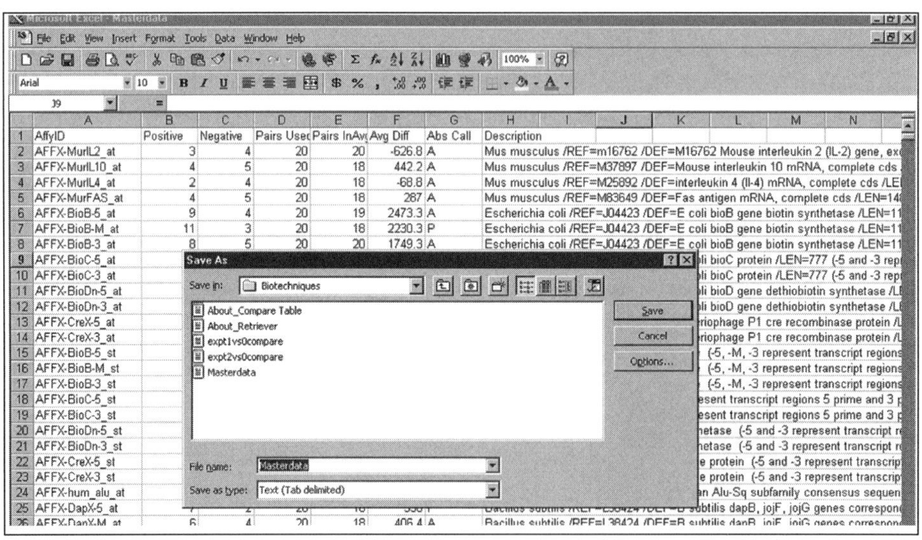

Figure 1. Representative input file for *Retriever*. The data-containing file (here called "Masterdata.txt") from the array hybridization experiment has been opened (in this case, using Excel). For correct input, the columns of the Masterdata.txt should be the same as this example. Column A is the Affymetrix ID string; column B is the number of positive probe pairs; column C is the negative probe pairs; column D is the numbers of probe pairs used in the average difference calculation; column E is probe pairs in the average; column F is the average difference, an indicator of RNA abundance; column G is the absolute call; and column H is the descriptive information. For input into *Retriever*, the Masterdata.txt file must be saved in the ".txt" extension (as illustrated using the "File/Save As" option) in the same folder as the Retriever.exe program. *Retriever* accepts any name for the Masterdata file (maximum 100 characters or shorter).

in an "absolute analysis" to determine various measures of signal intensity. For experiments in which treatment versus control hybridizations are performed, secondary "comparison analysis" allows the determination of how signal intensity changes for each gene on the microarray. The comparison analysis metric is frequently expressed as a "fold change" value, a signed number indicating the relative change in expression (up- or down-regulated) produced by the experimental treatment. These data are then examined to identify any biologically relevant genes in clusters or biological pathways for further hypothesis-generated experimentation (1,2,5,6,9).

A body of literature supports the use of high-density GeneChip arrays as highly specific, quantitative measures of steady-state mRNA abundance in complex RNA samples (15). These changes can be independently confirmed by northern analysis, RNase protection, or semi-quantitative RT-PCR assays (15,21). In applying this technique to analyze changes in gene expression in viral infected epithelium (21), we have needed to develop additional computational tools for data analysis that may be of utility to other investigators using the GeneChip methodology. Here, I describe two programs that can be run on desktop PCs (with Microsoft® Windows® 95/98/2K operating systems) to facilitate data extraction and multiple replicate analysis. The first program, *Retriever*, allows primary data extraction from an absolute analysis text file for individual probe sets. The second program, *CompareTable*, compares two lists of genes and generates a text file with genes common to the two original text files. The output files generated by these two programs can be opened and manipulated by any software application recognizing tab-delimited text files (e.g., Microsoft NotePad® or Excel®), adding to the tools for the analysis and interpretation of microarray data.

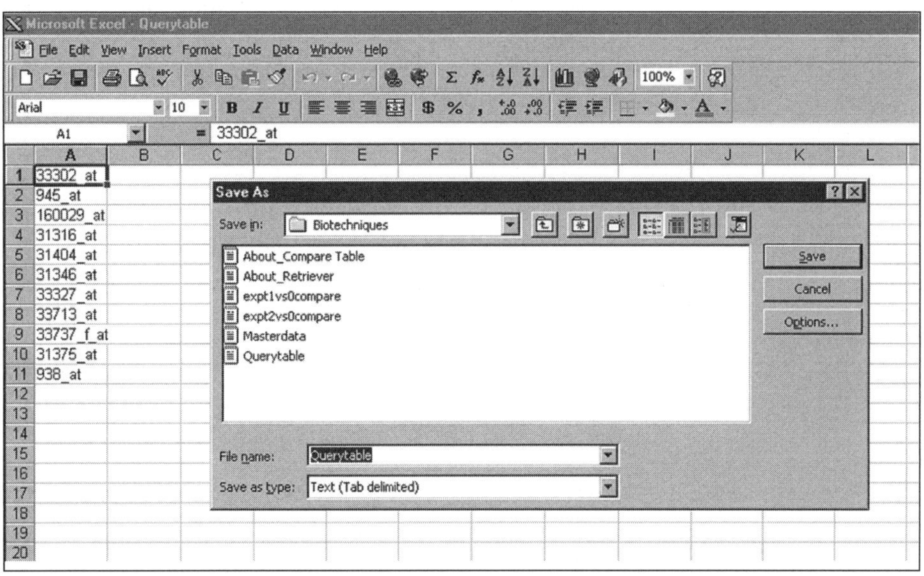

Figure 2. Representative query file for *Retriever*. The query list of Affymetrix ID probe set identifiers is entered as a single column. A representative Querytable file containing 11 entries as viewed by Excel is shown. *Retriever* will accept a maximum of 13 000 entries and will compare them until an empty row is encountered; as with the Masterdata.txt, Querytable must be saved as a ".txt" file (100 characters or shorter).

MATERIALS AND METHODS

GeneChip Microarray Analysis

RNA samples were hybridized to the Hu95A GeneChip (Affymetrix) containing 12 626 sequenced human genes under conditions as described by the manufacturer. After probe hybridization, washing, and staining, the GeneChip arrays were scanned using a Gene Array™ Scanner and analyzed using the GeneChip Analysis Suite 3.3 software (Affymetrix). For each gene, 16–20 probe pairs are immobilized as approximately 25-mer oligonucleotides that hybridize throughout the mRNA; each probe pair is represented as a perfect match oligonucleotide and a mismatch oligonucleotide as hybridization control. The average intensity of each probe cell is calculated after subtraction of the local background (the lowest 2% intensity of each sector—each probe cell is divided into 16 sectors). The normalized average intensity value is used to determine the number of positive and negative probe pairs. Based on the positive/negative ratio, positive fraction, and log average ratio of the perfect match to the mismatch, the absolute call [e.g., the gene is detected ("present" or "marginal") or not ("absent")] is determined (22). Finally, the average difference is determined by calculating the difference intensity between the perfect match and the mismatch of every probe pair and averaging the differences over the entire probe set. The aver-

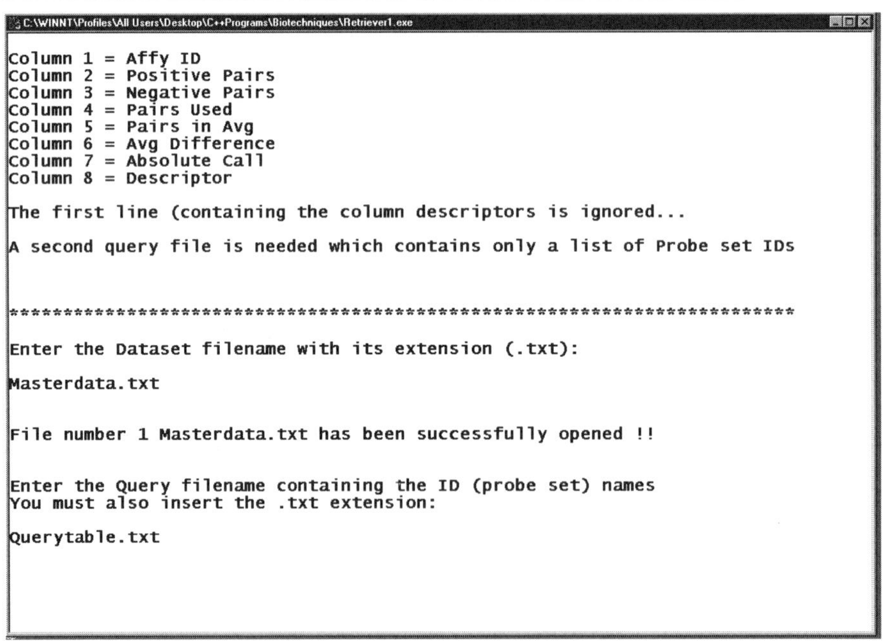

Figure 3. Operation of *Retriever*. When the executable *Retriever* program is launched (by double clicking on the program icon), the program runs as a virtual DOS machine on the Windows interface. The user is prompted for input file names with the ".txt" extension. If the input file is in the same folder as the executable program, then all that is necessary is the filename and extension; if the input file is in a separate folder, then the user must enter the file path and extension. Finally, the user specifies the name for the output file (with the ".txt" extension). A maximum of 100 characters, including the ".txt" extension, is accepted.

age difference statistic was retrieved for quantification of mRNA abundance in those samples in which the absolute call indicated that the gene was present. Probe set data were saved as ASCII tab-delimited text files and deposited into our data warehouse and relational database server (UTMB Bioinformatics Program).

Secondary Data Analysis

Retriever and *CompareTable* are two programs written in C^{++} programming language (compiled with Borland version 3.5). These executable programs are available for download by using the standard ftp anonymous protocol at the UTMB Bioinformatics ftp site: ftp://nmr.utmb.edu. To do this, download the two files *Retriever* and *CompareTable* into a data analysis folder. The path of the input data tables must be known or, more conveniently, copied into the same data analysis folder. Data output from the programs will be saved as user-specified names in this folder for subsequent inspection and analysis. The programs are launched by double clicking the program icon.

RESULTS AND DISCUSSION

After the acquisition of the array image, a number of methods can be employed to correct for relative differences in array intensity. These techniques include normalization to single (or multiple) housekeeping genes or normalization to a global variable, such as the median value of intensities of all the measured genes. Because it cannot be assumed *a priori* which transcripts are not affected by the treatment, nor necessarily if global transcription

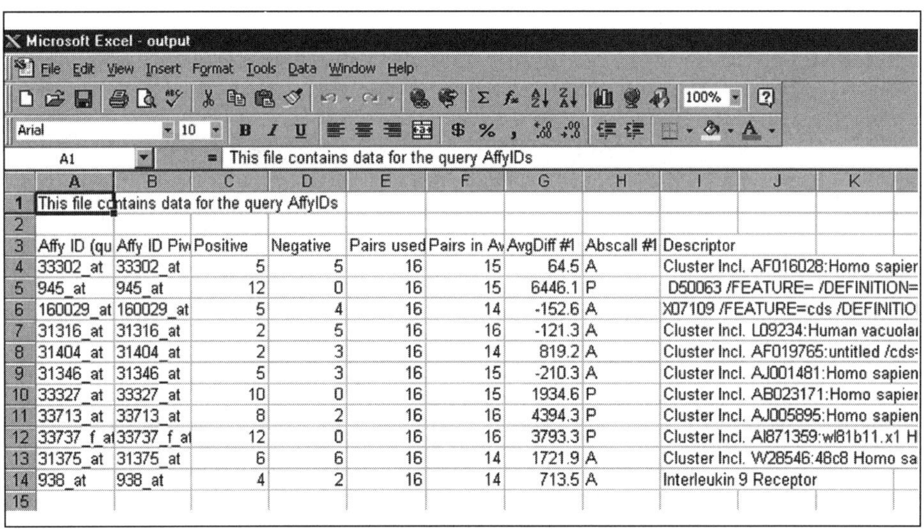

	A	B	C	D	E	F	G	H	I
1	This file contains data for the query AffyIDs								
2									
3	Affy ID (qu	Affy ID Piv	Positive	Negative	Pairs used	Pairs in Av	AvgDiff #1	Abscall #1	Descriptor
4	33302_at	33302_at	5	5	16	15	64.5	A	Cluster Incl. AF016028:Homo sapier
5	945_at	945_at	12	0	16	15	6446.1	P	D50063 /FEATURE= /DEFINITION=
6	160029_at	160029_at	5	4	16	14	-152.6	A	X07109 /FEATURE=cds /DEFINITIO
7	31316_at	31316_at	2	5	16	16	-121.3	A	Cluster Incl. L09234:Human vacuolai
8	31404_at	31404_at	2	3	16	14	819.2	A	Cluster Incl. AF019765:untitled /cds=
9	31346_at	31346_at	5	3	16	15	-210.3	A	Cluster Incl. AJ001481:Homo sapien
10	33327_at	33327_at	10	0	16	15	1934.6	P	Cluster Incl. AB023171:Homo sapiei
11	33713_at	33713_at	8	2	16	16	4394.3	P	Cluster Incl. AJ005895:Homo sapien
12	33737_f_at	33737_f_at	12	0	16	16	3793.3	P	Cluster Incl. AI871359:wl81b11.x1 H
13	31375_at	31375_at	6	6	16	14	1721.9	A	Cluster Incl. W28546:48c8 Homo sa
14	938_at	938_at	4	2	16	14	713.5	A	Interleukin 9 Receptor
15									

Figure 4. Sample data output. An output file generated by *Retriever* was opened in Excel and is shown. Column A is the Affymetrix ID probe set identifiers taken from the Querytable; column B is the Affymetrix ID string that it was matched to in the Masterdata file (serving as a visual check that the correct match was found); column C is the positive probe pairs; column D is the negative probe pairs; column E is the numbers of probe pairs used in the average difference calculation; column F is probe pairs in the average; column G is the average difference; column H is the absolute call; and column I is the descriptive information.

is not affected by the treatment, careful investigators are well advised to analyze their data by several methods before making any conclusions. For example, in the specific case of viral infection, global changes in gene expression can occur that may make normalization to median signal intensity potentially dangerous. Specifically, infection with Vesiculo Stomatitis, polio, or herpes simplex viruses can lead to the phenomenon of host cell "shut-off" where cellular RNA polymerase II-dependent genes are no longer transcribed (17,19,20). Conversely, the expression of standard housekeeping genes, such as β-actin or MHC class I genes, can be regulated by other viruses such as RSV, making normalization to these housekeeping transcripts problematic (13). For this reason, the retrieval of housekeeping genes from individual arrays can be extremely helpful in the initial inspection of array data. However, this retrieval can be time consuming, requiring multiple search/find queries for individual probe sets in spreadsheet programs (such as Excel). To facilitate multiple data point retrieval, I have developed *Retriever*, an executable program that can be run on desktop PCs that will rapidly extract data for multiple probe sets. The operation and application of *Retriever* is described below.

Retriever, A Computational Tool to Extract Primary Array Data from GeneChip Data Files

Retriever is an executable program that will retrieve primary data (from a ASCII ".txt" file) for any given probe set represented on the GeneChip. *Retriever* accepts two text files:

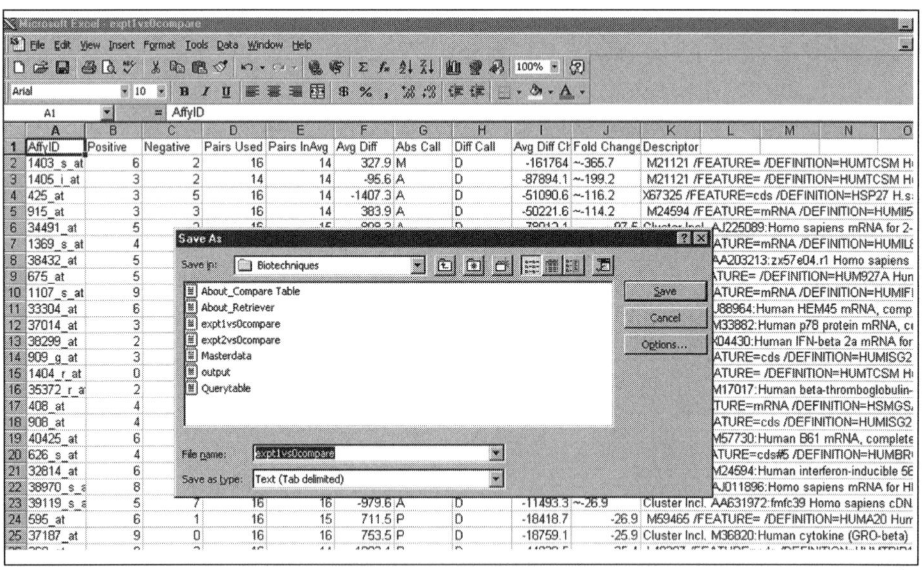

Figure 5. Input file format for *CompareTable*. An example input file for *CompareTable* as viewed by Excel. Both input files must match the file shown in this example and be saved as a tab-delimited ".txt" file using the "File/Save As" option (in this case, the file name is "expt1vs0compare.txt"). The first line containing the header names is ignored by *CompareTable*. For correct entry, the data should be arranged as the following: column A, the Affymetrix ID probe set identifiers; column B, the number of positive pairs; column C, the number of negative pairs; column D, the number of pairs used; column E, the number of pairs in the average; column F, the average difference; column G, the absolute call; column H, the difference call with "I" being increased and "D" decreased; column I, the average difference change; column J, the fold change; and column K, the gene descriptor.

a tab-delimited text file containing the Affymetrix primary data (indicated in this figure as Masterdata.txt) and a ".txt" file containing a query list of Affymetrix probe identification (ID) sets for which primary data is requested (indicated in this example as Querytable.txt). *Retriever* executes a simple loop where the first ID name from Querytable file is checked against the first ID entry in the Masterdata.txt file. When an exact match is found (partial matches are ignored), *Retriever* recovers the primary data from the Masterdata.txt file and outputs the information into the output file (indicated in this example as Output.txt). The next ID is then pulled from Querytable, and the search is repeated until the end of the list is reached. The required format of the two text files is shown on Figures 1 and 2. The text files can be opened and manipulated by data analysis software, such as Excel, as shown in Figures 1 and 2. Figure 3 shows the operation of the program. The user must either copy the input files into a folder containing the Retriever.exe program (in which case, the file name and extension are all that are needed to open the file) or specify the path to the data file. After launching *Retriever*, the user is prompted to enter the filename (including the ".txt" extension as shown; *Retriever* will accept file names of up to 100 characters in length). After program execution, the output file will be saved in the program folder. Figure 4 shows the output of the program after opening the Output.txt file by Excel. The data for each probe set ID are listed and include the number of positive probe pairs, the number of negative probe pairs, the number of probe pairs used, the average difference, and the absolute call (A, absent; P, present; or M, marginal). These data can then be manipulated, plotted, or merged with other datasets as desired by the user.

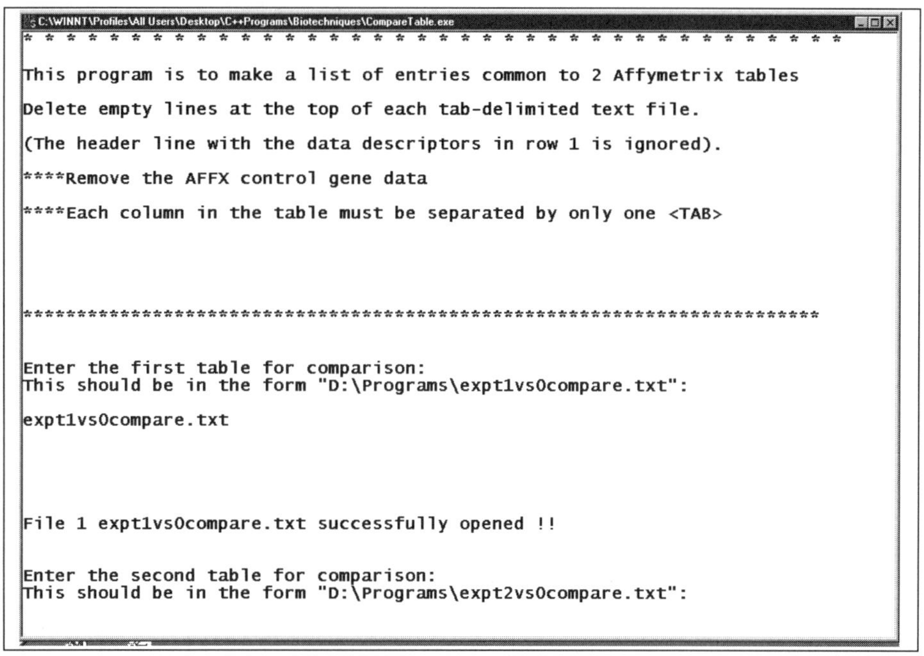

Figure 6. Operation of *CompareTable*. When the executable *CompareTable* is launched (by double clicking on the program icon), the program runs as a virtual DOS machine on the Windows interface. The user is prompted to input file names. If the input files are in the same folder as the executable program, then all that is necessary to enter is the filename and extension; if the input files are in a separate folder, then the user must enter the file paths and extensions. A limit of 100 characters (including the extension) is accepted.

Application

Retriever is useful for several applications. First, multiple independent probe sets representing the same gene are present on the Hu95A chip. Sometimes, the behavior of one probe set is aberrant from that of the others (21). *Retriever* allows data recovery for each probe set so that the investigator can analyze the individual hybridization statistics for each probe set. Second, should the investigator wish to normalize to individual or multiple housekeeping genes, *Retriever* can recover the hybridization signal for these values. Finally, it is sometimes necessary to extract data for selected genes on a given array, whether or not the gene expression is changed as a consequence of treatment (and therefore may not appear in the secondary comparison analysis of genes up- or down-regulated by a treatment).

High-density oligonucleotide arrays allow quantitative estimate of gene transcript levels in individual samples (11,15). From the absolute analysis data, GeneChip comparison analysis algorithms allow the comparison of gene expression levels between two samples. This typically involves the application of user-modified thresholds and results in a table of gene IDs, a “difference call” metric (increased, marginally increased, decreased, marginally decreased, and no change) and an estimate of the magnitude of the difference between the two samples (the fold change calculation). My colleagues and others have observed that 90% of probes showing a consistent 3-fold change (or greater) are significantly regulated by the experimental treatment using independent techniques [e.g., northern analysis, RNase Protection Assay (RPA) or RT-PCR (10,21,22)]. However, I have had difficulty easily identifying genes from long tables that are common to multiple-treatment replicates (many of the commercially available software programs that we tried are not enabled with this functionality). For this reason, *CompareTable* was developed; its operation and application are described below.

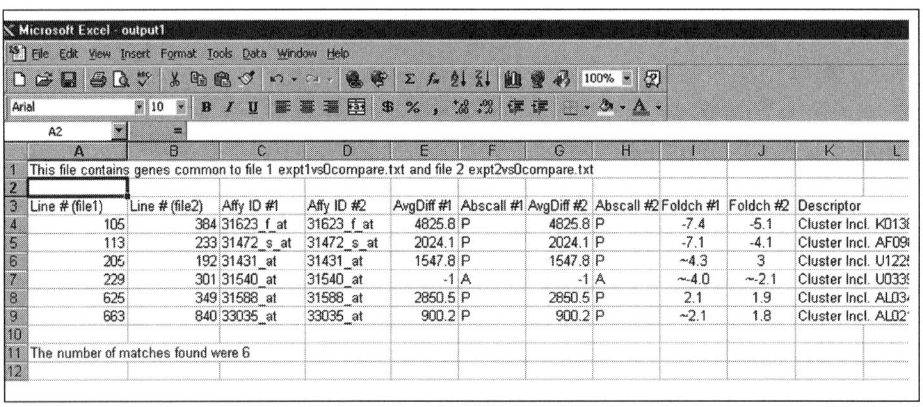

This file contains genes common to file 1 expt1vs0compare.txt and file 2 expt2vs0compare.txt

Line # (file1)	Line # (file2)	Affy ID #1	Affy ID #2	AvgDiff #1	Abscall #1	AvgDiff #2	Abscall #2	Foldch #1	Foldch #2	Descriptor
105	384	31623_f_at	31623_f_at	4825.8	P	4825.8	P	-7.4	-5.1	Cluster Incl. K013
113	233	31472_s_at	31472_s_at	2024.1	P	2024.1	P	-7.1	-4.1	Cluster Incl. AF09
205	192	31431_at	31431_at	1547.8	P	1547.8	P	~4.3	3	Cluster Incl. U122
229	301	31540_at	31540_at	-1	A	-1	A	~-4.0	~-2.1	Cluster Incl. U033
625	349	31588_at	31588_at	2850.5	P	2850.5	P	2.1	1.9	Cluster Incl. AL03
663	840	33035_at	33035_at	900.2	P	900.2	P	~2.1	1.8	Cluster Incl. AL02

The number of matches found were 6

Figure 7. Sample data output. Output file generated by *CompareTable* was opened in Excel. The output contains the following: column A, the line number of the first file from which the matching Affymetrix ID string was obtained; column B, the corresponding line from the second file where the match was made; columns C and D, the Affymetrix ID strings from each file to serve as a visual check that the correct match was made; columns E and F, the average difference and the absolute call (for the Affymetrix ID string) taken from the first table (indicated as Table I); columns G and H, the average difference and the absolute call taken from the second table (indicated as Table II); columns I and J, the fold change parameters from Tables I and II, respectively; and column K is the descriptor. The final row lists the number of matches found between the two tables.

CompareTable, A Computational Tool to Find Genes Common to Two Experimental Replicates

Experimental replicates of array experiments allow for the elimination of false positives and to control for biological variation between experiments. *CompareTable* was written to compare two randomly sorted lists of genes from two standard GeneChip comparison analysis tables and identify those genes that are present in both duplicates. *CompareTable* takes the Affymetrix ID from one table and sequentially queries all of the ID names in the second. If a match is found, then data from both tables are written to an output file; otherwise, *CompareTable* advances to the next entry and queries all of the entries in the second table. Figure 5 shows the format for *CompareTable* input files; these must be saved as a tab-delimited ".txt" files. In these tables, genes are represented that are significantly different in one chip compared to another (e.g., control). Typically, this data contains the ID, the signed fold change, and a descriptor (Figure 5 shows a representative comparison table). Figure 6 shows the operation of *CompareTable*, where the user is prompted for the file name of the tab-delimited input files and names the output file. Upon completion, the output file is saved in the folder containing the executable program. Figure 7 shows an example of the output generated by *CompareTable* (opened in Excel). For each gene found, the row where the gene was located on each file is printed, and the probe set identifier for both files is listed (as a visual check that the program is operating correctly), the average difference, the absolute call, and the fold change parameters from files I and II, respectively. The Affymetrix descriptor is contained on the last column. Finally, the output file will list the number of matches found in the two files as the last row of the output table.

In summary, a number of new bioinformatics tools are being and have been developed to facilitate the analysis and retrieval of data from microarray experiments. Here, two simple programs are described that can be run on desktop PCs (with Windows 95/98/2K operating systems) to facilitate data extraction and multiple replicate analysis. These programs will help research investigators more systematically evaluate this wealth of microarray data.

ACKNOWLEDGMENTS

The author thanks the UTMB Genomics Core Laboratory (T. Wood, Director) and the Bioinformatics Program (B.A. Luxon, Director) for advice and discussions. This project was supported by NAID grant nos. R21 AI48163 and AI40218 to A.R.B. and NIEHS grant no. P30 ES06676 (to R.S. Lloyd, UTMB).

REFERENCES

1. **Alon, U., N. Barkai, D.A. Notterman, K. Gish, S. Ybarra, D. Mack, and A.J. Levine.** 1999. Broad patterns of gene expression revealed by clustering analysis of tumor and normal colon tissues probed by oligonucleotide arrays. Proc. Natl. Acad. Sci. USA *96*:6745-6750.
2. **Brown, M.P.S. and W.N.L. Grundy.** 2000. Knowledge-based analysis of microarray gene expression data by using support vector machines. Proc. Natl. Acad. Sci. USA *97*:6745-6750.
3. **Chu, S., J. DeRisi, M. Eisen, J. Mulholland, D. Botstein, P.O. Brown, and I. Herskowitz.** 1998. The transcriptional program of sporulation in budding yeast. Science *282*:699-705.
4. **Der, S.D., A. Zhou, B.R.G. Williams, and R.H. Silverman.** 1998. Identification of genes differentially regulated by interferon α, β or γ using oligonucleotide arrays. Proc. Natl. Acad. Sci. USA *95*:15623-15628.

5. **Dopazo, J., E. Zanders, I. Dragoni, G. Amphlett, and F. Falciani.** 2001. Methods and approaches in the analysis of gene expression data. J. Immunol. Methods *250*:93-112.
6. **Eisen, M.B., P.T. Spellman, P.O. Brown, and D. Botstein.** 1998. Cluster analysis and display of genome-wide expression patterns. Proc. Natl. Acad. Sci. USA *95*:14863-14868.
7. **Fambrough, D., K. McClure, A. Kazlauskas, and E.S. Lander.** 1999. Diverse signaling pathways activated by growth factor receptors induce broadly overlapping, rather than independent, sets of genes. Cell *97*:727-741.
8. **Geiss, G.K., M.C. An, R.E. Bumgarner, E. Hammersmark, D. Cunningham, and M.G. Katze.** 2001. Global impact of influenza virus on cellular pathways is mediated by both replication-dependent and independent events. J. Virol. *75*:4321-4331.
9. **Golub, T.R., D.K. Slonim, P. Tamayo, C. Huard, M. Gaasenbeek, J.P. Mesirov, H. Coller, and M.L. Loh.** 1999. Molecular classification of cancer: class discovery and class prediction by gene expression monitoring. Science *286*:531-537.
10. **Harkin, D.P., J.M. Bean, D. Miklos, Y.-H. Song, V.B. Truong, C. Englert, F.C. Christians et al.** 1999. Induction of GADD45 and JNK/SAPK-dependent apoptosis following inducible expression of BRCA1. Cell *97*:575-586.
11. **Harrington, C.A., C. Rosenow, and J. Retief.** 2000. Monitoring gene expression using DNA microarrays. Curr. Opin. Microbiol. *3*:285-291.
12. **Iyer, V.R., M.B. Eisen, D.T. Ross, G. Schuler, T. Moore, J.C.F. Lee, J.M. Trent, L.M. Staudt et al.** 1999. The transcriptional program in the response of human fibroblasts to serum. Science *283*:83-87.
13. **Jamaluddin, M., S. Wang, R. Garofalo, T. Elliott, A. Casola, S. Baron, and A.R. Brasier.** 2001. INFβ mediates coordinate expression of antigen-processing genes in RSV infected pulmonary epithelial cells. Am. J. Physiol. Lung Cell Mol. Physiol. *280*:L248-L257.
14. **Lipshutz, R.J., S.P.A. Fodor, T.R. Gingeras, and D.J. Lockhart.** 1999. High density synthetic oligonucleotide arrays. Nat. Genet. *21(1 Supp)*:20-24.
15. **Lockhart, D.J., H. Dong, M.C. Byrne, M.T. Follettie, M.V. Gallo, M.S. Chee, M. Mittmann, C. Wang et al.** 1996. Expression monitoring by hybridization to high-density oligonucleotide arrays. Nat. Biotechnol. *14*:1675-1680.
16. **Ramsay, G.** 1998. DNA chips: state-of-the-art. Nat. Biotechnol. *16*:40-44.
17. **Spencer, C.A., M.E. Dahmus, and S.A. Rice.** 1997. Repression of host RNA polymerase II transcription by herpes simplex virus type 1. J. Virol. *71*:2031-2040.
18. **Watson, A., A. Mazumder, M. Stewart, and S. Balasubramanian.** 1998. Technology for microarray analysis of gene expression. Curr. Opin. Biotechnol. *9*:609-614.
19. **Yalamanchili, P., U. Datta, and A. Dasgupta.** 1997. Inhibition of host cell transcription by poliovirus: cleavage of transcription factor CREB by poliovirus-encoded protease 3Cpro. J. Virol. *71*:1220-1226.
20. **Yuan, H., B.K. Yoza, and D.S. Lyles.** 1998. Inhibition of host RNA polymerase II-dependent transcription by vesicular stomatitis virus results from inactivation of TFIID. Virology *251*:383-392.
21. **Zhang, Y., B.A. Luxon, A. Casola, R.P. Garofalo, M. Jamaluddin, and A.R. Brasier.** 2001. Expression of RSV-induced chemokine gene networks in lower airway epithelial cells revealed by cDNA microarrays. J. Virol. *75*:9044-9058.
22. **Zhu, H., J.-P. Cong, G. Mamtora, T. Gingera, and T. Shenk.** 1998. Cellular gene expression altered by human cytomegalovirus: global monitoring with oligonucleotide arrays. Proc. Natl. Acad. Sci. USA *95*:14470-14475.

Visual Genome Explorer™: A Comparative Visual Interface to Genome Data

E. Scott Archer, Lester W. Yee, and Frederick C. Leung

The University of Hong Kong, Hong Kong

Reprinted from BioTechniques 22:1164-1165 (June 1997)

ABSTRACT

In response to the deluge of genome data, we are developing Visual Genome Explorer™, an interactive graphical interface to genome data. Given is a description of the prototype program, which introduces the concept of visual comparative genomics for complete bacterial genomes.

INTRODUCTION

With the explosive growth of genomic databases, researchers face the increasingly overwhelming task of assimilating pertinent information. Retrieval of many kinds of data, such as sequences and their identified coding regions, is relatively simple because of the efforts made by several major institutions to simplify database access. However, it is often labor-intensive and time-consuming to extract important information because the data are typically only available in a noninteractive textual form. Although colorful illustrations have accompanied publications of complete genome sequencing projects (2,3,4), and sophisticated software has been developed for managing sequencing project data (References 5 and 6; see Availability), no software is readily available for simple visual interaction with the data. In response, we are developing Visual Genome Explorer™, an interactive graphical interface to genome data. Following is a description of Visual Genome Explorer *bacterial* (VGE*b*), the prototype of Visual Genome Explorer for use with bacterial genomes.

VGE*b* introduces the concept of visual comparative genomics, whereby the researcher is presented with connected global and local views of multiple genomes. This comparative presentation is highly integrated with the whole system, allowing the user to perform

BioComputing: Computer Tools for Biologists
Edited by Stuart M. Brown

all operations on a single genome or, comparatively, two genomes simultaneously. For example, when the user chooses to display a functional group, the program displays the selected group of coding regions on both genomes, providing a visual structural comparison of the gene locations involved in that group.

MATERIALS AND METHODS

The interface of VGE*b* is organized into two or four primary windows (depending on whether one or two genomes are loaded, respectively). Each genome is represented by both a circular global view and a linear local view, which displays a finer resolution of the genome and is synchronized (logically and visually) with the global view by the range selector (pie-slice-shaped shadow on global view). Several classes of information are associated with a sequence in VGE*b*: putative coding regions, which are color-coded by functional role classification (if role data are available); motifs, such as restriction enzyme cut sites; transfer RNAs; and ribosomal RNAs.

A genome is represented globally in VGE*b* as several concentric circles, or rings, as is customary for circular chromosomes. The outermost ring of the global view shows the locations of putative coding regions found on the plus (outer edge) and minus (inner edge) strands, color-coded by functional role. In the second or middle ring are motif (or restriction enzyme) sites. Up to four motifs can be displayed in the ring as equally scaled subrings of different colors. The innermost ring shows the locations of transfer RNAs (outer edge) and ribosomal RNAs (inner edge) in red and blue, respectively. The local view concurrently supports the same classes of information as the global view. It is drawn linearly below the circular global view and layered from bottom to top: scale, sequence (at sufficient magnification), putative coding regions (plus strand below minus strand), motifs, tRNAs and rRNAs.

In VGE*b*, the user can browse visually and perform queries on the data. To make interaction with data simple and intuitive, VGE*b* provides direct manipulation of views—zooming and scrolling. The user can change the zoom selection by dragging the ends of the range-selection area in the global view, by scrolling in the local view or using a direct input box for start- and end-sequence locations. As the cursor passes over objects in the local view, information about that object, such as the object's name and function, is displayed in the status bar at the bottom of the application window.

In addition to visual browsing, the user can perform simple queries on the associated data in VGE*b*. For example, one can search for any data objects that match a particular term, such as "metabolism". A list of all putative coding regions containing that term would then be displayed, allowing the user to selectively build a list of data items to display or hide. The user can also use a sequence scanning tool to perform searches for restriction enzyme cut sites and/or simple motifs.

VGE*b* uses data from several sources. The sequence and primary annotation tables associated with a genome (i.e., putative coding regions and RNAs) are read from a standard GenBank® flat-file (1). Currently, all of the published complete bacterial genomes are available in GenBank format. Functional role classifications (see Availability), however, are not generally included in the GenBank flat-file. Thus, functional role data for identified putative coding regions, if available, are assigned to loaded objects using an external table. Enzyme cut sites and other motifs are generated by VGE*b* and so are stored locally in custom files to improve performance.

AVAILABILITY

VGE*b* is currently available for PCs running 32-bit Windows® (Windows 95® or Windows NT™) software. It can be freely downloaded on the World Wide Web at http://www.hku.hk/zoology/vgeb. Check the above Web site for the latest information about VGE*b*.

The functional role classifications are made according to The Institute for Genomic Research's functional role classification in their publications of complete genomes in Science. e.g.: *H. influenzae* (3); *M. genitalium* (4); and *M. jannaschii* (2).

Documentation, code and data for the *C. elegans* database are available from anonymous file transfer protocol (ftp) servers at lirmm.lirmm.fr, cele.mrc-lmb.cam.ac.uk and ncbi.nlm.nih.gov.

ACKNOWLEDGMENTS

We thank Richard J. Douthart for his original idea of circular graphical chromosome representation used first in the CAGE/GEM software. This work is supported by the Hong Kong Research Grants Council (RGC), Hong Kong University (HKU) 196/95M.

REFERENCES

1. **Benson, D., M. Boguski, D.J. Lipman and J. Ostell.** 1996. GenBank. Nucleic Acids Res. *24*:1-5.
2. **Bult, C.J., O. White, G.J. Olsen, L. Zhou, R.D. Fleischmann, G.G. Sutton, J.A. Blake, L.M. FitzGerald et al.** 1996. Complete genome sequence of the methanogenic archaeon, *Methanococcus jannaschii*. Science *273*:1058-1073.
3. **Fleischmann, R.D., M.D. Adams, O. White, R.A. Clayton, E.F. Kirkness, A.R. Kerlavage, C.J. Bult, J.-F. Tomb et al.** 1995. Whole-genome random sequencing and assembly of *Haemophilus influenzae* Rd. Science *269*:496-512.
4. **Fraser, C.M., J.D. Gocayne, O. White, M.D. Adams, R.A. Clayton, R.D. Fleischmann, C.J. Bult, A.R. Kerlavage et al.** 1995. The minimal gene complement of *Mycoplasma genitalium*. Science *270*:397-403.
5. **Smith, S.W., R. Overbeek, C.R. Woese, P.M. Gilbert and P.M. Gillevet.** 1994. The genetic data environment: an expandable GUI for multiple sequence analysis. CABIOS. *10*:671-675.
6. **Sutton, G.G., O. White, M.D. Adams and A.R. Kerlavage.** 1995. TIGR assembler: a new tool for assembling large shotgun sequencing projects. Genome Sci. Technol. *1*:9-19.

Update to: Visual Genome Explorer™: A Comparative Visual Interface to Genome Data

K.H. Cheng, C.C. Hon, and F.C. Leung

Department of Zoology, The University of Hong Kong, Hong Kong (fcleung@hkucc.hku.hk)

Illustration of Enhanced Features of Visual Genome Explorer Beta 2.1 by comparative Analysis of *Helicobacter pylori* Strains 26695 and J99 Genomes

ABSTRACT

We have developed an updated version (beta 2.1) of the Visual Genome Explorer bacterial *(VGE*b*), a freely available genome analysis program for Windows® platform. The new version of VGE*b *has enhanced comparative features such as gapped nucleotide search, restriction enzyme cut-site support, (G+C) content variation chart, and online information reference. We have illustrated the advanced comparative features of VGE*b *beta 2.1 by performing a comparative analysis on* Helicobacter pylori *strains 26695 and J99. The analysis includes GC content comparison, pathogenicity-related loci comparison, and strain-specific loci distribution; major translocations between two genomes were graphically presented.*

INTRODUCTION

Helicobacter pylori is known to be associated with an increased number of cardiovascular, respiratory, extra-gastroduodenal digestive, neurological, skin, autoimmune, and growth diseases (5).

Several commercial computer programs have been developed for genome analysis, but most of them are costly or in-house software limited to well-equipped genome centers. Since 1997, VGE*b* has been released freely and enables individuals to perform bacterial genomic analysis and comparisons on their own personal computers (1). VGE*b* is notable for its highly integrated graphical interfaces. The program can support up to four genome projects concurrently. Each genome is provided with a global circular view and a local linear view. VGE*b* can also show explicitly the locations and distribution of the noncoding regions of all opened genome projects, as well as the variation of the (G+C) content throughout the genomes.

After the first release, VGE*b* has continued to grow and expand. Version beta 2.1 has enhanced comparative features such as gapped nucleotide search, restriction enzyme cut-site support, (G+C) content variation chart, and online information reference (2). These new comparative features were illustrated in this study by a comparative analysis on *H. pylori* strains 26695 and J99.

This study used *H. pylori* strains 26695 and J99, whose sequences were completed in 1997 (3) and 1999 (4), respectively. A global comparative genomic analysis between the two highly homologous pathogens was performed by Alm et al. in 1999 (4). *H. pylori* is known to be associated with an increased number of diseases (5). Strain-specific diversity may affect the development of diseases (6). The genomic organization, gene order, and predicted proteomes of *H. pylori* 26695 and J99 are quite similar (4).

METHODOLOGY

With VGE*b*, an Excel spreadsheet of all the genes in both strains is generated. The genes, except for strain-specific ones, are paired with their counterparts in the opposite genome. Based on this spreadsheet, the detailed comparison of the genomic organization of the two genomes was done, including the genomic rearrangements and strain-specific gene distribution. The GC content display function of VGE*b* allows us to compare the special GC content associated region of the two genomes. Restriction sites and short tandem repeat distributions were compared by the motif search function. Specific region comparison was done by gapped sequence queries on the opposite genome.

RESULTS

Minor Genomic Rearrangements

Single gene rearrangements between these two genomes can be compared using VGE*b*. When we searched the sequence of the HP0641 gene of 26695 against the J99 genome, we realized that the first half of the gene matches the first 120 base pairs of the JHP0584 gene of J99. However, the second half of the HP0641 gene matches the middle

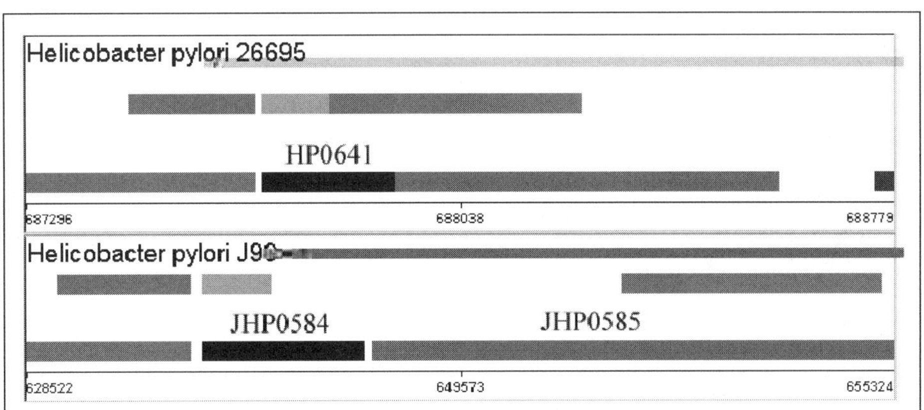

Figure 1. The rearrangement between HP0641, JH0584, and JHP0585. The first half of HP0641 matches the first 120 bp of JHP0584, while the second half matches the middle 110 bp of JHP0585. This reveals that deletion or insertion took place in one of the genomes at this locus. **(See color plate A6.)**

110 base pairs of the JHP0585 gene, which is supposed to be a strain-specific gene (Figure 1). A fragment of 1146 base pairs is supposed to be inserted or deleted from the genome. A search through BLAST revealed that the fragment matched some part of *H. pylori* strain J166 and strain 2600. Although the mechanism of such recombination remains to be studied, it can be demonstrated that VGE*b* can effectively help the sequence analysis between the genomes. In the same way, we can generate the sequence mapping of the HP0817-0818 genes to the JHP0753-0757 genes, in which a deletion of about 1500 base pairs is likely to have taken place.

Significant GC Content Regions

The average GC content of the whole genome is about 39%, and some regions have a significantly lower or higher GC content than the rest of the genome (Figure 2). The most notable regions are the pathogenicity island and the strain-specific "plasticity zone." There are five distinct GC regions in the 26695, while J99 has only four. In Figure 2, a region in 26695 cannot be found in the corresponding region of J99. In this region of 26695, the associated gene included an insertion sequence, called IS605, which cannot be found in the corresponding region J99. The extra low GC content of that region may possibly be related to the presence of IS605.

Virulence and the Pathogenicity Island

The virulence of indivdual *H. pylori* depends on their ability to produce a cytotoxin-associated protein (CagA) and an active vacuolating cytotoxin (VacA) (3). The cluster of CagA family genes that correlates to the severity of *H. pylori* related diseases was called the cag pathogenicity island, or cagPAI. It is located at a region of low GC content. VGE*b* analysis showed the cagPAI of both strains are very similar, with only minor differences. Moreover, IS605 was found to be in the CagPAI of other strains of *H. pylori*, such as NCTC11638, but absent in both J99 and 26695 (3,4).

Strain-Specific Genes

With the spreadsheet mentioned earlier, the strain-specific genes of the two genomes can be located and compared by VGE*b*. In the 1495 open reading frames (ORFs) in J99,

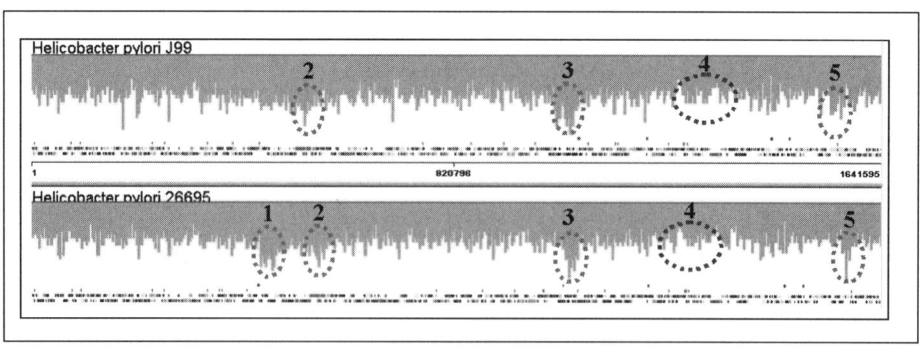

Figure 2. The distinct GC regions in both stains. The picture was taken from a local view of the VGE*b*, displaying the GC content of both genomes. The top of the orange region represents 50% of GC content, the longer the orange bars, the lower the GC content. Here we see the two genomes generally have GC content < 50%. The red circles indicated the regions with particularly low GC content, while the blue circle displays the regions with exceptionally high GC content. **(See color plate A7.)**

89 of them are J99 specific. While 177 out of 1590 ORFs in 26695 are strain specific. The most amazing fact is that these genes are not distributed evenly throughout the genome. They seem to appear as a cluster with relatively low GC content, called plasticity zone (Figure 3). The clustering of DNA with a lower GC content is suggestive of horizontal DNA transfer, and the strain-specific sequence differences are consistent with different origins of the DNAs (4).

Restriction Sites Polymorphism

We have compared the *Not*I sites between these two genomes using VGE*b*. The motif search function enables this comparison. There are 14 sites in J99 whereas 26695 has only 7, which is mainly due to the silent nucleotide mutation (6/7) within genes, except one mutation (1/7) resulting in alternation of one amino acid (4). The *Not*I restriction fragments size of both genomes was sorted by size and displayed as a histogram by VGE*b*. Analysis of *Nru*I site by VGE*b* showed similar results.

Homopolymeric Tracts and Dinucleotide Repeats

The presence of homopolymeric tracts and dinucleotide repeats in the beginning of genes led to the prediction that "slipped-strand repair" may modulate gene expression (3,4); also hypothesized is that it could lead to antigenic variation and adaptive evolution (4). The motif search function in VGE*b* can pinpoint the locations of particular types of short tandem repeats, e.g., homopolymeric and dinucleotide repeats. Five members of the

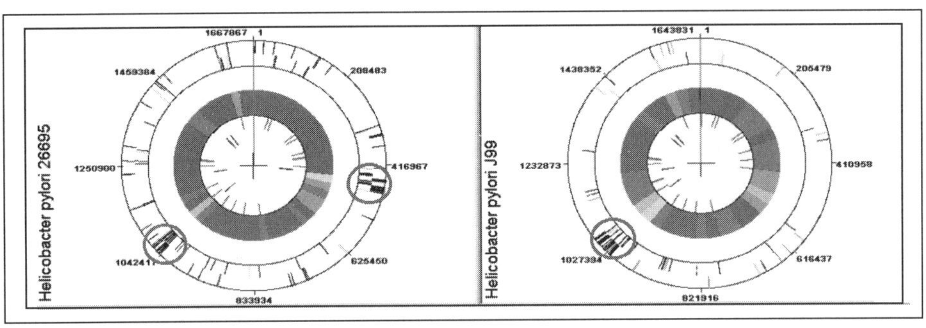

Figure 3. The distribution of strain-specific genes in both genomes. The red circles indicate that the strain-specific gene clusters called plasticity zones. These clusters are always relatively low in GC content. **(See color plate A7.)**

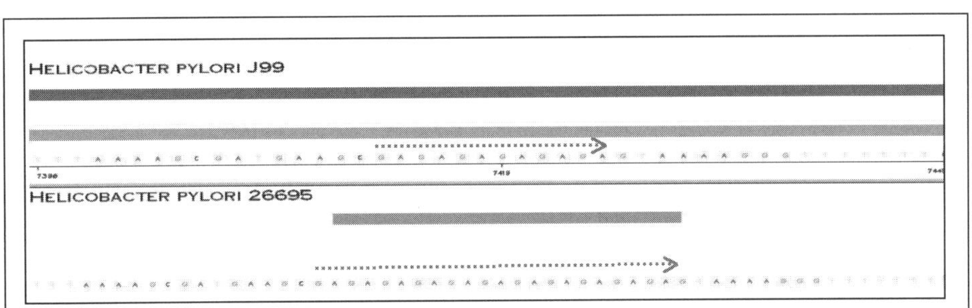

Figure 4. Both are in "on" status. In the beginning of HP0007 and JHP0009, there are different numbers of GA repeats, indicating an "on" expression status. **(See color plate A7.)**

OMP (outer membrane protein) family contain CT repeats in the beginning of the gene, with different numbers of repeats without affecting the predicted expression status. For example, both HP0007 and JHP0009 have CT repeats at the beginning but with different numbers of repeats. However, both of the predicted expression statuses of the genes are "on"(4) (Figure 4). But there are some cases that array with different numbers of repeats, turning the genes into an "on" or "off" status (4).

CONCLUSION

VGE*b* provides a good platform for bacterial genome analysis on Windows-based PCs. The full graphical presentation visualizes the bacterial genomic analysis on PCs, which enables easy and user-friendly manipulation and analysis of huge amounts of bacterial genome data. The newly enhanced features of Version beta 2.1 enable scientists to perform comparative bacterial genomic analysis as illustrated by the comparison of two *H. pylori* strains. VGE*b* is not a commercial software; it can be downloaded free of charge from the Internet. It is a useful tool for scientists with little background in genomic research to take advantage of the genomic data.

VGE*b* is available for PCs running Windows95/98/NT and is downloadable from the World Wide Web from http://www.hku.hk/zoology/vgeb. Please refer to the Web page for the latest information. The Excel spreadsheet of all genes in both strains can be downloaded directly at http://www.hku.hk/zoology/vgeb/HpyloriReport.xls.

REFERENCES

1. **Archer, E.S., L.W. Yee, and F.C. Leung.** 1997 Visual genome explorer: a comparative visual interface to genome data. BioTechniques *22*:1164-1165.
2. **Benson, D.A., I. Karsch-Mizrachi, D.J. Lipman, J. Ostell, B.A. Rapp, and D.L. Wheeler.** 2000. GenBank. Nucleic Acids Res. *28*:15-18.
3. **Tomb, J.F., O. White, A.R. Karlavage, R.A. Clayton, G.G. Sutton, R.D. Fleischmann, K.A. Ketchum, H.P. Klenk, S. Gill, B.A. Dougherty, et al.** 1997. The complete genome sequence of the gastric pathogen *Helicobacter pylori*. Nature *388*:539-547.
4. **Alm, R.A., L.S. Ling, D.T. Moir, B.L. King, E.D. Brown, P.C. Doig, D.R. Smith, B. Noonan, B.C. Guild, B.L. deJonge, et al.** 1999. Genomic-sequence comparison of two unrelated isolates of the human gastric pathogen *Helicobacter pylori*. Nature *397*:176-180.
5. **Tsang, K.W. and S.K. Lam.** 1999. *Helicobacter pylori* and extra-digestive diseases. J. Gastroenterol. Hepatol. *14*:844-850.
6. **Figura, N.** 1997. Identifiable *Helicobacter pylori* strains or factors important in the development of duodenal ulcer disease. Helicobacter Jul 2 (Suppl. 1):S3-12.
7. **Taylor, D.E.** 1992. Construction of a *Helicobacter pylori* genome map and demostration of the diversity at the genome level. J. Bacterol. Nov:174.

CHAPTER 35

DNA Array Analysis in a Microsoft® Windows® Environment

T. Conway, B. Kraus, D.L. Tucker, D.J. Smalley, A.F. Dorman, and L. McKibben

The University of Oklahoma, Norman, OK, USA

Reprinted from BioTechniques 32:110-119 (January 2002)

ABSTRACT

Microsoft® Windows®-based computers have evolved to the point that they provide sufficient computational and visualization power for robust analysis of DNA array data. In fact, smaller laboratories might prefer to carry out some or all of their analyses and visualization in a Windows environment, rather than alternative platforms such as UNIX. We have developed a series of manually executed macros written in Visual Basic for Microsoft Excel® spreadsheets, that allows for rapid and comprehensive gene expression data analysis. The first macro assigns gene names to spots on the DNA array and normalizes individual hybridizations by expressing the signal intensity for each gene as a percentage of the sum of all gene intensities. The second macro streamlines statistical consideration of the confidence in individual gene measurements for sets of experimental replicates by calculating probability values with the Student's t *test. The third macro introduces a threshold value, calculates expression ratios between experimental conditions, and calculates the standard deviation of the mean of the log ratio values. Selected columns of data are copied by a fourth macro to create a processed data set suitable for entry into a Microsoft Access® database. An Access database structure is described that allows simple queries across multiple experiments and export of data into third-party data visualization software packages. These analysis tools can be used in their present form by others working with commercial* E. coli *membrane arrays, or they may be adapted for use with other systems. The Excel spreadsheets with embedded Visual Basic macros and detailed instructions for their use are available at http://www.ou.edu/microarray.*

INTRODUCTION

With the publication of the first genome-wide expression data six years ago (13), the power of the DNA array was immediately evident. Since then, reviewers have predicted a revolution in the way biologists conduct their research (3–5,12). Today the trend toward widespread use of DNA arrays continues. Vast amounts of DNA array data are accumulating, and the need for standardized array annotation and data representation is being

BioComputing: Computer Tools for Biologists
Edited by Stuart M. Brown

addressed (http://www.ebi.ac.uk/microarray/MGED/). Relational databases are being developed to handle microarray data storage in a format that facilitates data processing and visualization, allowing researchers to analyze and interpret their experiments and disseminate the data (16). Generally, the database management software used is Oracle, Web interfaces are written in JAVA or XML, and scripts used for data processing, retrieval, etc. are written in Perl, C, or other programming languages. However, for the common microarray user, the implementation of such a system may be beyond their budget or exceed their actual needs. For these users, a central archive is an attractive alternative for data storage. An example of this, the EcoReg Consortium (http://gobi.lbl.gov/~ecoreg/index.html), is being established as a public database for storage and manipulation of *E. coli* microarray and proteome data. Still, it will be necessary for individual users of consortium databases to process, analyze, and format the data for submission.

DNA arrays have been used to examine the genetics and physiology of the comprehensive biological model, *E. coli*. Some research groups have used commercial membrane-based DNA arrays (2,17,18), while others have employed DNA microarrays (6,11,19,20) or oligonucleotide arrays (15). Several practical issues regarding the use of whole-genome arrays have been addressed, and the power of this technology as a means for deducing the physiological state of the bacterial cell is now well established.

In our laboratory, we routinely use membrane-based DNA arrays for *E. coli* gene expression profiling. Multiple replicates of each experimental condition are processed and analyzed with manually executed macros written in Visual Basic and run in Microsoft® Excel® spreadsheets on Microsoft Windows®-based computers. These macros, statistical analysis, and data processing protocols are described here and are freely available to the scientific community (http://www.ou.edu/microarray).

MATERIALS AND METHODS

Software and System Requirements

The analysis tools described here are written for Microsoft Excel and Microsoft Access®: Microsoft Office® 97 or higher is required. Optimal system requirements include a Windows-compatible computer with a PII processor and 128 Mb RAM or higher. Macintosh® computers running Microsoft Office 98 for Macintosh can also be used to run the macros in Excel.

Example *E. coli* Data Set

In this study, we compare a sample data set of *E. coli* MG1655 grown at pH 7.4 and pH 5.5 under otherwise identical conditions. Cultures were grown aerobically in 50 mL MOPS (pH 7.4) or MES (pH 5.5) minimal glucose (0.2%) medium (8) in 250-mL fleakers (Corning, Acton, MA, USA) at 37°C with 300 rpm agitation and harvested in mid-logarithmic growth phase. RNA isolation, radioactive labeling during cDNA synthesis, and hybridization to DNA array membranes were described previously (17).

Description of Raw Data

We routinely use Panorama *E. coli* Gene Arrays™ (Sigma-Genosys, The Woodlands, TX, USA) for gene expression profiling. Phosphorimaging of a hybridized membrane array produces a TIFF image file that must be further processed for data analysis. The image analysis software (ArrayVision™ version 5.1; Imaging Research, St. Catharines, Ontario, Canada) makes use of a customized template to accommodate three grid layers (3 × 1; 16 × 24; 4 × 4) according to the design of the Panorama *E. coli* gene arrays. The spot labeling protocol was edited such that each spot is named by its unique array coordinate, allowing the spot intensity measurements to be easily associated with the correct gene identifiers in subsequent processing steps. The customized ArrayVision template file for analysis of Panorama *E. coli* gene arrays, and detailed instructions for its use, are available at http://www.ou.edu/microarray. The spot intensities are represented in a row-column format and are exported into Excel spreadsheets for further analysis.

Data Processing

The macros and sample analyses can be downloaded from http://www.ou.edu/microarray. Follow the links to "Macroarray", "Data Analysis", and then "Spot-Finding and Image Quantitation" or "*E. coli* Data Analysis (software downloads)"; alternatively, the macroarray section of the site can be accessed directly at http://www.ou.edu/microarray/macroarray.htm. Raw DNA array data, exported from ArrayVision, are processed in a series of three Excel workbooks that are used to manually execute four macros written in Visual Basic (Table 1). These macros are designed to filter the data and calculate statistics to allow for further data analysis and interpretation. Detailed, step-by-step instructions for use of these analysis tools are provided on our Web site.

RESULTS AND DISCUSSION

The workbooks, macros, and subroutines used for DNA array data analysis are outlined in Table 1. The subroutines can be run in order individually, or the macro containing all relevant subroutines can be run once to execute all subroutines. In the following section,we provide an overview of important statistical considerations, the specific processing steps, and outcomes.

Statistical Significance

Various approaches for attaching significance to DNA array data have been published, including a simple "rule-of-thumb" criterion for the value of the expression ratio (2,20). Some researchers have used the standard deviation from the mean of the expression ratios as an indicator of confidence (6,17). Arfin et al. (1) applied the Student's *t* test to experimental replicates and considered the *P* value to be the most important indicator of significance. Richmond et al. (11) combined a confidence interval obtained with the Student's *t* test and a rule-of-thumb criterion for the expression ratio. Others prefer to consider the significance of a single experimental condition based on the coefficient of variation (18). A precedent for statistical analysis of array data has yet to be firmly established, and a standard is clearly needed. Whatever statistical approach is adopted, it is essential that DNA array experiments are properly replicated and the uncertainty that lies behind individual gene measurements be considered to attach significance to data sets.

We advocate the use of at least two replicates of each experimental condition. Membrane arrays typically have duplicate spots for each gene, and each spot is considered to be a separate determination, providing a total of four determinations for the two repli-

Table 1. Useful Purpose of Workbooks, Visual Basic Macros, and Subroutines Used in This Study

Workbook	Macro	Subroutine	Purpose
Image Data Cruncher	AllData Crunched6	ArvAllSort1	associates array coordinate with spot number
		Nameall2	associates spot number with unique identifier for gene
		CalcPercentage3	normalizes data by expressing each spot as percentage of sum of all spot intensities
		Cleanup4	reorganizes data and calculates average values for duplicate spots
		Statistics5	calculates averages of genomic DNA controls and blank spots
2-Replicate-Stats	AllAnalysis8	OrganizebySpotNo1	sorts each of the four data sets individually by spot number
		CalculateAverages2	calculates averages of volumes and percentage values for the control and test replicates
		copyvaluesintoPRaw3	copies and pastes percentage values into a separate spreadsheet for calculation of *P* values
		CalculateLn4	copies percentage values into a separate spreadsheet and natural log transforms data
		CalculatePRaw5	calculates the *P* value for the raw data by application of the Student's *t* test
		CalculatePLn6	calculates the *P* value for the log transformed data by application of the Student's *t* test
		CopyAllValues7	copies and pastes data used for ratio calculations into separate spreadsheet

Table 1. Continued.

Workbook	Macro	Subroutine	Purpose
Data Analysis			
	AllAnalysis6		
		SpotSort1	sorts control and test data sets by spot number
		DataSort2	copies data set into spreadsheet used for ratio calculations, sorts by total percentage value
		ThresholdRatios3	calculates ratio of Test/ Control using threshold of tot percentage value for 500th lowest gene
		Cleanup4	reorganizes data and calculates log (10) of ratio
		Stats5	calculates standard deviation of log ratio values and correlation between Test and Control
	(Manual Step)		copy and paste special values Praw and PLn values from 2-Replicate-Stats to Data Analysis
	MakeDBsheet		reorganizes data and copies into separate spreadsheet for entry into Access database

cates. Because membrane arrays are hybridized with a single labeled target mRNA and normalized independently (in effect a one-color experiment), the variation in the measurement is at the level of the raw data, not the measurement of the ratio (unlike the statistical approach that is popular with two-color microarrays and involves internal normalization of the measurements). The uncertainty that lies behind individual gene measurements can be variously calculated as the standard deviation of the determinations or the coefficient of variation. Since we are usually interested in the statistical significance of differences between an experimental condition and a control, we prefer the Student's *t* test as a means for calculating this probability, based on the uncertainty of the replicate measurements in both conditions. The Student's *t* test is best applied to natural log-transformed normalized data sets (7). Generally, a *P* value of less than 0.05 is chosen to indicate a 95% probability that the difference in gene expression between conditions is significant. However, it has been pointed out that with very large data sets (e.g., a bacterium with 5000 genes) choosing a value of $P < 0.05$ means that there could be up to 250 false positives in the data set (1). Thus, the researcher is left with two choices: to lower the *P* value to a level where no false positives are expected ($P < 0.0002$) or to consider a second statistical metric that, when combined with a reasonable *P* value ($P < 0.05$), is an excellent indicator of significance of a ratio value.

We use the standard deviation of the mean of the log ratios—within the context of the *P* value—to indicate significant up- or down-regulation of gene expression. This approach is meaningful where the expression level of the majority of genes does not change significantly between conditions and where the researcher is interested in genes that show substantially different expression. The standard deviation for the log ratios is calculated, and only those genes that differ by more than three standard deviations (99.9% confidence in each tail) from the mean of the log ratio (usually zero, or no change) are considered. In practice, emphasis is placed on those genes that have expression ratios greater than three standard deviations from the mean and have a reasonable probability of being significantly different between the conditions, based on a *P* value of less than 0.05. Where there are four or more determinations for each gene, the *P* value can be lowered to less than 0.005 with little change seen in the number of genes that are considered to vary significantly between conditions.

There may also be situations when the researcher is interested in changes in gene expression that are not a full three standard deviations from the mean but are still significant (i.e., where the differences in gene expression between conditions are subtle yet meaningful). In this case the Student's *t* test can be used as the sole measure of significance, but the *P* value must be adjusted to ensure that false positives are avoided. One approach for this is to apply the Bonferroni correction that describes a *P* value for significance in a large data set (7). Various strategies for implementing this correction factor have been described, and its proper use is somewhat controversial (14). The Bonferroni correction effectively lowers the *P* value to a point where false positives are avoided and consideration is given only to those genes for which there is a high degree of confidence in the ratio value. In this light, the Bonferroni correction seems to be a reasonable statistical tool but may be too stringent for some considerations.

"Image Data Cruncher" Workbook

To begin the data analysis process, the raw data from an experimental replicate is copied and pasted into a blank spreadsheet named "arvdata" in an Excel workbook named "Image Data Cruncher" containing the macro named "AllDataCrunched6" (Figure 1). The first subroutine in the macro, "ArvAllSort1", copies and pastes subsets from the "arvdata" spreadsheet into a second spreadsheet named "allfields" that contains information provided by the membrane manufacturer for associating the array location with a spot number that is unique to each target on the array. The second subroutine, "Nameall2", copies and pastes the data in the "allfields" spreadsheet into a third spreadsheet that associates the spot number with a unique identifier and associated genome annotation information for each gene.

Differences in spot intensities between replicate experiments arise from normal experimental variation, such as differences in growth conditions, radioactive nucleotide incorporation efficiency, hybridization conditions, or image acquisition. To compare experimental replicates (separate cultures) or technical replicates (same culture and same RNA sample), the data from each array must be normalized. The third subroutine, "CalcPercentage3", normalizes arrays; if the data are not normalized, then the values for replicate experiments, when plotted, will not pass through zero or be directly proportional (Figure 2). Array experiments can be normalized by expressing each gene-specific spot relative to an internal control, if a suitable set of control spots is present on the arrays. Unfortunately, the intensities of the genomic DNA control spots on the Panorama *E. coli* membranes vary significantly (data not shown) and are not reliable for normalization. In the absence of an internal standard, the preferred approach for normalization is to express each gene-specific spot as a fraction of the sum of all gene spots (n = 4290), a strategy that at once considers all variables that lie behind the array image (17,18). The "CalcPercentage3" subroutine normalizes the entire data set, expressing each spot as a percentage of the sum of all spots on the array.

The fourth subroutine, "Cleanup4", reorganizes the data to facilitate subsequent data

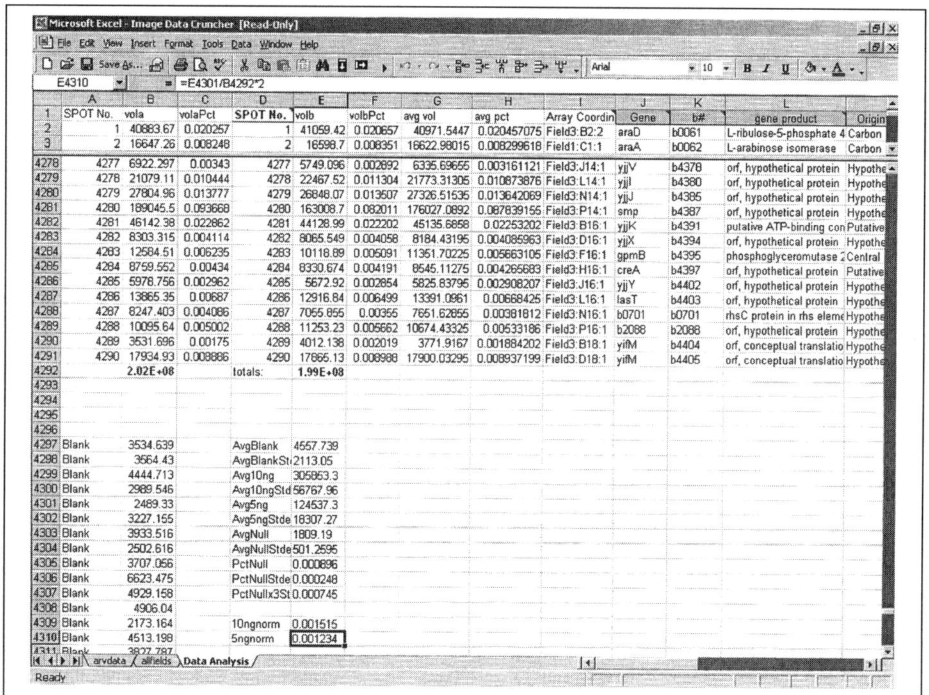

Figure 1. "Image Data Cruncher" workbook used to process raw gene expression profile data by normalizing expression values and assigning gene names to array spot coordinates.

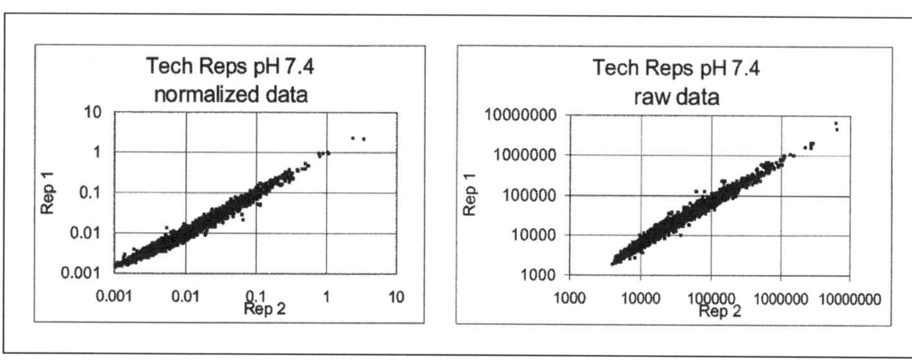

Figure 2. Scatter plot view of normalized (left) and raw (right) data showing linearity of experimental replicates for normalized data.

processing steps. The fifth subroutine, "Statistics5", calculates the averages and statistics for the blank spots (empty spots between genomic DNA control spots) and null spots (empty spots within array). However, we have not found these values to offer a reliable means for establishing background on the array or for empirical determination of a threshold value and therefore do not use them. The "AllDataCrunched6" macro in "Image Data Cruncher" runs all five subroutines at once and results in a file that contains the raw data, normalized data, and associated genome annotation information for each gene on the array.

"2-Replicate-Stats" Workbook

The second workbook, "2-Replicate-Stats", contains a macro named "AllAnalysis8" that is used to calculate the probability that the average of the experimental (test) replicates is significantly different from the average of the control replicates (Figure 3). The four replicate data sets are sequentially copied from the crunched data files, beginning with the first and then the second replicate of the control, followed by the first and then the second replicate of the experimental, and pasted into the "Enter Data (2 Replicates)" worksheet. The first subroutine, "OrganizebySpotNo1", sorts the four data sets by spot number, which aligns the gene-specific data in rows. The second subroutine, "CalculateAverages2", calculates the mean of the normalized (percentage) values for the four spot intensities from each experimental condition (two spots for each gene per membrane). The subroutine, "copyvaluesintoPRaw3", copies the percentage values into a

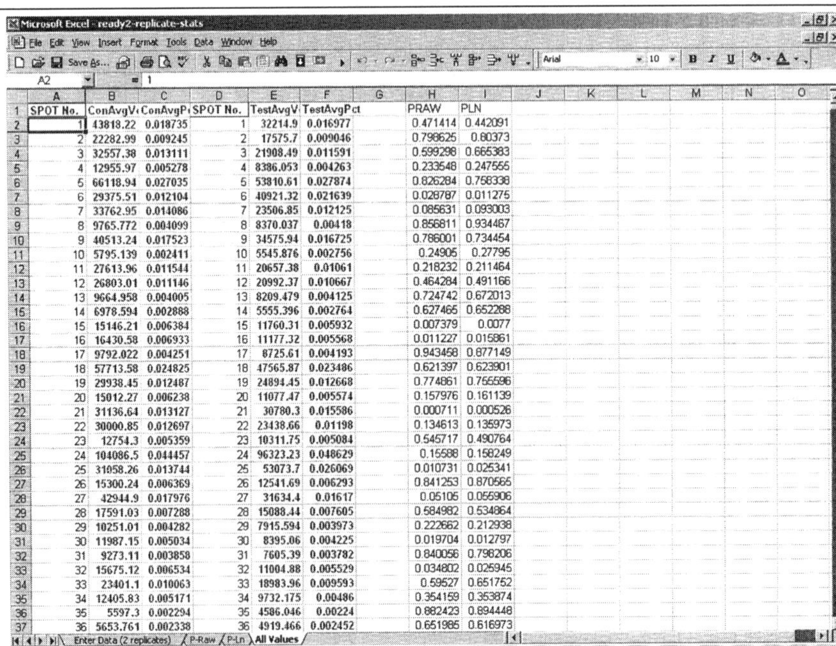

	A	B	C	D	E	F	G	H	I
1	SPOT No.	ConAvgV	ConAvgP	SPOT No.	TestAvgV	TestAvgPct		PRAW	PLN
2	1	43818.22	0.018735	1	32214.9	0.016977		0.471414	0.442091
3	2	22282.99	0.009245	2	17575.7	0.009046		0.798625	0.80373
4	3	32557.38	0.013111	3	21908.49	0.011591		0.599298	0.665383
5	4	12955.97	0.005278	4	8386.053	0.004263		0.233548	0.247555
6	5	66118.94	0.027035	5	53810.61	0.027874		0.826284	0.758338
7	6	29375.51	0.012104	6	40921.32	0.021639		0.028787	0.011275
8	7	33762.95	0.014086	7	23506.85	0.012125		0.085631	0.093003
9	8	9765.772	0.004099	8	8370.037	0.00418		0.856811	0.934467
10	9	40513.24	0.017523	9	34575.94	0.016725		0.786001	0.734454
11	10	5795.139	0.002411	10	5545.876	0.002756		0.24905	0.27795
12	11	27613.96	0.011544	11	20657.38	0.01061		0.218232	0.211464
13	12	26803.01	0.011146	12	20992.37	0.010667		0.464284	0.491166
14	13	9664.958	0.004005	13	8209.479	0.004125		0.724742	0.672013
15	14	6978.594	0.002888	14	5555.396	0.002764		0.627465	0.652288
16	15	15146.21	0.006384	15	11760.31	0.005932		0.007379	0.0077
17	16	16430.58	0.006933	16	11177.32	0.005568		0.011227	0.015861
18	17	9792.022	0.004251	17	8725.61	0.004193		0.943458	0.877149
19	18	57713.58	0.024825	18	47565.87	0.023486		0.621397	0.623901
20	19	29938.45	0.012487	19	24894.45	0.012668		0.774861	0.755596
21	20	15012.27	0.006238	20	11077.47	0.005574		0.157976	0.161139
22	21	31136.64	0.013127	21	30780.3	0.015586		0.000711	0.000526
23	22	30000.85	0.012697	22	23438.66	0.01198		0.134613	0.135973
24	23	12754.3	0.005359	23	10311.75	0.005084		0.545717	0.490764
25	24	104086.5	0.044457	24	96323.23	0.048629		0.15588	0.158249
26	25	31058.26	0.013744	25	53073.7	0.026069		0.010731	0.025341
27	26	15300.24	0.006369	26	12541.69	0.006293		0.841253	0.870565
28	27	42944.9	0.017976	27	31634.4	0.01617		0.05105	0.055906
29	28	17591.03	0.007288	28	15088.44	0.007605		0.584982	0.534864
30	29	10251.01	0.004282	29	7915.594	0.003973		0.222662	0.212938
31	30	11987.15	0.005034	30	8395.06	0.004225		0.019704	0.012797
32	31	9273.11	0.003858	31	7605.39	0.003782		0.840056	0.798206
33	32	15675.12	0.006534	32	11004.88	0.005529		0.034802	0.025945
34	33	23401.1	0.010063	33	18983.96	0.009593		0.59527	0.651752
35	34	12405.83	0.005171	34	9732.175	0.00486		0.354159	0.353874
36	35	5597.3	0.002294	35	4586.046	0.00224		0.882423	0.894448
37	36	5653.761	0.002338	36	4919.466	0.002452		0.651985	0.616973

Figure 3. "2-Replicate-Stats" workbook used to calculate probability by application of the Student's *t* test, based on uncertainty of measurements determined from replicate experiments, that gene expression is significantly different between the conditions analyzed.

separate spreadsheet, and the "CalculateLn4" subroutine transforms the raw percentage values by the natural log while copying them to an additional spreadsheet.

The subroutines "CalculatePRaw5" and "CalculatePLn6" are used to calculate the *P* values for the raw and log transformed data, respectively, by application of the Student's *t* test to the four determinations for each of the control and experimental conditions. The last subroutine, "CopyAllValues7", reorganizes the data by pasting the data columns to be used in subsequent steps into a separate spreadsheet named "All Values". Once the "All-Analysis8" macro is executed, the "2-Replicate-Stats" workbook serves as an archive for the raw and normalized data from the replicates being compared and contains the *P* values that are associated with the ratio calculations in the "Data Analysis" workbook.

"Data Analysis" Workbook

The third workbook, "Data Analysis", contains two macros. The first macro, "All-Analysis6", is used to calculate the log ratio of the expression levels in the experimental versus the control condition (Figure 4). The average normalized data from the "All Values" spreadsheet in the "2-Replicate-Stats" workbook are pasted into the "crunched data" spreadsheet in the "Data Analysis" workbook. The first subroutine, "SpotSort1", sorts the data by spot number such that the gene specific data are aligned in rows. The second subroutine, "DataSort2", copies the data and pastes them into the "DataAnalysis" spreadsheet to be used for ratio calculations.

The third subroutine, "ThresholdRatios3", is used to determine a threshold value for ratio calculation and calculates the absolute value of the ratio of the Test/Control such that genes that are more highly expressed in the test condition are given a positive value

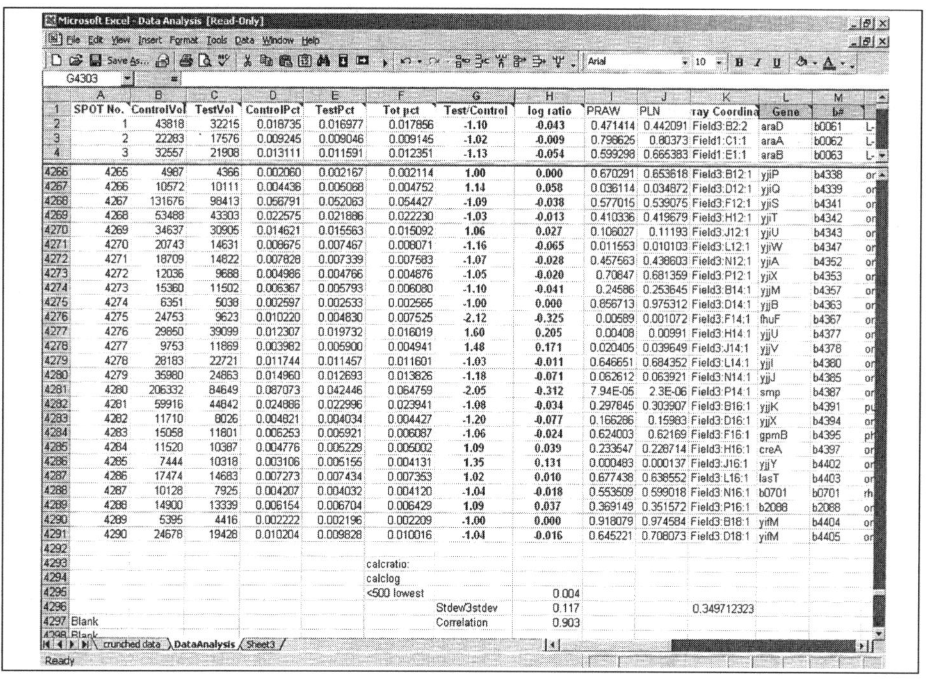

Figure 4. "Data Analysis" workbook used to calculate gene expression ratios between conditions.

and genes that are more highly expressed in the control are given a negative value. The threshold value is chosen to represent the limit of detection of an expressed gene (i.e., the signal intensity at which spots are considered to be significantly higher than the array background). Any spot intensity that falls below the threshold value is raised to that value to obtain a reasonable ratio in cases where a gene is expressed below the threshold value in at least one of the two experimental conditions. Ideally, the threshold would be determined independently for each gene, based on the local spot background and the known cross-hybridization to other expressed genes in the sample, but this is not possible because of the dense packing of some membrane arrays and the lack of prior knowledge as to the number and extent of gene expression in a given growth condition. These factors make determination of the threshold value difficult. We have chosen a conservative approximation of the threshold value corresponding to the 500th lowest expressed gene based on the average of the normalized expression levels in the two conditions. This threshold value is similar to that obtained by visual inspection of array images to determine the faintest of gene-specific spots and is reasonable in light of the predicted number of expressed genes based on the number of mRNA species in the *E. coli* cell and the arrangement of genes in operons (1380 mRNAs × average 2.5 genes per operon = 3450 expressed genes) (9). Others have calculated threshold values as three standard deviations above the mean of "blank" spots (10), which corresponds to approximately 214 in the example data set used in this study. If desired, researchers can write alternative approaches for threshold determination into the "ThresholdRatios3" subroutine, or the threshold level can be edited as described on the Web site.

The fourth subroutine, "Cleanup4", reorganizes the data and calculates the log (base 10) of the expression ratio. The fifth subroutine, "Stats5", calculates the standard deviation of the mean of the log ratios. The "AllAnalysis6" subroutine executes all five subroutines and results in a "DataAnalysis" spreadsheet that contains the averaged raw and normalized data, corresponding genome annotation information for each gene on the array, and the ratios. The "AllAnalysis6" macro concludes with creation of two empty columns that are used for manual pasting of the *P* value associated with each ratio calculation from the "2-Replicate-Stats" workbook. Finally, the "MakeDB" macro in the "Data Analysis" workbook can be executed to reorganize and paste key data columns into a separate spreadsheet that can be used for data entry into a suitable database (Figure 5).

Database Considerations

It is useful to build a database when multiple time points or multiple conditions are being compared. A sample Access database called "sampleDB" can be downloaded from our Web site. There are two tables in the sample database, one populated with annotation information such as b# (unique identifier), array coordinate, gene, gene product, functional groupings, and accession numbers, and the other table populated with data generated by the "MakeDB" macro in the "Data Analysis" workbook. Additional tables can be created and populated by copying the table (structure only), editing the design as necessary, and pasting the appropriate data from the spreadsheets created by "MakeDB". Queries are designed by linking tables by b# and can be run by manipulation of the parameters in query design. The query data can be copied or exported from Access into third-party presentation software packages. There are limitations on the total number of tables and data elements supported by Access that make it impossible to query more than 20 experiments. The macros and Excel workbooks described here are also suitable for processing data for entry into more robust database programs such as MYSQL or Oracle.

Modification and Adaptation of Visual Basic Macros

If desired, the macros described above can be modified in any number of ways to streamline the process or to tweak the statistical parameters. This is most easily accomplished using the Microsoft Visual Basic Editor within the Excel workbook environment. For example, the threshold value can be adjusted by simply changing the "cell" location that is used for the ratio calculation, or the four macros described here could be combined in a single workbook containing several spreadsheets. (We have chosen not to do so because this would create a very large, complicated workbook.) More importantly, these macros can be adapted for use with membrane systems other than the one described here. An example of this is the creation of an analogous set of macros for processing of data from Atlas™ Mouse 1.2 arrays (BD Biosciences Clontech, Palo Alto, CA, USA). These macros and associated Excel workbooks are also publicly available on our Web site.

Web Site and Public Access to Analysis Tools

A Web site has been created for users to download these analysis tools and protocols: http://www.ou.edu/microarray. The protocols on the site describe, in a simplistic way, the step-by-step implementation of the analysis tools.

Benefits of Semi-Automated Data Processing

Our earliest protocol for membrane array data acquisition and analysis required more than 10 man-hours for each experimental replicate (17). Thus, it was important to auto-

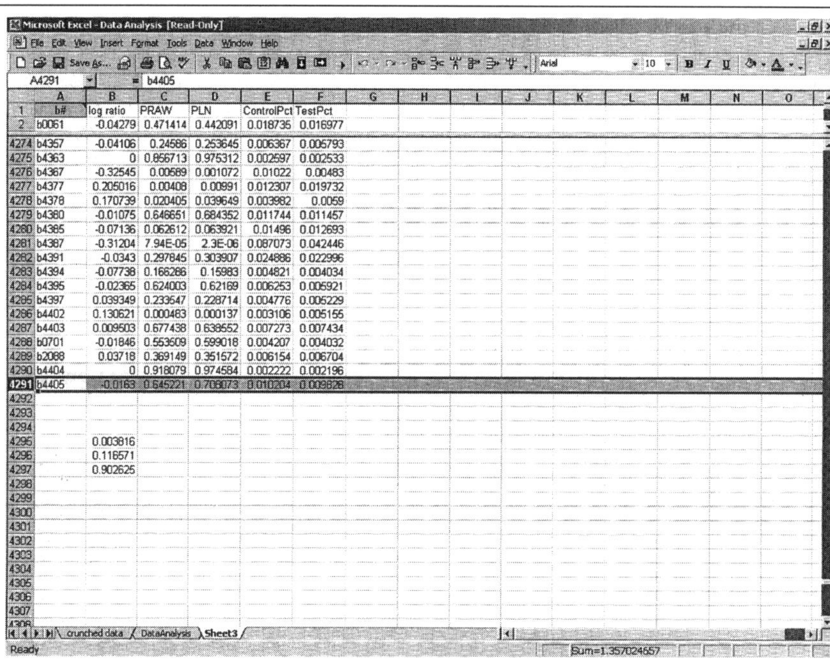

Figure 5. "MakeDB" worksheet containing selected data columns ready for entry into gene expression experiments database.

mate the data analysis process to the greatest extent possible. By taking advantage of improvements in commercially available software and by writing a series of macros in Visual Basic for semi-automated data processing in Microsoft Excel, the time was shortened to 5 min per experimental replicate. In addition, it is critical that the possibility of human error be eliminated when carrying out a large number of manipulations of massive data sets such as those involved in gene expression profiling. The macros described here effectively minimize the level of manual data processing by automating much of the process. The integration of four macros in three workbooks provides for a significant level of human supervision of an otherwise error-free, automated process.

ACKNOWLEDGMENTS

The authors wish to thank Bill Cuevas and Simon Sims for information before publication and helpful discussions. This work was supported by National Institutes of Health grant nos. to RO1 AI48945-01 and RR-01-005 to T.C.

REFERENCES

1. **Arfin, S.M., A.D. Long, E.T. Ito, L. Tolleri, M.M. Riehle, E.S. Paegle, and G.W. Hatfield.** 2000. Global gene expression profiling in *Escherichia coli* K12. The effects of integration host factor. J. Biol. Chem. *275*:29672-29684.
2. **Barbosa, T.M. and S.B. Levy.** 2000. Differential expression of over 60 chromosomal genes in *Escherichia coli* by constitutive expression of MarA. J. Bacteriol. *182*:3467-3474.
3. **Brown, P.O. and D. Botstein.** 1999. Exploring the new world of the genome with DNA microarrays. Nat. Genet. *21(1 Suppl)*:33-37.
4. **DeRisi, J.L. and V.R. Iyer.** 1999. Genomics and array technology. Curr. Opin. Oncol. *11*:76-79.
5. **Ferea, T.L. and P.O. Brown.** 1999. Observing the living genome. Curr. Opin. Genet. Dev. *9*:715-722.
6. **Khodursky, A.B., B.J. Peter, N.R. Cozzarelli, D. Botstein, P.O. Brown, and C. Yanofsky.** 2000. DNA microarray analysis of gene expression in response to physiological and genetic changes that affect tryptophan metabolism in *Escherichia coli*. Proc. Natl. Acad. Sci. USA *97*:12170-12175.
7. **Long, A.D., H.J. Mangalam, B.Y. Chan, L. Tolleri, G.W. Hatfield, and P. Baldi.** 2001. Improved statistical inference from DNA microarray data using analysis of variance and a Bayesian statistical framework. Analysis of global gene expression in *Escherichia coli* K12. J. Biol. Chem. *276*:19937-19944.
8. **Neidhardt, F.C., P.L. Bloch, and D.F. Smith.** 1974. Culture medium for enterobacteria. J. Bacteriol. *119*:736-747.
9. **Neidhardt, F.C., J.L. Ingraham, and M. Schaechter.** 1990. Physiology of the Bacterial Cell: A Molecular Approach. Sinauer Associates, Sunderland, MA.
10. **Pomposiello, P.J., M.H. Bennik, and B. Demple.** 2001. Genome-wide transcriptional profiling of the *Escherichia coli* responses to superoxide stress and sodium salicylate. J. Bacteriol. *183*:3890-3902.
11. **Richmond, C.S., J.D. Glasner, R. Mau, H. Jin, and F.R. Blattner.** 1999. Genome-wide expression profiling in *Escherichia coli* K-12. Nucleic Acids Res. *27*:3821-3835.
12. **Schena, M., R.A. Heller, T.P. Theriault, K. Konrad, E. Lachenmeier, and R.W. Davis.** 1998. Microarrays: biotechnology's discovery platform for functional genomics. Trends Biotechnol. *16*:301-306.
13. **Schena, M., D. Shalon, R.W. Davis, and P.O. Brown.** 1995. Quantitative monitoring of gene expression patterns with a complementary DNA microarray. Science *270*:467-470.
14. **Scherf, U., D.T. Ross, M. Waltham, L.H. Smith, J.K. Lee, L. Tanabe, K.W. Kohn, W.C. Reinhold et al.** 2000. A gene expression database for the molecular pharmacology of cancer. Nat. Genet. *24*:236-244.
15. **Selinger, D.W., K.J. Cheung, R. Mei, E.M. Johansson, C.S. Richmond, F.R. Blattner, D.J. Lockhart, and G.M. Church.** 2000. RNA expression analysis using a 30 base pair resolution *Escherichia coli* genome array. Nat. Biotechnol. *18*:1262-1268.
16. **Sherlock, G., T. Hernandez-Boussard, A. Kasarskis, G. Binkley, J.C. Matese, S.S. Dwight, M. Kaloper, S. Weng et al.** 2001. The Stanford Microarray Database. Nucleic Acids Res. *29*:152-155.
17. **Tao, H., C. Bausch, C. Richmond, F.R. Blattner, and T. Conway.** 1999. Functional genomics: expression analysis of *Escherichia coli* growing on minimal and rich media. J. Bacteriol. *181*:6425-6440.
18. **Tao, H., R. Gonzalez, A. Martinez, M. Rodriguez, L.O. Ingram, J.F. Preston, and K.T. Shanmugam.** 2001. Engineering a homo-ethanol pathway in *Escherichia coli*: increased glycolytic flux and levels of

expression of glycolytic genes during xylose fermentation. J. Bacteriol. *183*:2979-2988.

19. **Wei, Y., J.M. Lee, C. Richmond, F.R. Blattner, J.A. Rafalski, and R.A. LaRossa.** 2001. High-density microarray-mediated gene expression profiling of *Escherichia coli*. J. Bacteriol. *183*:545-556.

20. **Zimmer, D.P., E. Soupene, H.L. Lee, V.F. Wendisch, A.B. Khodursky, B.J. Peter, R.A. Bender, and S. Kustu.** 2000. Nitrogen regulatory protein C-controlled genes of *Escherichia coli*: scavenging as a defense against nitrogen limitation. Proc. Natl. Acad. Sci. USA *97*:14674-14679.

CHAPTER 36

ArrayExplorer©, a Program in Visual Basic for Robust and Accurate Filter cDNA Array Analysis

P.C. Patriotis[1,2], T.D. Querec[1], B.N. Gruver[1], T.R. Brown[3], and C. Patriotis[1]

[1]Fox Chase Cancer Center, Philadelphia, PA, [2]Penn State University, University Park, PA, and [3]Columbia University, NY, USA

Reprinted from BioTechniques 31:862-872 (October 2001)

ABSTRACT

Determining the dynamics in the global regulation of gene expression holds the promise of bringing a better understanding of the processes that govern physiological cell growth regulation and its disruption during the development of disease. The advent of cDNA arrays has created the possibility for the parallel analysis of expression of thousands of genes in a given cell population, simultaneously. The level of expression of a given set of genes within the studied tissue corresponds to the intensity of a labeled cDNA probe synthesized from the studied tissue RNA and bound specifically to the cDNAs of the genes spotted on the array. The accurate extraction of gene expression intensity values is essential for further data analysis and the interpretation of the obtained results. Here, we describe a new array image-processing software developed in Microsoft® Visual Basic, the ArrayExplorer©, which provides a user-friendly, multiple-window interface and a number of automatic and manual features that facilitate a reliable, robust, and accurate extraction of gene intensity values from filter-array images.

INTRODUCTION

The cDNA microarrays are an emerging new technology developing on the basis of recent technological achievements and the sequence information generated by the Human Genome Project. cDNA microarrays allow the parallel analysis of the expression of hundreds to thousands of genes in different cell populations, simultaneously (1–6). Gene expression arrays contain unique cDNA sequences corresponding to the genes of interest, which are spotted on the array substrate in a known configuration. Depending on the sub-

BioComputing: Computer Tools for Biologists
Edited by Stuart M. Brown

strate used, cDNA arrays may be glass or filter arrays. Arrays may be purchased pre-made from a number of commercial suppliers (e.g., Affymetrix, Clontech Laboratories, Research Genetics, Genome Systems) or may be individually spotted by in-house facilities with custom configurations. In the case of filter macroarrays, equivalent amounts (about 5 ng) of PCR-amplified DNA sequences (0.5–2.0 kb) of specific cDNA clones are immobilized at a density of 10–20 cDNAs/cm^2 on positively charged nylon membranes (3,4).

Complex cDNA probes are derived in vitro by reverse transcription of purified total or poly$(A)^+$ RNAs and an incorporated radioactive label (^{32}P or ^{33}P) and are then hybridized to the DNAs immobilized on the arrays. The detection of the hybridized DNA probe is usually carried out by means of phosphoimage scanning or direct autoradiography. Once captured, the high-resolution digital images are subjected to processing by densitometric analysis. The accurate measurement of the gene intensities from the array-generated analog images is quite challenging. One of the issues is the proper placement of the reference grid over the array image relative to the location and size of the hybridization signals. Another issue is the extraction of the hybridization intensities, particularly when they are overlapping and in the presence of nonspecific, background noise. A final but very important point is to present the results in a way that facilitates their further analysis and easy comprehension. However, most of the commercially available array image processing and analysis programs address some but not all of these issues, and their resolution contributes to a significant improvement of the degree of accuracy of the extracted gene intensity values, including a more linear detection of signals within a wide dynamic range, and the detection of a greater number of genes on each array.

ArrayExplorer© contains routines for flexible grid placement. The variable diameter of the hybridization signals, a common characteristic of array images obtained with radiolabeled probes, is incorporated in the gene intensity calculation. The problem of high spot-to-spot signal interference or overlapping hybridization spots is also solved by the ArrayExplorer. Automatic and manual routines facilitate the easy separation of such signals from each other and allow the accurate estimation of their intensities. Another feature of ArrayExplorer is flagging of signals that are solely due to higher local background noise or when an automatic routine fails to converge. The combination of such novel automatic and manual features allows for the robust and accurate extraction of gene intensity values, which are finally exported automatically in a text/spreadsheet format, where they are conveniently linked to information regarding array location, gene identity, and access to DNA and protein sequence databases.

MATERIALS AND METHODS

Program Design

ArrayExplorer is coded in the Microsoft® Visual Basic programming language. The interface is developed and compiled using Microsoft Visual Basic 6.0 (Service Pack 3) Professional Edition for 32-bit Windows™ applications. The code is optimized for execution speed and structured into object-classes for handling the storage, management, and analysis of gene information. These classes are subdivided into “whole grid”, “individual grid region”, “individual gene”, and “individual spot”. Each class contains flags and locations of the regions that it affects. The presentation of the data is reflected by a variety of flag-sets that give the hybridization signals different colors on the array image.

Many features set ArrayExplorer apart from other programs dealing with array image processing. The program uses circular scanning regions (CSRs), which may vary in location and size. The intensity of each array spot is calculated by integrating the pixel intensities within each CSR, which are corrected for the local background. For 8-bit images, the intensity value in each pixel may vary within a range of 0–255. The background is determined globally for the entire gridded array image or for each individual grid section. In either case, the background value is determined as the average intensity per pixel within regions that do not contain any spotted cDNAs and, therefore, with no hybridization signals. A number of regions have been incorporated as a default option in the program for background determination; however, the user may manually customize the size and location of these regions. Another important feature of ArrayExplorer is the ability to subdivide the CSR into selectable and sumable 1/8th-sectors, or "pie-sectors". This procedure is used at instances when signal bleeding from adjacent hybridization spots causes signal overlapping or when there is a distortion in the signal caused by hybridization artifacts. The user selects the pie-sector(s) that are free of signal interference at a given spot. The intensity of the selected pie-sector(s) is determined automatically and, since the hybridization spots are for the most part symmetrical and circular, the program calculates the intensity of the entire spot, assuming that it is proportional to the intensity determined within the selected pie-sector(s).

The most original feature of ArrayExplorer is its ability to locate and expand the CSRs to encompass the entire hybridization signal at each spot. Following grid alignment, a gradient method is used to locate the center of the hybridization signal within each square and align a CSR: two sums are taken by calculating the average intensity in the rim of the CSR, along the X and Y axes. As these sums are weighed trigonometrically, the program changes the coordinates of the CSR in the direction(s) that results in a greater sum. The difference of the sums between the top and bottom and the left and right sides of the CSR rim is calculated. The magnitude of the calculated sums in the CSR rim will instruct the program toward which direction and how much to move the CSR in each axis, for the difference in the sums between the left and right and the top and bottom of the rim to be close to zero. Once CSR rim intensity symmetry has been achieved, the CSR is expanded by default until the average intensity value in the rim reaches the sum of the average background value plus three standard deviations of the background. Additionally, the user may expand or reduce the size of the CSR using the operational window for manual fine-tuning. An array spot is automatically flagged when the automatic routines fail to detect the spot center or the CSR is expanded beyond the limits of the grid square.

ArrayExplorer exports the extracted intensity values in a text file, which also contains other relevant information for each gene included in the array. This information includes: (*i*) the gene's array location; (*ii*) where available, the gene's name, family, and functional classification; (*iii*) database access information, including GenBank® and SwissProt access numbers; (*iv*) the gene intensity value; and (*v*) custom or automatic flags. Each of these pieces of information is tab-delimited, and each of the genes is separated by carriage-return line-feed. This format allows for easy import into a custom program or a professional spreadsheet/data-analysis program such as Microsoft Excel®.

Simulated Array Images for Program Testing

Simulated array images were created with a custom program in Visual Basic that was designed to generate a bitmap image of a 1200-gene array. To reproduce the variability in the actual hybridization signals, the size, position within the grid square, and spot intensities were generated at random within the ranges of the observed values from actual arrays. The low-intensity hybridization spots were simulated by generating spots with a fixed diameter and variable height; the high-intensity spots were simulated by varying the diameter of the spot while keeping a constant, maximum height. Some of the high-intensity spots overlapped to varying degrees. The borders of the simulated spots were gradually blending into the background, much like the true hybridization signals. Furthermore, the simulated array gene signals were randomly offset from their center to test the accuracy of ArrayExplorer for CSR autoplacement. Finally, a global background and random noise was added to the entire image.

Hardware and System Requirements and Program Availability

ArrayExplorer requires an Intel® Pentium® II or later CPU (200 MHz or greater) with at least 64 MB RAM and 10 MB available disk space for installation. The program is compatible with Microsoft Windows 95, 98, NT4.0, or Windows ME, and it may be obtained upon request.

RESULTS AND DISCUSSION

General Program Layout

ArrayExplorer uses a multi-window interface to analyze array images in bitmap format (Figure 1). The main window displays the array image and its corresponding reference grid. Drop-down menu options *File*, *Edit*, *Mode*, and *Help* are displayed at the top of the main window. The menu options allow the user to initiate all available commands operating in the main window, as well as for opening other available operational windows. A selection of buttons and shortcut keys are available for commonly used commands, which increases the functionality of the program. Finally, the background and net intensity values for each selected grid location are displayed at the bottom right of the main window.

Four additional operational windows increase the flexibility of ArrayExplorer. The *Gene List* displays a searchable list of the genes spotted on the array and their grid locations. The *Fine Tuning* window allows the user to adjust manually the size and coordinates of the CSR around individual gene spots. The *Threshold* window provides a list of the gene intensities sorted in ascending order. This window allows the user to zero all points that have intensities below a given threshold value, which is caused by background noise. These operational windows are linked together such that selecting a gene at a grid location in the main window will display a magnified image of it in the *Fine Tuning* window, list its location number and name in the *Gene List*, and show its intensity in the *Threshold* list of intensities. A fourth window allows the user to look at the 2-D profile of the intensity values for a manually selected set of genes.

Input

ArrayExplorer uses bitmap-formatted images generated with a phosphorimager-linked application, Adobe® PhotoShop®, or other graphics-editing programs. Phosphorimager scanning at variable exposure times is carried out at 16-bit/50-μm resolution. Alternatively, autoradiographic images are obtained on BioMax® MS films (Eastman Kodak, Rochester, NY, USA) for variable exposure times and scanned at 16 bit/1200 dpi (25 μm) grayscale resolution using a flatbed scanner. The resolution of the obtained images is then converted to 8 bit/200 dpi using a standard graphics-editing program. Processing of higher-resolution images (400 or 600 dpi) with ArrayExplorer indicated that an increase in image resolution does not yield a higher degree of accuracy (data not shown). For a more precise grid alignment, before processing with ArrayExplorer, array images may be subjected to rotation and cropping using standard graphics-editing programs.

Features

Grid selection and placement. ArrayExplorer allows users to select from a list of standard reference grids for commercially available filter arrays. These grids are linked to corresponding files that contain gene identity and array location, as well as database access information (GenBank and SwissProt). Currently, the grids available with ArrayExplorer include all human, murine, and rat Atlas™ array series from Clontech Laboratories (Palo Alto, CA, USA). However, the versatility of ArrayExplorer allows users to design and save grids for arrays with custom configurations and size. Reference grids that correspond to new arrays can be easily incorporated in the program.

Once an appropriate configuration for the analyzed array is selected, the grid is aligned over the array image. This feature is quite flexible and allows the user to fine-tune the alignment of the grid over the array image. To address the inherent variability of the spotting process, ArrayExplorer contains three routines that facilitate rapid and accurate positioning of the grid: (*i*) the user positions the grid based on the known location of two genes in the grid; (*ii*) the grid can be stretched and dragged over the image with the mouse and cursor arrows; and (*iii*) individual grid sections can be selected and realigned, which allows for potential variability in the spacing between individual array sectors, as well as stretching of the filters after repeated hybridization. Since each gene intensity is calculat-

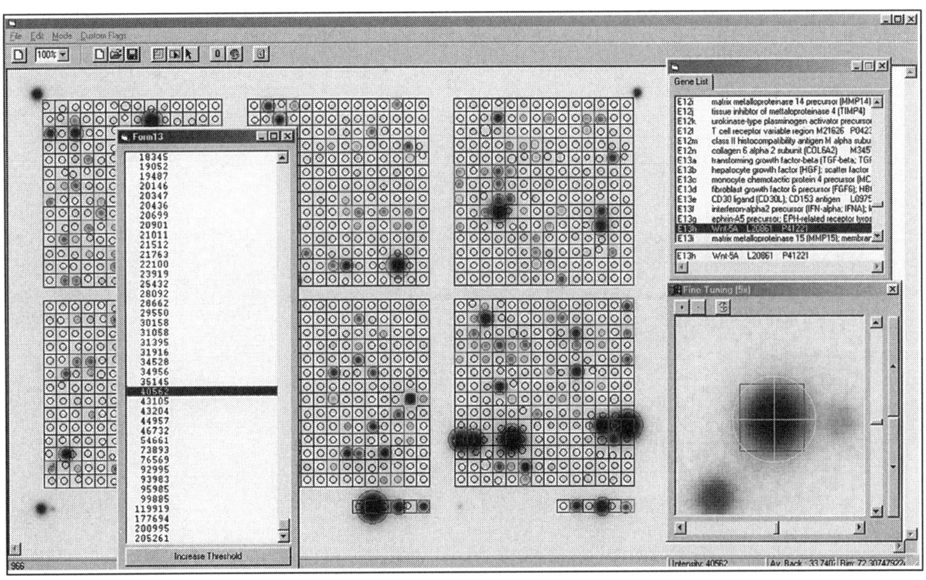

Figure 1. ArrayExplorer screen display. ArrayExplorer has four linked operational windows. Each window can be resized and repositioned on the screen. The main window displays the entire array image with the overlaid reference grid. Red boxes indicate signals above the background threshold; black boxes are below this threshold. The *Fine Tuning*, *Gene List*, and *Threshold* windows allow easy navigation and manual adjustment of the grid. Operations are facilitated through the menu items, the toolbar buttons, and the designated shortcut keys. **(See color plate A8.)**

ed within a CSR that is automatically aligned with the hybridization signal and further fine-tuned if necessary, the required level of precision for grid alignment is such that each gene spot grossly falls within a grid square.

Background estimation. ArrayExplorer has two different options for calculating the background around a particular gene. When the general background over the entire array image appears uniform, a global average background value can be determined automatically from areas within the borders of the grid. More often, however, the background varies between the different sections of the array image. In such cases, a sectional average background value can be calculated and then subtracted from the intensities of the genes included in each corresponding array section. In both cases, users can either choose a default background setting (Figure 2) or create a custom pattern of areas within the array image that are devoid of hybridization signals, which can then be used to calculate the average background value.

Intensity determination. Accurate extraction of the gene intensities is quite challenging because of irregularities in DNA spotting, signal bleeding beyond the coordinates of the grid square, and interference with adjacent gene signals.

ArrayExplorer addresses these problems through a combination of innovative features for flexible CSR placement and size. Once the CSRs are placed around each hybridization signal, their radius is increased until the circle encompasses the entire signal. This allows extraction of the intensity values from hybridization signals with variable levels of bleeding beyond the circumference of the spotted cDNAs (Figure 3). Gene spots whose hybridization signals cannot be determined automatically by ArrayExplorer because of signal interference and overlapping between adjacent gene spots are flagged as green boxes on the grid. Such spots can be easily selected and their CSR positioning can be adjusted using the Fine Tuning window (Figures 1 and 4). Two methods in ArrayExplorer allow the user to deal with intense hybridization signals that leak into adjacent grid

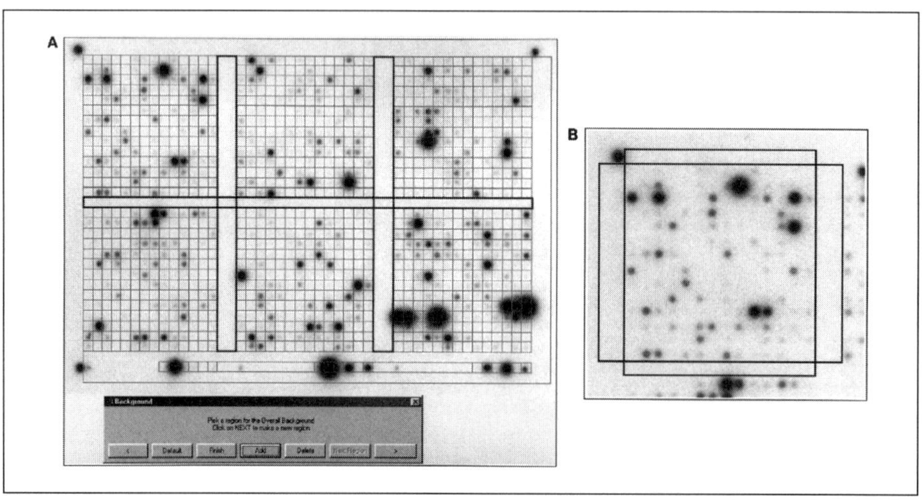

Figure 2. Background determination. Background can be calculated with ArrayExplorer in two ways: (A) The default overall background is calculated as the average intensity value per pixel in the central array areas, boxed in black, which are devoid of spotted cDNA/hybridization signals. (B) The default sectional background is calculated as above within signal-free regions along the borders of each array section. In both cases, the user can modify the size and location of the default background calculation regions.

squares. In simple cases, the program identifies and flags adjacent squares that do not contain signal from their own spotted gene cDNA. The user can manually select each of these boxes and annul the corresponding gene intensity. The result is color-coded in the output data file. A more complex situation occurs when one hybridization signal partially overlaps with another. To solve this problem, the *Pie-sectoring* option has been incorporated in ArrayExplorer, which allows the user to divide each of the overlapping hybridization signals into 1/8-slices and use only those slices that are free of signal interference (Figure 4). *Pie Auto* then readjusts the radius of the corresponding CSR on the basis of the calculated average intensity value per pixel in the selected set of 1/8-slices.

Four other options in ArrayExplorer increase its utility for image processing and further data analysis: The *Threshold* feature allows the user to determine a cutoff value for true hybridization signals and noise-related signals. With this option, all detected hybridization signals are automatically sorted in a descending order, and the threshold signal value is being determined through direct inspection of grid squares linked to corresponding values in the *Threshold* list. All signals below the threshold value are then automatically zeroed and specifically flagged in the output data file. Another option allows the user to select multiple grid squares that require the same operation so that they are manipulated simultaneously. The user also has the option to custom design up to five flags for labeling selected gene signals on the processed array image. Each flag is color-coded on the grid and has a designated label that appears adjacent to the corresponding gene intensity value in the output data file. These flags can help the user to easily locate selected groups of genes during the later stages of data processing. Finally, ArrayExplorer has the ability to display a 2-D profile of manually selected array areas (Figure 5). This option is particularly important for determining whether a detected intensity value is indeed due to a true hybridization signal or if a particular signal has reached its saturation point, therefore requiring processing from images acquired at shorter exposure times.

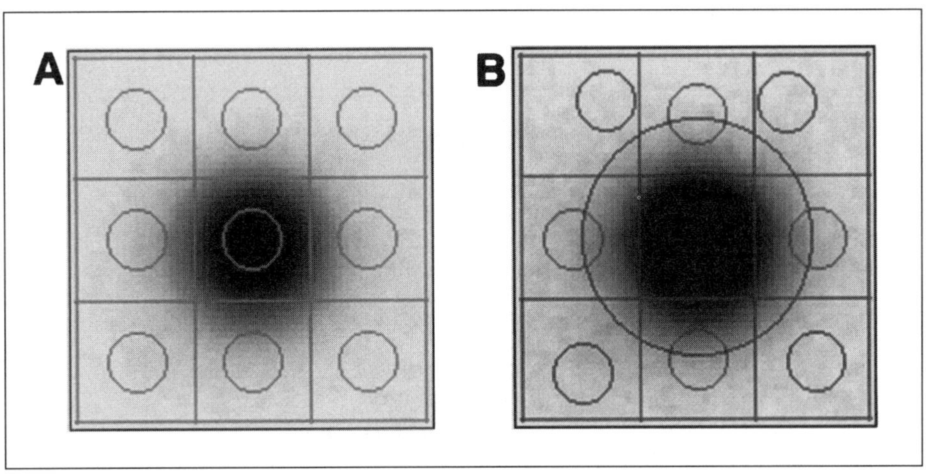

Figure 3. Automatic CSR expansion. The ArrayExplorer *AutoPlacement* option allows automatic, differential expansion of the grid CSRs to encompass hybridization signals with variable diameters. (A) Before CSR expansion, only a portion of the total gene hybridization signal is registered. (B) Automatic expansion of the CSR allows intensity calculation over the entire hybridization signal. Grid boxes flagged in gray represent areas that are manually zeroed by the user to avoid registration of false-positive signals caused by signal bleeding from adjacent boxes.

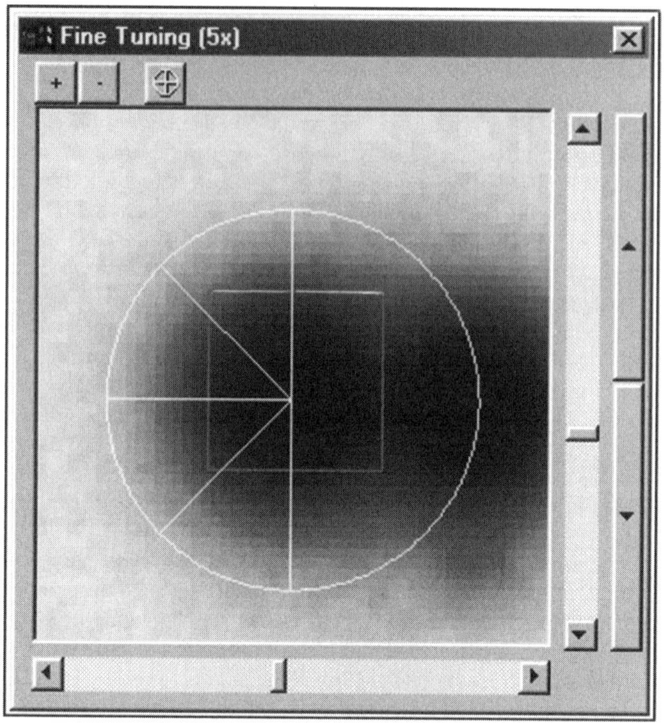

Figure 4. Fine Tuning and Pie-sectoring options. The Fine Tuning window displays an image centered on the selected gene spot. In this operational window, the user is able to manually adjust the diameter and positioning of the CSR. The Pie-sectoring option is used to determine the gene intensity value extrapolated from the selected portion (here from the left half) of the hybridization signal, which is free of signal interference from adjacent spots. The user may decide how many consecutive CSR sectors should be included for calculation of the total signal.

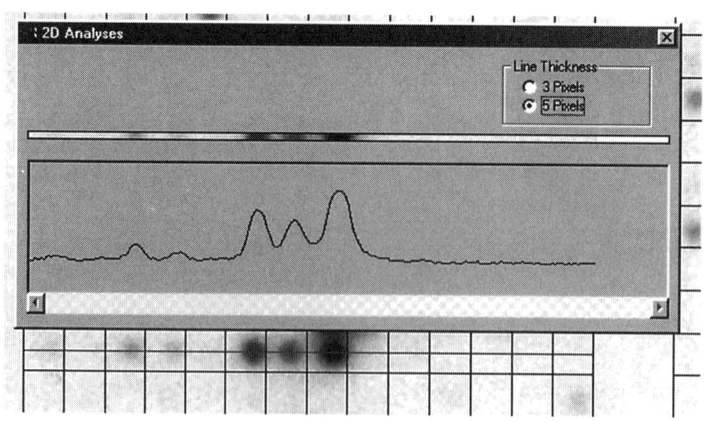

Figure 5. 2-D profile analysis. The user may choose sections of the gridded array image to obtain a 2-D display of the hybridization intensities. This may prove helpful during manual manipulation and fine-tuning of the grid, and for data processing.

Output

In addition to the gene intensity values extracted with ArrayExplorer, the generated output data file contains important relevant information regarding the location, identity, and database access entries for each gene and can be opened by a number of spreadsheet applications, including Microsoft Excel.

Simulated Images

The simulated array image (Figure 6B) was processed with ArrayExplorer similarly to actual array images (Figure 6A) obtained through autoradiography and scanning or with a phosphorimager. The results show that the intensity values extracted from the simulated array image with ArrayExplorer using standard automatic and manually adjusted CSRs are highly correlated to the actual input numbers, with an $r^2 = 0.999$ (Figure 6C). The flagged overlapping spots were processed using the *Pie-sectoring* option, and, again, the estimated values were in excellent agreement with the actual ($r^2 = 0.996$).

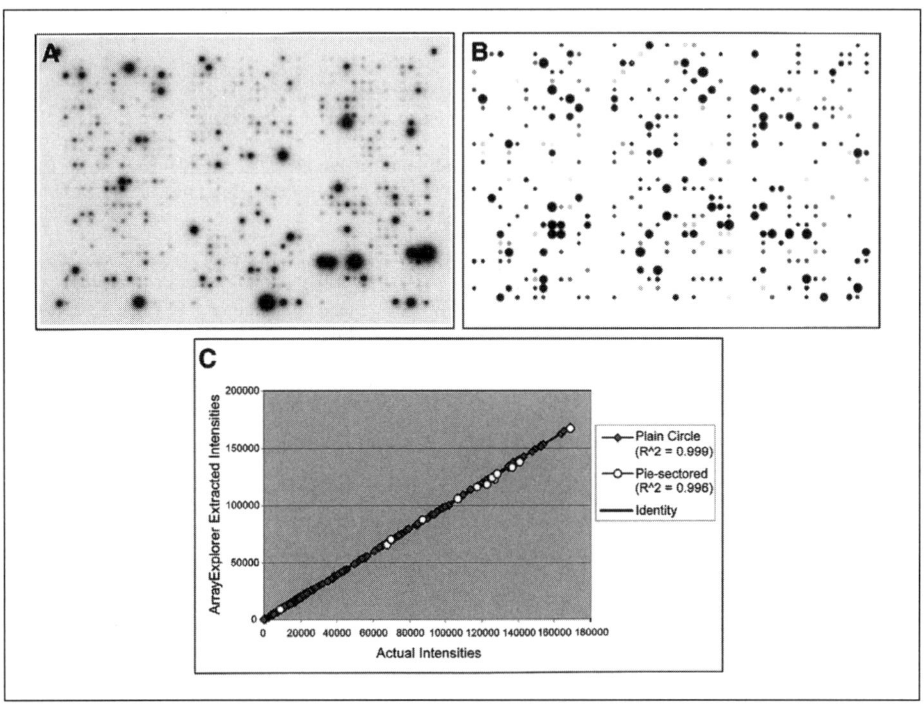

Figure 6. Accuracy of data extraction with ArrayExplorer. (A) Actual autoradiographic image of a hybridized Atlas 1.2 Human Cancer array. (B) Simulated array image: spots were generated randomly using a custom algorithm developed in the laboratory. The range of diameters and intensities found in the actual array images is duplicated in the simulated array. Plain spots in the simulated array were analyzed using a combination of automatic and manual CSR adjustment, while overlapping spots were analyzed using the Pie-sectoring option. (C) Correlation plot of the actual input intensity values versus the values extracted with ArrayExplorer from the simulated array image.

CONCLUSION

ArrayExplorer is an innovative program for the analysis of gene expression arrays obtained with radioactively labeled cDNA probes. The combination of automatic and manual features in ArrayExplorer yields highly accurate and robust array data extraction. Gene expression intensities are automatically exported in a spreadsheet/text file format, which is convenient for further analysis and data mining.

ACKNOWLEDGMENTS

The authors would like to thank Ms. R. Stoyanova for numerous suggestions for improving and testing the program, as well as Dr. E. Ross for critical review of the manuscript. The development of this software was supported by funds from the 5th District AHEPA Cancer Res. Foundation, Inc. awarded to C. Patriotis, and through NIH SPORE P50-CA83638 (PI: R. Ozols). C. Patriotis is a Liz Tilberis Scholar of the OCRF, Inc. P. Patriotis was a Fellow of the Howard Hughes Medical Institute High School Student Program.

REFERENCES

1. **DeRisi, J., L. Penland, P.O. Brown, M.L. Bittner, P.S. Meltzer, M. Ray, Y. Chen, Y.A. Su, and J.M. Trent.** 1996. Use of a cDNA microarray to analyse gene expression patterns in human cancer. Nat. Genet. *14*:457-460.
2. **Heller, R.A., M. Schena, A. Chai, D. Shalon, T. Bedilion, J. Gilmore, D.E. Woolley, and R.W. Davis.** 1997. Discovery and analysis of inflammatory disease-related genes using cDNA microarrays. Proc. Natl. Acad. Sci. USA *94*:2150-2155.
3. **Ramsay, G.** 1998. DNA chips: state-of-the art. Nat. Biotechnol. *16*:40-44.
4. **Schena, M., R.A. Heller, T.P. Theriault, K. Konrad, E. Lachenmeier, and R.W. Davis.** 1998. Microarrays: biotechnology's discovery platform for functional genomics. Trends Biotechnol. *16*:301-306.
5. **Schena, M., D. Shalon, R.W. Davis, and P.O. Brown.** 1995. Quantitative monitoring of gene expression patterns with a complementary DNA microarray. Science *270*:467-470.
6. **Schena, M., D. Shalon, R. Heller, A. Chai, P.O. Brown, and R.W. Davis.** 1996. Parallel human genome analysis: microarray-based expression monitoring of 1000 genes. Proc. Natl. Acad. Sci. USA *93*:10614-10619.

Update to: ArrayExplorer©, a Program in Visual Basic for Robust and Accurate Filter cDNA Array Analysis

Photis C. Patriotis[1], Troy D. Querec[2], Briana N. Gruver[1], Truman R. Brown[3], and Christos Patriotis[1]

[1]Division of Medical Science, Fox Chase Cancer Center, Philadelphia, PA; [2]Emory University, GDBBS, Atlanta, GA; [3]Department of Radiology & Biomedical Engineering, Columbia University, NY, USA (c_patriotis@fccc.edu)

The publication of the ArrayExplorer© program in the October 2001 issue of *BioTechniques* has generated heightened interest in this product, and more than 30 copies of the software have already been distributed in laboratories in the U.S. and elsewhere. The lack of any reported problems with the use of ArrayExplorer by these investigators indicates that the software is highly user-friendly and quite robust. In addition to our laboratory, a number of other investigators at Fox Chase Cancer Center use the software routinely, and some of the results obtained with ArrayExplorer have been included in a recent publication (1), while other have been submitted for publication or have created the backbone of several grant applications. To further expand the utility of the ArrayExplorer program in the gene expression array field, we have developed two new data analysis approaches, one for correction of saturated signals and the detection of a larger number of genes per array image, and a second for normalization of multiple single-channel array data sets on the basis principle component analysis (PCA). Two manuscripts describing these approaches have been submitted for publication.

REFERENCE

1. **Fashena, S.J., M.B. Einarson, G.M. O'Neil, C. Patriotis, and E.A.J. Golemis.** 2002. Cell Science Dissection of HEF1-dependent functions in motility and transcriptional regulation. J. Cell Sci. *115*:99-111.

CHAPTER 37

Computational Methods for Gene Expression-Based Tumor Classification

Momiao Xiong, Li Jin, Wuju Li, and Eric Boerwinkle

University of Texas—Houston Health Science Center
Houston, TX, USA

Reprinted from BioTechniques 29:1264-1270 (December 2000)

ABSTRACT

Gene expression profiles may offer more or additional information than classic morphologic- and histologic-based tumor classification systems. Because the number of tissue samples examined is usually much smaller than the number of genes examined, efficient data reduction and analysis methods are critical. In this report, we propose a principal component and discriminant analysis method of tumor classification using gene expression profile data. Expression of 2000 genes in 40 tumor and 22 normal colon tissue samples is used to examine the feasibility of gene expression-based tumor classification systems. Using this method, the percentage of correctly classified normal and tumor tissue was 87.0%. The combined approach using principal components and discriminant analysis provided superior sensitivity and specificity compared to an approach using simple differences in the expression levels of individual genes.

INTRODUCTION

Diagnostic pathology has traditionally relied on macro- and microscopic histology and tumor morphology as the basis for classifying tumors. Current classification frameworks, however, are unable to discriminate among tumors with similar histopathological features but vary in clinical course and response to treatment (12).

Cancer is a dynamic disease. It has been estimated that a new genetic alteration occurs in one out of 10 000 tumor cells at each cell division (15). This leads to an increasing number of genetic aberrations in cancer cells, some of which result in altered gene expression (14). These changes in gene expression are at least partially responsible for the characteristic of a tumor or tumor-type. Genome-wide gene expression profiles provide detailed "tumor signatures" and can potentially be used for molecular diagnosis and classification of tumors (13).

BioComputing: Computer Tools for Biologists
Edited by Stuart M. Brown

In the past decade, array technologies have been developed that can be used to simultaneously assess the expression of thousands of genes (3,8,11,12,16,17). An early use of human microarrays in cancer biology used an 870 gene array to identify gene expression differences between a malignant melanoma cell line and a nontumorigenic variant of this line (3). It was found that 17% of the genes were expressed at significantly lower levels in the normal cell line than in the tumor cell line and that 7.3% of the genes were expressed at significantly higher levels. Since then, several studies used arrays to analyze gene expression in colon, breast, and other tumors and demonstrated the potential power of molecular profiles for classifying tumors (1,4,5,9).

Information and data analysis are key components to the successful development of tumor classification systems based on gene expression profiles. Most existing gene expression data in cancer research focus on differential gene expressions in tumor and normal cells or clustering genes into groups based on similarity of gene expressions. However, the characteristics of gene expression profiles are yet to be fully explored for tumor classification.

Alon et al. (1) monitored expression of more than 6500 human genes in 40 tumor and 22 normal colon tissue samples. They used a deterministic annealing algorithm to separate tumor and normal tissues into two distinct clusters. Cluster analysis attempts to classify tissue samples in an unsupervised fashion and does not fully exploit prior knowledge of the tissue samples. In general, accuracy of classification by unsupervised clustering is lower than that by supervised learning algorithms. Golub et al. (4) proposed a neighborhood analysis method for class discovery and class prediction using gene expression profiles. Looking at the gene expression patterns from bone marrow samples of acute leukemia patients, Golub et al. (4) were able to distinguish between patients with acute lymphoblastic leukemia and those with acute myeloid leukemia. Their discriminant decision rules were based on the quantity of prediction strength (PS) and the threshold of the PS. However, they failed to give the procedure used for the automatic selection of the threshold of the PS for classification. Heuristic rules for selection of the PS threshold might have been used with a certain level of subjectivity. In addition, their method was not formulated in terms of standard statistical frameworks for classification. Therefore, the most widely used statistical methods and software for classification cannot be exploited.

The purpose of this report is to develop an analytic framework for a gene expression-based tumor classification scheme that can allow us to analyze data in a formal, systematic, and integrated manner and investigate the impact of some factors on the accuracy of the proposed tumor classification systems. Expressions of 2000 genes in each of 40 tumor and 22 normal colon tissue samples will be used to show that gene expression profiles provide a precise cancer classification system.

MATERIALS AND METHODS

Discriminant Analysis

We will use Fisher's linear discriminant function for classifying tumor and normal tissues. Fisher's approach does not assume that the observations are normally distributed. It does, however, implicitly assume that the population covariance matrices are equal (6). Tumor and normal tissues are classified on the basis of k selected feature variables. The feature variables can represent the level of expression of k genes or k summary measures representing the level of gene expression of any number of genes or other measures. Suppose that n_N normal and n_T tumor tissue samples are examined. For tissue sample i, we have the vector $Y_i' = (Y_{i_1}, Y_{i_2}, \ldots, Y_{i_k})$. The Y_is for normal (N) and tumor (T) samples constitute the following data matrix,

$$Y_N = [Y_{N1}, Y_{N2}, \ldots, Y_{Nn_N}]_{(k \times n_N)} \qquad \text{[Eq. 1]}$$

$$Y_T = [Y_{T1}, Y_{T2}, \ldots, Y_{Tn_T}]_{(k \times n_T)} \qquad \text{[Eq. 2]}$$

From these data matrices, the sample mean vectors and covariance matrices are determined by:

$$\bar{Y}_N = \frac{1}{n_N} \sum_{i=1}^{n_N} Y_{Ni},$$

$$S_N = \frac{1}{n_N - 1} \sum_{i=1}^{n_N} (Y_{Ni} - \bar{Y}_N)(Y_{Ni} - \bar{Y}_N)' \qquad \text{[Eq. 3]}$$

$$\bar{Y}_T = \frac{1}{n_T} \sum_{i=1}^{n_T} Y_{Ti},$$

$$S_T = \frac{1}{n_T - 1} \sum_{i=1}^{n_T} (Y_{Ti} - \bar{Y}_T)(Y_{Ti} - \bar{Y}_T)' \qquad \text{[Eq. 4]}$$

$$S = \frac{(n_N - 1)S_N + (n_T - 1)S_T}{n_N + n_T - 2} \qquad \text{[Eq. 5]}$$

Fisher's idea was to transform the multivariate observations Y_N and Y_T into univariate observations Z_N and Z_T such that the Zs were separated as much as possible. Fisher suggested taking linear combinations of the Ys to generate Zs, which can be easily manipulated mathematically. The midpoint, $\hat{m}$, between the two univariate sample means, $\bar{Z}_N = (\bar{Y}_N - \bar{Y}_T)' S^{-1} \bar{Y}_N$ and $\bar{Z}_T = (\bar{Y}_N - \bar{Y}_T)' S^{-1} \bar{Y}_T$, is given by:

$$\hat{m} = \frac{1}{2} (\bar{Y}_N - \bar{Y}_T)' S^{-1} (\bar{Y}_N + \bar{Y}_T) \qquad \text{[Eq. 6]}$$

The classification rule based on Fisher's linear discriminant function for an unknown sample, Y_0, is as follows:

Assign Y_0 to N if $(\bar{Y}_N - \bar{Y}_T)' S^{-1} Y_0 \geq \hat{m}$,

and

Assign Y_0 to T if $(\bar{Y}_N - \bar{Y}_T)' S^{-1} Y_0 \leq \hat{m}$.

To evaluate the precision of Fisher's method for classifying tissue type based on its gene expression profile, we will use a cross-validation scheme (10). The tissue samples for classification will be randomly divided into two sets. Fisher's linear discrimination function will be trained on the first set (training set) and then tested on the second set (test set). There is a tradeoff involved in any data partitioning. A small training set will compromise the accuracy of prediction, while a small test set will jeopardize the accuracy of validation. In this report, we will use three different proportional partitionings of the tissue samples: *(i)* 50% for training and 50% for testing; *(ii)* 68% for training and 32% for testing; and *(iii)* 95% for training and 5% for testing.

Q-Mode Analysis

A genome-wide gene expression profile, which involves thousands of genes, constitutes a data set of high dimension. A practical objective is to reduce the number of variables while preserving as much of the original information as possible. Although k variables (i.e., the expression of k genes) are required to reproduce the total system variability, often much of this variability can be accounted for by a small number, p, of principal components. Therefore, the p principal components can replace the initial k variables.

Due to the expense of each array experiment, the number of observations is generally smaller than the number of variables (genes). In addition, for tumor classification studies, biological sample characteristics rather than individual genes are of primary interest. Therefore, in this report, we use Q-mode principal component analysis (2), which is based on a $(n \times n)$ product—moment matrix of the observations, rather than the variables as preclassification data processing.

Let $X = Y' - Y'$, where Y' is the transpose of the matrix Y, as similarly defined in Equations 1 and 2, whose elements are the variables of gene expression for tumor and normal tissues. Y' is the sample mean of the matrix Y'. Principal component analysis intends to replace the k variables $x_1, x_2, \ldots, x_k$ by p principal components $z_1, z_2, \ldots, z_p$ where p is smaller than k, while minimizing loss of information. The variables $z_1, z_2, \ldots, z_p$ are the principal components if they are linear functions of $x_1, x_2, \ldots, x_k$ with the following properties (10):

1. z_1 has the maximum possible variance among all possible linear functions of $x_1, x_2, \ldots, x_k$
2. z_2 has the maximum possible variance among all possible linear functions of $x_1, x_2, \ldots, x_k$ subject to z_2 being uncorrelated with z_1.
3. In general, z_p has the maximum possible variance among all possible linear functions of $x_1, x_2, \ldots, x_k$, subject to z_p being uncorrelated with $z_1, z_2, \ldots, z_{p-1}$ for $2 \leq p \leq k$.

Some principal components may arise as a result of random sampling, measurement error, or individual variation. Such components contain no systematic variation in the original data and will not be used for classification. We include l principal components with the largest eigenvalues, which account for 90% of the summation of the eigenvalues (i.e., 90% of the total variance of gene expression) in the classification scheme.

Gene Expression Data of Tumor and Normal Colon Tissues

Feasibility of the aforementioned methods is explored using the expression profiles of 2000 genes using an oligonucleotide array (Affymetrix, Santa Clara, CA, USA) in 22 normal and 40 colon cancer tissues, which were originally retrieved from www.molbio.princeton.edu/colondata (1) and are now posted in www.sph.uth.tmc.edu/hgc.

RESULTS

Principal component analysis of the gene expression profiles of 2000 genes from the 62 colon samples revealed that the first seven principal components accounted for 90.44% of the total expression variation. The contribution of the remaining components was small and therefore ignored in subsequent analyses. The performance of Fisher's linear discriminant function for tumor classification was therefore evaluated by using the loading scores of the seven principal components for each sample as predictor variables.

The total tissue specimens were resampled using jackknifing. The resampled speci-

Table 1. Percentage of Correctly Classified Normal and Tumor Samples Using the First Seven Principal Components[a]

Proportion of Training-Test Set	Accuracy for Training Set	Accuracy for Cross-Validated Set	Accuracy for Test Set
50%–50%	89.5%	84.7%	85.7%
68%–32%	89.8%	87.0%	87.0%
95%–5%	90.2%	87.8%	88.2%

[a]Results are an average of 62 replicate partitions.

Table 2. Percentage of Correctly Classified Normal and Tumor Samples Using 42 Genes Showing Significant Differences in Gene Expression

Proportion of Training-Test Set	Accuracy for Training Set	Accuracy for Cross-Validated Set	Accuracy for Test Set
50%–50%	99.7%	54.7%	66.6%
68%–32%	99.8%	58.4%	64.7%
95%–5%	98.6%	65.2%	67.2%

mens were partitioned into three proportions of the training and testing sets: 50%:50%, 68%:32%, and 95%:5%. The result is presented in Table 1. The accuracy of classification is defined as the percentage of correct classification for both normal and tumor tissues. The accuracy of classification throughout this study was averaged over the 62 times. Using training and test proportions of 68% and 32%, respectively, the accuracy of classification for both the cross-validation set and the test set was 87.0%.

Comparison of expression patterns between normal and tumor tissue samples revealed that a large number of genes were expressed at similar levels. However, the expression profiles also revealed that 42 genes were expressed at significantly different levels ($P < 0.01$). We next determined whether these 42 genes, which show different expression between normal and tumor samples, could be used for reliable classification. Fisher's linear discriminant function was used for classification analysis (using SPSS software), and the results are presented in Table 2. As expected, the accuracies for the training sets using these 42 genes were generally higher than those using the seven principal components. The accuracies for both the cross-validated set and the test set based on the 42 genes were, however, much lower than the principal component analyses. The data presented here indicate that the combined approach of using principal components for data reduction and discriminant analysis for classification provides superior sensitivity and specificity compared to approaches that utilize simple differences in expression levels of individual genes.

DISCUSSION

An alternative approach to morphologic- or histologic-based tumor classification schemes is to use genomic expression profiles. Cancer occurs as a consequence of pro-

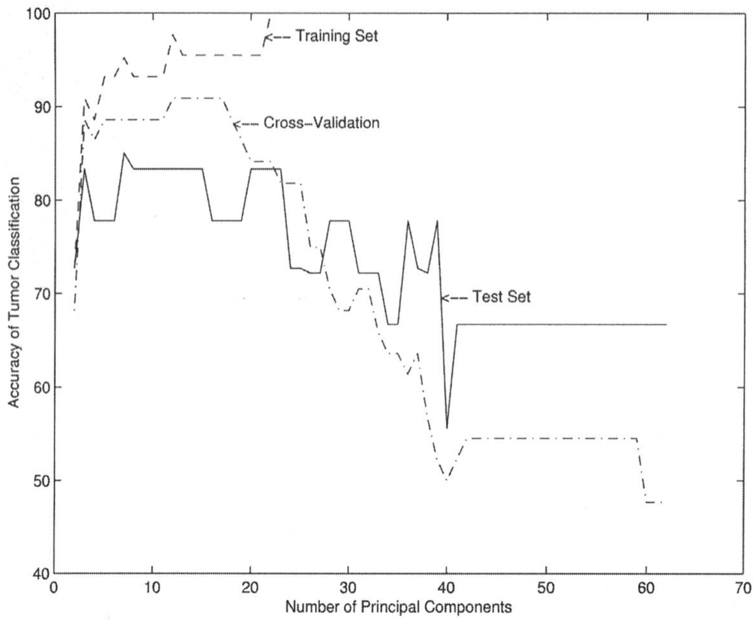

Figure 1. Classification accuracy for varying numbers of principal components. Training and test proportions were 68% and 32%, respectively.

gressive accumulation of genetic aberrations, which lead to a deviation of gene expression profiles from that of the normal cells (7). In addition, a large number of genes are expressed differentially in tumor versus normal cells. The pattern of total variations of gene expression across tissues distinguishes normal from tumor tissue and hence can be used for classification purposes. At present, the number of samples studied in a typical microarray gene expression study is much smaller than the number of genes whose expression is being monitored. From a mathematical point of view, it is infeasible to use the gene expression profiles of all of the genes for tumor classification using Fisher's linear discriminant function because the rank of the covariance matrix describing gene expression is equal to the number of samples, and hence, the covariance matrix is singular. Therefore, the maximum number of variables allowed for any classification study using a linear discriminant function is equal to the number of samples (n = 62 for the colon cancer study). To circumvent this problem, a data reduction algorithm is necessary.

There are two ways in which one might attempt to reduce the number of variables and hence the dimensionality of the data set. First, one could simply choose a subset of genes that show different expression between tumor and normal cells. In this study, we chose 42 genes that show significant difference in their expression between tumor and normal tissues using a simple t-test ($P < 0.01$). Using these 42 genes, it was found that the accuracy of tumor classification using Fisher's linear discriminant function was low. This finding suggests that the status of tissue samples is determined by expression of many genes that have a subtle yet systematic difference between tumor and normal tissues (13).

The level of expression of each gene across tissues was correlated (1). Therefore, it is likely that the information contained in a large number of genes could be captured by a much smaller number of factors without appreciable loss of information. It is well known that principal component analysis is a linear transformation of a set of correlated variables into a set of uncorrelated variables ordered by their variance. Thus, data reduction can be achieved for classification by including only those variables with the greatest variances. This approach is perhaps less appealing, intuitively, than that of retaining a subset of genes with significant difference in gene expression between normal and tumor tissues. However, it has the advantage that, for the same or less dimensionality, one could retain a greater proportion of the original gene expression information by using principal components. In this report, we demonstrate that, using seven principal components as classification variables, the accuracy of classification is more than 84% for all three partition schemes, which is better than that of using 42 genes selected based on expression difference between normal and tumor tissues.

It is interesting to note that misclassification may primarily be due to tissue sample composition. It has been reported that normal tissues have a high muscle index, while tumors have a low muscle index (1). For the partitioning of the tissue samples used in this report, three normal samples n_8, n_{34}, and n_{36} with a muscle index 0.2, 0.2, and 0.1, respectively, were classified as tumors, while three tumor samples t_{23}, t_{27}, and t_{34} with a muscle index of 0.2, 0.2, and 0.2, respectively, were classified as normal tissues. However, muscle index is not a unique factor that determines the type of tissue. Tumor samples with a much higher muscle index (e.g., t_{33} and t_{36} with a muscle index equal to 0.7) were correctly classified as tumors by the cluster algorithm proposed here.

One of the key issues in principal component analysis is to determine how great a reduction in dimensionality can be achieved while retaining much of the original variation. It was shown that seven principal components sufficed to describe the variance structure and accounted for 90.44% expression variations of 2000 genes. To evaluate the impact of the number of principal components used for classification on the percentage of correctly classified normal and tumor samples, we repeated the classification exercise using varying numbers of principal components. The results are plotted in Figure 1. Fisher's linear classifier using these seven principal components had the highest percentage of correctly classified normal and tumor samples, and using too many principal components for classification reduces the accuracy rather than improves the precision of tumor classification.

In summary, we have demonstrated that molecular classification of tumors is feasible and that a moderate number of tissue samples suffice to achieve reasonable precision of tumor classification. Development of microarray expression technology is revolutionizing tumor classification and may have far-reaching future implication for the diagnosis and treatment of cancer.

ACKNOWLEDGMENTS

M.M.X. is supported by National Institutes of Health grant nos. GM56515 and MH59518. The authors thank two anonymous reviewers for their helpful comments.

REFERENCES

1. **Alon, U., N. Barkai, D.A. Notterman, K. Gish, S. Ybarra, D. Mack, and A.J. Levine.** 1999. Broad patterns of gene expression revealed by clustering analysis of tumor and normal colon tissues probed by

oligonucleotide arrays. Proc. Natl. Acad. Sci. USA *96*:6745-6750.

2. **Basilevsky, A.** 1994. Statistical factor analysis and related methods. Theory and Applications, p. 278. John Wiley & Sons, New York.

3. **DeRisi, J., L. Penland, P.O. Brown, M.L. Bittner, P.S. Meltzer, M. Ray, Y. Chen, Y.A. Su, and J.M. Trent.** 1996. Use of a cDNA microarray to analyse gene expression patterns in human cancer. Nat. Genet. *14*:457-460.

4. **Golub, T.R., D.K. Slonim, P. Tamayo, C. Huard, M. Gaasenbeek, J.P. Mesirov, H. Coller, M.L. Loh et al.** 1999. Molecular classification of cancer: class discovery and class prediction by gene expression monitoring. Science *286*:531-537.

5. **Iyer, V.R., M.B. Eisen, D.T. Ross, G. Schuler, T. Moore, J.C.F. Lee, J.M. Trent, L.M. Staudt et al.** 1999. The transcriptional program in the response of human fibroblasts to serum. Science *283*:83-87.

6. **Johnson, R.A. and D.W. Wichern.** 1982. Applied Multivariate Statistical Analysis, p. 503. Prentice-Hall, Englewood Cliffs, NJ.

7. **Khan, J., L.H. Saal, M.L. Bittner, Y. Chen, J.M. Trent, and P.S. Meltzer.** 1999. Expression profiling in cancer using cDNA microarrays. Electrophoresis *22*:223-229.

8. **Lockhart, D.J., H. Dong, M.C. Byrne, M.T. Follettie, M.V. Gallo, M.S. Chee, M. Mittmann, C. Wang et al.** 1996. Expression monitoring by hybridization to high-density oligonucleotide arrays. Nat. Biotechnol. *14*:1675-1680.

9. **Perou, C.M., S.S. Jeffrey, M.V.D. Rijn, C.A. Rees, M.B. Eisen, D.T. Ross, A. Pergamenschikov, C.F. Williams et al.** 1999. Distinctive gene expression patterns in human mammary epithelial cells and breast cancers. Proc. Natl. Acad. Sci. USA *96*:9212-9217.

10. **Roecker, E.B.** 1991. Prediction error and its estimation for subset-selected models. Technometrics *33*:459-468.

11. **Spellman, P.T., G. Sherlock, M.Q. Zhang, V.R. Iyer, K. Anders, M.B. Eisen, P.O. Brown, D. Botstein, and B. Futcher.** 1998. Comprehensive identification of cell cycle-regulated genes of the yeast *Saccharomyces cerevisiae* by microarray hybridization. Mol. Biol. Cell *9*:3273-3297.

12. **Stephenson, J.** 1999. Human genome studies expected to revolutionize cancer classification. JAMA *282*:927-928.

13. **Strausberg, R.L. and M.J.F. Austin.** 1999. Functional genomics: technological challenges and opportunities. Physiol. Genomics *1*:25-32.

14. **Theillet, C.** 1998. Full speed ahead for tumor screening. Nat. Med. *4*:767-768.

15. **Tlsty, T.D., B.H. Margolin, and K. Lum.** 1989. Differences in the rates of gene amplification in nontumorigenic and tumorigenic cell lines as measured by Luria-Delbruck fluctuation analysis. Proc. Natl. Acad. Sci. USA *86*:9441-9445.

16. **Wodicka, L., H. Dong, M. Mittmann, M.H. Ho, and D.J. Lockhart.** 1997. Genome-wide expression monitoring in *Saccharomyces cerevisiae*. Nat. Biotechnol. *15*:1359-1367.

17. **Yang, G.P., D.T. Ross, W.W. Kuang, P.O. Brown, and R. Weigel.** 1999. Combining SSH and cDNA microarrays for rapid identification of differentially expressed genes. Nucleic Acids Res. *27*:1517-1523.

CHAPTER 38

Software for Quantitation and Visualization of Expression Array Data

David Strehlow

Boston University Medical Center, Boston, MA, USA

Reprinted from BioTechniques 29:118-121 (July 2000)

ABSTRACT

Software is described that facilitates the analysis of phosphoimages from large array hybridizations. The Macintosh® PowerPC™-compatible application and its manual are available at no charge from http://people.bu.edu/strehlow. The software is compatible with both custom formats and array filters from three commercial manufacturers. It allows the rapid quantitation of every spot on images of hybridizations to large arrays. The user drags grids of squares over the spots on the image to define the coordinates of each spot, then aligns and edits the position of the grid. The software then corrects the positions as necessary and quantitates up to 27 000 spots per image. It stores the numerical values for each signal in a format called the fingerprint file. Fingerprint files can be directly averaged or compared, allowing the user to find mean values or differences in data from independent hybridization experiments. Data can be recalled from the fingerprint file and can be output in a variety of spreadsheet formats with several options for background correction. Finally, the software offers an output format that allows the convenient visualization of data points using animated, three-dimensional graphs.

INTRODUCTION

Large filter membranes containing thousands of expressed sequence tags or known cDNAs have recently been made available from several commercial vendors. Hybridization to these filters allows the determination of gene expression patterns. However, the full use of this information requires the ability to catalog numerical values for the expression of each clone on the filter. With proper computer software, it is possible to rapidly compare the expression levels of tens of thousands of genes in a given tissue. Also, it should be possible to compare expression levels of thousands of genes at many stages during developmental programs, during oncogenesis and during progression of other disease states.

BioComputing: Computer Tools for Biologists
Edited by Stuart M. Brown

Commercial software is generally expensive, does not facilitate comparisons of whole expression patterns using pixel data and is limited in the filter types it will analyze. Many investigators perform only a small number of hybridizations and so find it difficult to justify the purchase of expensive software that might be used only once or a few times. Other free or less expensive software requires that the user compare images by superimposing one image on top of another and subtracting the pixel values. This superposition method requires that the images be identical trapezoids and that probe-specific activity be the same in all experiments. This is rarely the case, however.

A better method for comparing data from different hybridizations is not to compare images directly, but rather to quantitate each signal and compare the relative mathematical values for each clone on a pairwise basis. Since array filters typically contain thousands of data points, such software should accommodate large arrays of numerical values. DataMachine handles these large arrays and stores the clone pixel values with corresponding position data in a format called fingerprint files.

DataMachine Software

DataMachine allows the user to (a) open 16-bit phosphoimage data files; (b) quantitate total numerical pixel values representing any individual clone; (c) export total numerical pixel values of a user-determined number of clones with a variety of background correction methods; (d) find average and difference data from independent hybridizations; (e) export pixel values in a format that allows visualization of the data in 2-D spreadsheet formats or animated 3-D graphs; and (f) generate mapping functions to group data in 3-D space Mapping functions can then be applied to independent hybridization images to distinguish and display differentially expressed clones.

For (b), the total numerical pixel value is the sum of all pixels that contribute to the

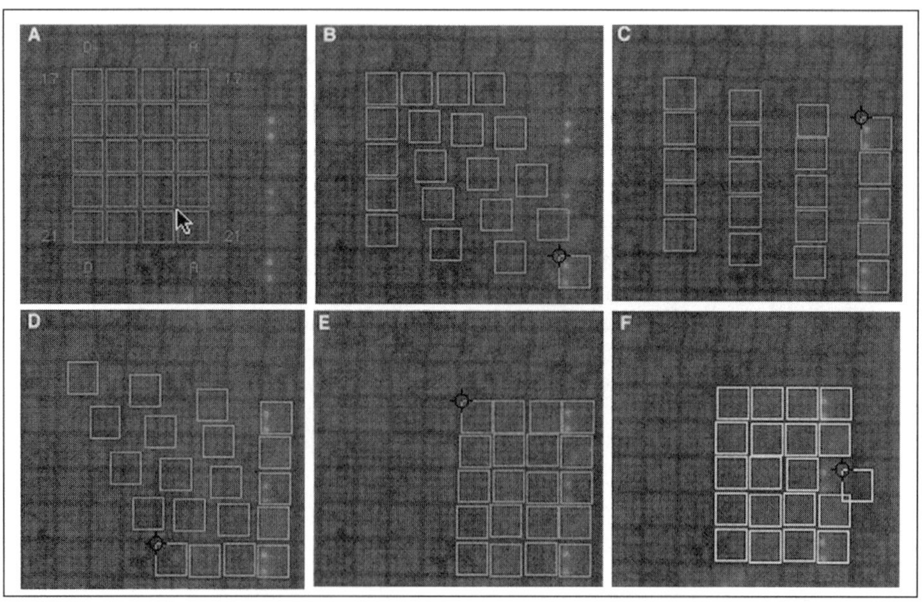

Figure 1. Dragging and aligning a DataMachine grid over an array phosphoimage (see text for details).

spot image of a single clone in a phosphoimage. This value is limited to the maximum size of an imaginary rectangle that surrounds each spot but does not overlap neighboring spots. The actual size of the rectangle depends on the filter manufacturer. If a unique clone is represented by two spots on the filter, the total numerical pixel value for each spot is found, and then the two values are averaged. See the Manual for further details.

The DataMachine version 1.0 software package will open and analyze industry-standard 127-dpi, 16-bit TIFF phosphoimages. [This includes Molecular Dynamics (.GEL) images (Amersham Pharmacia Biotech, Piscataway, NJ, USA) as well as Packard Cyclone images and others.] DataMachine will accommodate several types of filters. It has preset values that allow it to catalog clones from filters made by Genome Systems (Incyte Genomics, St. Louis, MO, USA), Clontech Laboratories (Palo Alto, CA, USA) and Research Genetics (now Invitrogen, San Diego, CA, USA). It can also be tailored to accommodate filters with a user-defined number of fields, rows, columns and spot sizes. Thus, it should work with nearly every phosphoimage and filter array configuration.

The DataMachine Method

DataMachine relies on a combination of user input and digital image recognition to identify the positions of each clone spot. The user first defines the type of filter being analyzed. If it is one of the three commercially available types recognized by DataMachine, then the software automatically identifies the number of fields, rows and columns of clones as well as the size of each spot on the filter. Alternatively, the user can customize these values to allow the analysis of filters of up to 30 fields, each containing 30 rows and 30 columns. Next, the user drags a grid of boxes (up to 30 × 30 boxes at one time) over the clones to define their positions (Figure 1A). The user can stretch the grid at each corner (Figure 1, B–E) or alter the positions of grid boxes individually (Figure 1F). Once the grid is defined, it can be fixed in position (Figure 2) and the process continued until the coordinates of all clones have been entered. Fixing a grid in position can, at the user's option, invoke computer coordinates correction. If this function is invoked, the user-designated clone positions may be altered by as much as ± 3 pixels horizontally or vertically to optimize the signal from any significant clones in the image. This function can be used to generate more consistent coordinates from several similar images.

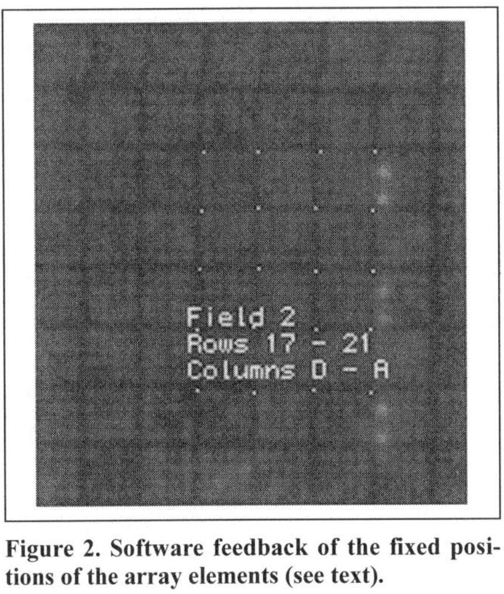

Figure 2. Software feedback of the fixed positions of the array elements (see text).

Once all grid overlays have been positioned, the user can generate the fingerprint file. A new window will then appear that represents the clones in the image that have been entered into the fingerprint file as well as any that are not represented. The user can add clones or redefine their position by editing this file at any time. When fingerprint files have been generated for each phosphoimage, they can be combined to generate average or difference fingerprint files. In the former case, as many as nine images can be combined to find the average expression data at each clone position for all experiments. In the latter case, relative local values for each clone on each filter are automatically generated, and then differences between the relative values on any two fingerprint files can be found. The user can optionally ask that only clones that are significant on one of two filters be output. This option allows rapid identification of differentially expressed clones.

The data from either a mean comparison or a difference comparison is saved in a new fingerprint file. Values from original, mean or difference fingerprint files can be output in spreadsheet format. The user can output any number of values, from a sample of the top values to all 27 000 values in a custom image. The output is automatically formatted so that each clone pixel value is presented next to its field, row and column designations. Output can be ranked according to either pixel value or position in the filter (field 1, row 1, column 1 being first and so on). Formatting by position allows the user to directly compare all clones on two filters by pasting the data to neighboring columns in a spreadsheet.

The user can also specify the type of background correction requested. Background correction options include using a blot average (or global average), a local average or no correction (raw data). Local average is useful for finding corrected values of clones on blots with nonuniform backgrounds. Raw data can be output and then corrected in spreadsheet format, if desired, using a selected clone or clones as standards on the filter. Commercial services generally find the average value of all pixels in the phosphoimage and use this value as the background. As long as the blot has a uniform background, the preference in our laboratory is to output and save the raw data, which we can then manipulate with background correction later. The user can also choose an arbitrary position on the blot as background. If the user samples the value of an arbitrary position on the blot, this value can then be subtracted from all raw values using a spreadsheet program.

Comparing Expression Fingerprinterprints in 3-D

In addition to spreadsheet format, the user can request that the data be exported in a form suitable for display in a 3-D graph (Figure 3). The 3-D format is useful for representing data from a large data set. DataMachine will output the top clone values in a file ready for importing into a free, limited version of the shareware Pro Fit. Viewing data in 3-D requires Pro Fit version 5.1, a trial version of which can be downloaded free from http://www.quansoft.com/index.html. The required 3-D module that plugs into Pro Fit to give it 3-D capabilities is available free at ftp://ftp.quansoft.com/pub/customer/quansoft/modules/Plot3D_210.hqx. Pro Fit will import the data and display it as an animated 3-D graph. DataMachine automatically sets the color of each clone in the 3-D graph according to the clone's pixel values. Red spheres in the graph are the most intense clones, blue are intermediate and green are the weakest. The position of each clone in the 3-D graph is determined by the position of the clone in the filter (e.g., filter position: field 1, row 1, column 1 translates to graph coordinates X, Y, Z = 0, 0, 0; filter position: field 1, row 1, column 2 translates to graph coordinates X, Y, Z = 1, 0, 0).

DataMachine additionally allows comparison of expression data sets in 3-D. At the user's option, DataMachine will group the highest value clones from one hybridization experiment and represent these near the origin in a 3-D graph. The functions used to group the highest clones, the mapping functions, can be saved and applied to a second hybridization experiment. High clones that are the same in both experiments will be displayed near the origin for both experiments. High clones in the second experiment that are different from high clones in the first experiment will be displayed at isolated positions away from the origin (compare Figures 4A and 4B). Thus, differentially expressed clones are readily visualized in the animated 3-D space. DataMachine also contains a utility for converting from the XYZ coordinates in 3-D space to the field, row and column designations on the original filter. Therefore, the positions of clones visualized in 3-D space can be readily identified on the hybridization filter.

DISCUSSION

DataMachine has three valuable characteristics. First, it simplifies the task of quantifying array data. Second, it inserts numerical values for expression data in a file format that allows calculation of averages and differences in independent experiments. Third, it facilitates visualization of array data from either 2-D spreadsheet formats or animated 3-D formats.

ACKNOWLEDGMENT

The author would like to thank the Scleroderma Foundation for its generous ongoing support.

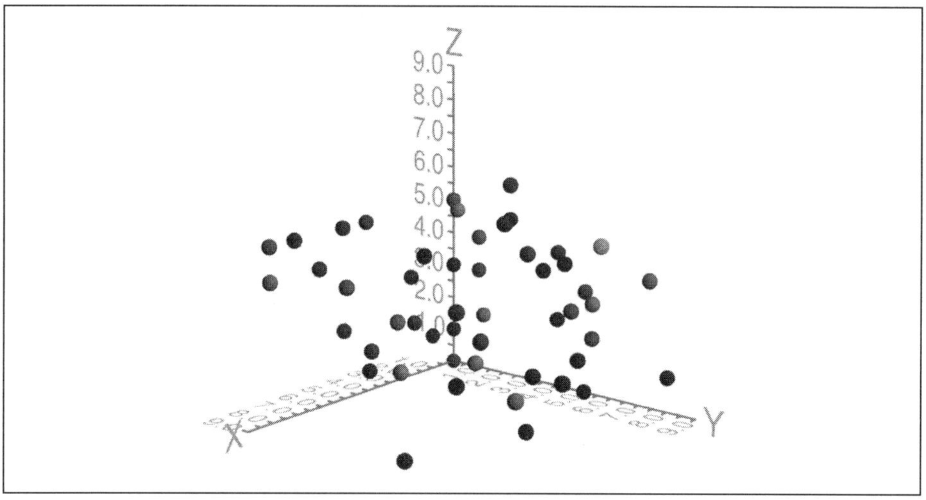

Figure 3. Representation of the high clones from an array hybridization in three dimensions (see text).

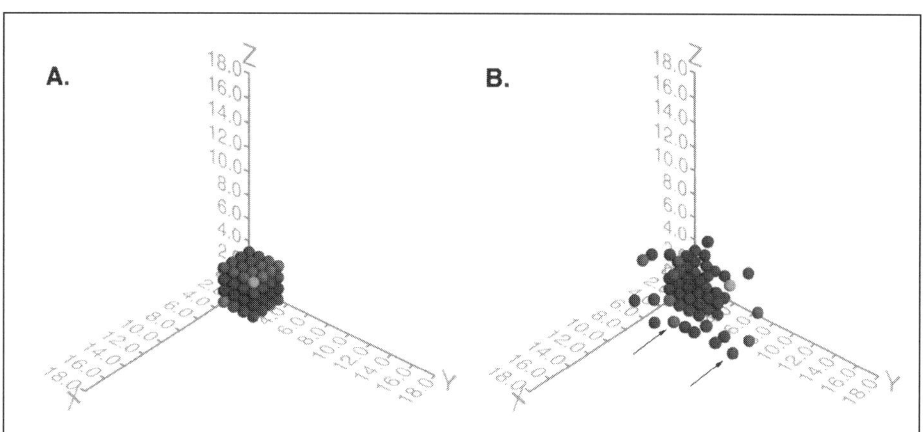

Figure 4. Comparison of high clones from two different array hybridizations after applying the same grouping function to each array and plotting in three dimensions (see text).

Update to: Software for Quantitation and Visualization of Expression Array Data

David Strehlow

Boston University Medical Center, Boston, MA, USA (strehlow@bu.edu)

DataMachine has been downloaded several hundred times, and many users have offered generous feedback. In response to their remarks, the program has improved substantially, and two new releases were posted since the original publication. The most substantial change was the ability to import a wide variety of tagged image format (tif) files. Thus, since a large number of file types can be imported into Adobe® Photoshop® and then saved in 16-bit tif format, nearly any image format can be dealt with using DataMachine after conversion to a tif file. A number of smaller bug fixes have been incorporated. These include fixes to the statistical calculations for background subtraction. Finally, a new version of the manual was written to reflect the changes and clarify some of the features of DataMachine.

CHAPTER 39

Microarray Assistant Clone Organizer and Array Simulator

Ramaswamy Anbazhagan

The Johns Hopkins University School of Medicine, Baltimore, MD, USA

Reprinted from BioTechniques 32:1398-1402 (June 2002)

ABSTRACT

Microarrays are extensively used in molecular biology experiments. While several vendors offer microarrays on a variety of platforms, many researchers prefer to use custom microarrays with a selected list of clones for their experiments. Many research centers have established core facilities for the production of custom microarrays. Microarray production involves a number of steps, including maintaining a master list of stock clones, selecting required clones for custom microarrays, subculturing selected clones, amplifying inserts, recording results, and identifying the orientation of clones in the microarray. We have created a simple, user-friendly, and versatile Microsoft® Excel® spreadsheet-based software, Microarray Assistant, *which can assist the user in all the steps of microarray design and synthesis. In addition, the program gives options to insert, delete, or interchange clones during various steps. The program also gives a visual picture of the locations of the clones in the plates, as well as in the microarray. The program can also be used to assist in the transfer of clones between plates of different configuration.*

INTRODUCTION

Microarrays are extensively used in molecular biology experiments. Commercial vendors market different kinds of microarrays. While many vendors incorporate only named genes in their microarrays, others include a mixture of named genes and expressed sequence tags. Microarrays with a small number of targeted lists of genes from different gene families or specific biological pathways have been marketed by some other vendors. However, many researchers like to use their own custom list of clones in their microarray based on the area and scope of their study. Many research centers have established core facilities for the production of custom microarrays.

The major tasks of a microarray facility include maintaining a master list of stock

BioComputing: Computer Tools for Biologists
Edited by Stuart M. Brown

clones, selecting required clones for custom microarrays and subculturing them (inoculating the stock cultures into new culture tubes or plates and regrowing them), performing PCR amplification of the inserts, recording the subculture and PCR results, and figuring out the orientation of clones in the microarray to correlate with hybridization data.

We have developed a number of programs for molecular biology experiments based on Microsoft® Excel® (1,2). Here we describe a simple, user-friendly, and versatile spreadsheet-based software, *Microarray Assistant*, which can assist the user during different stages of microarray design, synthesis, and analysis. In addition, the program gives options to insert, delete, or interchange clones during various steps of the microarray design. The program can also be used to assist in the transfer of clones between plates of different configuration.

MATERIALS AND METHODS

Obtaining and Opening the Program

The program works with Microsoft Excel 2000 and is available free of charge for nonprofit use under a Material Transfer Agreement (MTA) with the Johns Hopkins University School of Medicine. It is also available for exclusive licensing for commercial use. Please e-mail the author to obtain a copy of MTA. The program files will be e-mailed after receiving the completed copy of MTA.

The program files, *maa.xls* and *maa.hlp*, need no special installation. Simply copy these two files to a new directory in your hard drive. The program can be started by opening the program file *maa.xls*. A dialogue box appears, asking to either enable or disable macros. Click on the *Enable Macros* button. When the program opens, a new menu, named *Microarray Assistant*, is added to the program menu bar. This menu has 11 menu items, namely *New Microarray Workbook*, *Selected List*, *Subculture Plates*, *Subculture List, PCR Plates*, *PCR List*, *Microarray Plates*, *Microarray List*, *Make Microarray*, *Final List*, and *Microarray Help*. These menu items can be executed to perform various functions of the program.

Opening a New Workbook

When the program is opened, it automatically creates a customized workbook with 14 sheets. You can also create new customized workbooks by executing the command *New Microarray Workbook* from the *Microarray Assistant* menu. The first 10 sheets of this customized workbook are used for recording, manipulating, and visualizing clones. These sheets are named *Master List*, *Selected List*, *Subculture Plates*, *Subculture List*, *PCR Plates*, *PCR List*, *Microarray Plates*, *Microarray List*, *Microarray*, and *Final List*. The program automatically creates column headings in the *Master List* sheet. The sheet named *Configuration Data* provides a simple form for entering information about plate, pin, and array configuration. The last three sheets, named *Subculture Plate Template*, *PCR Plate Template*, and *Microarray Plate Template*, are meant to provide templates for use. The program automatically enters default data in these template sheets.

Usually, many clones are rearranged and/or replaced in various steps during microarray design, and they may also be transferred from 96-well plates to 384-well plates. The worksheets in the *Microarray Assistant* program are organized in a sequential order to help through these different stages of microarray design and synthesis.

Master List Sheet

This first worksheet, named *Master List*, is used to maintain a list of stock clones (Figure 1). This sheet also assists in selecting clones for making custom microarrays. The first cell in the first column of this sheet (A1) displays the label "Select" and the cell below that shows a "0". The remaining cells in this column are meant for marking the clones that would be selected for subculturing. The next three columns (B, C, and D) are meant for recording information about plate, row, and column numbers of the clones in the stock plates, and column E is for recording clone identification numbers (Clone ID). In addition to these five columns, the user can fill up any number of additional columns with data pertaining to the clones, but they should be added only after the Clone ID column. Also, it is mandatory that Clone IDs should be unique for each clone in the master list. In case there are duplicate clones in the list, please make sure that each one is given a unique Clone ID.

The user can select the clones for a custom microarray by putting a mark in column A in the corresponding row. Either a single character or multiple characters (alphanumerical) can be used to mark the selected clones. The cell A2 initially shows a "0", indicating that no clones have been selected. This cell is automatically updated by the program each time a clone is marked for selection or unmarked by the user. This cell indicates the total number of clones marked by the user at any given time.

Configuration Data and *Plate Template* Sheets

Before proceeding to the next step, have a brief look at the *Configuration Data* sheet

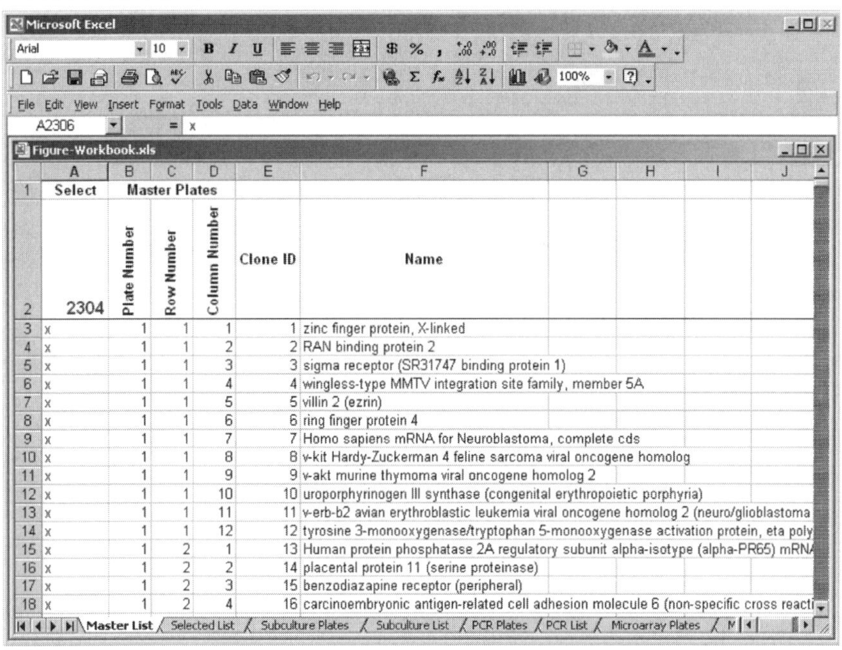

Select	Plate Number	Row Number	Column Number	Clone ID	Name
x	1	1	1	1	zinc finger protein, X-linked
x	1	1	2	2	RAN binding protein 2
x	1	1	3	3	sigma receptor (SR31747 binding protein 1)
x	1	1	4	4	wingless-type MMTV integration site family, member 5A
x	1	1	5	5	villin 2 (ezrin)
x	1	1	6	6	ring finger protein 4
x	1	1	7	7	Homo sapiens mRNA for Neuroblastoma, complete cds
x	1	1	8	8	v-kit Hardy-Zuckerman 4 feline sarcoma viral oncogene homolog
x	1	1	9	9	v-akt murine thymoma viral oncogene homolog 2
x	1	1	10	10	uroporphyrinogen III synthase (congenital erythropoietic porphyria)
x	1	1	11	11	v-erb-b2 avian erythroblastic leukemia viral oncogene homolog 2 (neuro/glioblastoma
x	1	1	12	12	tyrosine 3-monooxygenase/tryptophan 5-monooxygenase activation protein, eta poly
x	1	2	1	13	Human protein phosphatase 2A regulatory subunit alpha-isotype (alpha-PR65) mRN
x	1	2	2	14	placental protein 11 (serine proteinase)
x	1	2	3	15	benzodiazapine receptor (peripheral)
x	1	2	4	16	carcinoembryonic antigen-related cell adhesion molecule 6 (non-specific cross reacti

Figure 1. *Master List* sheet of *Microarray Assistant*. Note the clones selected with an "x" in column A and the total number of selected clones displayed in cell A2.

and the three template sheets towards the end of the workbook. The *Configuration Data* sheet displays a user form that has to be filled up by the user with plate, pin, and array configuration data (Figure 2). The total number of subculture plates, PCR plates, and microarray plates are to be manually entered by the user in the appropriate columns. The program automatically enters a set of default row and column values, 8 and 12 respectively for these plates. These values can be left as such or can be modified by the user appropriately. For example, if the user performs PCR in a 384-well plate, then the data regarding PCR plate in the *Configuration Data* sheet and the corresponding template in the *PCR Plate Template* sheet should be modified accordingly. The form also provides space for entering data related to pin and array configuration.

The plate template sheets (*Subculture Plate Template*, *PCR Plate Template*, and *Microarray Plate Template*) are meant to provide templates to the program to indicate the order in which the clones should be transferred from one plate to the other. Each of these templates represents four 96-well plates, and they are automatically filled up with a set of numbers from 1 to 384. If the user performs all the steps in 96-well plates, then these templates can be left unchanged. If the user transfers the clones from a 96-well plate to either 384-well plate or any other plate format, then the corresponding template should be modified to indicate the order of transferred clones in the new plate. When the program performs transfer of clones between different plates, it reads this template and performs the transfer in the same orientation as indicated in the template. Some of the most commonly used templates are included in the help file. These can be copied and pasted into the template sheet, or the user can create new templates.

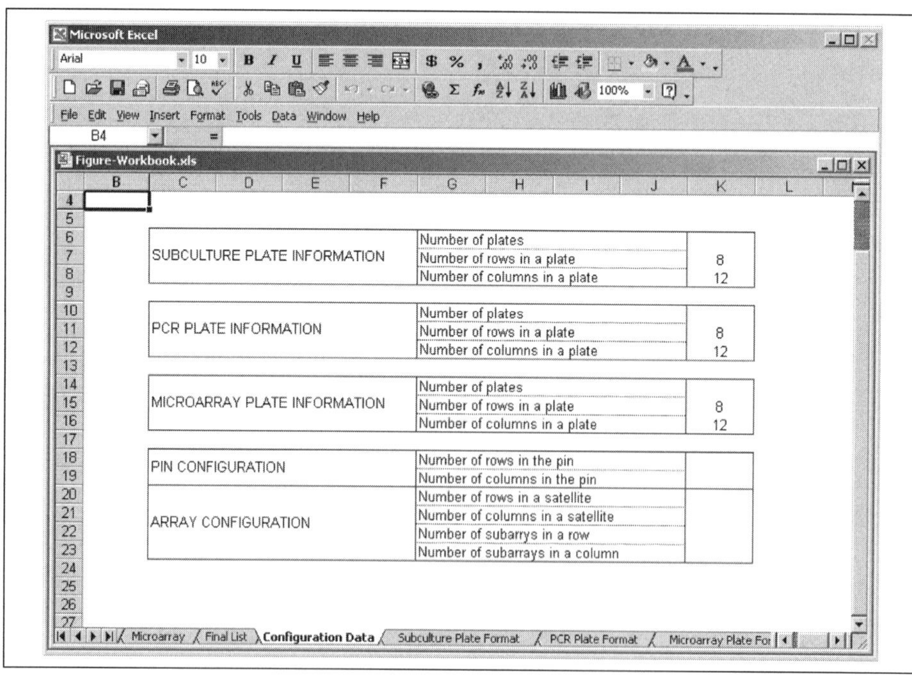

Figure 2. ***Configuration Data*** **sheet showing the user form filled up with plate, pin, and array configuration data.**

Selected List Sheet

After marking the required number of clones in the *Master List* sheet, go to the *Microarray Assistant* menu and execute the menu item *Selected List*. After a brief delay, the program activates the *Selected List* sheet. This sheet now displays all the selected clones copied from the *Master List* sheet. In addition, the program inserts four new columns just before the Clone ID column. Three of these new columns (E, F, and G) are meant for recording the position of clones in the subculture plates (plate, row, and column numbers). The program automatically fills up these columns based on the subculture plate configuration data from the *Configuration Data* sheet. The user can organize the selected clones in the subculture plates using the values in these columns. After the list of selected clones is finalized, execute the command *Subculture Plates* from the *Microarray Assistant* menu. The program copies all Clone IDs from the *Selected List* sheet into the *Subculture Plates* sheet.

Subculture Plates Sheet

The *Subculture Plates* sheet now displays the Clone IDs of the selected list of clones in a row and column format reflecting their position in real plates. In addition, adjacent plates are colored differently and marked with an outline around the plates (Figure 3). In addition, plate numbers are displayed on the right side of each plate. This sheet can be printed out and kept on the bench side as a guide while inoculating clones for subculture.

The user can add, delete, or interchange clones in this *Subculture Plates* sheet. How-

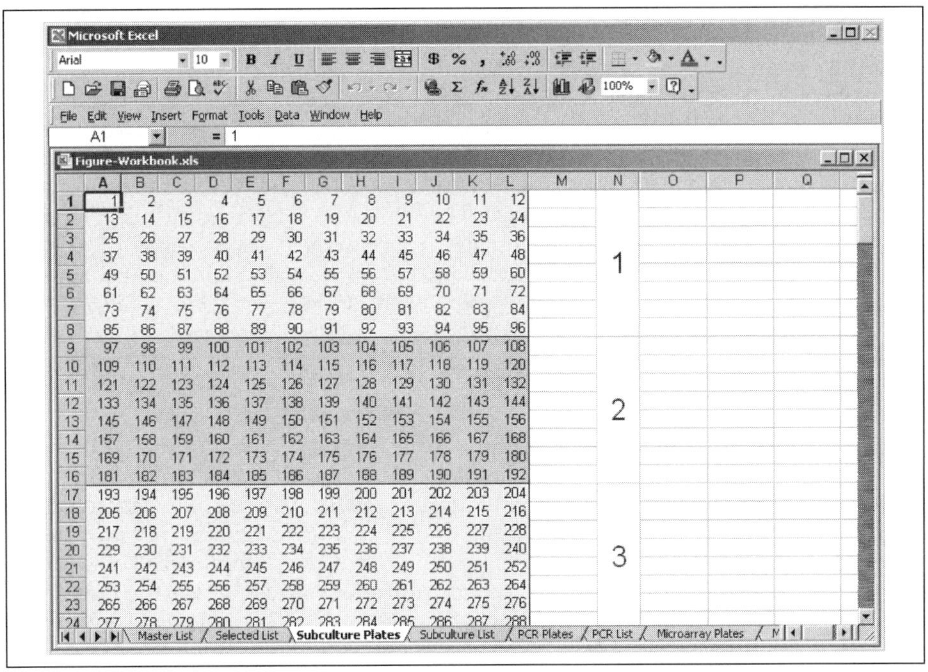

Figure 3. *Subculture Plates* sheet showing the Clone IDs displayed in a plate configuration. Note the different background shades for adjacent plates and the plate numbers on the right side.

ever, make sure that the plates maintain their continuity to each other (i.e., there are no empty rows between any of the plates). None of the wells in the plates should be left empty, and also there should be no duplicate IDs. If you want to leave some wells empty, fill up the wells with Clones IDs such as Empty1, Empty2, etc. After culturing the clones, execute the command *Subculture List* from the *Microarray Assistant* menu. After a brief delay, you will see that the Clone IDs from the *Subculture Plates* sheet and their corresponding plate, row, and column information from the *Master List* sheet is copied into the *Subculture List* sheet.

Subculture List Sheet

Before copying data into the *Subculture List* sheet, the program checks to make sure that there are no duplicate Clone IDs or empty wells in the *Subculture Plates* sheet. If there are duplicate Clone IDs or empty wells, then the program displays an error message. The program reads each Clone ID in the *Subculture Plates* sheet and copies the relevant data from the *Selected List* sheet into the *Subculture List* sheet. In addition, the program adds a new column named "Subculture Result" just before the Clone ID column. This column is meant for recording the results of the subculture.

PCR Plates and *PCR List* Sheets

If the configuration of the subculture plates and PCR plates are the same (i.e., if both the plates are of 96-well format), then the user can proceed with the default values in the *Configuration Data* sheet and *PCR Plate Template* sheet. In other situations (using 384-plate or any other plate format), these values should be modified appropriately. You can use the templates provided in the help file or create a new template for this purpose. Then execute the command *PCR Plates* from the *Microarray Assistant* menu.

The program reads the Clone IDs from the *Subculture Plates* sheet and transfers them to the *PCR Plates* sheet. The transfer is performed in accordance with the information provided in the *Configuration Data* sheet and *PCR Plate Template* sheet. After transferring the Clone IDs, the program marks the plates with different colors and draws an outline around each plate as described earlier. At this stage, the user again has the option to add, delete, or interchange the clones in the *PCR Plates* sheet. However, precautions should be taken to avoid duplicates and empty wells, as described previously. After performing PCR amplification, execute the command *PCR List*. The program reads the Clone IDs in the *PCR Plates* sheet and copies all relevant data from the *Subculture List* sheet into the *PCR List* sheet. In addition, the program inserts three more columns for PCR plate data and another column for recording the PCR results.

Microarray Plates and *Microarray List* Sheets

While in most cases the PCR plates with their products are directly processed for arraying, the PCR products are occasionally transferred to new plates. The two sheets *Microarray Plates* and *Microarray List* are meant to facilitate recording data relevant to the final plates used for arraying. The corresponding commands *Microarray Plates* and *Microarray List* can be executed to perform these functions. The program uses the data provided in the *Configuration Data* sheet and *Microarray Plate Template* sheet to perform these functions. All options for altering the Clone IDs in the *Microarray Plates* sheets and restrictions are similar to the ones described above.

Microarray and *Final List* Sheets

Before proceeding to the next step, make sure that relevant pin configuration (number of rows and columns in the spotting pin) and array configuration data (number of rows and columns in a satellite, number of subarrays in a row, and number of subarrays in a column) are properly entered in the *Configuration Data* sheet. Then, execute the menu item *Make Microarray* from the *Microarray Assistant* menu. After a brief delay, the *Microarray* sheet displays the clone IDs of the PCR products in the final array format. The position of each Clone ID in this sheet represents its position in the microarray. In addition, the program marks the satellites with different colors and draws an outline around each subarray. Before executing this command, the program checks to see if the plate, pin, and array configurations match with each other and displays an error message if there is any mismatch.

To get a final list of all the clones in the microarray and data regarding their location, execute the menu item *Final List* from the *Microarray Assistant* menu. This will create a list of all clones in the microarray in the *Final List* sheet with complete information about each spot on the microarray, including the location of the respective clones in the master plate, subculture plate, PCR plate, microarray plate, and microarray (Figure 4).

The program comes with its own help file *maa.hlp*, which can be opened by executing the *Microarray Help* command from the *Microarray Assistant* menu or by directly clicking on the help file. The first page of the help file shows a list of topics for which help is available, and individual topics can be visualized by clicking on the topic names.

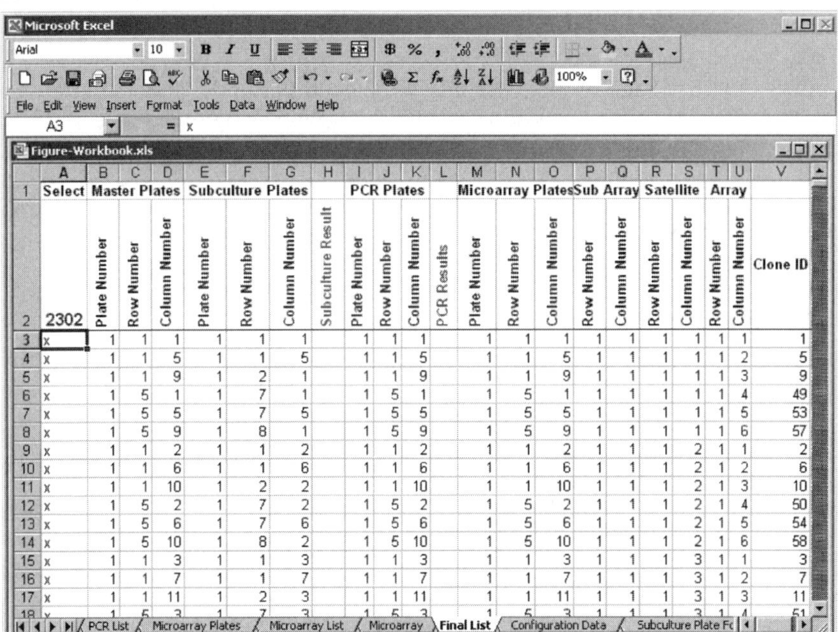

	A	B	C	D	E	F	G	H	I	J	K	L	M	N	O	P	Q	R	S	T	U	V
1	Select	Master Plates			Subculture Plates				PCR Plates				Microarray Plates			Sub Array		Satellite		Array		
2	2302	Plate Number	Row Number	Column Number	Plate Number	Row Number	Column Number	Subculture Result	Plate Number	Row Number	Column Number	PCR Results	Plate Number	Row Number	Column Number	Row Number	Column Number	Row Number	Column Number	Row Number	Column Number	Clone ID
3	x	1	1	1	1	1	1		1	1	1		1	1	1	1	1	1	1	1	1	1
4	x	1	1	5	1	1	5		1	1	5		1	1	5	1	1	1	1	1	2	5
5	x	1	1	9	1	2	1		1	1	9		1	1	9	1	1	1	1	1	3	9
6	x	1	5	1	1	7	1		1	5	1		1	5	1	1	1	1	1	1	4	49
7	x	1	5	5	1	7	5		1	5	5		1	5	5	1	1	1	1	1	5	53
8	x	1	5	9	1	8	1		1	5	9		1	5	9	1	1	1	1	1	6	57
9	x	1	1	2	1	1	2		1	1	2		1	1	2	1	1	1	2	1	1	2
10	x	1	1	6	1	1	6		1	1	6		1	1	6	1	1	1	2	1	2	6
11	x	1	1	10	1	2	2		1	1	10		1	1	10	1	1	1	2	1	3	10
12	x	1	5	2	1	7	2		1	5	2		1	5	2	1	1	1	2	1	4	50
13	x	1	5	6	1	7	6		1	5	6		1	5	6	1	1	1	2	1	5	54
14	x	1	5	10	1	8	2		1	5	10		1	5	10	1	1	1	2	1	6	58
15	x	1	1	3	1	1	3		1	1	3		1	1	3	1	1	1	3	1	1	3
16	x	1	1	7	1	1	7		1	1	7		1	1	7	1	1	1	3	1	2	7
17	x	1	1	11	1	2	3		1	1	11		1	1	11	1	1	1	3	1	3	11
18	x	1	5	3	1	7	3		1	5	3		1	5	3	1	1	1	3	1	4	51

Figure 4. *Final List* sheet showing data regarding location of each spot in the microarray. In addition, data regarding the location of each clone in the master plate, subculture plate, PCR plate, and microarray plate are also displayed in this sheet.

DISCUSSION

The major advantages of the program are that it employs a well-known interphase of the Excel spreadsheet, and it is very user friendly. Thus, the learning curve required to master the program is very much reduced. The program combines all relevant data required for the microarray design and synthesis in a single workbook. The program helps to prepare different lists and also gives a visual display of all plates, as well as the microarray. The program also gives options to insert, modify, delete, or replace the clones during various steps.

In addition to the commands provided by the *Microarray Assistant* menu, the user can also make use of all the built-in Excel commands. For example, if the *Microarray* sheet is too large to view, the user can execute the built-in "*zoom*" command from the Excel menu to visualize a large area. Additional enhancements for easy visualization of the data in different worksheets can be achieved using different background colors, font colors, or by using special outlines.

The program might be a little slow in executing commands that create Subculture List, PCR List, and Microarray List. This is because the program searches and combines the relevant data while executing these commands. Users are advised to try these commands with a small number of clones (1000 or 2000) to get an idea about the time it takes to complete the execution of these commands. Attempts will be made to improve the speed of these commands in future versions.

Researchers can utilize the Microarray Assistant to integrate their work related to microarray design and synthesis. They can use the program if they find it easy to use when compared with commercial software or if some of the features found in *Microarray Assistant* are not available in their commercial software. This program also will be very useful to researchers who perform their procedures manually (e.g., transfer of clones from 96-well plates to 384-well plates).

Further improvements of the program will depend on the feedback from the users. While critical suggestions would be incorporated, the emphasis on a simple, user-friendly interphase will be maintained.

ACKNOWLEDGMENTS

I thank Dr. Edward Gabrielson for his support and encouragement. I also thank Drs. Qingbin Guo and Ie-Ming Shih for useful discussions.

REFERENCES

1. **Anbazhagan, R. and E. Gabrielson.** 1999. Spreadsheet-based program for alignment of overlapping DNA sequences. BioTechniques *26*:1180-1185.
2. **Anbazhagan, R., J. Herman, K. Enika, and E. Gabrielson.** 2001. Spreadsheet-based program for the analysis of DNA methylation. BioTechniques *30*:110-114.

Update to: *Microarray Assistant* Clone Organizer and Array Simulator

Ramaswamy Anbazhagan and Edward Gabrielson

Johns Hopkins University School of Medicine, Baltimore, MD, USA (anba@jhmi.edu)

Microarray production involves a number of steps including maintaining a master list of stock clones, selecting required clones for custom microarrays, subculturing selected clones, amplifying inserts, recording results, and indentifying the orientation of clones in the microarray. Although commercial software products are available to assist these functions, they are expensive and also require additional training for use. I have created a simple user-friendly versatile Excel spreadsheet-based software, *Microarray Assistant*, which can assist the user in all the steps of microarray design and synthesis. In addition, the program gives options to insert, elete, or interchange clones during the various steps. The program also gives a visual picture of the location of the clones in the plates, as well as in the microarray. The program can be used to assist in the transfer of clones between plates of different configuration. Te program works with Microsoft Excel 2000 and is available free of charge for faculty members of academic institutions under a material transfer agreement with the Johns Hopkins University School of Medicine. Instructions for getting the program can be obtained by sending an e-mail (anba@jhmi.edu) to the author.

Section V

Phylogenetics and Genotyping

Along with the boom in gene expression studies, databases of complete genome sequences have also given a huge boost to the study of genetic variation, both between species (phylogenetics) and within a single species (genotyping). Once again, these studies require computer analysis of large data file. For these types of studies, free software developed by scientists is as popular as commercial programs. Articles in this section include descriptions of specific software for building phylogenetic trees and other forms of similarity indexes from sequence data, DNA fingerprints and VNTR probes, detection of single base sequence variants in sequence databases, as well as general overviews of the use of free phylogenetic tools on the Internet.

CHAPTER 40

Phylogenetics—A Web of Trees

Tim Littlejohn

eBioinformatics, Sydney, Australia

Reprinted from BioTechniques 29:482-484 (September 2000)

The quantity and diversity of genome sequence data continues to grow rapidly, thanks to ever-cheaper data production technologies and the relentless drive of the global genome effort. As a consequence, opportunities continue to expand around the use of these resources for richer comparative sequence analysis to discover the mode and nature of evolutionary change. In its latest release (June 15, 2000), the GenBank® database contained over 8.6 billion bases of sequence from over 75 000 different organisms (http://www.ncbi.nlm.nih.gov). While this information is by no means equally distributed across the species represented in the database, it does represent a rich repository of data ripe for phylogenetic analysis. The tools for such analysis are not just of interest to evolutionary biologists; the patterns of similarities between sequences they can reveal are also used in population genetics and molecular diagnostics, for example. Numerous "anonymous" Internet resources (sites that can be accessed without a username or password) exist for phylogenetic analysis. Also, Internet-based application service providers (ASPs) are particularly suited to phylogenetic analysis because they store the data and the results of analyses—along with the tools and databases necessary to analyze the data—with the often intense computing resources needed to run phylogenetic investigations.

Collecting Data

What are the best ways to collect this data so that evolutionary information can be mined from it? It is critical that the data being analyzed are homologous, that is, related by evolutionary descent. Database records can be retrieved using powerful text-searching engines such as Entrez (http://www.ncbi.nlm.nih.gov/Entrez/) and SRS (http://srs.ebi.ac.uk/) on the Web, or similar systems in bioinformatics' ASP systems such as the TextSearch function in BioNavigator (http://www.bionavigator.com) or similar academic ASPs such as those in the EMBnet (http://www.embnet.org). The pitfall with this approach is that it can be difficult to identify homologous sequences using keyword searches because databases frequently contain errors and inconsistencies in how homologous genes are named. Some databases, such as the Organelle genome database GOBASE (http://megasun.bch.umontreal.ca/gobase/gobase.html), Hobacgen (http://biom3.univ-lyon1.fr/databases/ hobacgen.html) and Hovergen (http://biom3.univ-lyon1.fr/databases/hovergen.html) are ideal for phylogenetic analysis as they provide collections of homol-

BioComputing: Computer Tools for Biologists
Edited by Stuart M. Brown

ogous sequences for organelles, bacteria and vertebrates for such in-depth research.

Sequence-similarity searching methods can also be used to retrieve homologous sequences. Tools like BLAST are ideally suited to this function (see BLAST on the Web by Bruno A. Gaëta, BioTechniques, Vol. 28. No. 3, March 2000). This tool can be accessed at sites such as the NCBI (http://www.ncbi.nlm.nih.gov), the Japanese GenomeNet server (http://www.blast.genome.ad.jp), the European Bioinformatics Institute (http://www.ebi.ac.uk) and ASPs such as http://www.bionavigator.com. Organism-specific collections of sequence data—for example, the *Arabidopsis* sequence database at http://www.arabidopsis.org—are an excellent way to collect specific data ready for phylogenetic analysis.

Multiple Sequence Alignments

Once collected, the next step for the analysis of phylogenetic information in molecular sequence collections is to perform a multiple sequence alignment (MSA). This step is critical in phylogenetic analysis because data is typically analyzed down columns of residues across aligned taxa under the assumption that residues in each column are homologous, that is, they share a common history. (See BioTechniques Vol. 29, No. 1, July 2000, for the article, Multiple Sequence Alignment Tools on the Web, by Giles J. Gaskell). The ClustalW multiple sequence alignment tool is one of the most popular and can be reached on the Internet at several sites including the European Bioinformatics Institute (EBI) (http://www2.ebi.ac.ukclustalw/), the Baylor College of Medicine (http://dot.imgen.bcm.tmc.edu:9331/multi-align/Options/clustalw.html) and the Institut Pasteur Software Environment (PISE) in France (http://bioweb.pasteur.fr/intro-uk.html) or ASPs like BioNavigator. Other software tools for aligning and refining alignments can be found elsewhere (see Gaskell's article in Vol 29, No. 2, cited above). Frequently, MSA files need manual fine-tuning as none perform a perfect job at aligning homologous sites. Internet-based tools like CINEMA (http://www.bioinf.man.ac.uk/dbbrowser/CINMA2.1/) and JalView (http://www2.ebi.ac.uk/~michele/jalview/contents.html) offer Java tools for such functions.

Phylogenetic Tree Inference

When a reliable multiple sequence alignment (MSA) file has been generated, the real phylogenetic analysis can begin. The three major methods for phylogenetic inference include maximum likelihood, parsimony and distance methods. Maximum likelihood methods calculate the probability that a given tree will result in the observed MSA. Programs using this approach are typically computer-intensive and have not been implemented on the Internet as widely as those using other methods. Parsimony methods determine the minimum number of changes required to produce the observed MSA. Unlike maximum likelihood and distance trees, parsimony trees do not include branch lengths. Distance methods work by pairwise comparisons of sequences, which infers an evolutionary distance between them, and then derive a tree based on these distances using methods like neighbor joining.

Some MSA programs produce dendrograms (trees indicating how the sequences are clustered, indicating their overall relationships with each other). In fact, ClustalW even allows the user to create Neighbor Joining trees with a bootstrap option. As described above, many Internet resources allow access to ClustalW. Other sequence alignment tools include TreeAlign, a multiple sequence alignment application that builds trees even as it aligns DNA or protein sequences using a combination of distance matrix and approximate parsimony methods. TreeAlign can be accessed through the PISE system (http://bioweb.pasteur.fr/seqanal/phylogeny/intro-uk.html).

The next step in phylogenetic analysis is the application of these MSA files as input data to phylogenetic tree construction software. Integrated ASP systems platforms such as BioNavigator (http://www.bionavigator.com) allow all these analyses to be conducted on one site using a seamless flow of data through pipelines offering data collection from public and private data collections, pruning these sets, aligning and then phylogenetic tree inference. Other sites such as the Pasteur Institute's PISE system (http://bioweb.pasteur.fr/intro-uk.html) also facilitate sequence collection and alignment and then pass on these alignments for phylogenetic analysis. The BioNavigator system (http://www.bionavigator.com) further permits these analyses to be recorded as histories and then re-run automatically as autoprotocols, using other query sequences.

There are over 175 phylogenetic programs available (further discussion is beyond the scope of this review) that can be downloaded from the Internet or run from their Web sites directly. Many of these programs are specific to MacintoshÒ or MicrosoftÒ Windows, and rely heavily on their operating system-specific user interfaces and so are not suitable for integration into a standard Web interface. However, as pointed out in an earlier On-Ramp article, all tree-building programs make a certain number of assumptions and it is critical that users understand these limitations before launching into any analyses.

Some packages (for example, PHYLIP) contain applications that allow analysis using all the major methods for molecular (nucleotide and protein) sequence data but also other data types. The PHYLIP package allows analysis of restriction site, gene frequency, quantitative characters and discrete characters, and can be found at http://evolution.genetics.washington.edu/phylip/programs.html. An excellent summary of these resources can be found at the University of Washington site at http://evolution.genetics.washington.edu/phylip/software.html. Web-based sites allowing online analyses through the anonymous Web are listed at http://evolution.genetics.washington.edu/phylip/software.serv.html. Other lists for phylogenetic software exist at EBI in the BioCatalog (http://corba.ebi.ac.uk/Biocatalog/Phylogeny.html), the "Tree of Life" Project (http://phylogeny.arizona.edu/tree/programs/programs.html) and at the Willi Hennig Society site (http://www.cladistics.org/education.html).

A hallmark of most phylogenetic programs is that they are computationally intensive: as a consequence, there are few public sites that allow phylogenetic analysis because of the large computational requirements. Indeed, as Joe Felsenstein, author of the popular PHYLIP package (http://evolution.genetics.washington.edu/phylip.html) says, "Some sites have also been bold enough to make PHYLIP available on a server basis... I think they're very brave considering how slow some of our programs are". ASP Web providers such as BioNavigator offer the PHYLIP and other packages (e.g., GCG: http://www.gcg.com and fastDNAml:ftp at ftp://rdp.life.uiuc.edu) for phylogenetic analysis and have no restrictions on usage of these packages. PHYLIP is also available on the Web through the WebPhylip interface developed at the National University of Singapore (http://sdmc.krdl.org.sg:8080/~lxzhang/phylip). The PISE system allows access to PHYLIP, fastDNAml, MOLPHY, BIONJ, TREE-PUZZLE, Seq-Gen, TreeAlign, QR2, PhyloQuart, LVB, and BAMBE. Other sites such as the Computational Biochemistry Group of the ETH in Zürich, Switzerland (http://cbrg.inf.ethz.ch/) has the TreGen, AllAll, Multalign and Darwin programs available.

Tree Evaluation and Visualization of Results

When a phylogenetic tree is created, the next step is to evaluate the tree for robustness. The most common method for doing this is bootstrap analysis. Some sites allow bootstrapping using the PHYLIP package (e.g., BioNavigator, WebPhylip and PISE), which essentially involves resampling the dataset and then analyzing the resampled data. Phylogenies that are well supported by the alignment will tolerate variations in the data brought about by the resampling process, although alignments that do not have a consistent evolutionary history along their length will give rise to different trees when small changes are introduced through resampling. In addition, it is often useful to provide a user tree as a guide for phylogenetic analyses. This permits a narrowing down of the tree space that has to be evaluated, an advantage that greatly speeds up the analysis.

User trees can also be used to compare different trees and to evaluate whether they are significantly different from each other, given the experimental evidence. This is particularly important when the "best" tree found by a program appears to contradict another tree derived from different evidence. In such a case, it is important to compare the two trees to ensure that the data clearly supports one tree over the other using a Kishino-Hasegawa-Templeton (KHT) test. The BioNavigator system's Web implementation of PHYLIP allows phylogenetic analysis with user trees as well as tree comparison using the KHT test.

While some phylogeny programs create rooted trees in which the location of the common ancestor is inferred from the data, this inference requires assumptions that are often regarded as unacceptable, such as that of a molecular clock. Without assumptions like these, most programs will generate unrooted trees where only the relative topology of the taxa is known. After producing such a tree it is often necessary to "root" it, based on alternative evidence (for example, morphological evidence).

When trees are created they need to be manipulated and visualized. The FragTree (http://193.50.234.246/~beaudoin/anrs/Tree.html) and Phylodendron (http://iubio.bio.indiana.edu/treeapp/treeprint-form.html) servers allow such visualizations, as does the BioNavigator system. BioNavigator allows online graphics viewers, relabeling and drawing of both rooted and unrooted trees in a range of formats, and basic manipulation of trees (including flipping branches and re-positioning of the root of the tree). BioNavigator also allows the launching of helper applications such as TreeView (http://taxonomy.zoology.gla.ac.uk/rod/treeview.html), which can be launched directly from the browser to add significantly improved interactive tree viewing and manipulation.

Conclusion

There are many steps to phylogenetic analysis requiring databases, local data repositories and applications. Phylogenetic analysis tends to be an iterative, complicated process, and it is essential that the researcher doing the analysis has a good understanding of the methods to use and their potential pitfalls. One of the strengths of the PHYLIP package is its extensive documentation that provides a detailed description of the assumptions that are behind the various approaches. Integrated systems such as BioNavigator are ideal for phylogenetic analysis because they allow a flow of information all the way from database query through to a refined, visualized tree. In addition, sites that provide step-by-step protocols such as those found on BioNavigator greatly assist the new or infrequent user in identifying and choosing the right approach for such complicated analyses.

CHAPTER 41

Phylogenetics on the Web

Stuart M. Brown,

NYU School of Medicine, New York, NY, USA

Reprinted from BioTechniques 27:1146-1148 (December 1999)

Anyone who spends a little bit of time working with DNA and protein sequences quickly starts wondering how these sequences are related. This often leads to the more fundamental question of how the organisms from which these sequences come are related, which is the basis for molecular taxonomy or phylogenetics.

The essence of taxonomy is choosing heritable traits (characters) that accurately represent groups of organisms, i.e., all organisms with feathers and wings are classified as birds. Taxonomy has the mystique of an old discipline, practiced in musty rooms in the sub-basement of museums where white-haired scientists spend their time staring through magnifying glasses at jars of pickled specimens. However, taxonomists were early adopters of the new technologies of molecular biology because DNA (and protein) sequences have many advantages over more classical types of visible taxonomic characters.

- Character states can be scored unambiguously (genotypes can be read directly rather than relying on phenotypes);
- Huge numbers of characters can be scored for each individual;
- Information can be obtained on both the extent and the nature of divergence between sequences (nucleotide substitutions, insertion/deletions or genome rearrangements);
- Comparisons can be made between groups of organisms (populations) that have minimal phenotypic differences;
- Phylogenies can be built for groups of organisms that are so widely diverged that they do not share many morphological traits.

It seems logical that for a given gene, closely related organisms have similar sequences and more distantly related organisms have more dissimilar sequences. However, it is necessary to consider phylogenetic analysis, conducted with molecular data, in the context of the full body of taxonomic and biological knowledge. If calculated correctly, relationships derived from sequence data actually represent the relationships between genes, which is not necessarily the same as relationships between whole organisms. The sequences that were used for data may not have had the same phylogenetic history as the species within which they were contained. Different genes (and different species) evolve at different speeds, and there is the possibility of horizontal gene transfer either by hybridization, vector-mediated DNA movement or direct uptake of DNA.

Within the field of phylogenetics, there are two different, yet hotly contested and philosophically opposed methods of building trees—phenetics and cladistics. These two

BioComputing: Computer Tools for Biologists
Edited by Stuart M. Brown

different approaches to the interpretation of sequence data lead to different computational algorithms and, ultimately, to different phylogenies computed from the same data.

The phenetic approach constructs trees by considering the overall phenotypic similarities among species without trying to understand the evolutionary history that has brought the species to their current phenotypes. Sequence data is used to classify organisms into trees based on the absolute number of characters that they share. Distances between DNA sequences are simply the sum of all base pair differences between two aligned sequences. Distances between aligned protein sequences are calculated using both amino acid identities and similarities (calculated from a scoring table such as the PAM or BLOSUM matrix).

Computer phylogeny programs based on the phenetic model use a matrix of pairwise distances and simple clustering algorithms such as the unweighted pair group method using arithmetic averages (UPGMA) and Neighbor Joining to build trees. These methods build clusters by first grouping pairs of sequences with the fewest differences, adding the next most distant and then repeating this process until all sequences are included.

The cladistic approach is based on the study of clades, which are groups of related organisms. Evolutionary trees are "reconstructed" by considering the various possible pathways of evolution (branching points where groups diverged from a common ancestor) and choosing the best possible tree from among these. Known ancestral relationships are considered as well as current data. Instead of looking at all the phenotypic characters of organisms, cladistics uses only a subset of characters that are uniquely shared by a group of organisms and not present in their ancestors. Computer algorithms based on the cladistic model generally rely on parsimony (the least number of evolutionary changes) or maximum likelihood (the most probable ancestors) methods for the calculation of relationships and the building of trees.

For data such as classical morphological characters and for higher (or perhaps we should say deeper) levels of taxonomy, the cladistic approach is almost certainly superior. However, cladistic methods are often difficult to implement, and require assumptions that are not always satisfied with molecular data. Molecular data make pure "phenotypic" characters since there is no inherent reason for a researcher to think that one mutation is more important or more "ancestral" than another. Phenetic approaches lead to faster algorithms and often have nicer statistical properties for molecular data.

The starting point of a phylogenetic analysis is usually a set of related proteins. It is possible to make phylogenies from DNA sequence, but unless you are dealing with either noncoding sequences or a set of very closely related sequences (intra-species), it is much more informative to work with proteins; much more distant relationships can be analyzed.

Once a list of related proteins is collected, the next step is to build a multiple alignment. It is important to align only homologous regions of the sequences. In some cases, distantly related genes may share homology in some regions, but not in others. Phylogenetic analysis programs require a multiple alignment with no large gaps. It is also extremely important to have all the sequences in the alignment be of the same length. If a list of proteins includes some partial sequences, the decision will have to be made whether to eliminate these shorter sequences or to truncate all of the other sequences and make an alignment of only the region that is shared.

Once the sequences have been aligned, the multiple alignment file becomes the input for a phylogenetic analysis program. Phylogenetics programs are available for both desktop computers (Macs and PCs) and for mainframes. A few computing centers have recently begun to offer free phylogenetics servers on the Web. This is quite brave, considering the potential for these programs to consume huge amounts of computing power if large numbers of long sequences are used. The amount of computation is also dependent on the type of algorithm used: distance methods (UPGMA and Neighbor Joining) are quick, parsimony methods take longer, and maximum likelihood methods are very slow.

The Institut Pasteur, Paris has a PHYLIP server at http://bioweb.pasteur.fr/seqanal/phylogeny/phylip-uk.html

Joaquin Dopazo at the Centro Nacional de Biotecnologia, Madrid, Spain has made a nice server interface for PHYLIP at http://www.cnb.uam.es/cgi-bin/dopazo/PHYLIP/phylip

The Belozersky Institute at Moscow State University in Russia has its own "GeneBee" phylogenetics server at http://www.genebee.msu.su/services/phtree_reduced.html

The last challenge for phylogenetic analysis is to produce a publication-quality printout of the results. The PHYLIP drawtree program is rather poor in this regard, both in terms of the user interface and in the quality of the final product. The Phylodendron Web site is a tree drawing program with a better user interface and many formatting options, including PDF. Whatever program is used, the final tree may need to be redrawn in a graphics program: http://iubio.bio.indiana.edu/treeapp/treeprint-form.html

Phylogenetic analysis of protein families is one of the most challenging and interesting tasks in bioinformatics. This type of analysis is rarely done and, therefore can often make a valuable contribution to a research area. For example, new genes are being discovered on a daily basis, but little is being done to establish the evolutionary relationships between these genes and their homologs. A biologist with an interest in phylogenetic analysis could make a significant contribution to the annotation of these sequences.

CHAPTER 42

Convenient Method to Improve the Graphical Quality of Phylogenetic Trees Computed by the MEGA Program

Rémi N. Charrel[1,2], Philippe De Micco[1], and Xavier de Lamballerie[1]

[1]Faculté de Médecine, Marseille, France and [2]University of Texas, Medical Branch, Galveston, TX, USA

Reprinted from BioTechniques 27:702-704 (October 1999)

Over the last two decades, analyses of nucleotide and amino acid sequences have been increasingly used for the study of evolution and phylogenetic relationships among macro- and microorganisms. Subsequently, numerous phylogenetic programs were developed to help analyze this data by computer-assisted methods.

Among these programs, molecular evolutionary genetics analysis (MEGA) (Pennsylvania State University, University Park, PA, USA) (1) is one of the most popular. Its simple use permits the calculation of genetic distances from nucleotide or protein alignments using pairwise distance, Jukes-Cantor, Tamura, Tamura-Nei, Tajima-Nei, Kimura and gamma distribution algorithms. The construction of phylogenetic trees can be performed using neighbor-joining, unweighted pairgroup method with arithmetic mean (UPGMA) and parsimony methods and tested by bootstrap resampling analysis. These different phylogenetic tools are available in a compact package that can now be downloaded at the following Web site address: http://evolgen.biol.metro-u.ac.jp/MEGA/.

MEGA functions under the MS-DOS environment (or in a DOS window of the Microsoft® Windows® [Microsoft, Redmond, WA, USA] environment). The constructed phylogenetic trees can be either directly printed or exported under a unique ASCII format. In the first case, the trees cannot be saved as computer files nor can they be modified by word- or graphic-processors. In the second case, the ASCII file can be printed using the DOS editor. This permits minor modifications of the alphanumeric characters. The absence of different export formats is limiting, because treatment of the phylogenetic trees is usually necessary to clarify them for the reader of a scientific paper or the audience of a lecture. An excellent program, TreeExplorer, which offers rich tree-handling options and permits the printing of trees under the Windows environment, has recently been made available by Koichiro Tamura and can be downloaded at the aforementioned MEGA Web site address. However, this program is difficult to use if the MEGA program is not run in a DOS window of the Windows environment (which is often necessary to optimize the computer random-access memory [RAM] and work with large input data files).

BioComputing: Computer Tools for Biologists
Edited by Stuart M. Brown

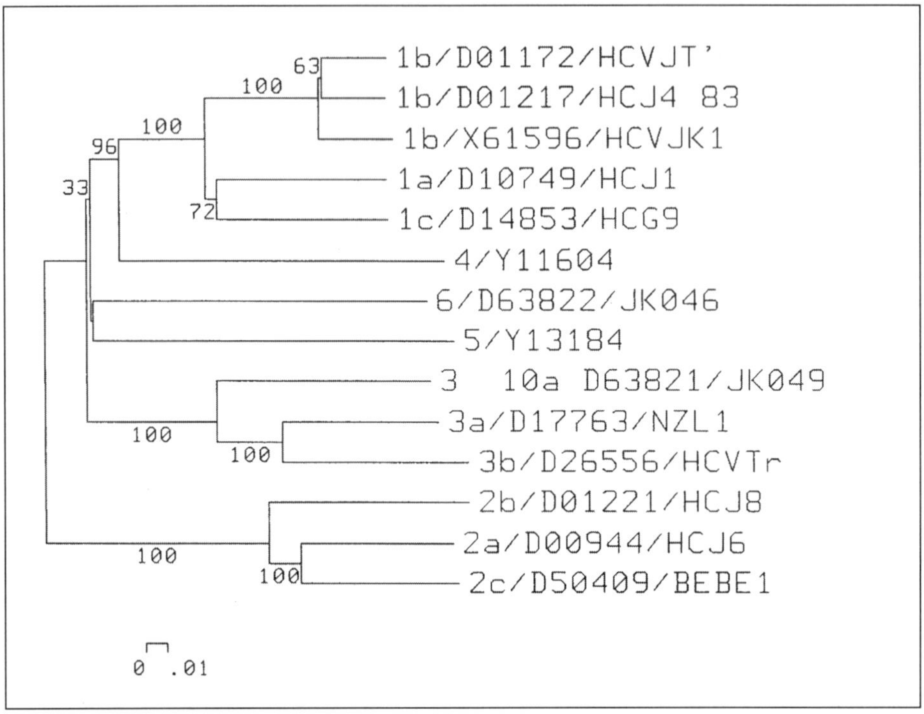

Figure 1. Tree printed directly from MEGA.

We propose here a very simple and rapid method to obtain high-quality phylogenetic trees from the MEGA ASCII exported files and to save tree data on disks. This method is based on the execution of a custom Macro program in Microsoft Word® 7.0 processor (Microsoft) under PC or Macintosh® (Apple Computer, Cupertino, CA, USA) environment. It consists of the replacement of the ASCII characters Ä, Ú, ³, À, ´, Ã and Å by the following characters: ChrW(9472), ChrW(9484), ChrW(9474), ChrW(9492), ChrW(9508), ChrW (9500) and ChrW(9532), respectively. The resulting tree can be edited using Courier, Courier New or LinePrinter fonts. It can be enlarged or reduced by changing the font size. The labels of taxa can be easily modified. Different word processor options can then be used to improve the editing quality, and the resulting trees can be used to prepare scientific manuscripts or slides.

Figures 1 and 2 show phylogenetic trees constructed from the same data set. Fourteen complete polyprotein sequences of hepatitis C isolates representing the 6 viral genotypes were used. Alignments were performed by Clustal W Version 1.7 (Metrowerks, Austin, TX, USA) (2) using the default parameters and the output format compatible with MEGA. Phylogenetic analysis was performed using the gamma ($\alpha = 2$) algorithm for distance calculation, the neighbor-joining method and 500 replications for the bootstrap test.

Figure 1 represents the phylogenetic tree printed directly from MEGA. Figure 2 represents the same tree, obtained after opening an exported ASCII file in Microsoft Word 7.0, before (Figure 2A) and after (Figure 2B) conversion using the Macro described in this study.

This custom Macro and the instructions for complete installation are available by

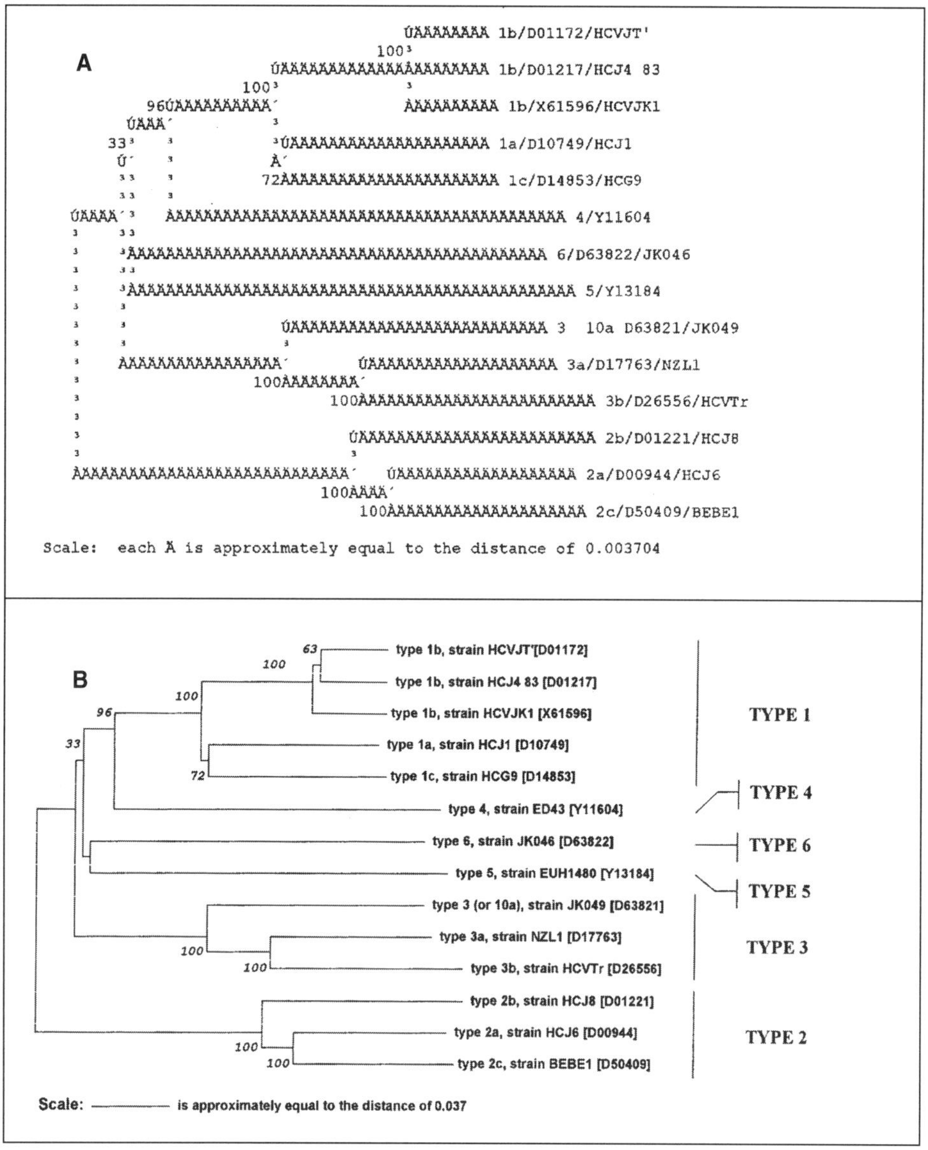

Figure 2. Trees obtained from the MEGA-exported ASCII file. (A) Tree printed from Microsoft Word 7.0 before any modification. (B) Tree printed after the Tree Macro and several editing improvements using Microsoft Word 7.0.

e-mail at the following Internet address (virophdm@lac.gulliver.fr). They are free of charge and will be sent as two attached text files that were tested virus-free by Norton AntiVirus 4.0 (Symantec). The only software requirements are MEGA versions 1.01 or 1.02 and Microsoft Word 7.0.

We believe that this simple Macro program will allow the users of MEGA to easily improve the graphic quality of their phylogenetic trees and that of their scientific manuscripts and slides under the familiar and convenient environment of their word processor. When more sophisticated modifications of their phylogenetic trees are required, we recommend the use of the TreeExplorer program.

REFERENCES

1. **Kumar, S., K. Tamura and M. Nei.** 1993. MEGA: Molecular Evolutionary Genetics Analysis, Version 1.02. Pennsylvania State University, University Park, PA.
2. **Thompson, J.D., D.G. Higgins and T.J. Gibson.** 1994. CLUSTAL W: improving the sensitivity of progressive multiple sequence alignment through sequence weighting, position-specific gap penalties and weight matrix choice. Nucleic Acids Res. *22*:4673-4680.

CHAPTER 43

Computer Program for Automatically Calculating Similarity Indexes from DNA Fingerprints

E. Scott Archer and Frederick C. Leung
The University of Hong Kong, Hong Kong

Reprinted from BioTechniques 25:252-254 (August 1998)

ABSTRACT

DNA Simdex™ Version 2.1 is a third-generation Windows® computer program that uses sophisticated image-analysis techniques to automatically locate bands on electrophoresis gel images and calculate similarity indexes. Here we give a functional description of the program.

INTRODUCTION

DNA fingerprinting is useful in studies on the relatedness between individuals and species, as it can be used in estimating individual and population homozygosity and relatedness. DNA sampled from an individual is digested by one or more restriction enzymes, producing a collection of DNA fragments of varying lengths, which are resolved using gel electrophoresis. The similarity index between DNA fingerprints, measured as a function of the number of matching vs. non-matching bands, as described by Lynch (2,3) can then be used to make inferences about the genetic structure of populations.

Although helpful in its ability to quickly calculate similarities if band positions were known, the initial version of DNA Simdex™ (5) had no built-in image analysis and thus relied on outside means for determining band locations. Several other image-analysis systems are available that are capable of automated band detection and subsequent calculation of similarity, but the cost of these systems is somewhat prohibitive since they typically require specialized hardware and costly software. In response, we have developed DNA Simdex Version 2.1 (Simdex2.1), a user-friendly graphical analysis program that requires no more expensive hardware than a PC running Windows 95® or NT™ and an image-acquisition device such as a desktop scanner or camera/digitizer.

BioComputing: Computer Tools for Biologists
Edited by Stuart M. Brown

MATERIALS AND METHODS

A desktop scanner can be used to digitize an autoradiogram or fluorescent photograph, as described previously (1,6,7). Alternatively, a charge-coupled device (CCD) camera-based, image-analysis system can be used, but it must allow saving of digitized images in standard image file formats. For Simdex2.1, the image must be stored as a 256 gray-scale bitmap, in bitmapped (BMP) format, with the wells of the gel located along the top of the image and the lanes running vertically down.

Upon loading the gel image file, two windows are created: one that contains the digitized gel image and one that will show the corrected band positions when they are available. The user is prompted to use the mouse to designate a rectangular region on the gel image, which defines the locations of the wells, the migration front and the outermost lanes. The user can then designate the lanes' locations, such that the lane-vector runs approximately through the middle of the entire lane.

When the user is finished editing the lane locations, the band-editing dialog pops up. This dialog acts as a sort of control box, from which the user can initiate an automated search for bands, which is performed using a template-matching technique, similar to that described previously (4). In this technique, a small "window" of intensity values along the lane is checked for a match against a relative-value template. The window is iterated along the entire length of the profile, one pixel at a time. For example, if the window size is set at five, call each profile value in the window a, b, c, d and e. A "peak" is detected if the values match the template: $a \leq b \leq c \geq d \geq e$. The lane's background is used as a filter during peak detection on the profile, such that any detected peaks, or local maxima, with values above background are designated as bands. The user can adjust the filter value of the background to compensate for an excessively faint, dark or noisy gel image.

Because the effectiveness of finding bands using template-matching is about 85%, the user can also add, remove and edit band locations manually using the mouse. To aid the user in editing bands, a graph of each lane's intensity profile can selectively be drawn along the lane. This allows the user to make sure that bands are accurately positioned.

The length of a DNA fragment's migration through a gel, and consequently its band position, is determined by a combination of factors, including (but not necessarily limited to) its molecular weight and its charge and conformation. Migration length can be influenced by other factors too, such as inconsistencies in the gel matrix and the quantity of DNA loaded into the lane. Variations in migration length can cause mis-

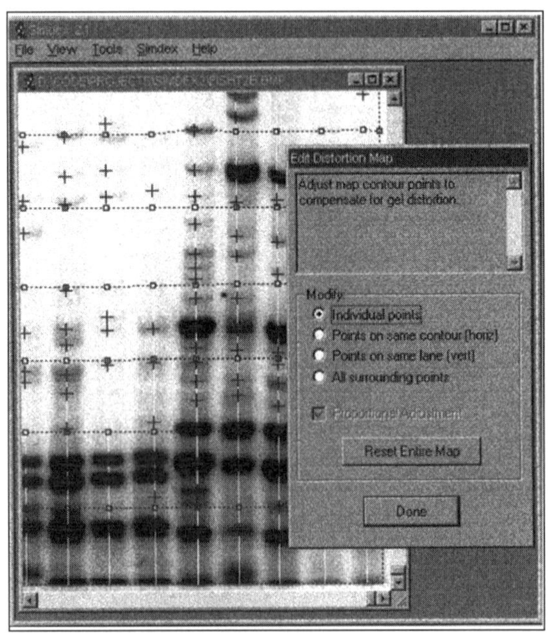

Figure 1. DNA Simdex2.1 distortion map editing.

matches in bands that should otherwise match. While the human observer easily compensates for these distortions, the computer must base its band-matching decisions on discrete pixel locations.

To this end, Simdex2.1 provides a distortion map (Figure 1). It acts as a set of visible contours that run across all of the lanes, representing equal migration length. Using the mouse, the user can modify the location of any contour-lane intersection, which re-maps the migration length of any nearby bands. This turns out to be an effective method of compensating for most gel distortions.

The molecular weight (MW) of a particular band on a gel image is not directly implied by its visible migration position, which in Simdex2.1 is measured in pixels. Simdex2.1 assigns approximate molecular weights to bands using references to lanes that contain a MW standard. The bands of a MW standard are of known molecular weight and thus provide a set of reference points for mapping corrected migration lengths to molecular weights.

First, the user must select the MW standard lanes. After assigning a MW standard from the modifiable MW standard database, the user must assign a molecular weight from the standard to one of the bands in a standard lane. With the user's assistance, Simdex2.1 maps the rest of the bands in the standard lanes to corresponding molecular weights in the MW standard. Simdex2.1 then calculates the molecular weights of the bands in the rest of the lanes, based on their migration length, as corrected by the distortion map.

Simdex2.1 can calculate the similarity indexes for any set of lanes on a single gel or on different gels. The similarity index is given by Equation 1:

[Eq. 1]
$$S_{ab} = \frac{2n_{\text{ab}}}{n_a + n_b}$$

where S_{ab} is the similarity index between lanes a and b, n_{ab} is the number of bands common to both lanes and n_a and n_b are the number of bands in lanes a and b, respectively. For similarity index calculations on lanes contained in a single gel, the bands' corrected migration distances are used. However, when comparing lanes on more than one gel, the molecular weights of the bands involved must be known.

This is an obvious requirement, as molecular weight is the only unambiguous quantity available between gels.

Simdex2.1 displays band migration lengths, MWs and any calculated similarity indexes in tabular displays (Figure 2). These data can also be printed or exported to files as American Standard Code for Information Interchange

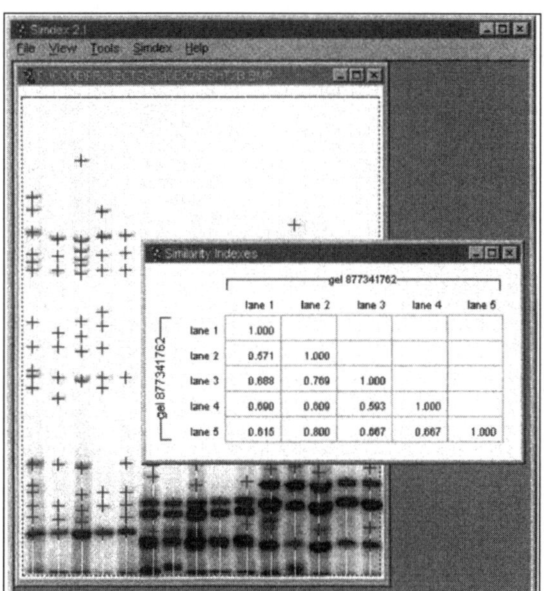

Figure 2. DNA Simdex2.1 output - similarity indexes.

(ASCII) text for import into other software, such as a spreadsheet program. Additionally, work can be saved to a project file at any time for later retrieval. Slated for future releases are the additions of TWAIN scanner support and an open data base connectivity (ODBC)-compliant database connection to support similarity calculations across very large data sets.

ACKNOWLEDGMENTS

We thank Dr. Darrell P. Chandler and Mr. Ryan D. Quesenberry for providing DNA fingerprint autoradiograms used in the testing and development of DNA Simdex Version 2.1. This work was partly supported by the US Department of Agriculture (USDA) under Related Service Agreement with the US Department of Energy (DOE) under Contract No. DE-AC06-76RLO 1830 at Pacific Northwest Laboratory (PNL). PNL is operated for DOE by Battelle Memorial Institute. This work was also supported by the Hong Kong Research Grants Council, The University of Hong Kong 196/95M.

REFERENCES

1. **Donovan, R.S., C.W. Robinson and B.R. Glick.** 1994. Detection enhancement of computer images of bands from Western blots using a grayscale scanner. BioTechniques *17*:660-661.
2. **Lynch, M.** 1990. The similarity index and DNA fingerprint. Mol. Biol. Evol. *5*:584:599.
3. **Lynch, M.** 1991. Analysis of population genetic structure by DNA fingerprinting, p. 113-126. *In* T. Burke, G. Dolf, A.J. Jeffreys, and R. Wolff (Eds.), DNA Fingerprinting: Approaches and Applications. Birkäuser Verlag, Switzerland.
4. **Parker, J.R.** 1994. Practical Computer Version C. John Wiley & Sons, New York.
5. **Scott, D.A., M. Welt and F.C. Leung.** 1993. A computer program to aid in calculating similarity indexes from DNA fingerprints. BioTechniques *14*:980-983.
6. **Shea, T.B.** 1994. An inexpensive densitometirc analysis system using a Macintosh™ computer and a desktop scanner. BioTechniques *16*:1126-1128.
7. **Velleman, S.G.** 1995. Quantifying immunoblots with a digital scanner. BioTechniques *18*:1056-1058.

CHAPTER 44

Fingerprinting Method for Phylogenetic Classification and Identification of Microorganisms Based on Variation in 16S rRNA Gene Sequences

Gajendra P.S. Raghava, Rajesh J. Solanki, Vishal Soni, and Pushpa Agrawal

Institute of Microbial Technology, Chandigarh, India

Reprinted from BioTechniques 29:108-116 (July 2000)

ABSTRACT

The paper describes a method for the classification and identification of microorganisms based on variations in 16S rRNA sequences. The 16S rRNA is one of the most conserved molecules within a cell. The nature of the variable and spacer regions has been found to be specific to a given organism. Thus, the method presented here can be very useful for the classification and identification of microorganisms for which very little information is available. To automate the method, a comprehensive computer program called FPMAP has been developed for the analysis of restriction fragment pattern data. The method involves the restriction digestion of genomic DNA, preferably using four-cutters that may recognize 6–9 sites within the 16S rDNA. The fragments are separated on a polyacrylamide gel along with a suitable marker, then transferred into a nylon membrane and hybridized with a radiolabeled 16S rDNA probe. After autoradiography, the fragment sizes are calculated, and the data are analyzed using the FPMAP software. We demonstrate that the method can be used for identification of strains of Streptomyces *and* mycobacteria. *The software is available from our ftp site ftp://www.imtech.res.in/pub/com/fpmap/unix.*

INTRODUCTION

The identification and classification of microorganisms are of fundamental importance in microbiology. Classical methods are mainly emperical and depend on phenotypic data. They are also time consuming and expensive. More recently, various fingerprinting methods using some unique property of an organism have been described. The existing approaches to fingerprinting can be broadly classified as: *(i)* classical DNA fingerprinting

BioComputing: Computer Tools for Biologists
Edited by Stuart M. Brown

such as restriction enzyme analysis of the genomic DNA on pulse field gel electrophoresis, *(ii)* the use of a repetitive DNA sequence as a probe for a particular region of the genome, *(iii)* DNA-DNA hybridization, *(iv)* PCR amplification of a particular section of the genome followed by restriction analysis of the amplified fragment and (v) the use of an insertion sequence.

In all of these methods, genomic DNA has been the subject of analysis (6,10,22). A serious limitation of these approaches is that alterations in the DNA structure and composition occuring during the processing of the DNA are completely overlooked. Further, it is a well-known fact that many organisms undergo large deletions in chromosomal DNA, or loss or gain of plasmids, acquisition of transposons or an *IS* element either during different growth condition or in response to environmental conditions. An organism may exhibit slightly different characteristics depending on the environment from which it was isolated. Thus, the properties related to the presence or absence of plasmids, transposons or *IS* elements are unsuitable as tools for phylogenetic identification. However, almost all of these methods have been used for epidemiological studies or to differentiate between strains of the same species or to confirm the species of the genus.

Ribosomes are a part of the translational machinery of a cell, and rRNA is vital for cellular growth, function and survival. Consequently, the primary, secondary and tertiary structures of rRNA molecules have been conserved during evolution (8). Analysis by classical methods of comparative oligonucleotide cataloguing and consideration of full or partially complete sequences of 16S rRNA have revealed that the primary structure of rRNAs consists of highly conserved regions interspersed by regions of moderate to low homology within related species (4). Despite the highly conserved nature of rRNAs, they vary in size and in the organization of the spacer as well as variable regions within the rRNA. This results in variation of the cleavage sites for restriction endonucleases. The small size of 5S rRNA and extensive secondary and tertiary structures present in the 23S rRNA render these molecules unsuitable for sequencing and further analysis. However, the moderate size of the 16S rRNA, sequence conservation during evolution and proposed and confirmed importance in classification has proven that 16S rRNA sequences can be used effectively for phylogenetic identification of the microorganisms at various levels of heirarchy. Most of the earlier studies have used rRNA sequences for some kind of epidemiological studies or to confirm the earlier data of the phylogeny of a particular group. No earlier studies in the literature have concentrated on using the 16S rRNA sequence for the phylogenetic identification of organisms.

In the present work, the variation within 16S rRNA is exploited for phylogenetic identification and classification of microorganisms. An important fingerprinting technique is the analysis of the restriction fragment length polymorphism (RFLP). Several computer programs have been developed to compare the RFLP of a test organism with the RFLP database generated in the same laboratory, but not in different laboratories (3,5,20,21). However, in this study, we have developed a comprehensive software named FPMAP (fingerprint mapping) that not only compares the RFLPs but also facilitates *(i)* the selection of suitable restriction enzymes for the sequences obtained from GenBank® or the ribosomal database project (RDP), *(ii)* the graphical presentation of the positions of restriction sites, *(iii)* the graphical presentation of the positions of the DNA fragments according to the molecular weights, *(iv)* RFLP analysis and comparison with the subject strain RFLP and *(v)* the creation of a PHYLIP format of the complete data and the use of the PHYLIP software package to generate an unrooted tree. The phylogenetic tree obtained indicates the identity of the test strain.

To demonstrate the use of a fingerprinting technique in the identification and classification of microorganisms, we selected two genera, *Streptomyces spp.* and *Mycobacterium spp.* The restriction enzyme sites of the 16S rRNA sequences of the type strains of these genera available in GenBank and RDP were analyzed using the FPMAP program. Using a 16S rRNA sequence as a probe and FPMAP as a pattern analysis tool, one can phylogenetically identify a strain. Because the maximum size of the 16S rRNA is 1500 bp, many restriction enzymes will have no sites. Methylase-sensitive enzymes may not be a good choice. An enzyme that does not cut frequently may also not be a good choice because conservation in the 16S rRNA sequences is high and only those enzymes that have small recognition sequences are likely to produce good RFLP analyses. Success of the method depends, therefore, on the selection of suitable restriction enzymes, suitable molecular weight markers and good electrophoretic resolution. We recommend the selection of restriction enzymes that cut frequently and produce DNA fragments in the range of 20–700 bp, and the use of 7%–8% nondenaturing polyacrylamide gel for electrophoresis. The FPMAP software has been developed in standard C language so that it can be ported to a variety of other platforms. Executable and source codes of the FPMAP are available at ftp://www.imtech.res.in/pub/com/fpmap/unix.

MATERIALS AND METHODS

Media, Chemicals and Strains

Type strains of *Mycobacterium smegmatis* (ATCC 14468), *M. vaccae* (DSM 43292) and *Streptomyces rimosus* (from M. Goodfellow's collection, University of Newcastle upon Tyne, UK) were obtained from the Microbial Type Culture Collection (IMTECH, Chandigarh, India) and *M. bovis* BCG was obtained from the Tuberculosis Vaccine Laboratory, Madras, India. *S. coelicolor* B385 was a gift from Prof. D.A. Hopwood (John Innes Centre, Norwich, UK). A new isolate of *Saccharomonospora sp.* (PA136, classified and identified by chemotaxonomy) and *Saccharomyces cerevisiae* DNA were used as controls. *M. bovis* BCG and *M. smegmatis* and *M. vaccae* were grown on Sauton's medium. (One liter contains 0.05 g ferric ammonium citrate, 4 g L-asparagine, 2 g citric acid, 0.5 g K_2HPO_4, 0.5 g $MgSO_4$, 60 mL glycerine, 2.5 mL 0.1% $ZnSO_4$ and 6 mL 5 M NaOH to make pH 7.2.) *Nocardia lurida* (DSM 43134) and *N. asteroides* (DSM 43757) were grown on the CM medium (2 g yeast extract, 10 g malt extract, 4 g glucose and *S. coelicolor* B385 and *Saccharomonospora sp.* were grown as described by Hopwood et al. (9). All chemicals were obtained from Sigma (St. Louis, MO, USA), and the medium components were obtained from Difco, Detroit, MI, USA. Restriction enzymes and labeling kits were purchased either from Promega (Madison, WI, USA) or from New England Biolabs (Beverly, MA, USA).

Preparation of Genomic DNA and Southern Hybridization

Genomic DNA from *Streptomyces sp.* was isolated essentially as described by Hopwood et al. (9) but with some modification. Fifty milligrams of wet mycelium was suspended in 0.5 mL lysozyme solution (2 mg/mL and 50 µg/mL RNase from Promega) and incubated at 37°C until the cells became translucent. In this mixture, 0.25 mL 2% SDS was added and vortex mixed until the viscosity of the solution decreased noticeably. To this solution, 0.25 mL neutral phenol-chloroform was added and vortex mixed for 30 s.

The solution was centrifuged for 2 min, and the supernatant was collected in a fresh tube. The procedure was repeated twice more before final chloroform extraction. To this supernatant, 0.1 vol 3 M sodium acetate (pH 4.8) (0.3 M as a final concentration) was added and mixed. DNA was precipitated with 2 vol cold absolute alcohol by slow mixing and cooling in ice. We found that by intermittent cooling and slow mixing, it was possible to obtain high molecular weight DNA as a pellet that could be removed by a glass rod or pipet tip.

Mycobacterial DNA was prepared by a method developed in the author's laboratory. The 21-day-old culture was grown in Sauton's medium at 37°C at 200 rpm, Tween® 80 was added aseptically at a final concentration of 1%, and the culture was grown for another 24 h. The cells were harvested by centrifugation at 10 000× *g* at 4°C and washed twice in TE (10 mM Tris-HCl and 1 mM EDTA). The resuspended cells were again incubated at 45°C for 30 min in a Tris buffer (25 mM Tris-HCl, pH 7.5). This step seems to remove a lot of the waxy layer present outside the mycobacteria cell wall and made the cells more accessible to lysozyme treatment. The cells were resuspended in 25 mM Tris-HCl and 10 mM EDTA, pH 8.0. Lysozyme was added to the final concentration of 10 mg/mL (about 30 mg wet weight of cells) and incubated at 37°C for 2 h. To this mixture, 10 µg/mL proteinase K and 1 µg/mL RNase were added, and incubation was continued for another 2 h at 37°C. The preparation was extracted twice with phenol and then twice with chloroform. DNA was precipitated in the presence of 0.3 M sodium acetate (pH 4.5) and 2 vol cold absolute ethanol. The DNA pellet was washed twice in 70% ethanol, air-dried and dissolved in TE.

Three micrograms of genomic DNA was digested to completion by *Hha*I, *Mbo*I, *Rsa*I and *Bst*NI. Lambda phage DNA digests (*Hin*dIII, *Eco*RI + *Hin*dIII and *Pst*I) and the *Hin*fI and the *Bst*NI digest of the Bluescript® plasmid (SK^+, SK^-, Stratagene, La Jolla, CA, USA) were used as molecular weight markers. The digests were separated on a 7% polyacrylamide gel, and the separated fragments were electrotransferred onto a nylon membrane (HybondN®; Amersham Pharmacia Biotech, Little Chalfont, UK). Because of the high gel concentration and the nature of polyacrylamide, capillary transfer and gel pretreatment is not recommended. DNA was denatured in the membrane by placing it in 0.4 M NaOH for 10 min (DNA side up). The membrane was rinsed in 2× SSC for 2 min, air-dried and then fixed at 80°C for 30 min (the DNA can also be UV fixed). The 16S rDNA from five different newly isolated and unidentified actinomycetes strains was amplified using the following primers: 27f-5′-AGAGTTTGATC**M**TGGCTCAG-3′ and 5′-AAGGAGGTG**W**TCCA**R**CC-3′ (13) where M = A + C, W = A + T and R = A + G.

The PCR product (approximately 1.5 kb) was purified from the gel by using a DEAE-cellulose membrane (NA-45; Schleicher and Schuell, Keene, NH, USA) (19). The ends of the PCR products were phosphorylated by T4 polynucleotide Kinase, cloned into the dephosphorylated pGEM5Zf(+) vector (both from Promega) at the *Eco*RV site and transformed into the *E. coli* strain JM109. The white clones from ampicillin-X-Gal-IPTG plates were collected, and the plasmid was checked for the 16S rDNA inserts by *Nco*I and *Spe*I digestions. The entire 16S rDNA fragment was sequenced to confirm the chemotaxonomy results (to be published separately) and to use the cloned fragments as a probe. The GenBank accession numbers for these strains are as follows: AF 223346, AF 223347, AF 223348, AF 223349 and AF 223350. For use as a probe, the cloned fragment was recovered by digesting the recombinant vector with *Nco*I and *Spe*I enzymes and was purified from the agarose gel as described by Sambrook et al. (19) using DEAE-cellulose membrane. The purified 16S rDNA fragment was radiolabelled with $\alpha[^{32}P]$dCTP using a NEBlot® random priming kit (New England Biolabs) as described by the manufacturer. Hybridization was carried out at 50°C using a hybridization bag in a buffer containing

30% formamide, 5× SSPE containing 0.9 M NaCl, 50 mM sodium phosphate, 5 mM EDTA (pH 7.7), 0.5% SDS and 5× Denhardt's reagent containing 0.05% BSA, 0.05% Ficoll® Type 400 (Sigma) and 0.05% polyvinylpyrrolidone. The membrane was incubated in a hybridization buffer at 50°C for 2 h before adding the radiolabeled (α[^{32}P]dCTP) and denatured probe in the same solution. After removing all the air bubbles and sealing the bag, hybridization was carried out for 22 h. The membranes were washed in 2× SSC with 0.1% SDS at 67°C for 30 min and in 1× SSC with 0.1% SDS at 40°C for another 30 min before autoradiography with Kodak® X-Omat™ films (Eastman Kodak, Rochester, NY, USA) for 10–24 h. The sizes of the hybridizing bands were calculated with the program DNASIZE (16).

RESULTS

Figure 1 suggests that a good-quality, high-molecular-weight DNA can be obtained from mycobacteria without using a commercial kit or guanidine hydrochloride.

To test the FPMAP identification software, we selected species of the genera *Mycobacteria* and *Streptomyces*. Both genera are medically, industrially and phylogenetically important. Genomic DNA of various species of these genera were digested with the restriction enzymes *Hha*I, *Mbo*I, *Rsa*I and *Bst*NI. Figure 2 shows the result of *Hha*I digests obtained after hybridizing with 16S rRNA as probe. The selection of enzymes was based on the restriction enzyme analysis of the published 16S rRNA sequences (17). The DNA fragments present in the Southern blot were measured using the program DNASIZE (16). The RFLPs of test strains were compared with those of databases created for the strains. The result is shown in the Figure 3, A and B, in which, for *S. rimosus* and *M. bovis*, the RFLP generated by the restriction enzymes *Rsa*I and *Hha*I, respectively, is identical to the one generated using the published sequence. The RFLP data was then processed with PHYLIP, and an unrooted evolutionary tree was generated (Figure 4, A and B). Figure 5 shows, however, the phylogenetic tree that would be obtained from the sequence data. It is evident from the tree that the subject strains have formed a closed cluster with the type strain of *S. rimosus* and *M. bovis* and other related species of *Streptomyces* and *Mycobacteria*. The restriction enzymes *Mbo*I and *Bst*NI produced distinctive RFLPs during sequence analysis, but neither of the enzymes worked well during digestions. Both *Mbo*I and *Bst*NI enzymes are methylase sensitive. Perhaps methylation of the DNA at the *Mbo*I and *Bst*NI sites is the cause of a slightly different RFLP of the test strain than expected. Most of the strains tested produced good RFLP, but it is beyond the scope of this report to show large data.

Figure 1. DNA prepared from *M. bovis* BCG using in-house method. (1) Undigested *M. bovis* BCG DNA. (2) λ *Hin*dIII marker.

DISCUSSION

Evolutionary conservation and the uniqueness of the ribosomal RNA allow a broad spectrum of applications. Grimont and Grimont (6) described the importance of the 16S rRNA-based probes as taxonomic tools. The method is known as ribotyping (the generation of characteristic fragment patterns by hybridization of restriction endonuclease fragments of total DNA with labeled standard rRNA). This method has shown considerable discriminatory power in the identification of organisms up to the species level and in the typing of a number of bacteria, for example, *Aeromonas spp.* (1), *Campylobacter spp.* (15), *Gordona spp.* (18), *Legionella spp.* (7), *Mycobacterium spp.* (12), *Mycoplasma spp.* (23), *Rhodococcus spp.* (14) *Saccharomonospora spp.* (24), *Staphylococcus spp.* (11) and *Streptomyces spp.* (2).

The recent review by Gurtler and Stanisich (8) emphasizes the importance of the spacer regions of 16S rDNA and 23S rDNA genes for typing and identification. In their method, the spacer regions of the 16S–23S rDNA are amplified and then the strain differentiations are made either by digesting the PCR fragments with RFLP or by single strand conformation polymorphism (SSCP). However, these methods rely on rather complicated procedures with large resource requirements that involve several steps before one obtains a RFLP. It is apparent that the complete sequence of the 16S rRNA of an organism is enough to determine the phylogenetic position of a strain. Despite automation of DNA sequencing, it is beyond the means of many laboratories to sequence large numbers of microorganisms. However, the need to protect biodiversity from the negative impact of globalization has revived the demand for simple yet comprehensive and reliable methods of identification. In the present study, we approach the use of the properties of the 16S

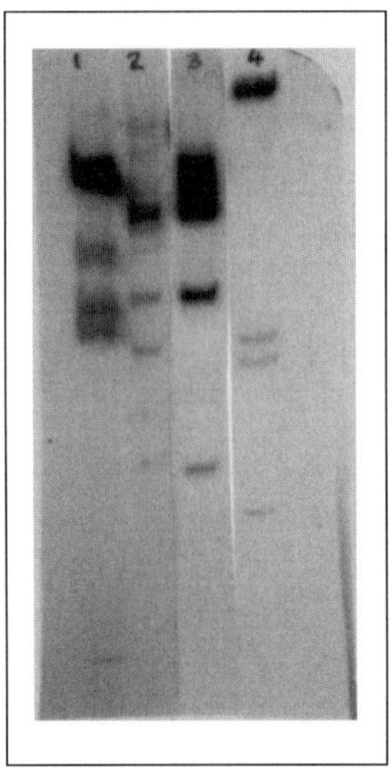

Figure 2. 16S rDNA *Hha*I restriction pattern hybridizing with 16S rDNA probe. (1 and 4) pBluescript plasmid SK+ and SK- digested with *Hin*fI and *Bst*NI, respectively, as molecular size markers. Lane 1, 1074, 517, 456, 396 and 75 bp; lane 4, 1849, 314, 288 and 121 bp. Lane 2, *M. bovis* BCG and lane 3, *S. rimosus*. For experimental detail, please see Materials and Methods.

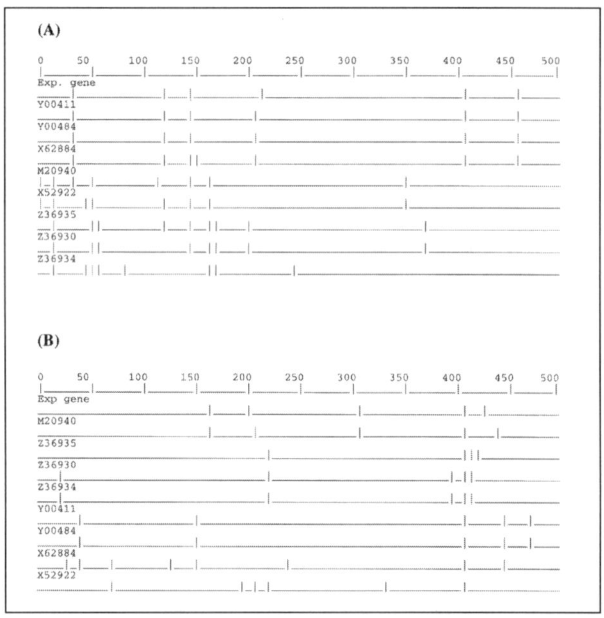

Figure 3. Screen output of FMAP comparing the fragment maps of the experimental genes as obtained by the Southern hybridization with that obtained from the 16S rDNA sequence present in the database. (A) Fragment map of *S. rimosus* (Exp. gene) and experimental gene for restriction enzyme *Rsa*I. (B) Fragment map of *M. bovis* (Exp. gene) and experimental gene for restriction enzyme *Hha*I where GenBank accession numbers X62884, Y00411, Y00484, Z36934, Z36935, Z36930, M20940 and X52922 represent *S. rimosus*, *S. coelicolor*, *S. lividans*, *N. asteroides*, *N. brasiliensis*, *N. nova*, *M. bovis* and *M. smegmatis*, respectively.

rDNA for phylogenetic identification from a different perspective.

We argue that if one selects a set of restriction enzymes that cut frequently within the 16S rRNA genes and use them to digest the genomic DNA of a strain, the variability in the spacer regions and in the whole operon can be detected by hybridization using 16S rDNA as a probe. Since the size of the 16S rRNA is about 1.5 kb, many restriction enzymes will have either no recognition sites or less than three. Conservation within 16S rRNA is such that long stretches of homologous sequences are common among species of the same genus, and the GC content of the rDNA does not vary much between species. Therefore, enzymes that have short recognition sequences have more sites and can cut the DNA more frequently to reveal the RFLP within a small region of variability.

The two important factors in classification and identification using the method described here are the selection of restriction enzymes and the availability of correct and complete 16S rRNA sequences. The restriction enzymes should be such that *(i)* they cut frequently to produce distinctive RFLPs, *(ii)* the DNA fragments produced by the enzymes should separate well during electrophoresis (should not generate doublets or triplets) and *(iii)* the fragments produced should be within the working limits of the methods of electrophoresis. The partial sequence of many organisms

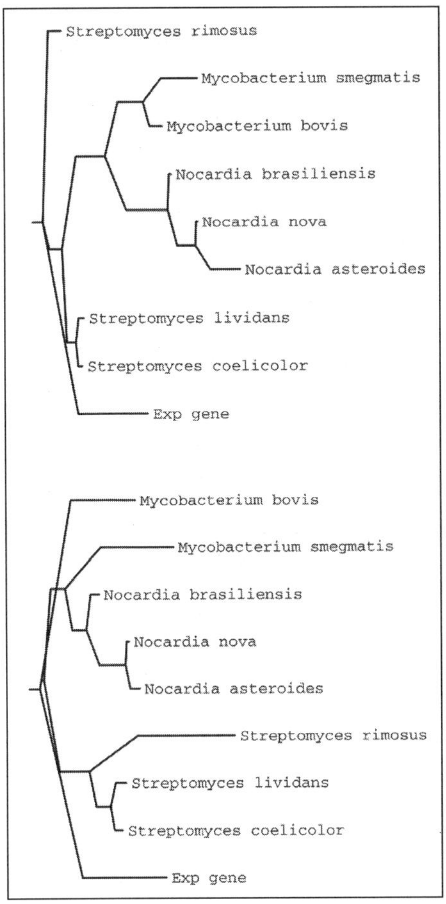

Figure 4. Phylogenetic tree was generated by using FPMAP and PHYLIP (restml and drawgram). (A) *S. rimosus* (Exp. gene) and experimental genes for restriction enzyme *Rsa*I. (B) Fragment map of *M. bovis* (Exp. gene) and experimental genes for restriction enzyme *Hha*I.

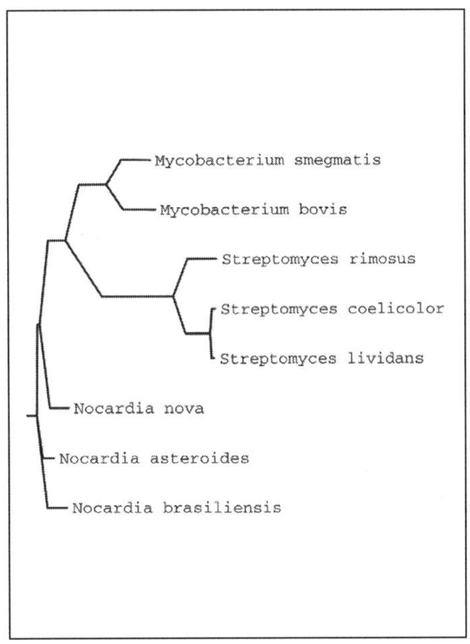

Figure 5. Phylogenetic tree obtained using 16S rDNA sequences available in the database. All sequences were compared using the multiple sequence alignment program CLUSTAL W. The phylogenetic tree was obtained using the program drawgram of the PHYLIP.

Table 1. FPMAP Menu as Displayed on the Computer Screen

FPRINT 1.0
Search restriction sites in nucleodide sequence
Display restriction sites in a sequence
Display of size of fragments in a sequence
Comparison of restriction sites in sequences
Comparison of fragments of sequences
Identification of an unknown gene
Generate fragment mapping in PHYLIP format (input)
Exit from program

that are available in the databases are not useful because the partial sequence does not provide a true picture of the variation within the system. The size of DNA fragments generated by the restriction enzymes is not really a limitation of this method; however, a fragment size of 6–10 bp and 1 kb and above creates some practical problems such as poor resolution, very faint signals even with the radioisotopes and difficulties obtaining appropriate molecular weight markers and selecting the optimal electrophoresis gel concentration. It is important to use a molecular weight marker that covers the entire range of fragments produced by the restriction enzyme. Because FPMAP presents restriction sites and fragment maps graphically, the selection of appropriate restriction enzymes is facilitated. However, the identification of a strain may not be straightforward. As stated earlier, it is important to obtain distinct RFLPs. In a situation when there is some doubt, the identification option of the software allows one to visualize the fragment maps of the test and similar strains, which may aid the user in identification.

The objective of the present study was to provide a quick identification tool but not to carry out a taxonomic study. It is therefore beyond the scope of this article to present large amounts of data that could provide a real feeling for mismatches between the RFLP patterns inferred from the sequence data and the patterns obtained through experiments. Integration of FPMAP with PHYLIP permits processing of the RFLP data by PHYLIP to produce a phylogenetic tree.

The menu displayed on the computer screen is shown in Table 1.

ACKNOWLEDGMENTS

This work was funded by the Council of Scientific and Indusrial Research (CSIR), New Delhi, India. We are grateful to Mr. Rajeev Kumar for his assistance in database creation. This is IMTECH communication no. 26/99.

REFERENCES

1. **Altwegg, M., R. Altwegg-Bissig, A. Demarta, R. Peduzzi, A.W. Reeves and B. Swaminathan.** 1988. Comparison of four typing methods for *Aeromonas* species. J. Diarrheal Dis. Res. *6*:88-94.
2. **Doering-Saad, C., P. Kampfer, S. Maulis, G. Kritzman, J. Schneider, J. Zakrzewska-Czerwinska, H. Schrempf and I. Barash.** 1992. Diversity among *Streptomyces* strain causing potato scab. Appl. Environ. Microbiol. *58*:3932-3940.

3.**Gartler, V. and V.A. Stanisich.** 1996. New approaches to typing and identification of bacteria using the 16S-23S rDNA spacer region. Microbiology *142*:3-16.

4.**Gopo, J.M., R. Melis, E. Filipska, R. Meneveri and J. Filipski.** 1988. Development of a *Salmonella*-specific biotinylated DNA probe for rapid routine identification of *Salmonella*. Mol. Cell. Probes *2*:270-271.

5.**Graves, L.M. and B. Swaminathan.** 1993. Universal bacterial DNA isolation procedure, p. 617-621. *In* D.H. Persing, T.H. Smith, F.C. Tenover and T.J. White (Eds.), Diagnostic Molecular Microbiology: Principles and Applications. American Society for Microbiology, Washington DC.

6.**Grimont, F. and P.A.D. Grimont.** 1986. Ribosomal ribonucleic acid gene restriction patterns as potential taxonomic tools. Ann. Inst. Pasteur/Microbiol. *137*:165-175.

7.**Grimont, F., M. Lefevre, E. Ageron and P.A.A. Grimont.** 1989. rRNA gene restriction patterns of *Legionella* species: a molecular identification system. Res. Microbiol. *140*:615-626.

8.**Gutell, R.R.** 1992. Evolutionary characteristics of 16S rRNA and 23S rRNA structures, p. 243-309. *In* H. Hartman and K. Matsuno (Eds.), The Origin and Evolution of the Cell. World Scientific Publishing, Singapore.

9.**Hopwood, D.A., M.J. Bibb, K.F. Chater, T. Kiesser, C.J. Bruton, H.M. Kieser, D.J. Lydiate, C.P. Smith et al.** 1985. Genetic manipulation in *Streptomyces*: A Laboratory Manual. John Innes Foundation, Norwich, UK.

10.**Huys, G., R., Coopman, P. Janssen and K. Kersters.** 1996. High-resolution genotypic analysis of the genus *Aeromonas* by AFLP fingerprinting. Int. J. Syst. Bacteriol. *46*:572-580.

11.**Izard, N.C., H. Hachler, M. Grehn and F.H. Kayser.** 1992. Ribotyping of coagulase-negative staphylococci with special emphasis on intraspecific typing of *Staphylococcus epidermidis*. J. Clin. Microbiol. *10*:817-823.

12.**Kauppinen, J., J. Pelkonen and M.L. Katila.** 1994. Random amplified polymorphic DNA genotyping of *Mycobacterium malmoense*. J. Microbiol. Methods *19*:261-267.

13.**Lane, D.J., B. Pace, G.J. Olsen, D.A. Stahl, M.L. Sogin and N.R. Pace.** 1991. Rapid determination of 16S ribosomal RNA sequences for phylogenetic analysis. Proc. Natl. Acad. Sci. USA *82*:6955-6959.

14.**Lasker, B.A., J.M. Brown and M.M. McNeil.** 1992. Identification and epidemiological typing of clinical and environmental isolates of the genus *Rhodococcus* with use of a digoxigenin-labeled rDNA gene probe. Clin. Infect. Dis. *15*:223-233.

15.**Moureau, P., I. Derclaye, D. Gregoire, M. Janssen and G.R. Cornelis.** 1989. *Campylobacter* species identification based on polymorphism of DNA encoding rRNA. J. Clin. Microbiol. *27*:514-1517.

16.**Raghava, G.P.S.** 1994. Improved estimation of DNA fragment length from gel electrophoresis. BioTechniques *17*:100-104.

17.**Ragahva, G.P.S. and G. Sahni.** 1994. GMAP: a multi-purpose computer program to aid synthetic gene design, cassette mutagenesis and introduction of potential restriction sites into DNA sequences. BioTechniques *16*:1116-1123.

18.**Richet, H.M., P.C. Craven, J.M. Brown, B.A. Lasker, C.Cox, M.M. McNeil, A.D. Tice, W.R. Jarvis and O.C. Tablan.** 1991. A cluster of *Rhodococcus* (*Gordona*) *bronchialis* sternal-wound infections after coronary-artery bypass surgery. N. Engl. J. Med. *324*:104-109.

19.**Sambrook, C.J., E.F. Fritsch and J. Maniatis.** 1989. Molecular Cloning: A Laboratory Manual. CSH Laboratory Press, Cold Spring Harbor, NY.

20.**Schneider, M. and F.J. De Bruijn.** 1996. Rep-PCR mediated genomic fingerprinting of rhizobia and computer-assisted phylogenetic pattern analysis. World J. Microbiol. Biotechnol. *12*:163-174.

21.**Schimd, J., Y.P. Tay, I. Wan, M. Curr, D. Parr and W. Mckinney.** 1995. Evidence for nosocomial transmission of *Candida albicans* obtained by Ca.3 fingerprinting. J. Clin. Microbiol. *33*:1223-1230.

22.**Skuce, R.A., D. Brittain, M.S. Hughes, L.-A. Beck and S.D. Neil.** 1994. Genomic fingerprinting of *Mycobacterium bovis* from cattle by restriction fragment length polymorphism analysis. J. Clin. Microbiol. *32*:2387-2392.

23.**Yogev, D., D. Halachmi, G.E. Kenny and S. Razin.** 1988. Distinction of species and strains of *Mycoplasmas* (*mollicutes*) by genomic DNA fingerprints with an rRNA gene probe. J. Clin. Microbiol. *26*:1198-1201.

24.**Yoon, J.-H., H. Kim, S.-B. Kim, W.Y. Kim, S.T. Lee, M. Goodfellow and Y.H. Park.** 1996. Identification of *Saccharomonospora* strains by the use of genomic DNA fragments and rRNA gene probe. Int. J. Syst. Bacteriol. *46*:502-506.

Update to: Fingerprinting Method for Phylogenetic Classification and Identification of Microorganisms Based on Variation in 16S rRNA Gene Sequences

Pushpa Agrawal

Institute of Microbial Technology, Chandigarh, India (pushpaagrawal@lycos.com)

A Web site has been created for the computer program FPMAP, unlike in the original version, in which the software was available from IMTECH's file transfer protocol (FTP) site. In its present format, an on-line documentation is provided via Internet from the Web site (http://imtech.res.in/raghava/fpmap/). The document consists of complete information about FPMAP, including downloading and installation instructions. To make the software user-friendly and widely available in its present form, the software can run on a personal computer (PC). It has also been compiled for a number of other operating systems and computer hardware. FPMAP binaries and/or executables are now available for following major operating systems and/or platforms: (*i*) Linux™; (*ii*) Digital UNIX®; (*iii*) SUN Solaris™; (*iv*) Windows®/DOS®; and (*v*) SGI® IRIX® systems.

CHAPTER 45

GELSTATS: A Computer Program for Population Genetics Analyses Using VNTR Multilocus Probe Data

Steven H. Rogstad and Stephan Pelikan
University of Cincinnati, Cincinnati, OH, USA

Reprinted from BioTechniques 21:1128-1131 (December 1996)

ABSTRACT

GELSTATS, a computer program for population genetics analyses utilizing genetic markers revealed with variable number tandem repeat (VNTR) multilocus probes, is described and made available (both as C++ source code and as an executable DOS program). The program calculates several population genetics parameters, including: (i) *individual and population band numbers;* (ii) *population bands exhibiting complete linkage (redundant examples of such bands can be removed in subsequent analyses);* (iii) *similarity (fraction of bands shared) between individuals and average similarity within and between designated groups;* (iv) *estimated probability that two individuals chosen at random will have identical band profiles;* (v) *heterozygosity estimates for designated groups; and* (vi) *Fst estimates. Nonparametric permutation methods are used to assess the significance of differences in both within- and between-group similarity. A jackknife test for heterozygosity differences between groups is also computed. Examples of GELSTATS analyses illustrate some features of the program.*

INTRODUCTION

Loci for which alleles differ primarily due to variable number of tandem repeats (VNTR; Reference 14) of a "core" DNA sequence [including minisatellite (8,9), microsatellite (11) and simple sequence repeat (21) loci] are often extremely polymorphic. VNTR multilocus probes simultaneously reveal alleles at several different loci (8,17) and have been used to examine population genetics characteristics of several organisms (e.g., References 1, 3, 5 and 19). The GELSTATS computer program was designed to facilitate such studies by providing selected population genetics parameters based on VNTR multilocus probe data. GELSTATS calculates a wider range of parameters than

BioComputing: Computer Tools for Biologists
Edited by Stuart M. Brown

previous programs [e.g., SIM (1) for analyzing similarity or ThumbPrint (3) for heterozygosity] and also uses, as described in the next section, nonparametric analyses for assessing the statistical significance of differences in similarity between populations. It offers an alternative to the approaches of Lynch (12,13).

GELSTATS analyses depend on several assumptions (detailed in the README file accompanying the program), including: *(i)* bands with the same migration distance are identical alleles at one locus; and *(ii)* bands are transmitted in a Mendelian fashion [several studies have now shown that, for the most part, bands are transmitted this way (e.g., References 2, 4, 18 and 23)]. In calculations involving estimated allele frequencies and heterozygosities, it is assumed that populations are in Hardy-Weinberg equilibrium at each locus.

MATERIALS AND METHODS

Traits for multilocus VNTR marker analyses are population bands (probed endonuclease DNA fragments appearing in autoradiographs or chemilumigraphs), and each individual is scored for the presence [1] or absence [0] of each population band. Thus, data sets submitted to GELSTATS include population band profiles for all individuals, each of which is coded as to population membership. The README file includes information on data collection, on American Standard Code for Information Interchange (ASCII) file data input, on creation of an ASCII output file that can be converted for use with statistical and word-processing programs, and on other practical and theoretical details of the program.

Data analyses include the following features. In the initial GELSTATS analysis, a list is provided of polymorphic population bands that are redundant and thus may be completely linked (all and only those individuals having one population band also have the "linked" band). Redundant cases of such putative, completely linked bands can be removed in subsequent analyses to examine how inclusion of such bands affects results (retention would give extra weight to such a locus). Care must be taken to determine whether reported completely linked bands are in fact due to clonal growth (e.g., in plants, multiple samplings of ramets of one clonally spreading genet in which a unique population band occurs).

The frequency of each population band (number of individuals in which a population band occurs divided by the total number of individuals in the group of interest) in each group and across all individuals is calculated. The number of bands for each individual, mean number of bands per individual ($\bar{x}$) by group and across the entire sample (with s.d. and s.e.) and permutation tests (Reference 6; see below) of whether these mean values differ between groups are given. A table of all possible pair-wise comparisons of similarity is provided [$s = (2*N_{XY})/(N_X + N_Y)$, where N_{XY} is the number of bands shared between individuals X and Y, and N_X and N_Y are the number of bands in individuals X and Y, respectively; References 12, 13, 16 and 24]. This similarity matrix can be imported into other statistical packages (i.e., see example below). The program computes, for each designated group, for all possible pair-wise comparisons of individuals between groups, and for the whole data set, both the average interindividual similarity ($\bar{s}$) and an estimate of the probability ($\bar{s}^{\bar{x}}$) that two individuals randomly chosen from a group will have identical banding patterns (9).

Allele frequency (p) for each population band is estimated (according to References 10 and 20) both within each group and across all individuals. These frequencies are used to estimate the number of loci (L) and heterozygosities (H) for groups and across all individuals (calculated according to both References 10 and 20).

Permutation tests (6) are conducted by GELSTATS to explore whether groups differ in the following: *(i)* average number of bands per individual; *(ii)* average within-group interindividual similarity; and *(iii)* average within-group interindividual similarity vs. average between-group interindividual similarity. All of these tests take the same form, and more details are given in the README file. In general, for a particular permutation test of whether two groups differ, the value of the statistic being tested is computed for one group (the target group). Then, individuals are drawn, without replacement, at random from both groups to construct a "randomized" group the same size as the target group. For each randomized group, the "statistic of interest" is computed. By default, 5000 randomized groups are so analyzed (this default value can be changed), and the fraction of all the randomized groups that yield values of the statistic more extreme than that observed for the target group is used as a "p-value" for assessing whether the two groups differ in the "statistic of interest".

For any two populations, the permutation tests for similarity comparisons include three types of tests. First tested is whether similarity within one group differs from that in another group. Note that it is possible for two groups to have exactly the same level of similarity but simultaneously be completely genetically differentiated in population markers (e.g., within each group, all members share all bands, but no bands are shared between the groups). To test the degree of genetic differentiation, two further permutation tests are conducted: one in which the level of similarity within one group is compared to similarity for intergroup comparisons, and another in which similarity for the other group is compared to the intergroup comparisons. In the latter two tests, the degree to which within-group similarities exceed between-group similarities reflects the degree to which populations are genetically differentiated.

To examine whether the levels of heterozygosity estimated for two populations differ, a jackknife (22) resampling test is conducted in which heterozygosity is calculated for subgroups of each population. Subgroups are of size n-2, where n is the size of the population, randomly drawn from each population without replacement. The n-2 subgroup size is used to increase the number of possible representative subgroup comparisons possible. Such resampling comparisons between populations are conducted 5000 times (by default; a different number can be specified) with a "p-value" reported that is the fraction of the comparisons where one population has a heterozygosity greater than the other.

The nonparametric tests used by GELSTATS offer advantages over several other statistical approaches for the analysis of VNTR data sets (6). Tests using complex statistics with unknown sampling distributions (for example, H, which estimates heterozygosity) can be performed without requiring large sample sizes. Groups of unequal size or with unequal numbers of bands can be analyzed. Permutation tests avoid the problems of dependency discussed by Lynch (12,13), are frequently more powerful than rank-based, nonparametric tests and generally produce exact probabilities (6).

RESULTS AND DISCUSSION

A data set was analyzed with GELSTATS to demonstrate selected features of the program and to show how GELSTATS permutation tests of similarity compare with analyses of the same similarity matrices using the nonmetric multidimensional scaling (MDS) module of SYSTAT (Reference 25; the README file gives instructions for importing

Table 1. GELSTATS Similarity Tests for Humpback Whale Groups (Data from Reference 3)

Permutation Test of Similarity	"p-value"
Within group G > within group C	0.643
Within group G > between groups G and C	0.001*
Within group C > between groups G and C	0.004*
Within group G > within group A	0.634
Within group G > between groups G and A	0.000*
Within group A > between groups G and A	0.000*
Within group G > within group H	0.559
Within group G > between groups G and H	0.000*
Within group H > between groups G and H	0.000*
Within group C > within group A	0.534
Within group C > between groups C and A	0.553
Within group A > between groups C and A	0.498
Within group C > within group H	0.735
Within group C > between groups C and H	0.497
Within group H > between groups C and H	0.151
Within group A > within group H	0.904
Within group A > between groups A and H	0.537
Within group H > between groups A and H	0.224

G = Gulf of Maine, C = central California, A = southeastern Alaska and H = Hawaii (average similarity within these groups = 0.306, 0.284, 0.290 and 0.315, respectively). An asterisk indicates accept the test at the "p-value" ≤0.05 level. See text for further explanation.

similarity matrices calculated by GELSTATS into SYSTAT). The data set includes the combined data from Tables 4–6 in Reference 3 (data from VNTR multilocus probes 33.15, 3′HVR and M13, respectively). This data set consists of band profiles from 20 humpback whales (five from each of four groups, here denoted G = Gulf of Maine, C = central California, A = southeastern Alaska and H = Hawaii) with a total of 168 population bands. An initial GELSTATS analysis of this whale data detected no monomorphic population bands (considering all 20 individuals), although 35 non-monomorphic population bands redundant with other bands (completely linked?) were found, and these redundant bands have been removed from the following analyses of the remaining 133 population bands.

Results of GELSTATS within and between group similarity tests are given in Table 1. Note that although average similarities within groups differ slightly, these values are not significantly different. However, as noted above, groups can have equal levels of within-group similarity and yet be genetically differentiated. An examination of the within-group vs. between-group similarity permutation tests reveals that only those tests involving the

G group indicate genetic differentiation has occurred (note tests with asterisks), while differentiation is not detected in any comparisons among the A, C or H groups.

The latter points are supported by SYSTAT MDS analysis of the same GELSTATS similarity matrix. Figure 1 shows the placement of individual whales in the first two dimensions of the MDS analysis (accounting for 81.8% of the variance after 20 iterations with stress stabilizing at 0.15). Note that, in agreement with the permutation tests of similarity, the only group that is clearly differentiated is the G group.

GELSTATS provides two different estimates (10,20) of heterozygosity (h). The bias-corrected h values (10) for the combined data sets with redundant examples of possibly linked bands removed are: Gulf of Maine h = 0.8922; California h = 0.8076; Alaska h = 0.7685; and Hawaii h = 0.7445. Resampling tests indicate that none of these values are significantly different. GELSTATS estimates Fst over the entire data set by two methods: *(i)* Fst as described by Lynch (References 12 and 13; using similarity values) = 0.096; and *(ii)* Fst as described by Nei (References 7 and 15; using heterozygosity values) = 0.073.

The above examples demonstrate some of the types of data analyses possible with GELSTATS and show that the permutation tests can be used in conjunction with previously utilized tests (e.g., MDS) to further interpret data. Hopefully, GELSTATS will facilitate analyses of VNTR multilocus probe genetic data at the population level. The program and the supporting README file are available as **GELSTATS.ZIP** at **ftp.uc.edu**.

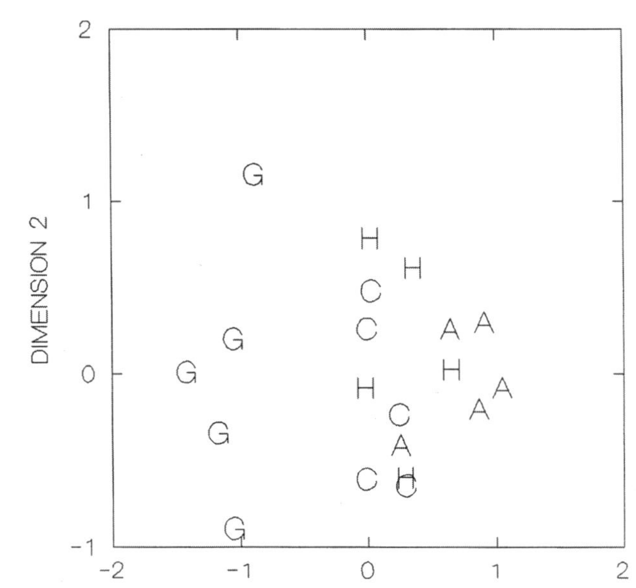

Figure 1. Nonmetric multidimensional scaling analysis of VNTR genetic markers in humpback whale individuals. Data taken from Reference 3. Placement of individuals in the first two dimensions based on analysis of a similarity matrix generated by GELSTATS is shown. Sampled individuals were from four groups: G = Gulf of Maine; C = central California; A = southeastern Alaska; and H = Hawaii. More details are given in the text.

ACKNOWLEDGMENTS

Both authors contributed equally to this effort; author order was determined by coin toss. We thank D. Busemeyer, B. Keane, H. Lim, H. Mills and 3 anonymous reviewers. A portion of this research was supported by funds from NSF grant DEB 9096317 (S.H.R.) and from the University of Cincinnati.

REFERENCES

1. **Alberte, R.S., G.K. Suba, G. Procaccini, R.C. Zimmerman and S.R. Fain.** 1994. Assessment of genetic diversity of seagrass populations using DNA fingerprinting: implications for population stability and management. Proc. Natl. Acad. Sci. USA *91*:1049-1053.
2. **Arens, P., P. Odinot, A.W. van Heusden, P. Lindhout and B. Vosman.** 1995. GATA- and GACA-repeats are not evenly distributed throughout the tomato genome. Genome *38*:84-90.
3. **Baker, C.S., D.A. Gilbert, M.T. Weinrich, R. Lambertsen, J. Calambokidis, B. McArdle, G.K. Chambers and S.J. O'Brien.** 1993. Population characteristics of DNA fingerprints in humpback whales (*Megaptera novaeangliae*). J. Hered. *84*:281-290.
4. **Dow, B.D., M.V. Ashley and H.F. Howe.** 1995. Characterization of highly variable $(GA/CT)_n$ microsatellites in the bur oak, *Quercus macrocarpa*. Theor. Appl. Gen. *91*:137-141.
5. **Gilbert, D.A., N. Lehman, S.J. O'Brien and R.K. Wayne.** 1990. Genetic fingerprinting reflects population differentiation in the California Channel Island fox. Nature *344*:764-766.
6. **Good, P.** 1993. Permutation Tests. Springer-Verlag, New York.
7. **Hartl, D.L. and A.G. Clark.** 1989. Principles of Population Genetics, 2nd ed. Sinauer Associates, Sunderland, MA.
8. **Jeffreys, A.J., V. Wilson and S.L. Thein.** 1985. Hypervariable 'minisatellite' regions in human DNA. Nature *314*:67-73.
9. **Jeffreys, A.J., V. Wilson and S.L. Thein.** 1985. Individual-specific 'fingerprints' of human DNA. Nature *316*:76-79.
10. **Jin, L. and R. Chakraborty.** 1993. A bias-corrected estimate of heterozygosity for single-probe multilocus DNA fingerprints. Mol. Bio. Evo. *10*:1112-1114.
11. **Litt, M. and J.A. Luty.** 1989. A hypervariable microsatellite revealed by in vitro amplification of a dinucleotide repeat within the cardiac muscle actin gene. Am. J. Hum. Genet. *44*:397-401.
12. **Lynch, M.** 1990. The similarity index and DNA fingerprinting. Mol. Bio. Evo. *7*:478-484.
13. **Lynch, M.** 1991. Analysis of population genetic structure by DNA fingerprinting. *In* T. Burke, G. Dolf, A.J. Jeffreys and R. Wolff (Eds.), DNA Fingerprinting: Approaches and Applications. Birkhauser Verlag, Basel.
14. **Nakamura, Y., M. Leppert, P. O'Connell, R. Wolff, T. Holm, M. Culver, C. Martin, E. Fujimoto, M. Hoff, E. Kumlin and R. White.** 1987. Variable number of tandem repeat (VNTR) markers for human gene mapping. Science *235*:1616-1622.
15. **Nei, M.** 1973. Analysis of gene diversity in subdivided populations. Proc. Natl. Acad. Sci. USA *70*:3321-3323.
16. **Nei, M. and W-H. Li.** 1979. Mathematical model for studying genetic variation in terms of restriction endonucleases. Proc. Natl. Acad. Sci. USA *76*:5269-5273.
17. **Rogstad, S.H.** 1993. Surveying plant genomes for variable number of tandem repeats loci. Methods Enzymol. *224*:278-294.
18. **Rogstad, S.H.** 1994. Inheritance in turnip of variable number tandem repeat genetic markers revealed with synthetic repetitive DNA probes. Theor. Appl. Gen. *89*:824-830.
19. **Scribner, K.T., J.W. Arntzen and T. Burke.** 1994. Comparative analysis of intra- and interpopulation genetic diversity in *Bufo bufo*, using allozyme, single-locus microsatellite, minisatellite, and multilocus minisatellite data. Mol. Bio. Evo. *11*:737-748.
20. **Stephens, J.C., D.A. Gilbert, N. Yuhki and S.J. O'Brien.** 1992. Estimation of heterozygosity for single-probe multilocus DNA fingerprints. Mol. Bio. Evo. *9*:729-743.
21. **Tautz, D. and M. Renz.** 1984. Simple sequences are ubiquitous repetitive components of eukaryotic genomes. Nucleic Acids Res. *12*:4127-4138.
22. **Weir, B.S.** 1990. Genetic Data Analysis. Sinauer Associates, Inc., Sunderland, MA.
23. **Weising, K., H. Nybom, K. Wolff and W. Meyer.** 1995. DNA Fingerprinting in Plants and Fungi. CRC Press, Boca Raton.
24. **Wetton, J.H., R.E. Carter, D.T. Parkin and D. Walters.** 1987. Demographic study of a wild house sparrow population by DNA fingerprinting. Nature *327*:147-149.
25. **Wilkinson, L.** 1990. SYSTAT Manual. SYSTAT Inc., Evanston, IL.

Update to: GELSTATS: A Computer Program for Population Genetics Analyses Using VNTR Multilocus Probe Data

Steven H. Rogstad and Stephan Pelikan

University of Cincinnati, Cincinnati, OH, USA (steven.rogstad@uc.edu)

Since version 2 of GELSTATS, an exact permutation test of hypotheses about similarities has been included. This test produces "exact" p-values (as opposed to the p-values reported in our paper) by permuting individuals among the groups. This modification is described in detail in the updated README file that accompanies the program. In addition, several minor changes to the GELSTATS and README files have been made, so we recommend using the latest versions.

We continue to recommend, after a first GELSTATS analysis of population data, the removal of redundant examples of potentially completely linked bands, since retaining these will impart disproportionate weight to the bands of that linkage group. In analyzing data from several plant species, we have, as expected, detected extensive clonal reproduction in some populations (2,4,7,8). Removal of redundant examples of co-clonal individuals from the initial data set is recommended in calculating the final population genetics parameters for such populations. In addition, we are advocating that, in the final calculation of population genetics parameters, all monomorphic bands be retained. Although GELSTATS reports redundant examples of such bands as completely linked, it seems likely that fixed monomorphic bands may occur with some regularity throughout the genome and, thus, may not be completely linked. The retention of these bands may better reflect the degree of minisatellite variation and thus provide more accurate calculation of the various population genetics parameters. Retention of monomorphic bands should also render population genetics calculations more comparable across studies.

A partial list of examples of utilization of GELSTATS includes: (*i*) examining whether two or more populations are genetically homogeneous (1–4,6,7); (*ii*) examining populations for extent and effects of clonality (2,4,7,8); (*iii*) assessing whether pollution affects population genetics (2,3); (*iv*) determining whether a genetic bottleneck has occurred in the formation of two closely related species (6); and (*v*) examining whether differences in genetic variation can be detected between size–age classes (6).

Version 2.63 of GELSTATS is available on the Web at (http//math.uc.edu/~pelikan/GELSTATS/gsindex.html). The source code for additional programs to manipulate GELSTATS format data sets is also available there. These programs include DEDUP (to remove redundant cases of identical columns [e.g., when multiple co-clonal individuals are present] from a data set), PASTE (to combine two or more data sets), JACK (to perform jackknife tests on equality of pairwise similarities), and MAND (to enumerate all possible parent–offspring relationships in a data set).

We note that another approach to making inferences based on similarity data was developed by Leonard et al. (5) using a bootstrap procedure to estimate the variance of the similarities. Statistical Analysis System (SAS) programs for these procedures are available at (http//www.epa.gov/nerleerd/gendivsoftware.htm). The new e-mail address for Steven Rogstad is steven.rogstad@uc.edu.

ACKNOWLEDGMENTS

Continuing work on utilization of GELSTATS has been partially supported by the U.S. Environmental Protection Agency grant R826602-01-0 and funding from the Ohio Plant Biotechnology Consortium.

REFERENCES

1. **Busemeyer, D.T., S. Pelikan, R.S. Kennedy, and S.H. Rogstad.** 1997. Genetic diversity of Philippine *Rubus moluccanus* L. (Rosaceae) populations compared with synthetic DNA VNTR loci probes. J. Trop. Ecol. *14*:867-884.
2. **Keane, B., M.K. Smith, and S.H. Rogstad.** 1998. Genetic variation in red raspberries (*Rubus idaeus* L.; Rosaceae) from sites differing in organic pollutants compared with synthetic tandem repeat DNA probes. Environ. Toxicol. Chem. *17*:2027-2034.
3. **Keane, B., S. Pelikan, M.K. Smith, G. Toth, and S.H. Rogstad.** 1999 Genetic diversity and the effect of pollutants in *Typha latifolia* examined with synthetic tandem-repetitive DNA probes. Am. J. Bot. *86*:1226-1238.
4. **Kumar, A. and S.H. Rogstad.** 1998. A hierarchical analysis of minisatellite DNA genetic diversity in Gambel oak (*Quercus gambelii* Nutt.; Fagaceae). Mol. Ecol. *7*:859-869.
5. **Leonard, A.C., S.E. Franson, V.S. Hertzberg, M.K. Smith, and G.P. Toth.** 1999. Hypothesis testing with the similarity index. Mol. Ecol. *8*:2105-2114.
6. **Lim, H.W., S. Pelikan, and S.H. Rogstad.** 2002. Genetic diversity among populations and size classes of buckeyes (*Aesculus*: Hippocastanaceae) examined with multilocus VNTR DNA probes. Plant Systematics Evolution *230*:125-141.
7. **Rogstad, S.H., B. Keane, and J. Beresh.** Genetic variation across VNTR loci in central North American *Taraxacum* surveyed at different spatial scales. Plant Ecol. (In press.)
8. **Rogstad, S.H., B. Keane, M.H. Collier, and J.R. Shann.** 2001. Plant biomonitors: pollution, dandelions, and mutation rates, p. 350-362.. *In* Lipnick, R L., R.P. Mason, M.L. Phillips, and C.U. Pittman, Jr. (Eds.), Chemicals in the Environment: Fate, Impacts, and Remediation. ACS Books, Washington, D.C.

CHAPTER 46

Increased Throughput for Fragment Analysis on an ABI Prism® 377 Automated Sequencer Using a Membrane Comb and STRand Software

Robert J. Toonen and Shayne Hughes
University of California, Davis, Davis, CA, USA

Reprinted from BioTechniques 31:1320-1324 (December 2001)

ABSTRACT

This manuscript outlines our protocol for using a freely downloadable fragment analysis software package (STRand) together with a 96+4 RapidLoad membrane comb to increase throughput of samples for fragment analysis on ABI sequencers without costly upgrades from the manufacturer. We outline how using these products allows one to score 90 lanes of sample per gel on an ABI PRISM® 377XL (64-lane sequencer), saving both time and money in the processing of samples. This protocol is a major modification to those suggested by the manufacturer. This protocol gives more consistent results that are easier to score than standard protocols, and it reduces reagent costs. Interest in fragment analysis (primarily microsatellites and AFLPs) is steadily increasing among both population ecologists and geneticists, and methods that simultaneously increase sample throughput while reducing costs associated with these analyses by over 50% per gel should prove useful to anyone using an ABI, MJ Basestation™, or LI-COR® automated sequencer for fragment analysis.

INTRODUCTION

Interest in microsatellites and AFLPs as genetic markers of choice for population and ecological genetics studies is steadily growing, and the advantages of fragment analysis on automated sequencers has increased demands on available systems concomitantly. Given increased demands on available sequencers, issues of optimization of multiplex reactions, sample throughput and cost have become the focus of many researchers when deciding which detection technique to use for scoring fragment polymorphisms. Upgrades

BioComputing: Computer Tools for Biologists
Edited by Stuart M. Brown

of the ABI Prism® 377 DNA Sequencer (Applied Biosystems, Foster City, CA, USA) can increase sample throughput to 96 lanes per gel but are expensive and often difficult to justify because of the substantial cost of matching peripheral equipment to the 96-well format. This article describes how we use STRand software (5) and RapidLoad membrane combs (4) from The Gel Company (San Francisco, CA, USA) to score up to 90 samples per gel using GeneScan® (Applied Biosystems) collection software on an ABI 377XL (64-lane) DNA sequencer. This method will work with either ABI 373 or 377 sequencers and provides a cheap alternative to increase the throughput of fragment analysis with only slight modification to existing protocols and without requiring any expensive equipment upgrade.

MATERIALS AND METHODS

PCR and Multiplex Optimization

Using 5′-end labeled Phophoramidite (6-FAM, HEX, and NED) custom primers from Applied Biosystems, dye-labeled amplification products are generated via PCR. PCR protocols were modified from the GeneScan Reference Guide provided by Applied Biosystems (1) to optimize amplification efficiency for marine crabs. ABI slab gel sequencers read four-colored dyes simultaneously, but one dye (ROX) is used for the internal size standard. Hence, three overlapping loci can be scored simultaneously in an individual via multiplex PCR amplification. Where loci with non-overlapping allelic ranges exist within the limitations of the size standard used (Applied Biosystems markets a variety of size standards), two (or more) loci can be labeled with the same dye to increase the number of loci scored in a single reaction with multiple primer pairs. Because the allelic range in the marine crabs *Petrolisthes cinctipes* and *Cancer magister* is unusually large (up to 190 bp), a maximum of seven loci can be simultaneously amplified in a single multiplex reaction depending on the degree of overlap among the allelic range of amplification products.

Although the primer front can be cut from the gel in STRand or GeneScan, the presence of unincorporated primer often obscures small amplification products. This requires either the removal of unincorporated primer from the PCR products via a clean-up step or optimization of the PCR to minimize the amount of unincorporated primer remaining after amplification. We have chosen the latter strategy to eliminate the cost of PCR clean up and increase the speed of scoring samples following PCR. By sequentially halving the primer concentration until the amount of unincorporated primer is minimized, amplification products can often be improved, and the template-to-primer ratios of each locus can be adjusted as necessary to equalize the intensity of amplification among loci. Elimination of the majority of the primer front reduces interference with scoring small products (<100 bp) while saving money on dye-labeled primer use, ROX size standard, and PCR clean-up kits. This procedure also reduces the steps involved and reduces background, thereby making the resulting gel file easier to score.

Gel Preparation

Long Ranger™ Singel packs (BMA, Rockland, ME, USA) are cast according to the manufacturer's protocol (3) using a casting comb (0.2 mm mylar no. 401913; Applied Biosystems). Once the gel is completely polymerized, the casting comb is removed and the well is washed with deionized (DI) water to remove any remaining bits of polymerized acrylamide. The well is visually inspected for a sharp, clean, and straight interface, the gel is held at a 45° angle, and a Kimwipe® is then used on one corner to wick all remaining DI water from the well (while avoiding contact with the gel surface). To ensure the well is completely dry before loading, the gel is left in a rack upside-down (with the well at the bottom) while preparing samples for loading.

Although unnecessary, we have found that pre-warming the equipment produces more consistent and easily scorable gels. We pre-warm the sequencer by using the gel and buffer from the previous run; the sequencer is simply restarted in PreRun mode during sample preparation to bring the system to temperature in advance of gel loading. Once the sequencer is pre-warmed, the new gel comes to running temperature very quickly after being inserted into the machine, which precludes a lengthy pre-run.

Sample Preparation

Dye-labeled amplification products are generated in advance as outlined above, and 1 μL of the PCR amplification is combined with 1.1 μL loading buffer, which consists of 0.16 μL GeneScan-500 ROX size standard (Applied Biosystems), 0.68 μL deionized formamide (Sigma, St. Louis, MO, USA), and 0.26 μL blue dextran/EDTA loading buffer (Applied Biosystems; supplied with GeneScan-500 ROX size standard). This loading mixture is greatly diluted relative to the recommendations of the Applied Biosystems protocol provided with the GeneScan-500 ROX size standard and is the lowest amount of size standard per lane we could load with reliable results. At this loading volume, the size standard is detected at 600–1500 relative fluorescence units (rfu). If higher signal strength is needed, then the volume of the ROX standard in the loading buffer can be easily increased. Some experimentation is required to minimize unincorporated primer in the reactions and to optimize the template-to-primer ratios for each locus to produce an amplification product detected at between 1000 and 4000 rfu (as described above). Once accomplished, this minimum loading volume (0.16 μL/lane) of the ROX size standard works very well and uses only about one-third the suggested amount of this expensive reagent per gel.

After the PCR and loading buffer are mixed, the PCR plate is sealed and the samples are denatured at 95°C for 3 min on a GeneAmp® 9700 thermal cycler (Applied Biosystems) before being transferred directly onto ice. Samples are kept on ice while transferring 0.9 μL of the denatured mixture from a 96-well PCR plate into the 96+4 well RapidLoad membrane comb dipping tray (The Gel Company). The company recommends loading only 0.4–0.7 μL sample onto each tooth of the comb, but the extra volume is sometimes desirable, and the volume of sample loaded can be easily controlled during dipping (explained below). Despite the increased volume transferred to the dipping tray—among evaporation, denaturing, and pipettor error—very little of the denatured samples will be left in the PCR plate after transfer, and reloading is generally not possible in case of sample loss. If multiple loads are frequently necessary, then you may wish to increase your sample preparation volume.

Gel Loading and Running

The denatured samples in the RapidLoad tray are loaded onto the 96+4 membrane comb immediately before being loaded onto the gel. Samples are applied to the comb by dipping the comb into the numbered wells on the tray until the sample is absorbed. If the sample wicks up beyond three-quarters of the height of any tooth, then decrease either the time that you allow samples to wick into the comb (we simply withdraw the comb as soon as the samples cover approximately three-quarters of the tooth) or the amount of the denatured sample transferred to the dipping tray for future loads. In practice, three-quarters of the tooth on the 96+4 RapidLoad membrane comb translates to approximately 0.7 μL sample and ensures no mixing of samples that is due to wicking between lanes.

Without the 96-lane upgrade on the ABI 377XL, the path of the laser is not wide enough to take advantage of the entire 100-lane capacity of the RapidLoad 96+4 membrane comb. The laser detection zone reads only the lanes numbered 4–93 on the RapidLoad dipping tray. However, using STRand, it is possible to take advantage of the full read path (maximum of 90 lanes) of the laser on the ABI 377XL sequencer. The drawback of increasing the number of lanes scored per gel is a reduced number of channels per sample (7). However, in comparisons of 128 multiplex reactions run with the method outlined here to those same samples run using a traditional 64-lane GeneScan protocol (1), we found not a single difference in allele sizes called for a given sample. Unfortunately, the 96+4 RapidLoad combs do not fit the entire width of the gel as do the standard shark's-tooth combs; this is problematic because 90 lanes are the absolute maximum width of the data-gathering region, and any slight variation in centering the comb in the read region will cause either the first or last lane to be missed on the gel file. Therefore, it is useful to mark the comb position on the plate to ensure the same position in every run and to run positive controls in lanes 1 and 90 of each gel so that a slight shift in the comb position does not result in lost data. With the two positive controls, 88 lanes of data can be collected reliably on each gel.

Although it does not appear to be necessary, we leave the RapidLoad dipping tray on ice throughout sample transfer and comb loading because in our experience it increases band resolution and uniformity slightly. Once the samples are transferred into the dipping tray, the gel from the previous run (used to pre-warm the sequencer, as described above) is removed and the new gel is loaded onto the sequencer. If any additional liquid has appeared in the well while the plate is inverted, then the gel is again tilted at a 45° angle and the liquid is removed as previously described. After confirming a flat baseline with a plate check, the front heating plate is attached and buffer is added to the correct level in the bottom buffer tray only. The upper buffer chamber is attached but left empty. The gel run (module GS 36D-1200, 36-cm WTR XL Scan, 2.5-h collection time) is started using a "dummy" sample sheet (we use a 64 sample sheet each time, but the sample sheet information and lane number collected for the gel are irrelevant) and then immediately paused (a No EP Current Error will pause the machine automatically if you fail to do it manually).

A pipettor is used to add 400–500 μL of 5% Ficoll® loading solution (diluted with PCR-quality water from The Gel Company's 20% Ficoll loading solution) to completely fill the well to the point where the solution just begins to bead between the plates. However, it is particularly important to avoid getting any gel loading solution on the back plate while filling the well. Any liquid on the back plate will cause the comb to stick and make loading it awkward, greatly increasing the likelihood of damaging the comb teeth during insertion. Bubbles in the loading solution should be removed with a gel-loading tip or micro-syringe. Once the well is filled with loading solution, we try to minimize the amount of time taken to apply the samples to the comb, insert the comb into the well, and start the electrophoresis run (it should not take more than a minute or two).

After samples have been applied to the membrane comb, it is ready for immediate loading into the loading solution-filled well. This is most easily accomplished by holding the comb at a 45° angle to the back plate and allowing the teeth to first penetrate the bead of loading solution left between the plates (as described above). As the teeth penetrate the bead and then come into contact with the back plate, the angle of the comb can be increased toward 90° to allow the teeth to easily slide down between the plates until they rest just above or on the gel surface. Over-insertion of the comb can damage the gel matrix and lead to irregular spacing between lanes or "bent" lanes that are much more difficult to score. Although the dilute loading solution can have extra blue dextran dye added to increase the resolution of the gel interface, if the gel was poured using the casting comb recommended above (Applied Biosystems), then the teeth of the membrane comb will just barely reach the gel surface when the upper edge is flush with the top of the back plate. After only a few loads to familiarize ourselves with the technique, we found that this method of loading becomes much faster and easier than traditional methods of loading a shark's-tooth comb using a multichannel syringe.

Although the company protocol (4) states that samples will not diffuse into the loading solution, we try to start the run as quickly as possible after inserting the comb. That is why we start and then pause the electrophoresis run before filling the well and loading the samples onto the comb. Once the comb is inserted, the upper buffer chamber can be quickly (but gently) filled. As soon as the door is closed, the run can be resumed, thereby minimizing the time that samples sit in the comb. After restarting the run, we allow 2 min electrophoresis (at 200 W) to transfer the samples into the gel. The run is then paused, and the comb is removed and discarded. We have heard of some people reusing the membrane combs after soaking in DI water to remove any residual sample, but we have not tried this technique. The 5% Ficoll loading solution is then gently rinsed from the well with 1× TBE from the upper buffer chamber using a 200-µL pipettor. It is essential that the loading solution be completely rinsed from the well, but it is also important that the rinsing be relatively gentle to avoid irregularly spaced or wavy gels. The standard method of vigorous rinsing with a 60-cc syringe invariably produced unscorable gels. The 5% dilution is much more easily dislodged than the original 20% Ficoll solution, so gentle rinsing several times across the width of the well is all that is needed to completely displace the loading solution. Once the well is thoroughly rinsed, the lid to the upper buffer chamber is put into place and the run is resumed.

RESULTS

After the gel run is finished, the gel file is transferred to a PC running Microsoft® Windows 98 SE®, on which STRand (5) has been installed. The STRand software package is freely downloadable from the *BioTechniques* Software Library (www.BioTechniques.com) or from the University of California, Davis Veterinary Genetics Laboratory Web page (http://www.vgl.ucdavis.edu/STRand; note that the server is case sensitive) and was installed and configured following the instruction manual (6). The unanalyzed gel file produced by the GeneScan collection software is copied from a Macintosh®-format zip disk onto a PC using MacDrive 2000 for Windows™ (Mediafour, West Des Moines, IA, USA), and the raw gel file is imported into STRand directly. Users can predefine the number of lanes to be found or let the software determine the number of visible lanes on the gel. The tracked gel is then analyzed using a set of user-defined rules for each locus included in the multiplex reaction. These locus rules are fully customizable for specific applications. The program calls both the size standards and the allele/band sizes for each lane of data and allows the user to examine and alter any of the peak or size designations. The program displays a list of peak heights and alleles called for each locus. A separate window that provides user alerts (for problems such as having more than two alleles called for a locus) is also included along with the electropherogram for each sample in a single window, eliminating the need to switch between windows or programs to complete the analysis. Individual dye colors can be hidden as necessary, and both the x- and y-axes of the electropherogram are easily adjustable with sliders for cases when zooming in on the electropherogram can help with scoring.

DISCUSSION

In our experience, results from gels that are run using the protocol we outline here are generally indistinguishable from those using a traditional shark's-tooth comb (Figure 1). We believe that the RapidLoad comb offers three primary advantages over the shark's-tooth comb that are unrelated to data quality. First, all 90 lanes are loaded simultaneously with the RapidLoad comb, which not only decreases load times but also eliminates the need for stagger-loading typically required with multichannel syringes used with a shark's-tooth comb. Second, there is no possibility of samples leaking between wells with a RapidLoad comb. Finally, less product is required for running with the RapidLoad combs, which proves to be a significant cost savings over standard protocols using a shark's-tooth comb.

Although the use of the RapidLoad comb simplifies loading and reduces the cost of reagents per lane on the gel, it is the analysis software, STRand, that allows us to use the full width of the laser path on the gel and get 88–90 lanes of scorable data from the 64-lane ABI 377XL. STRand was originally designed to address the needs of the University of California, Davis Veterinary Genetics Laboratory (VGL) while developing its equine DNA test using 14 STR loci multiplexed into a single lane. The software available at the time was inefficient for that level of high-throughput genotyping and had a tedious user interface without any relational database integration. The goal for STRand was to maximize accuracy while minimizing the analysis time. Flexibility was also important; STRand is ready for new technologies through the easy addition of any gel image format,

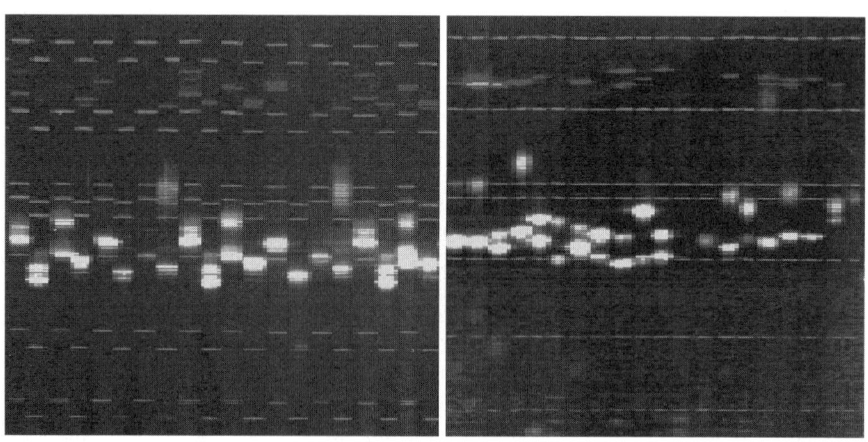

Figure 1. Magnified sections of a typical 64-lane shark's-tooth comb (left) and 64+4 RapidLoad membrane comb (right). Each lane is a four-locus multiplex reaction amplified from individual anomuran crabs, *Petrolisthes cinctipes*, and uses the GeneScan-500 ROX size standard from Applied Biosystems. We have zoomed in on the 200–450 bp portion of the center of each gel (lanes 22–42) for comparison. **(See color plate A8.)**

any number of lanes, any number of dyes, etc. Currently, ABI 373, 377, 310, 3100, and 3700, MJ BaseStation™ (MJ Research, Waltham, MA, USA), and LI-COR IR2® (LI-COR, Lincoln, NE, USA) formats are all supported.

STRand performs the full analysis from the original collection file to genotypes as opposed to using two programs such as GeneScan and Genotyper (Applied Biosystems). VGL found that STRand could cut analysis time by a factor of 4–6 because lane tracking is done automatically, sample sheets are pulled from a database, and the analysis window shows the electropherogram along with the other data with no need to switch between windows or programs. This year VGL expects to complete 140 000 DNA types using 13 markers each for a total of almost 1.7 million genotypes, all analyzed using STRand. A 96-lane gel with 13 STR loci per lane takes a skilled VGL analyst about 45 min to complete. Over the years, STRand may have lost a little of it lead as Applied Biosystems product literature claims 60 min to analyze 96 lanes of 14 markers for GeneScan/Genotyper and 10 min using their new GeneMapper™ software (2). New features on the horizon for STRand should continue to keep it competitive in terms of performance. More attractive to individual research laboratories is the fact that STRand is licensed under the GNU Public License (www.fsf.org) from The Regents of the University of California, which means that it is freely available. The source code is also available so that you can modify and improve the program as needed. The STRand homepage is located at http://www.vgl.ucdavis.edu/STRand (server is case sensitive), from which it is possible to download the software and a manual as well as to subscribe to the users e-mail list.

To manage all of the data generated, STRand stores all results in a relational database. The database can be either a Microsoft Access® database file or Microsoft SQL Server (other database systems may also work but have not been tested). Using a relational database server allows STRand users to reliably share the same database across a network so that all researchers have access to the same samples. Keeping the complete data set in a single location also provides easy access for data checking and for collecting statistics such as allele frequencies. STRand exports results in a text file format similar to that used by Genotyper, or results can be accessed directly from the database.

Although the protocol detailed here is specific to the ABI PRISM 377 automated sequencer, it should be readily adaptable to any automated sequencer that STRand supports. In summary, the method outlined here allows users to easily increase STR sample throughput on an automated sequencer without expensive equipment upgrades and, more importantly, without sacrificing data quality. The additional cost of the RapidLoad membrane combs over a reusable mylar shark's-tooth comb is easily offset by the savings on the GeneScan-500 ROX size standard consumed with each gel. Aside from gaining 24 additional lanes of data per gel, this protocol also reduces our cost; overall, our cost per gel is reduced by over 50% with this protocol relative to the standard 64-lane GeneScan protocol we followed originally (1). Together with the freely downloadable STRand fragment analysis software, the added benefits of the ease of sample loading, and the increased sample throughput of the RapidLoad membrane combs, this protocol should prove useful to anyone using an automated sequencer for fragment analysis.

ACKNOWLEDGMENTS

We are grateful to the staff of the Division of Biological Sciences Sequencing Facility, the Veterinary Genetics Laboratory, and The Gel Company for their advice and assistance in developing and troubleshooting this protocol. We thank M. Locke, A. Wilson, and R.

Grosberg for their assistance and the helpful comments they provided to improve the manuscript. This research was funded in part by a grant from the National Sea Grant College Program, National Oceanic and Atmospheric Administration (NOAA), U.S. Department of Commerce, under grant no. NA06RG0142, project no. R/F-177 through the California Sea Grant College System, and in part by the California State Resources Agency. The views expressed herein are those of the author and do not necessarily reflect the views of NOAA or any of its sub-agencies. The U.S. Government is authorized to reproduce and distribute for governmental purposes.

REFERENCES

1. **Applied Biosystems.** 1998. 377 DNA Sequencer: microsatellite analysis. Applied Biosystems Technical Document, Foster City, CA.
2. **Applied Biosystems.** 2000. ABI PRISM® GeneMapper high-performance genotyping software for linkage mapping applications. Applied Biosystems Technical Document, Foster City, CA.
3. **BioWhittaker Molecular Applications.** 1998. Long Ranger Gel Protocols. BMA, Rockland, ME.
4. **Gel Company.** 1999. RapidLoad Membrane Comb Protocol for Sample Kit KAT96. The Gel Company, San Francisco.
5. **Hughes, S.** 1998. STRand Nucleic Acid Analysis Software. Regents of the University of California, Davis (http://www.vgl.ucdavis.edu/STRand).
6. **Locke, M., E. Baack, and R.J. Toonen.** 2000. The STRand Manual (http://www.vgl.ucdavis.edu/STRand/docs/STRand_Manual.pdf).
7. **Velickovic, Z.M., M.B. Velickovic, and S.K.G. Grebe.** 1999. Increasing sample throughput on a standard 36-lane model 377 DNA sequencer to 48 or 64 samples per run. BioTechniques *27*:652-654.

Update to: Increased Throughput for Fragment Analysis on an ABI Prism® 377 Automated Sequencer Using a Membrane Comb and STRand Software

Robert J. Toonen[1] and Shayne Hughes[2]

[1]University of Hawaii–Manoa, Kane'ohe, Hawaii (toonen@hawaii.edu), and [2]University of California–Davis, Davis, CA, USA

A useful companion program to STRand is ResultViewer. This utility provides several ways to extract genotype data from the STRand database and calculates several useful statistics on that data. The results can be exported to a text file allowing import into spreadsheet programs.

INSTALLATION

ResultViewer is now a part of the STRand package and is automatically installed along with STRand. Visit (http://www.vgl.ucdavis.edu/STRand) to download the latest version of STRand.

USAGE

When you run ResultViewer, you are presented with one main window where you configure the results to display on the left and view those results in a spreadsheet on the right. In the top-left, there are four tabs. The Single and Gel tabs are probably the most useful to people outside of the Veterinary Genetics Laboratory (VGL). The Serial and Project tabs are specific to VGL, and while the Serial tab will work for anyone, the Project tab will not function correctly with the default STRand configuration.

The Single tab allows the user to add sample names one at a time or to load a text file containing a list of sample names. The Gel tab accepts a gel name and loads all the samples on that gel. It gives the option of loading only data found on the one gel specified or using the sample list from the gel to load all data for those samples across all gels.

Below the tabbed area are a few other settings that are seldom used. The Newest Results Only check box tells ResultViewer to only look at data for a particular marker from the most "recent" gel (see below for notes on how this is determined). Turning this setting off is useful only for debugging.

The selection of Type Results and Marker Stats toggles the spreadsheet on the right between these two views of the data. It can be used once ResultViewer has loaded the data to switch back and forth between the views. Type Results shows markers in the columns and samples in the rows. This gives a full display of the genotyping results for the samples displayed. The Marker Stats display also shows markers in the columns, but the rows contain several statistics. First, the number of alleles found for a marker is listed. Below that are observed heterozygosity (OH), predicted heterozygosity (PH), probability of exclusion (PE), polymorphic information content (PIC), and combined probability of exclusion (CPE) is calculated for the combination of all the markers. Below those, all of the alleles configured in STRand's loci rules are listed along with their frequency. Next, the observed genotypes are listed along with frequency and chi-squared values. This last bit of data is still under development.

The Min Gel Name setting allows limiting the range of gel names allowed for the data. This is meant to specify a limit on the age of data used if you use the gel naming scheme that I will describe below.

Finally, the Show button tells ResultViewer to go get the data specified. Use the save button to write the data out to a text file. You will need to save each view of the data, Type Results, and Marker Stats, individually.

Types displayed in red indicate that ResultViewer has found more than two alleles for this sample and marker. Also, single alleles in the STRand data are assumed to be homozygous and are duplicated to show the two alleles.

ResultViewer makes an assumption about the naming of the gels to determine the chronological ordering of the gels. When the data is retrieved, the gels are sorted in reverse alphabetical order, and it is assumed that the most recent gels will be at the top of the list. At VGL, the gels are given a name that starts with the date (i.e., 20020331). Users of ResultViewer will need to do something similar to assure that the correct results are considered when a sample is typed more than once with the same marker.

EXAMPLE

From a fresh installation of STRand, you can use ResultViewer to look at the results of one of the included example gels. Start ResultViewer by clicking the Start button, then Programs, then STRand, and click on ResultViewer. Choose the Gel tab and type in 20000113-C for the gel name. Click Show, and the program will think for a moment before displaying types for all 96 lanes of the gel. This gel has only been automatically analyzed by STRand, so there are several red boxes showing more than two alleles and blank types. You can now click the Marker Stats button to toggle to the statistics view. Clicking Done exits the program.

Section VI

The Internet

This is a catch-all collection of articles which describe a variety of Internet tools that are useful to molecular biologists. Online sequence analysis tools are presented for PCR primer design, multiple alignment, molecular modeling, SNP analysis, automated keyword searching of sequence and literature databases, laboratory products, and biology-related news. While some of the URLs cited in these articles have changed or become obsolete (many of these changes are mentioned in the authors' updates), the majority of these services are still functional. The failure of many Internet-based businesses has had minimal effect on the free services available to biologists.

CHAPTER 47

Designing PCR Primers on the Web

Mildad Biñas

eBioinformatics, Sydney, Australia

Reprinted from BioTechniques 29:988-990 (November 2000)

PCR is perhaps the single most significant advancement in molecular biology in over a decade. Not only has PCR greatly influenced the amount of sequence information currently available, but it is also changing the face of disease diagnosis, food microbiology and even forensic science.

Among all the parameters employed in PCR, perhaps the most critical is primer design. The length of the amplification product, its melting temperature and the total product yield of a reaction are all influenced by the quality of the primer sequences. Several guidelines can be followed when designing primers (5). There are numerous online tools dedicated to helping molecular biologists design effective PCR primers. Most sites have been designed to allow users to easily modify any parameter that affects the design of effective primers.

Primer3

Primer3 (9) is one of the most popular primer design tools available online (see below) and is based on programs Primer 0.5, written by S. Lincoln, M. Daly and E. Lander (6), and Primer v2, written by R. Resnick. For any given sequence, Primer3 determines which subsequence is most suitable for amplification by first using the information about target region and/or product length as specified by the investigator. Once the target region(s) has been determined, Primer3 then looks for regions that may be significantly similar to repeat elements, if any are specified, and further restricts potential target regions for amplification. The T_m for potential primers are then calculated as previously described (2,10). Primers that pass these criteria are paired, and the potential product length for each pair is calculated. The primer pairs' self and pairwise complementarity are assessed, and their sequences are again checked for similarity to repeat sequences using a slower, more sensitive method than what was initially used. The program then chooses the optimal primers from the list generated; one pair of primers is suggested for each range of product length requested.

Primer3 is available at the Center for Genome Research at the Whitehead Institute for Biomedical Research (http://www.genome.wi.mit.edu/cgi-bin/primer/primer3_www.cgi), where an older Web implementation of the same program (http://www.genome.wi.mit.edu/cgi-bin/primer/primer3.cgi) and the Primer3 source code are also available. Local

BioComputing: Computer Tools for Biologists
Edited by Stuart M. Brown

copies of this older implementation are located at several other sites, including the University of Cambridge's Department of Pathology (http://www.path.cam.ac.uk/cgi-bin/primer3.cgi), Northwestern University Medical School (http://www.basic.nwu.edu/biotools/Primer3.html), Marquette University Biomathematics Computational laboratory (http://eve.mscs.mu.edu/RESEARCH/PPICK/pp_info.html) and at the Department of Biological Services at Weizmann Institute of Science, Israel (http://bioinformatics.weizmann.ac.il/cgi-bin/primer/primer3.cgi), while a beta version of the latest implementation is found at the Oxford University Bioinformatics Centre (http://www.molbiol.ox.ac.uk/cgi-bin/primer3_www.cgi). These sites allow users to cut and paste a single nucleotide sequence into a form and set all the various parameters including melting temperature, size, GC content, PCR product size, positional restrictions within the source sequence and other miscellaneous restrictions such as sequence quality information. These sites also allow investigators to change the default penalty weights of each criterion that Primer3 uses to find the most optimal primers. The latest Web implementations found at the Whitehead Institute for Biomedical Research and the Oxford University Bioinformatics Centre also allow investigators to enter sequences of potential primers and test them on the specified criteria.

Primer3 is also available through the Web-based sequence analysis service provider, BioNavigator® (http://www.bionavigator.com) as one of several PCR primer design tools. Sequences can either be cut and pasted, uploaded from the investigator's computer in a variety of known file formats or obtained directly from one of the databases maintained onsite. The output generated by Primer3 at this site allows investigators to select the sequences and save them as query sequences to BLAST and other database similarity searching programs. Moreover, investigators can use multiple sequences to design primers by using tools that extract the consensus sequence from a multiple sequence alignment.

Prime

Prime, which is distributed as part of the Wisconsin Package from the Genetics Computer Group, is also a well-known program for primer design. This program calculates primer melting temperatures using the nearest neighbor model (1) calculation determined by Santalucia (11), while product melting temperatures are determined using the formula as modified by Rychlik et al. (10). Like Primer3, this program also checks primer self and pairwise complementarity as described by Hillier and Green (4). Prime allows investigators to set values for the standard design parameters, which include primer and product length, primer GC content, primer melting and annealing temperatures. This program also allows investigators to test potential primer sequences.

Prime is found in many bioinformatics Web servers, including BioNavigator, the Human Genome Mapping Project, UK (http://www.hgmp.mrc.ac.uk/), the Danish Biotechnological Database, BioBase, at the University of Aarhus, Denmark (http://biobase.dk/index.html) and the Biotechnology Computing Facility at the University of Arizona (http://bcf.arl.arizona.edu/gcg.html).

CODEHOP

CODEHOP is a useful degenerate primer design tool for designing primers to amplify genes encoding members of specific protein families, starting from the sequence of other protein family members. It produces a set of primers with each primer consisting of a short degenerate 3′ core region and a longer 5′ consensus clamp region (8). CODEHOP accepts a set of blocks as input, where a block is an aligned array of amino acid sequence segments without gaps representing a highly conserved region of homologous proteins (7). A position-specific scoring matrix is then computed for each block using the odds-ratio method, which computes weights on diversity observed at each alignment position, rather than traditional weighting method assumptions (3). One consensus residue is selected for each position in the block, and the most common codon corresponding to each consensus residue is selected using the codon usage table specified by the investigator. The DNA position scoring matrix, which is calculated using the position-specific amino acid matrix, is then used to determine the most likely codon for each DNA position, and the high-scoring nucleotide residues are considered for the consensus clamp regions. Possible degenerate regions are identified by scanning the DNA matrix generated in the 3′→5′ direction, looking for an invariant 3′ nucleotide position, a length of 11–12 positions ending on a codon boundary and have a maximum degeneracy, the product of the number of possible bases at each position, of 128. These candidate degenerate regions are then extended using the 5′ consensus clamp regions previously identified; the length is controlled by T_m calculations (8).

CODEHOP is available at the Fred Hutchinson Cancer Research Center (http://www. blocks.fhcrc.org/), its mirror site at the Weizmann Institute of Science (http://bioinformatics.weizmann.ac.il/blocks/index.html) and at BioNavigator. Programs for generating the sequence blocks required as input by CODEHOP are also available at the same sites.

Other Online Primer Design Sites

There are several other sites dedicated to primer design. GeneFisher is a primer design tool (http://bibiserv.techfak.uni-bielefeld.de/genefisher/) that allows investigators to use as input either single sequences or multiple closely related ones. Multiple sequences can be aligned before using this tool, but if sequences are entered unaligned, they are first aligned using ClustalW (13) or DCA (Divide-and-Conquer Multiple Sequence Alignment), a program for producing fast, high-quality multiple sequence alignments using the DCA algorithm, a heuristic approach to sum-of-pairs optimal alignment (12). GeneFisher is intended for those unfamiliar to primer design because it guides the users through each step. The Primer Generator (14) is particularly useful for designing primers appropriate for site-directed mutagenesis experiments. It analyzes the nucleotide sequence and its corresponding amino acid sequence and designs a primer that either has a new restriction enzyme site or is missing an old one, which helps in differentiating between mutated and nonmutated sequences. Primer Generator is located at http://www.med.jhu.edu/medcenter/primer/primer.cgi. Other online primer design tools include the Virtual Genome Center's Web Primers, (http://alces.med.umn.edu/webprimers.html), Primers! for the World Wide Web at Williamstone Enterprises, (http://www.williamstone.com/), DoPrimer at the Virtual Laboratory at Interactiva (http://doprimer.interactiva.de/), Stanford University's Web Primer (http://genome-www2.stanford.edu/cgi-bin/SGD/web-primer/) and Primer Selection at the Baylor College of Medicine Search Launcher (http://dot.imgen.bcm.tmc.edu:9331/seq-util/seq-util.html).

The large number of programs for PCR primer design reflects the central role of PCR in modern molecular biology. It is important to remember that all these programs are designed to assist in the primer design process and are not meant to replace the eye of the experienced researcher, especially considering the sometimes unpredictable nature of PCR experiments. When planning important PCR experiments, it is usually worthwhile to compare the predictions of several different programs and to use common sense and laboratory experience to evaluate the suggested primers before committing to their synthesis.

REFERENCES

1. **Borer, P.N., B. Dengler, I. Tinoco, Jr. and O.C. Uhlenbeck.** 1974. Stability of ribonucleic acid double-stranded helices. J. Mol. Biol. *86*:843-853.
2. **Breslauer, K.J., R. Frank, H. Blocker and L.A. Marky.** 1986. Predicting DNA duplex stability from the base sequence. Proc. Natl. Acad. Sci. USA *83*:3746-3750.
3. **Henikoff, J.G. and S. Henikoff.** 1996. Using substitution probabilities to improve position-specific scoring matrices. CABIOS *12*:135-143.
4. **Hillier, L. and P. Green.** 1991. OSP: a computer program for choosing PCR and DNA sequencing primers. PCR Methods Appl. *1*:124-128.
5. **Innis, M.A. and D.H. Gelfand.** 1994. Optimization of PCRs, p. 13-20. *In* M.A. Innis, D.H. Gelfand, J.J. Sninsky and J.J. White (Eds.), PCR Protocols: A Guide to Methods and Applications. CRC Press, London.
6. **Lincoln, S.E., M.J. Daly and E.S. Lander.** 1991. PRIMER: a computer program for automatically selecting PCR Primers (http://www.genome.wi.mit.edu/ftp/distribution/software/primer.0.5/).
7. **Posfai, J., A.S. Bhagwat, G. Posfai and R.J. Roberts.** 1989. Predictive motifs derived from cytosine methyltransferases. Nucleic Acids Res. *17*:2421-2435.
8. **Rose, T.M., E.R. Schultz, J.G. Henikoff, S. Pietrokovski, C.M. McCallum and S. Henikoff.** 1998. Protein consensus-degenerate hybrid oligonucleotide primers for amplification of distantly related sequences. Nucleic Acids Res. *26*:1628-1635.
9. **Rozen, S. and H.J. Skaletsky.** 1996-1998. Primer3. Code available at http://www.genome.wi.mit.edu/genome_software/other/primer3.html.
10. **Rychlik, W., W.J. Spencer and R.E. Rhoads.** 1990. Optimization of the annealing temperature for DNA amplification in vitro. Nucleic Acids Res. *18*:6409-6412.
11. **Santalucia, J., Jr.** 1998. A unified view of polymer, dumbbell, and oligonucleotide DNA nearest-neighbor thermodynamics. Proc. Natl. Acad. Sci. USA *95*:1460-1465.
12. **Stoye, J., V. Moulton and A.W. Dress.** 1997. DCA: an efficient implementation of the divide-and-conquer approach to simultaneous multiple sequence alignment. CABIOS *13*:625-626,
13. **Thompson, J.D., D.G. Higgins and T.J. Gibson.** 1994. CLUSTAL W: improving the sensitivity of progressive multiple sequence alignment through sequence weighting, position-specific gap penalties and weight matrix choice. Nucleic Acids Res. *22*:4673-4680.
14. **Turchin, A. and J.F. Lawler, Jr.** 1999. The primer generator: a program that facilitates the selection of oligonucleotides for site-directed mutagenesis. BioTechniques. *26*:672-676.

CHAPTER 48

Multiple Sequence Alignment Tools on the Web

Giles J. Gaskell

eBioinformatics, Sydney, Australia

Reprinted from BioTechniques 29:60-62 (July 2000)

The immense number of nucleotide and protein sequences that can be accessed through public databases on the Internet is an invaluable resource to scientists working in the fields of molecular biology, protein chemistry and molecular diagnostics. Tools such as BLAST (1) (also discussed in the March 2000 Internet On-Ramp) and FASTA (7) are used mainly to search for database sequences that are similar or potentially homologous to query sequences supplied by the investigator. However, in many circumstances, this is only the first stage of a more complex series of sequence analyses. Multiple sequence alignment analyses are fundamental in allowing investigators to draw conclusions about the similarity, function, structure and the potential evolutionary relationships between sequences. This column will focus on a number of different multiple sequence alignment tools available as services over the Web.

ClustalW

ClustalW (10), one of the most popular multiple nucleotide or protein sequence alignment programs, uses a progressive alignment approach (4). Progressive multiple alignments are created by first aligning the most similar of a set of sequences and then incorporating less similar sequences successively into the alignment. ClustalW aligns sequences using a global alignment algorithm (6) that generates an alignment over the entire length of the sequences. A neighbor-joining method (8) is first used to construct a guide tree. This determines the order in which the sequences are incorporated into the alignment. A comparison of multiple sequence alignment programs revealed that ClustalW performs well when aligning equidistant sequences of a similar length and when aligning small to large families of similar sequences, in which a few divergent sequences are also included in the alignment (11). In both cases, the performance of ClustalW was maintained from short sequences of less than 100 residues to those of over 400 residues.

ClustalW is available on the Internet through several locations: the European Bioinformatics Institute (EBI) (http://www2.ebi.ac.uk/clustalw/), the Baylor College of Medicine (http://dot.imgen.bcm.tmc.edu:9331/multi-align/Options/clustalw.html), the GenomeNet

BioComputing: Computer Tools for Biologists
Edited by Stuart M. Brown

server in Japan (http://www.clustalw.genome.ad.jp/), the Institut Pasteur Software Environment (PISE) in France (http://bioweb.pasteur.fr/intro-uk.html), the Pôle Bio-Informatique Lyonnais (http://pbil.ibcp.fr/) also in France, the Transfac group at the GBF in Germany (http://transfac.gbf.de/programs.html) and the Web-based sequence analysis service provider, BioNavigator™ (http://www.bionavigator.com). These servers allow investigators to cut and paste their sequences into forms on their Web sites and set various parameters, such as penalty values associated with the insertion of gaps into the sequences, to optimize the overall alignment. After submitting the job, the server computes the alignment, which is then displayed on a new page. Note that most ClustalW Web services that present the alignment directly to the investigator often time out on long alignments and lose the results. These instances occur when many and/or long sequences are submitted to these servers. Therefore, they are only suitable for aligning small numbers of short sequences. Services that get around this limitation are those that send their results by e-mail and those that store the results on the server, such as BioNavigator.

PileUp

Like ClustalW, PileUp uses both a progressive approach and the global alignment algorithm (6) when it aligns sequences. However, the order in which the sequences are incorporated into the alignment is determined by a guide tree that is constructed using the UPGMA method (9). PileUp is usually distributed as part of the Wisconsin Package from the Genetics Computer Group (GCG) (http://www.gcg.com), which is licensed to numerous bioinformatics services on the Internet, including BioNavigator, NIH Helix Systems (http://helix.nih.gov/newhelix.html), the Human Genome Mapping Project, UK (HGMP) (http://www.hgmp.mrc.ac.uk/) and the Biotechnology Computing Facility at the University of Arizona (http://bcf.arl.arizona.edu/gcg.html). The Wisconsin Package also includes tools for improving the appearance of multiple sequence alignments, which will be discussed later in this paper.

Dialign

Dialign (5) is a multiple sequence alignment program that takes an approach that is different from ClustalW and PileUp. Dialign uses a local alignment algorithm that compares sequence segments with each other to generate a multiple alignment. Sequences to be aligned are first broken down into gap-free segments of high similarity. These segments are then built up into a multiple alignment using an iterative procedure. In the same study comparing multiple sequence alignment programs (11), Dialign performed well when aligning sequences in which large terminal extensions or large insertions were present. In contrast to ClustalW, this program is useful for finding blocks of highly conserved regions within a set of sequences. Dialign is available on the Internet from Genomatix (http://genomatix.gsf.de/), Bielefeld University's Faculty of Technology (http://bibiserv.techfak.uni-bielefeld.de/), PISE and the HGMP.

Match-box

Match-box (3) is a multiple sequence alignment tool that is designed to align blocks of conserved or similar regions within sets of protein sequences. The Match-box server is located at the University of Namur in Belgium (http://www.fundp.ac.be/sciences/biologie/bms/matchbox_submit.html). Its method of alignment does not take into account gap penalty values, and the program requires that only a few parameters be set. A study that compared Match-box to six other multiple sequence alignment programs (2) revealed that Match-box was able to predict, with a high degree of reliability, structurally conserved regions between sets of protein sequences of known structure. Match-box scores each column in an alignment block to indicate the reliability of the aligned sequences within that block. This feature is not available with many other programs.

A comprehensive listing of multiple sequence analysis tools available from the Internet can be found at the VSNS BioComputing Division at Bielefeld University's Faculty of Technology (http://www.techfak.uni-bielefeld.de/bcd/Curric/MulAli/welcome.html).

Tools for Improving the Appearance of Multiple Sequence Alignments

So far, tools that generate multiple sequence alignments have been discussed. However, many of these programs display their results in their own formats, which are often difficult to read and may not be suitable for publication. This section will discuss a few tools that take multiple sequence alignments and reformat them into publication-quality figures. Note that these programs do not actually generate the multiple sequence alignments themselves.

BoxShade (written by Kay Hofmann and Michael Baron) is a tool that reformats multiple sequence alignments. It takes an alignment and shades or colors the nucleotides or amino acids of each sequence according to their degree of similarity in the alignment. This shading by similarity is either based on an identity threshold for residues in each column of the alignment or according to a specific sequence. The program also provides the option of displaying a consensus sequence beneath the alignment. The resulting file can be saved in a variety of different formats that can be imported into common word processors, image editors or desktop publishing packages. BoxShade is available on the Web through the Swiss Node of the European Molecular Biology network (http://www.ch.embnet.org/software/BOX_form.html) and PISE. Both servers allow multiple alignments to be pasted or uploaded into the program and are capable of shading alignments in varying tones of gray (depending on the level of similarity).

Two other programs that format multiple alignments are Pretty and PrettyBox from the Wisconsin Package (GCG). Pretty generates and displays consensus sequences in an alignment, but it does not shade residues according to levels of similarity. PrettyBox works essentially the same way as Pretty, though the program also provides shading features that make it function similarly to BoxShade. However, unlike BoxShade, PrettyBox gives the investigator additional control over the shading thresholds for different levels of similarity. Integrated bioinformatics Web sites can allow the investigator to align multiple sequences using a program such as ClustalW and to improve the appearance of the resulting alignment using PrettyBox or BoxShade. Both functions can be accomplished with relative ease in contrast to several other stand-alone multiple sequence alignment services that require several manual steps between different Web sites to achieve the same result.

The Internet is a significant resource to the bioscientist and is one of the easiest means of gaining access to a wide range of powerful bioinformatics tools for the analysis of sequence data. While there are several multiple sequence alignment programs available on the Web, it is important to realize that the results they generate can be used in other analytical tools, such as those designed for molecular phylogenetic or protein molecular modeling studies. Integrated Web servers such as BioNavigator greatly facilitate the ability to conduct such studies.

ACKNOWLEDGMENT

I greatly appreciate the advice and help of Dr. Bruno Gaëta, Senior Scientist at eBioinformatics, in the preparation of this manuscript.

REFERENCES

1. **Altschul, S.F., W. Gish, W. Miller, E.W. Myers and D.J. Lipman.** 1990. Basic local alignment search tool. J. Mol. Biol. *215*:403-410.
2. **Briffeuil, P., G. Baudoux, C. Lambert, X. De Bolle, C. Vinals, E. Feytmans and E. Depiereux.** 1998. Comparative analysis of seven multiple protein sequence alignment servers: clues to enhance reliability of predictions. Bioinformatics *14*:357-366.
3. **Depiereux, E., G. Baudoux, P. Briffeuil, I. Reginster, X. De Bolle, C. Vinals and E. Feytmans.** 1997. Match-Box_server: a multiple sequence alignment tool placing emphasis on reliability. Cabios *13*:249-256.
4. **Feng, D.F. and R.F. Doolittle.** 1987. Progressive sequence alignment as a prerequisite to correct phylogenetic trees. J. Mol. Evol. *25*:351-360.
5. **Morgenstern, B., A. Dress and T. Werner.** 1996. Multiple DNA and protein sequence alignment based on segment-to-segment comparison. Proc. Natl. Acad. Sci. USA *93*:12098-12103.
6. **Needleman, S.B. and C.D. Wunsch.** 1970. A general method applicable to the search for similarities in the amino acid sequence of two proteins. J. Mol. Biol. *48*:443-453.
7. **Pearson, W.R. and D.J. Lipman.** 1988. Improved tools for biological sequence comparison. Proc. Natl. Acad. Sci. USA *85*:2444-2448.
8. **Saitou, N. and M. Nei.** 1987. The neighbor-joining method: a new method for reconstructing phylogenetic trees. Mol. Biol. Evol. *4*:406-425.
9. **Sneath, P.H. and R.R. Sokal.** 1973. Numerical Taxonomy. Freeman, San Francisco, CA.
10. **Thompson, J.D., D.G. Higgins and T.J. Gibson.** 1994. ClustalW: improving the sensitivity of progressive multiple sequence alignment through sequence weighting, position-specific gap penalties and weight matrix choice. Nucleic Acids Res. *22*:4673-4680.
11. **Thompson, J.D., F. Plewniak and O. Poch.** 1999. A comprehensive comparison of multiple sequence alignment programs. Nucleic Acids Res. *27*:2682-2690.

CHAPTER 49

Molecular Modeling on the Web

Shoba Ranganathan

eBioinformatics, Sydney, Australia

Reprinted from BioTechniques 30:50-52 (January 2001)

With the worldwide genomic sequencing efforts producing a steady stream of data, the sequences of tens of thousands of proteins can now be determined. Understanding protein function requires knowledge of protein structure. Information on the 3-D structure of a protein can provide insight to the protein's interactions with other molecules; this information can also be exploited in such endeavors as protein engineering, mutagenesis, and computer-aided drug design. Detailed structural information is available for only 10% of all known protein sequences. Of the remaining 90%, a homologous protein of known structure has been identified for 10%–25%. For this subset, a rapid in silico approach to 3-D structure is accomplished using homology or comparative modeling. For another 25% of the available protein sequences, structural models can be generated using secondary and tertiary structure prediction methods.

Until recently, many biologists have been reluctant to use molecular modeling as a computational tool because of the high costs of modeling software, its user "unfriendliness", and the expensive hardware (usually UNIX® platforms) dictated by the software. Biologists have thus been hindered from making critical associations relating biological structure to function. So, if you have a protein sequence (referred to as the query sequence) and would like to build a molecular model using Internet-based tools, here is a route map.

UNDERSTANDING PROTEIN STRUCTURE

An understanding of protein structure is a prerequisite to molecular modeling. The Protein Databank (PDB; www.rcsb.org/pdb) is the single worldwide archive of structural data on biological macromolecules, primarily proteins. The structural information in the PDB is derived primarily from experimental studies using electron microscopy, X-ray crystallography, and NMR. On the structural scale, a protein can span four levels: primary, secondary, tertiary, and quaternary structures. A brief introduction to these and similar terms used in describing protein structure is available on the Web (http://molmod.angis.org.au/~shoba/asbmb/). A more detailed introduction to proteins and structures can be accessed at the PDB site itself (http://www.rcsb.org/pdb/education.html). A tip for first-time visitors to the PDB is to start with a structure itself and follow all the Web pages describing that structure. My favorite is 1SHA (the PDB identifier or ID code), an example of the phosphotyrosine recognition domain. Simply enter "1SHA" in the box next to

BioComputing: Computer Tools for Biologists
Edited by Stuart M. Brown

"Enter a PDB ID" under the "Search" heading at the left of the PDB Web page. When you do not know the PDB ID, you can search the PDB using "SearchLite" or SearchFields" and look for structures containing the keywords that you have specified. The PDB database entry itself is a very large test file that contains a descriptive section called the header and the x, y, and z coordinates for each atom of the macromolecule. Normally, only the "heavy" atoms are listed, excluding the hydrogen atoms. It is nearly impossible to read this text file and figure out how the protein is folded without the aid of graphic viewers such as Rasmol (www.bernstein-plus-sons.com/software/rasmol) or the Web browser plug-in, Chime (http://www.mdli.com/support/chime/default.html).

IS THERE A STRUCTURAL HOMOLOG?

Identifying a structural homologue simplifies the molecular modeling problem to one of homology model building. BLAST and PSI-BLAST searches of a new protein sequence against the sequences in the PDB database can be used to locate structural templates. One of the problems resulting from database matching is figuring out which of the hits to base the model on. A match spanning the entire length of a sequence and with a good score can be considered good. A poor score over a small match length (less than 25% of the query sequence) is not considered significant, and secondary or tertiary prediction methods will then have to be tried.

The homology modeling approach works well when the query protein has close homologues of known structure and is at the basis of the SWISS-MODEL (2) automated structure prediction system (http://www.expasy.ch/swissmod/SWISS-MODEL.html). The Prodom database of protein domains (http://www.toulouse.inra.fr/prodom.html) provides a BLAST server to search for domains contained in a query sequence and allows this information to be passed directly to the SWISS-MODEL server for subsequent protein structure modeling. Another completely automated service for homology models is the CPH Model server (http://www.cbs.dtu.dk/services/CPHmodels/), which uses a FASTA search to locate homologues within the PDB and then performs rapid homology modeling using distance constraints predicted by neural networks and probability density functions (5). The procedure works well for single domain sequences that are highly homologous to a structure but fails for multidomain protein sequences. For a quick look (the "rough and ready" approach), this server will generate a homology model complete with coordinates in the PDB format.

The integrated Web-based bioinformatics analysis system BioNavigator (http://www.bionavigator.com) takes a more "manual" approach to structure prediction using the same principle: the researcher can run a BLAST search of the PDB protein structure database and then enter the query sequence and its matching structure into other programs for building a structural model of the query protein based on the structure. The latter method gives the researcher complete control over the entire model-building process.

Proteins often contain several modules or domains, each with a distinct evolutionary origin and function. The new CD-Search service from NCBI (http://www.ncbi.nlm.nih.gov/Structure/cdd/cdd.shtml), based on the reverse position-specific BLAST algorithm, is very useful in identifying the conserved domains present in a protein sequence. Since these conserved domains correspond to compact structural units, it is possible to identify individual domains of multidomain proteins. If you do not find any significant hits with the default parameters, you might be able to locate more remote homologues by following the link for "advanced options" and increasing the expectation value.

TEMPLATE SELECTION

When there is more than one structural homologue available, the template for building the model will have to be selected carefully, in accordance with rules for comparative protein modeling (7). Generally, an X-ray structure is preferred to an NMR structure, and among X-ray structures, the one with the lowest resolution is chosen. Normally, more than one template is used only when the extra templates contribute to significant new structural information.

LINING UP THE QUERY AND THE TEMPLATE SEQUENCE

Model-building techniques depend on the alignment of the query sequence with those of the templates. The matched blocks in the alignment usually represent the structurally conserved regions (SCRs) and include secondary structural elements, while the insertion-deletion segments (indels) in the alignment represent the structurally variable regions (VRs), usually loops. Alignments are usually improved by using a set of homologous sequences, and the many programs available for this step have been reviewed recently (3).

BUILDING THE MODEL

Based on the alignment, the SCRs of the query sequence are constructed by simply copying the backbone and building new side-chains. The VRs are then modeled by joining the appropriate backbone atoms for deletion regions or by building in new segments for the insertion regions, normally using loop generation methods (4).

Besides automated model building available from the SWISS-MODEL server, the WHAT IF server (8) (www.cmbi.kun.nl:1100/WIWWWI) can generate a homology model based on the researcher's alignment with a single template structure. However, WHAT IF does not handle VRs in the alignment, resulting in structural models with chain breaks. Nevertheless, the WHAT IF structural model is very useful especially for side-chain conformations and can be used for in silico experiments such as mutations and disulfide bridge location before wet-lab experimentation.

A high-quality structural model with all VRs and loop refinements using Monte Carlo simulations can be built using the ICM (1) methodology via the BioNavigator interface. The input required here are the query sequence and an alignment with a single structural template.

HOW GOOD IS THE MODEL STRUCTURE?

The quality of the model is estimated using structural quality analysis tools. Complete analysis can be done using WHAT CHECK (6), which is available as part of the WHAT IF package from the WHAT IF server. Ramachandran plots for checking the backbone torsions can be generated using the ICM package described above.

BEYOND MOLECULAR MODELING

The structural model is only the beginning of a series of in silico experiments that can be performed. With a model, it is possible to look for functional residues at an active site, based on the location of key residues, which can then be mutated in the laboratory. Stabilizing a protein by incorporation of disulfide bridges is a common technique used in molecular biology and biochemistry. Instead of mutating pairs of residues at random, the structural model can be used to scan for the best location of a disulfide bridge, serving to direct experimental research. Such computer-based experiments are easily accessible through the previously mentioned WHAT IF and BioNavigator Web sites.

ACKNOWLEDGMENTS

The author's current address is Shoba Ranganathan, Bioinformatics Centre, National University of Singapore, Singapore. e-mail: shoba@bic.nus.edu.sg.

REFERENCES

1. **Cardozo, T., M.M. Totrov, and R.A. Abagyan.** 1995. Homology modeling by the ICM method. Proteins Struct. Funct. Genet. *23*:403-414.
2. **Guex, N., A. Diemand, and M.C. Peitsch.** 1999. Protein modeling for all. Trends Biochem. Sci. *24*:364-367.
3. **Gaskell, G.J.** 2000. Multiple sequence alignment tools on the Web. BioTechniques *29*:60-62.
4. **Johnson, M.S., N. Srinivasan, R. Sowdhamini, and T.L. Blundell.** 1994. Knowledge-based protein modeling. Crit. Rev. Biochem. Mol. Biol. *29*:1-68.
5. **Lund, O., K. Frimand, J. Gorodkin, H. Bohr, J. Bohr, J. Hansen, and S. Brunak.** 1997. Protein distance constraints predicted by neural networks and probability density functions. Protein Eng. *10*:1241-1248.
6. **Rodriguez, R., G. Chinea, N. Lopez, T. Pons, and G. Vriend.** 1998. Homology modelling, model and software evaluation: three related resources. Bioinformatics *14*:523-528.
7. **Sali, A. and J.P. Overington.** 1994. Derivation of rules for comparative protein modeling from a database of protein structure alignments Protein Sci. *3*:1592-1596.
8. **Vriend, G.** 1990. WHAT IF: a molecular modeling and drug design program. J. Mol. Graph. *8*:52-56.

Update to: Molecular Modeling on the Web

Shoba Ranganathan

National University of Singapore, Singapore (shoba@bic.nus.edu.sg)

INTRODUCTION

In the era of comparative genomics, building a molecular model based on a structural template has become the first approach to predicting and localizing protein function. Here is a brief update to the earlier paper (6) on how to use Web-based tools to build a 3-dimentional (3D) protein model by homology or comparative modeling. A single integrated interface as provided by BioNavigator (6) is no longer available, but a judicious combination of several freely available Web-servers may be used in its place.

UPDATE ON UNDERSTANDING PROTEIN STRUCTURE

The Protein Databank (PDB) (www.rcsb.org/pdb) is the single worldwide archive of structural data on biological macromolecules, primarily proteins containing 11 737 structures (as of 1 April 2002). The structural information in the PDB derives primarily from experimental studies using X-ray crystallography and electron microscopy (82.5%) and nuclear magnetic resonance (NMR) (15.3%). About 3% of the entries are from theoretical molecular modeling studies. A brief introduction to the protein structure is available on the Web (lecture 6 from the S* Life Science Informatics Alliance: http://www.s-star.org/schedule.html).

NEW WAYS TO LOOK FOR STRUCTURAL HOMOLOGUES

Protein Basic Local Alignment Search Tool (BLAST) and Position-Specific Iterated (PSI)-BLAST against the sequences in the structural database PDB still remain the most rapid methods to locate structural templates, using the BLAST server at the National Center for Biotechnology Information (NCBI) (http://www.ncbi.nlm.nih.gov/BLAST/). Since most proteins of approximately 150 residues contain >1 structural units, it is possible to identify individual domains of multidomain proteins. For multidomain proteins, the search for Conserved Domains on the BLAST server now offers two options: reverse position-specific (RPS)-BLAST and the domain architecture (DART) search.

Remote homologues may be detected using the 3D-Position-Specific Scoring Matrix (PSSM) (5) server (http://www.sbg.bio.ic.ac.uk/~3dpssm/), based on a fast fold recognition method using 1-dimensional (1D) and 3D sequence profiles coupled with secondary structure and solvation potential information. FUGUE (8) (http://www-cryst.bioc.cam.ac.uk/~fugue/prfsearch.html) is another fold recognition server that can provide low similarity templates from sequence-structure profile alignments.

LINING UP THE QUERY AND THE TEMPLATE SEQUENCE

This aspect is still the Achilles' heel of model building. Two different algorithms: CLUSTALW and MULTALIN, are conveniently available at the Network Protein Sequence Analysis server (3) (http://npsa-pbil.ibcp.fr/cgi-bin/npsa_automat.pl?page=/NPSA/npsa_server.html). It is worthwhile trying out both alignment algorithms and then selecting the best alignment based on conservation of functional residues, the lowest number of gaps, and their occurrence in loop or turn segments of the structural templates.

BUILDING THE MODEL

3D-JIGSAW (1) (http://www.bmm.icnet.uk/servers/3djigsaw/) is a new automated server for generating 3D models, with both interactive as well as automated modes of template selection. This methodology performed extremely well (10) in the homology modeling category at the fourth Critical Assessment of Structure Prediction (CASP) competition. 3D-JIGSAW is suitable for multidomain proteins, since templates are identified for each domain separately and then assembled together, and is freely available.

Another server that is freely available from the San Diego Supercomputer Center (SDSC) is SDSC1: SDSC Protein Structure Homology Modeling Server (http://cl.sdsc.edu/hm.html), based on Hidden Markov method for improving sequence alignments (9). This server is provided mainly for testing and benchmarking purposes and is not optimized for high quality template selection or model generation. It may be used to generate a "quick-and-dirty" model.

HOW GOOD IS THE MODEL STRUCTURE?

The ERRAT (2) server (http://www.doe-mbi.ucla.edu/Services/ERRAT/) provides a new tool for rapid identification of hot-spots in modeled structures. It is also useful comparing the quality of template and modeled structures, as well as the improvement in model quality as a function of iterative model building.

BEYOND MOLECULAR MODELING

Besides mutation analysis and protein engineering (4), molecular modeling can be used to predict antigenic sites on the surface of viral capsids (7), as well as peptide segments for vaccine design.

REFERENCES

1. **Bates, P.A., L.A. kelley, R.M. MacCallum, and M.J.E. Sternberg.** 2001. Enhancement of protein modelling by human intervention in applying the automatic programs 3D-JIGSAW and 3D-PSSM. Proteins *45 (Suppl 5)*:39-46.
2. **Colovos, C. and T.O. Yeates.** 1993. Verification of protein structures: patterns of nonbonded atomic interactions. Protein Sci. *2*:1511-1519.
3. **Combet, C., C. Blanchet, C. Geourjon, and G. Deléage.** 2000. NPS@:network protein sequence analysis. Trends Biochem. Sci. *25*:147-150

4. **Giannakis, E., D.A. Male, R.J. Ormsby, C. Mold, T.S. Jokiranta, S. Ranganathan, and D.L. Gordon.** 2001. Multiple ligand binding sites on domain seven of human complement factor H. Int. Immunopharm. *1*:433-443.
5. **Kelley, L.A., R.M. MacCallum, and M.J.E. Sternberg.** 2000. Enhanced genome annotation using structural profiles in the program 3D-PSSM. J. Mol. Biol. *299*:499-520.
6. **Ranganathan, S.** 2001. Molecular modeling on the web. BioTechniques *30*:50-52.
7. **Ranganathan, S., S. Singh, C.L. Poh, and V.T.K. Chow.** 2002. The hand, foot and mouth disease virus capsid: sequence analysis and prediction of antigenic sites from homology modeling. Appl. Bioinformatics *1*:43-52.
8. **Shi, J., T.L. Blundell, and K. Mizuguchi.** 2001. FUGUE: sequence-structure homology recognition using environment-specific substitution tables and structure-dependent gap penalties. J. Mol. Biol. *310*:243-257.
9. **Shindyalov, I.N. and P.E. Bourne.** 2000. Improving alignments in HM protocol with intermediate sequences, p. A-92. *In* Fourth Meeting on the Critical Assessment of Techniques for Protein Structure Prediction.
10. **Tramontano, A., R. Leplae, and V. Morea.** 2001. Analysis and assessment of comparative modeling predictions in CASP4. Proteins *45(Suppl 5)*:22-38.

CHAPTER 50

Get Your Bioinformatics on the Web!

Stuart M. Brown
NYU School of Medicine

Reprinted from BioTechniques 28:244-246 (February 2000)

A new kind of Web-based bioinformatics service is emerging to aid bench scientists. Integrated suites of programs are being hosted on remote servers and being offered as a package for a membership fee, or for free with the option to purchase added value services. The companies providing these Web tools are following a new Internet business strategy known as the Application Service Provider (ASP) model.

The basic idea of an ASP is that it is more efficient to have complex and expensive software hosted centrally on the well-maintained computers of an expert service provider rather than to buy the software and run it on one's own computer. From the perspective of a lab manager or PI, this is not so different from the central computing centers that most large universities and pharmaceutical research divisions provide for their scientists. However, these new bioinformatics ASPs offer this level of professional service to individual investigators and small labs that do not have access to high-quality central computing facilities. An ASP can be considered an extreme form of outsourcing of computer technology—where the vendor not only provides the computer hardware, software and system maintenance, but the actual computers that are located off-site as well.

Bioinformatics software plays a similar central role at a pharmaceutical company or biomedical research institution as does enterprise software for a bank or large retail company. Forrester Research has estimated that large companies spend about 80% of the initial cost of their major enterprise software applications each year on maintenance. This large investment also means that it is hard to change these applications quickly. In the ASP model, the costs and hassles of maintenance and upgrades are the responsibility of the ASP vendor, not the scientist. If new or better software becomes available, the ASP vendor can add it, or the scientist can move to another vendor that offers better service. Another advantage of an ASP is that it is possible for users to pay for access to software as they use it, and to have access to more different types of software that offer richer functionality than they might be able to afford on their own.

The availability of bioinformatics tools on the Web is nothing new—NCBI has been providing BLAST searches on the Web for over five years. The primary advantages of these new integrated packages offered by the ASP vendors are a permanent place to store your data and the results of your analysis and a wide selection of programs available from a single Web interface. There are also various automated tools that can regularly search

BioComputing: Computer Tools for Biologists
Edited by Stuart M. Brown

new sequence data with a set of your own query sequences or automatically perform a series of analyses on a batch of data files. Some of the ASP vendors also offer various forms of technical support and user training, but this tends to be a more costly value-added service.

Fee for Service Bioinformatics ASPs

The Genetics Computing Group (GCG) is a prime mover in the area of bioinformatics ASPs with its new SeqWeb™ interface to the Wisconsin Package of DNA analysis programs. GCG has recently entered a strategic partnership with Viaken Systems, Inc. to provide secure Internet access to SeqWeb™ as well as GCG's new SeqStore™ data management product, so that labs can have a central database to store sequence data, primers, plasmids, genotypes and data from gene expression microarrays. The target for these services are labs at smaller universities that do not provide central molecular biology computing resources, and small companies. However, Viaken may be able to offer better service and technical support than some larger university "computing centers" that often provide an out-of-the-box GCG installation and out-of-date local databases with minimal technical support from a systems manager with no knowledge of molecular biology or bioinformatics.

The University of Sidney has spun off a private company called eBioinformatics, Inc., based on technology developed as part of the Australian National Genomic Information Service (ANGIS) to provide commercial bioinformatics services to academic and corporate scientists. eBioinformatics offers access to its own comprehensive suite of Web-based bioinformatics tools called BioNavigator™ with a usage-based fee for time used and amounts of data stored. Registration on BioNavigator provides access to GCG SeqWeb as well as eBioinformatics' own proprietary software and databases. Subscribers are offered a comprehensive system that includes storage of files, software, databases, documentation, training and support. BioNavigator tools include PCR primer design, restriction mapping, protein secondary structure prediction, database similarity searching (and scheduled automatic searches), pairwise and multiple alignment and construction of phylogenetic trees. BioNavigator provides "Protocols" that are step-by-step tutorials that guide users through common bioinformatics procedures such as making multiple alignments or running a similarity search. BioNavigator also provides the feature "SMARTlink™" that features an intelligent application service link to any IMAGE clone that is identified in a bioinformatics operation, such as a BLAST search of GenBank®, and an immediate connection to GenomeSystems Inc. for acquisition of the clone.

The Canadian Bioinformatics Resource provides Web and Telnet access to GCG and other bioinformatics applications for a small yearly fee. However, accounts are only available to researchers who are located in Canada and are affiliated with a university, hospital, government department or nonprofit organization. This service is a replacement for the sale of outsourced GCG accounts, which was formerly offered by the bioinformatics company, Base4, Inc.

Hyseq Inc. has set up its own version of a Web-based bioinformatics/e-commerce tool called GeneSolutions™. GeneSolutions offers pay-as-you-go access to Hyseq's own proprietary databases of over 12 million DNA sequences (primarily ESTs). The Hyseq databases include many rarely expressed genes, extended sequence of cDNAs that are found in public databases and sequences of the same gene from multiple individuals that may aid in the characterization of polymorphisms. The GeneSolutions Web site allows users free access to store query sequences and to run scheduled searches against new data, but there are fees to see the sequences discovered by the searches and much larger fees to obtain the clones from Hyseq. Hyseq has also recorded tissue-specific expression profiles for thousands of genes based on actual counts of the numbers of molecules found in RNA extracted from approximately 50 libraries of tissue sources. Tissue-expression data may be added to a sequence profile for yet another fee.

"Free" ASPs and Integrated Bioinformatics Web Sites

DoubleTwist, Inc. (formerly Pangea Systems of Oakland, CA, USA) is in the process of creating a Web-based bioinformatics tool called DoubleTwist™. DoubleTwist is a "free integrated Web portal for online genetic research" that will include automated "research agents" to simplify common bioinformatics operations. Users will be able to store their own sequences on the site, set up automated searches and store search results. The site will make use of proprietary databases such as DoubleTwist's EST consensus sequences, the ProNet™ database of protein interactions from Myriad Genetics and a set of 50000 proprietary full-length cDNA sequences from AlphaGene. A researcher using DoubleTwist who finds a match between a query sequence and an AlphaGene clone will be able to license or purchase the clone. DoubleTwist will also provide "contextual" e-commerce links from life science vendors such as Clontech Laboratories that will offer products for sale that target a researcher's specific needs. In a sense, DoubleTwist is a portal Web site for biologists, offering them free bioinformatics tools to attract them and then sell them items such as clones and reagents. Other portal sites such as the BioMedNet HMS Beagle (www.biomednet.com/hmsbeagle) have targeted biologists by providing interesting magazine-like content or science-related news headlines (www.biotaq.com).

The Computational Biology Group at the National Center for Supercomputing Applications (NCSA), University of Illinois, has developed a free Web-based program called The Biology Workbench, a computing environment that integrates many standard protein and nucleic acid sequence databases and a wide variety of sequence analysis programs into a single interface. The WorkBench allows users to perform database searches, multiple alignments, phylogenetic analyses, protein secondary structure prediction and motif searches, all without worrying about file formats or often-confusing command line options and flags. Sequences, projects and results are saved in a password-protected area for each user, and encryption is used to transmit data between the NCSA computer and the Web browser on the user's computer.

The UK Human Genome Mapping Project Resource Centre offers free access (for bona fide academic researchers) to a comprehensive range of programs and databases to aid genomic and proteomic research. Online tools include GCG, Staden, GDE, PHYLIP and a bewildering array of other software. They have also developed their own graphical Web-based interfaces for a number of common molecular biology tasks including DNA sequence assembly (using Phred and Phrap), BLAST searches, identification of genes in genomic sequences, protein identification, multiple alignment, phylogenetic analyses and genetic linkage analyses.

CuraGen Corporation of New Haven, CT, USA has created a suite of genomics tools, which it calls CuraTools™, available for free on a Web site called the GeneScape Portal™. CuraTools is an integrated suite of bioinformatics tools that allows users to easily perform analyses on collections of DNA and protein sequences. Users can upload their own sequence data that can be edited and stored indefinitely in a password-protected personal file storage area on the CuraGen Web site. Using a simple Web interface, users can rapidly analyze their data and associate it with many other forms of genomic data following similarity searches, motif analyses, structure-function predictions, etc. Results from the various analyses are displayed with the aid of specifically tailored viewers and interpreters. Results can be saved together with personal data sets.

CuraTools includes BLAST similarity searches, ClustalW multiple alignment, PCR primer design, DNA sequence assembly with CAP2, DNA restriction analysis and ORF prediction, various protein analyses, protein domain analyses with the PROSITE, BLOCKS, PRODOM, PRINTS, SBASE and Pfam databases and phylogenetic tree analyses calculated with the PHYLIP neighbor-joining algorithm. Most of these tools are implemented by sending queries out to a wide variety of other Web servers and capturing the results that are reformatted into a consistent "report" format. While not as powerful as GCG, the CuraTools Web site offers a simpler interface and better integration of results than other sites that provide a single point of access to aggregations of Web tools such as the BCM Search Launcher, the Biologists Search Palette and Emmanuel Skoufos' Gene Discovery Page. In addition, the ability to store personal sequences and search results makes CuraTools by far the best free integrated bioinformatics Web site.

There are also risks associated with outsourcing crucial bioinformatics software. If a business or lab does not own the software applications they use, they may find it difficult to maintain a depth of knowledge about them, which could lead to a risky level of dependence on the ASP. There is also a vulnerability inherent in depending on someone else to maintain a computer system—slowdowns or failures at the ASP could delay critical research. A relationship with an ASP for bioinformatics software is like a strategic partnership, a concept that is quite familiar to many biotechnology and pharmaceutical companies.

Update to: Get Your Bioinformatics on the Web!

Stuart M. Brown

NYU Medical Center, New York, NY, USA (browns02@med.nyu.edu)

The landscape of bioinformatics vendors offering remote services has shifted a bit over the past 2 years. The financial failure of many "dot com" companies has led to a much greater emphasis on short-term profitability for companies providing Internet-based services and some consolidation of smaller companies. Looking forward, scientists need to consider the possibility that an ASP partner may go out of business or discontinue a service which they have come to rely on (and may contain lots of valuable data).

- Celera Genomics is the new 800-pound gorilla in the online bioinformatics world.
- The Genetics Computer Group (GCG) is now part of Accelrys, which also provides cheminformatics and simulation software for the pharmaceutical and chemical industries.
- Viaken Systems still provides remote access to GCG programs over the Web as an ASP. Viaken also offers fully managed on-site installations of its databases and software tools.
- The company formerly known as eBioinformatics has changed its name to Entigen, but it still provides a comprehensive bioinformatics ASP platform known as BioNavigator, which has over 150 programs including all GCG programs. Entigen provides a free trial of BioNavigator at www.BioNavigator.com, which includes free BLAST searches, private file storage space, and a tryout incentive of 100 units of usage.
- Hyseq no longer offers the GeneSolutions Web-based access to its database. The www.genesolutions.com Web site no longer exists.
- DoubleTwist has discontinued its free "web portal" service. A similar service called GenomeZone™ is now being offered over the Internet to individual scientists or larger organizations on a fee for service basis. This includes access to the proprietary DoubleTwist Annotated Human Genome Database as well as sequence similarity (BLAST) and clustering/alignment tools. DoubleTwist also offers fully managed on-site installations of its databases and software tools.
- Curagen still operates its CuraTools™ integrated bioinformatics service on the web, free for academic users.
- The NCSA Biology Workbench and the UK Human Genome Mapping Project Resource Center are still operating their free bioinformatics services on the Web. In uncertain economic times, government and academic services may be more reliable than private companies.

CHAPTER 51

Snapping Up SNPs

Stuart M. Brown

NYU Medical Center, New York, NY

Reprinted from BioTechniques 26:1090-1095 (June 1999)

Single nucleotide polymorphisms (SNPs) are a hot item in current discussions about the practical applications of genomics and Human Genome Project research. SNPs are simply DNA point mutations—single base-pair changes or insertion/deletions—that are present at measurable frequencies in the population. SNPs are the most common variations in the human genome; the current estimate is that SNPs occur once every 100–300 bp. SNPs exist at defined positions within genomes (sequence tagged sites [STS]) and can be used for gene mapping, defining population structure and performing functional studies. SNPs are expected to greatly facilitate large-scale genetic studies concerned with determining linkage between sequence variations and heritable phenotypes. SNPs might also become an efficient tool for genetic identification for legal and forensic applications.

SNPs are particularly interesting as markers because many known genetic diseases, such as sickle cell anemia, are in fact caused by single base mutations. Therefore, an assay for an SNP marker could directly identify the disease-causing mutation. Some genetic diseases, such as cystic fibrosis, are caused by any of a large number of different mutations in a single gene. SNPs can be used to quickly assay for all known alleles in a large gene.

SNPs have been proposed as an ideal tool for the emerging discipline of pharmacogenomics. This new discipline hopes to use genetic information to fine-tune patient diagnosis and treatment. Large-scale analysis of the associations between the effects of drugs and genetic markers may allow physicians to match drugs to the genetic makeup of individual patients to better predict beneficial and harmful effects.

Some advantages of SNPs over other types of genetic markers such as isozymes, restriction fragment length polymorphisms, variable number tandem repeats and simple sequence repeats include: very large numbers of polymorphic loci; loci distributed throughout the genome; markers present within coding regions, introns and regions that flank genes; simple and unambiguous assay techniques; high levels of polymorphism in the population; stable Mendelian inheritance; and low levels of spontaneous mutation. Kwok et. al (2) have observed that most of the SNPs identified from sequence databases are highly heterozygous in the population.

SNP Databases on the Web

In January 1998, Francis Collins, director of the National Human Genome Research Institute (NHGRI), launched a $30 million project to create a collection of 100000 SNP markers to be "calibrated" against a standard set of human DNA samples and made available to any interested scientist. In September 1998, the National Center for Biotechnolo-

BioComputing: Computer Tools for Biologists
Edited by Stuart M. Brown

gy Information (NCBI) in collaboration with NHGRI established a database of SNPs (dbSNP) on its Web site (http://www.ncbi.nlm.nih.gov/SNP) to serve as a central repository of SNP sequences. This database contains sequence information flanking each polymorphism, links to related GenBank® accessions and to the GenBank database of STS (dbSTS), the experimental conditions used to assay the polymorphism and the frequency of alleles in various ethnic groups. The dbSNP has 3953 accessions (as of February 1999), and it is growing at a rate of 100–500 new entries per month.

The NHGRI and the Centers for Disease Control have developed a set of "standard" human DNA samples (NIH Polymorphism Discovery Resource [NIHPDR]) that can be used to evaluate polymorphic markers in a consistent manner and to facilitate comparisons of markers developed by different research groups using different methods. This resource will comprise cell lines and DNA from 450 unrelated individuals, female and male. It is designed to reflect the range of diversity in the human population. The material in the resource will be available to any investigator from the Coriell Institute for Medical Research (Camden, NJ, USA; http://arginine.umdnj.edu/).

The National Cancer Institute (NCI) has taken an alternate approach to assembling a database of useful SNP markers. The NCI project, know as the Genetic Annotation Initiative (GAI), looks for SNPs by sequencing the coding regions and flanking regions of several thousand genes suspected of contributing to cancer susceptibility or resistance. SNPs discovered by the GAI are added to the NCBI's public dbSNP. This will complement the NCI's Cancer Genome Anatomy project, which collects STS data from a broad set of cancerous and noncancerous tissues.

The Whitehead Institute/MIT Center for Genome Research, has a Human SNP Mapping Project funded by a consortium consisting of Affymetrix, Bristol-Myers Squibb and Millennium Pharmaceuticals. As of May 1998, they released data on 2227 mapped SNPs and 3358 total SNPs. These data are available on their Web site (http://www.genome.wi.mit.edu/SNP/human/index.html). This Web site also includes a nice graphical map locating its SNP markers on all human chromosomes.

Washington University at St. Louis (WUSTL) has its own SNP database on the Web (http://www.ibc.wustl.edu/SNP). The WUSTL database is organized by cytogentic location (chromosome arm). Each SNP has a polymerase chain reaction (PCR) primer and assay conditions associated with it.

A large number of SNPs have already been identified using public data in GenBank from the genome sequencing projects and expressed sequence tag (EST) sequences. SNPs have been found by scanning for regions of overlapping sequence in bacterial artificial chromosome (BAC) and P1-derived artificial chromosome (PAC) clones that come from different individuals or from homologous chromosomes in a single individual (4). It is also possible to discover SNPs by scanning for variation among homologous EST sequences (1,3). The EST approach is particularly attractive since these sequences are by definition derived from expressed genes, and in fact, some SNPs may represent mutations with detectable phenotypes. An additional advantage of this approach is that many of these database sequences have already been mapped.

Several private companies are compiling their own sets of SNPs by large-scale sequencing of identical genomic regions from genetically diverse individuals. CuraGen (New Haven, CT, USA) announced in November of 1998 that it had collected over 60000 SNPs located in expressed genes. Celera Genomics (Rockville, MD, USA) (a collaboration between Perkin-Elmer and The Institute for Genomic Research) and Incyte Pharmaceuticals (Palo Alto, CA, USA) both plan to sequence the entire human genome before the federally funded Human Genome project is completed, and they hope to pay for this sequencing effort by each compiling private databases of approximately 100000 SNPs. Other companies involved in developing their own private SNP databases include Genset (Paris, France), Orchid Biocomputer (Princeton, NJ, USA) and Variagenics (Cambridge, MA, USA).

Ideally, these sets of SNPs should be evenly distributed across the genome, and have high levels of polymorphism in a wide range of different human subpopulations. If the public dbSNP grows as rapidly as predicted by Francis Collins at the NHGRI, then the value of private SNP databases will primarily be in the head start that they provide to private companies in developing technologies for large-scale SNP assays.

SNP Assays

To make these SNP markers useful, it will be necessary to develop simple, rapid, unambiguous methods that allow large numbers of SNPs to be assayed on a single DNA sample. SNPs can be used as genetic markers by creating PCR primers, ligase chain reaction primers, DNA chips or other assay systems that allow rapid identification of alternate alleles in DNA samples. DNA chip microarrays are particularly well suited to SNP assays, since large numbers of SNPs can be assayed simultaneously on a single sample.

Prototype genotyping chips have been developed by Affymetrix (Santa Clara, CA, USA), which were used by Wang et al. (5) at the MIT Whitehead Institute to simultaneously genotype 500 SNPs. Another new technology for assaying SNP markers is being developed by Nanogen (San Diego, CA, USA). Nanogen has developed a semiconductor microchip array that utilizes electrical currents to improve the speed and specificity of binding of nucleotide-specific probes to DNA samples and then assays the bound probes using a fluorescent imaging detector. This technology can also perform multiplex assays using a variety of different probes simultaneously on a single DNA sample.

In March of 1999, Orchid Biocomputer and the University of Washington (Seattle) School of Medicine announced a collaborative agreement to establish a new high-throughput SNP genotyping facility. The center will utilize a proprietary automated genotyping system developed by Orchid that will enable researchers to score up to 30000 SNP genotypes per day for pharmacogenetic analysis of clinical samples.

A key question facing researchers and diagnostic laboratories is how soon to adopt SNP-based assays and which SNP assay technology to choose. Each technology comes as a complete package of relatively expensive equipment, including devices for loading samples, imaging, computers, software and modules that allow different genes to be assayed. It will not be possible to mix and match equipment from different manufacturers, so an initial choice of technology will “lock in” a laboratory with a particular supplier. If a superior technology is later developed by a competitor, it will be expensive to make a change. Therefore, in the short term, individual investigators are likely to rely on large-scale service laboratories to perform SNP assays on a contract basis, rather than making a commitment to a particular technology.

REFERENCES

1. **Gu, Z., L. Hillier and P.Y. Kwok.** 1998. Single nucleotide polymorphism hunting in cyberspace. Hum. Mutat. *12*:221-225.
2. **Kwok, P.Y., Q. Deng, H. Zakeri, S.L. Taylor and D.A. Nickerson.** 1996. Increasing the information content of STS-based genome maps: identifying polymorphisms in mapped STSs. Genomics *31*:123-126.
3. **Picoult-Newberg, L., T.E. Ideker, M.G. Pohl, S.L. Taylor, M.A. Donaldson and M. Boyce-Jacino.** 1999.

Mining SNPs from EST databases. Genome Res. *9*:167-174.
4. **Taillon-Miller, P, Z. Gu, Q. Li, L. Hillier and P.Y. Kwok.** 1998. Overlapping genomic sequences: a treasure trove of single-nucleotide polymorphisms. Genome Res. *8*:748-754.
5. **Wang, D.G., J.B. Fan, C.J. Siao, A. Berno, P. Young, R. Sapolsky, G. Ghandour and N. Perkins.** 1998. Large scale identification, mapping and genotyping of single-nucleotide polymorphism in the human genome. Science *280*:1077-1082.

Update to: Snapping Up SNPs

Stuart M. Brown

NYU Medical Center, New York, NY, USA (browns02@med.nyu.edu)

The Human Genome Project has had great success in producing sequence data more rapidly and at lower cost than even the most optimistic scientists involved in the project could have predicted a few years ago. The discovery of SNPs — single base differences in the genomes of different people — has exceeded expectations even more dramatically. In January of 1998 Francis Collins set the objective of collecting 100 000 SNPs in a public database at NCBI (dbSNP). By February of 1999 this database contained just under 4000 SNPs. In November of 2001, dbSNP contained 3 895 421 human SNPs!

Part of this tremendous surge in the discovery of SNPs is a direct result of the final stages of assembly of the human genome. As the sequences of large BAC clones are completed and dropped into place on the contiguous sequence of the chromosomes, overlaps between clones from different libraries can be scanned for any sequence differences. Also, as EST and complete mRNA sequences are aligned onto the genome, additional sequence variations can be found.

However, the contributions of another group known as the SNP Consortium has been at least as important in driving the growth of the public SNP database. The SNP Consortium consists of 14 pharmaceutical and technology companies (APBiotech, AstraZeneca, Aventis, Bayer AG, Bristol-Meyers Squibb, Hoffmann-LaRoche, Glaxo Wellcome, IBM, Motorola, Novartis, Pfizer, Searle, SmithKline Beecham, and the Wellcome Trust). The data is managed by Lincoln Stein at Cold Spring Harbor Laboratory and it is publicly available on the SNP Consortium Web site http://snp.cshl.org. As of September 2001, there are 1.2 million SNPs in the SNP Consortium database. All of these have been mapped onto the consensus human genome sequence (the UCSC Golden Path) and submitted to dbSNP at GenBank.

It is quite amazing that all of these pharmaceutical companies, who are usually fiercely possessive of their data, have agreed to share SNP sequences with each other and the worldwide scientific community. It is clear that the value of the collective database is greater for all of them than the value of any "private" SNPs that they might discover in isolation. Furthermore, raw SNP data does not have any commercial value. It is only when a polymorphism is linked to a medically important phenotype that it becomes valuable. This is very similar to the process that brought huge amounts of EST data into the public domain, rather than locking them up in patents and proprietary databases.

A SNP is initially discovered as a single base sequence difference between two overlapping fragments of genomic sequences. These may be clones from different libraries, or possibly different clones derived from a library made from the DNA of a single person who is heterozygous at that locus. Each SNP must be validated (shown to be a real difference, not just a sequencing error) and allele frequencies must be determined in various human populations. SNPs will only be useful markers if they occur at sufficient frequencies in target populations. Orchid BioSciences (Princeton, NJ, USA) is in the process of evaluating 60 000 SNPs selected from The SNP Consortium database. Each SNP is being genotyped on DNA samples from 42 genetically diverse individuals assembled by the

Coriell Institute for Medical Research.

While more than 3 million SNPs have been discovered, the majority of these are not within the protein coding sequences of genes. Only about 1% of the SNPs in NCBI's dbSNP affect the coding regions of genes, and only half of those cause a change in an amino acid (the rest are silent mutations). Several papers have reported surveys of a small number of gene sequences across 50–100 different people. In each case, an average of about 2 coding SNPs that affect the translated protein (the so called cSNPs) were found per gene. In a broader sample of people, it is likely that cSNPs can be found for virtually every gene.

It is also possible for SNPs located within introns to affect intron splicing or impact on mRNA function in some other way. It is estimated that an equal number of SNPs which are located in regulatory regions outside the coding sequence also have a direct impact on gene function. However, SNPs that have no direct functional impact may still be valuable as genetic markers if they are tightly linked to some other medically significant allele.

CHAPTER 52

Be Alert: Free Sequence Alerting Services

Stuart M. Brown

NYU School of Medicine, New York, NY, USA

Reprinted from BioTechniques 29:1210-1212 (July 2000)

It has always been a challenge to keep up with all of the new data flowing into biological databases. Now, as the stream of DNA sequences from the Human Genome Project has increased to a torrent, the chances have drastically increased that something new has rendered obsolete the database searches that you did last week. You can spend lots of time obsessively repeating your searches or look for an automated service that will perform regular searches and send you notification when relevant new sequences have entered the database. There are also some free services that will check for new additions to the PubMed®/MEDLINE® database and notify you when relevant new journal articles appear.

Several free automated "sequence alerting" services exist on the Web in various forms. These can serve as a method to attract scientists to a multifunctional Web site, or merely be an altruistic contribution from a bioinformatics group. There are two basic types of sequence alerting services: those that perform similarity searches based on a user's query sequence and those that perform keyword searches of title/annotation information. In addition, various services are designed to monitor different databases, such as the SwissShop service (http://www.expasy.ch/swiss-shop), which only tracks updates to the SwissProt database.

A crucial aspect of an alerting service is that it only sends you NEW finds from the database, not merely repeating a search at some interval and sending you a complete listing of all matches, which would lead to huge amounts of time wasted sorting through the same sequences in every report.

The European Molecular Biology Laboratory (EMBL) in Heidelberg, Germany operates a Sequence Alerting System (http://www.bork.embl-heidelberg.de/Alerting/). The system searches each day for sequences similar to your query sequence and will inform you by e-mail if it has detected a new match. However, the system only searches protein databases (e.g., TREMBL, PDB, SwissProt and PIR), so it will not detect new ESTs or genomic sequences that contain sequences that have not been automatically detected and submitted to a translated protein database such as TREMBL.

The ExPASy server at the University of Geneva (home of the SwissProt database) runs the alerting service called SwissShop. New protein sequences that are added to the SwissProt database can be scanned on a weekly basis for a user-defined query. The service provides for both keyword searches and sequence-based searches (using either the BLAST

BioComputing: Computer Tools for Biologists
Edited by Stuart M. Brown

similarity algorithm or pattern searches use exact matching of PROSITE type patterns).

GeneNet (http://brac.postech.ac.kr/eng/) is a meta-search system for the analysis of sequence similarity via the Web that is developed and maintained by the Biological Research Information Center, Pohang University, Korea. For protein sequences, searches are performed on GenBank®, PDB, BLOCKS and KEGG; for DNA sequences, only GenBank is searched. The summarized results are provided by e-mail. Users can get full results on a Web page by following an embedded HTML link. Once an initial analysis is made, users can choose to have the search repeated on a monthly basis and the results sent by e-mail.

PubCrawler (http://www.pubcrawler.ie/) is a free alerting service that scans daily updates to the NCBI MEDLINE (PubMed) and GenBank databases, so it can alert users both to new sequences and new journal articles. PubCrawler can keep scientists informed of the current contents of MEDLINE and GenBank by listing new database entries that match their research interests. "It goes to the library. You go to the pub."™

PubCrawler maintains a database of previous matches for each user, so you are only alerted when new sequences and/or journal articles appear that you have not seen before. Full results can be sent by e-mail notification when new results are available, or the user can just retrieve them directly on the PubCrawler Web site whenever he or she wishes to check them.

PubCrawler was developed in the Department of Genetics, Trinity College, Dublin, Ireland. The PubCrawler software (UNIX®, Microsoft® Windows™ and Macintosh® versions) is also available for free to install on any computer to manage an update service for one user or an entire institution.

BioNavigator (http://www.bionavigator.com) is a commercial Web-based bioinformatics service operated by the eBioinformatics company (Eveleigh, NSW, Australia), which provides a free notification service for BLAST and keyword searches. BLAST searches of GenBank and text/keyword searches are free and can be set to be repeated weekly using the "Notify" function. Keyword searches use the "TextSearch" program, which utilizes the SRS database engine developed by Lion Bioscience and allows the user to search GenBank, BLOCKS, Pfam, PDB, Prosite, SwissProt and TREMBL. Registration for the BioNavigator service is also free and does not require the user to enter billing information. All results are stored on the Web in individual user's project areas.

DoubleTwist (http://doubletwist.com) is a commercial service focused on genomics. It provides free daily sequence updates via monitor agents for up to 10 query sequences (DNA or protein). DoubleTwist uses multiple similarity searching algorithms to compare the user's query sequences against about 20 public and proprietary databases including DNA, protein, ESTs, structures, motifs, domains, patented sequences, etc. When new matching sequences enter these databases, DoubleTwist notifies you by e-mail. The service offers e-mail alerts, but the actual results are retrieved via the Web. For a monthly subscription fee, additional query sequences can be searched and additional databases are available such as DoubleTwist's proprietary computationally generated clustered EST and annotated human genome databases.

The Biotechnology Computing Facility at the University of Arizona has created an automated Entrez Query Tool (http://bcf.arl.arizona.edu/query/) that makes regular searches of the Entrez database at NCBI and notifies users when new items are found. The user can customize an account for creating and managing individual queries. In each query, the search term is specified, along with the database (PubMed/MEDLINE, Protein, Nucleotide, Structure and Genome), frequency of the query (Daily, Weekly or Monthly) with dates for starting and terminating the automated search. If new results are found for a query, then the information is e-mailed to the user. The service is free and open to the public.

BioMail (http://www.biomail.org and http://biomail.sourceforge.net/biomail/) is a free

Web application that regularly (weekly by default) searches for articles that have recently appeared in the PubMed/MEDLINE database using user-customized search terms. It then e-mails lists of the found articles to the user. The HTML-formatted e-mails generated by BioMail can be used to view selected references in MEDLINE format (which is compatible with most reference manager programs). BioMail is developed and maintained at SUNY Stony Brook. BioMail source code was written in Perl for Linux; it is free software released under GNU GPL license.

JADE (Journal Articles Delivered Electronically) (http://www.biodigital.org/jade/) is another automated MEDLINE search agent developed and maintained by The National Center for Emergency Medicine Informatics. JADE makes weekly searches of MEDLINE with user-specified search terms and sends the results by e-mail. JADE also has a "Personal Archive" feature that allows each user to store citations to articles found in the searches in a personal database. Individual articles can also be added to the archive using the MEDLINE UID.

DBWatcher (http://www-igbmc.u-strasbg.fr/BioInfo/LocalDoc/DBWatcher/) is a free UNIX program that makes automated BLAST searches. It is developed and distributed by the Institut de Genetique et de Biologie Moleculaire et Cellulaire, Strasbourg, France. This is NOT a Web-based service, but rather a program intended to be installed on a UNIX computer to serve a number of scientists. The program is extremely customizable both in terms of query terms and databases searched (GenBank's BLAST databases or a local BLAST formatted database). Searches can be set to execute at any interval.

Only novel similarities are reported, thus saving the time of browsing through bulky results files. Results are sent by e-mail to one or several addresses, making DBWatcher particularly suitable for collaborative database surveys and for users who do not have access to locally maintained databases.

There is also a vast array of journal article/table of contents-based alerting services. Some will e-mail you the table of contents when a new issue of a specific journal is released. Others will scan titles (or abstracts) for keyword matches from a group of journals and notify you by e-mail when a relevant article is published. Many of these services are offered by specific publishers for their own journals or by document delivery services that have distribution agreements with a number of publishers.

By making use of one or more of these alerting tools, researchers can stay on top of newly discovered sequences and new journal articles relevant to their research with no need to constantly repeat the same database searches. However, for those who prefer to repeat searches manually, rather than receive automatic updates by e-mail, the NCBI has recently added a new feature called the Cubby to their Entrez Web site (http://www.ncbi.nlm.nih.gov:80/entrez/cubby/login.fcgi?call=so.SignOn..Login).

The Cubby is a password-protected personal storage space for Entrez queries (free registration is required). Stored search information in the Cubby includes the search name, date and time last updated, database searched, search terms and field limits—but not actual sequences or journal citations. Up to 100 different search queries can be stored. This can be very useful because it is difficult to remember from month to month the exact combination of keywords and field limits that were used to find a useful set of sequences or journal articles. A stored search can be updated with a single click, which will retrieve only those matching database records that have been added to the database since this search was last performed. Alternatively, an entire search can be repeated to bring up all matching records regardless of whether they have been seen before. Stored searches can also be modified before repeating them. Use these Internet tools to power your research with the flow of genome information rather than being swamped by it.

CHAPTER 53

Image Processing the Internet Way

Michael J. Herron

Department of Dermatology, University of Minnesota,
Minneapolis, MN, USA

Reprinted from BioTechniques 20:626-628 (April 1996)

Once upon a time, software programmers conspired with armies of tree choppers, printers, disk duplicators and long-distance haulers in an effort to ship you thick, paper manuals and a small heap of floppies. Soon after reading the manuals, you invested the rest of your afternoon listening to a soothing prerecorded voice explain that there were only thirteen callers ahead of you on what was euphemistically called a "help line". Once connected to a real human you soon realized that "Clyde" had never heard of electro for rhesus or any other subject that interests you.

Those days are over. This month's **Internet On-Ramp** introduces our readers to a free program that is a real scientific jewel. The manner in which this program is distributed and supported entirely through the Internet is a positive example of how the Internet is reshaping the way we work. If you make the small effort necessary to investigate the program featured this month, you will learn a bit about the Internet and gain a very useful program for your laboratory.

The program in question is **NIH Image**, a free image-processing and analysis program for the Macintosh® written by Wayne Rasband (wayne@helix.nih.gov) of the National Institutes of Health (NIH).

Most laboratories have some need for imaging, and NIH Image will run on a computer you may already have. But don't despair if your laboratory is an all-PC operation; a number of alternatives area also available for the PC.

NIH Image can grab an image from video, charge-coupled device (CCD), scanners or other means. It will enhance, analyze and animate your images. The program can read most common image formats, count objects, measure lengths and intensities, edit and enhance images. In addition, it can control cameras, microscope stages, focus mechanisms, VCRs, and much, much more. All of these tasks can be done on the widely used (and famously easy) Macintosh (Apple Computer, Cupertino, CA, USA). Using NIH Image you can analyze and document electrophoretic gels, project three-dimensional data (i.e., from confocal microscopes), extract data from old graphs and prepare images for presentation and publication.

To get your copy of NIH Image, start up your Web browser and enter this uniform resource locator (URL) http://rsb.info.nih.gov/nih-image/ into the "Location" box and hit the RETURN key. When you connect, you will find that NIH Image, like many other

BioComputing: Computer Tools for Biologists
Edited by Stuart M. Brown

Macintosh programs, is available in three versions. The 68xxx version is for older Macintosh computers using the Motorola 68020 or higher processor. The PPC version has been optimized for speed on the newer PowerPC® Macintosh computers. Finally, the FAT version is a little larger program that will run on both platforms. If you're not sure which version you need, click on the FAT version to download this program to your hard drive.

Besides the basic NIH Image program, there are many related files available on this Web Site. Be sure to read the Frequently Asked Questions (FAQ) and check out the related Web Sites that are "hotlinked" there.

When I first used NIH Image it seemed to be a simple image-editing program that would also allow a few interesting image manipulations and measurements. However, looking closer, I saw that by combining the various operations, complex image manipulation and analysis is possible.

There is solid documentation included with the program as it comes, and more documents and tutorials are available at the Web Site. The program also includes *Balloon Help*, which explains the purpose of a tool or menu item when you point at it with the mouse.

Even more useful than the documentation, tutorials and *Balloon Help* (and much better than "help-line Clyde") is the free **NIH Image mailing list**. This electronic mailing list is maintained by John Ladwig, Department of Soil Science, University of Minnesota (owner-listproc@soils.umn.edu) and allows you to ask very specific questions that will be read worldwide by the hundreds of other NIH Image users who subscribe to it. Typically, answers to your inquiry will begin to appear within a few hours after you post your question.

This e-mail list has been quite active since its inception in 1992, and you may sense that your questions have been asked and answered before. If so, there is an archive of all the mail list letters that you can search at http://www.soils.agri.umn.edu/infoserv/lists/nih-image/. I often find it useful to search this archive to formulate a specific question for the e-mail list.

To subscribe to the mailing list, send a mail message to listserv@soils.umn.edu with the following in the body of the message: **subscribe nih-image Firstname Lastname**. The computer at this address will automatically subscribe you to the mail list and send you a message acknowledging your subscription. From then on you will begin to receive copies of all the messages sent to the list. If you find that you need to stop these messages, send the command **unsubscribe nih-image** to the same address.

The first hurdle that you will face when actually using NIH Image is getting images into the computer. One method is by way of a video or CCD camera. This is fast, and with the wide variety of cameras available, it is often quite flexible. An excellent review of the options available can be found in Shaw et al. 1995. Digital Photography for the Light Microscope: Results with a Gated, Video-Rate CCD Camera and NIH-Image Software. BioTechniques *19*:946-955.

Another common method to bring an image into the computer is by using a desktop scanner. This method is less flexible, but can inexpensively yield images of very high xy resolution. Information explaining the use of desktop scanners with NIH Image can be found in the documentation.

The third method of introducing images involves machines that already produce images as digital data. This includes equipment such as confocal microscopes, magnetic resonance imagers and surface scanning microscopes, such as the Atomic Force Microscope. Thanks to the network of users on the NIH Image mail list, many of these image formats can be easily imported into NIH Image by macros (small user-written programs) that are available at the NIH-Image Web Site.

Once you introduce your visual into NIH Image, you may want to measure the intensity of bands on electrophoretic gels, blots or other images of interest.

Measuring intensities can be a bit tricky, so be sure to read the excellent document *NIH Image Engineering* assembled by Mark Vivino of the NIH Division of Computer Research and Technology. This document is available from http://rsb.info.nih.gov/nih-image/InsideImage/inside.html.

If you would like to investigate the utility of NIH Image in viewing and processing confocal images, be sure to download *Use of NIH-Image 1.57 with Confocal Images* prepared by Harvey J. Karten, University of California at San Diego from http://rsb.info.nih.gov/nih-image/confocals.html.

Don't give up if you find that NIH Image does not seem to be capable of performing the operations that your project requires. Instead, post a letter explaining your disappointment to the mail list. The chances are pretty good that someone on the list knows a way to solve your problem.

As you can see, it is possible to distribute a program through a file transfer protocol (FTP) site on the Internet, support that program with documents and tutorials located at World Wide Web sites and provide specialized help with an e-mail list and the support of interested users. This rapid and cost-effective method of distribution and support will become increasingly common as laboratories continue to adopt the Internet as an integral tool.

Update to: Image Processing the Internet Way

Michael J. Herron

Department of Entomology, University of Minnesota, St. Paul, MN, USA (herro001@umn.edu)

The most recent version of NIH Image version 1.62 was completed in 1999. The program is still in wide use and is supported by a very active user base. After 1999, Wayne Rashbad has devoted his time to development of ImageJ, an all new high-performance Java-based imaging program, which runs on Macintosh®, Windows®, and Unix® platforms. ImageJ has most of the capabilities of NIH Image, with many powerful new features of its own.

The Web site (http://rsb.info.nih.gov/nih-image/) contains links to NIH Image, ImageJ, and the listservs that serve users of both programs. This excellent site is worth exploring for many related links and software imaging tools.

ImageTool for the personal computer (PC) can now be found at (http://ddsdx.uthscsa.edu/dig/itdesc.html) and Strangers in Paradise: A Measurement Training Exercise Using NIH Image, is now hosted by (http://davem2.cotf.edu/ete/teacher/tmodules/tstrange.html).

CHAPTER 54

LabVelocity™: Online Tools for Life Science Products, Protocols, Technical Information, MEDLINE® Searches, and Laboratory Calculations

David A. Riethof and Rama Balakrishnan

LabVelocity, Inc., San Francisco, CA, USA

(Reprinted from BioTechniques 30:1310-1315 (June 2001)

ABSTRACT

As the pace of life science discovery increases, so do the demands on researchers. To remain competitive in the life science industry, researchers must use every tool at their disposal to keep up with new products, protocols, news, and literature in their field. While there are now myriad Web sites that assist researchers with this problem, many suffer from confusing user interfaces, poorly designed search engines, and a narrow information focus. Here, we present LabVelocityÔ, a user-friendly Web site that provides a free multidisciplinary information-gathering service for the life science research community. Using LabVelocity, a researcher can quickly find the products, protocols, technical references, news, MEDLINEÒ abstracts, and interactive software tools necessary for an experiment. This aggregation of information can streamline experimental planning and is especially useful when researchers want to set up a new laboratory or to venture outside their field of expertise.

INTRODUCTION

While the savvy research community has embraced advances in technology to improve the accuracy and speed of almost every step of the research process, the critical and time-consuming information-gathering steps of experimental planning are still often done by cumbersome traditional methods. Any innovation that improves on the use of laborious paper catalog, reference book, and journal searches will significantly enhance the scientist's ability to collect the vital information necessary to stay competitive at the bench.

The innovations of the Internet promise to streamline such information gathering, and

BioComputing: Computer Tools for Biologists
Edited by Stuart M. Brown

recent studies suggest that most scientists have incorporated various Web site services into their research routine, spending an average of 66 minutes each day using the Internet at work. Among the most common uses of the Internet in research, besides e-mail, are searches for technical information on products (90% of Internet-using life scientists) and for literature (77% of Internet-using life scientists) (4). Today, it is almost inconceivable that a major product supplier or journal would not have an online presence. A small but growing number of suppliers and journals have no paper version at all (3).

Ironically, the recent proliferation of life science Web sites has, in many ways, made information gathering for the scientist more confusing, not less, for now there is too much information in too many places and of often uncertain quality. Unfortunately, many current search engines are unable to filter usable life science information out of the ever-increasing morass of online data. The low signal-to-noise ratio of keyword searching can be particularly frustrating when researchers want specific life science information. In one study, over 60% of Internet-using life scientists complained they were often unable to find useful information online because of the limitations of current search engines (4). As a result, today's harried scientists must often search several Web sites of variable quality before they find the information they need to advance their research.

To address these problems, we created LabVelocity™ (www.labvelocity. com), a free and easy-to-use Web site that serves as a central online repository for the information life scientists need to conduct experiments. LabVelocity builds and maintains its own universal on-site databases of commercial products and protocols, technical references, news, and MEDLINE® abstracts. To make it easy for researchers to find complex life science products, LabVelocity's product database is structured in a tree-like hierarchy by discipline and technique, and its data are tagged so that products can be found by their attributes. Additionally, LabVelocity's on-site databases are integrated so that a researcher who finds a product from the product database can link instantly to that product's protocol within the protocol database, without having to perform a separate search. This integration of information can significantly streamline information gathering and experimental planning. Researchers can also take advantage of LabVelocity's many labor-saving interactive software tools such as automated MEDLINE agents and buffer and conversion calculators. Finally, researchers can create personalized libraries of all the technical information found on the site using a password-protected and customizable file system called the Lab Manager.

MATERIALS AND METHODS

The LabVelocity Web site is compatible with all currently available Microsoft® Windows™- and Macintosh®-based operating systems and major Internet browsers. To view PDF protocols and technical references on the site, researchers need to install Adobe® Acrobat® Reader™ 4.0 (freely available at http://www.adobe. com/products/acrobat/readstep. html). To play the site's interactive QuickHelp tutorials, researchers need to install Macromedia® Flash™ (freely available at http://www.macromedia.com/shocwave/download/index.cgi?P1_Prod_Version=Shockwave Flash). Site registration is required to see product prices and to use some customizable functions, including MEDLINE search agents and the archival file system called the Lab Manager. LabVelocity does not distribute registered-user information without permission. Researchers can read LabVelocity's privacy policy at http://www.labvelocity.com/about/index.jhtml? podsection= Privacy+Policy. All sections of the site are free.

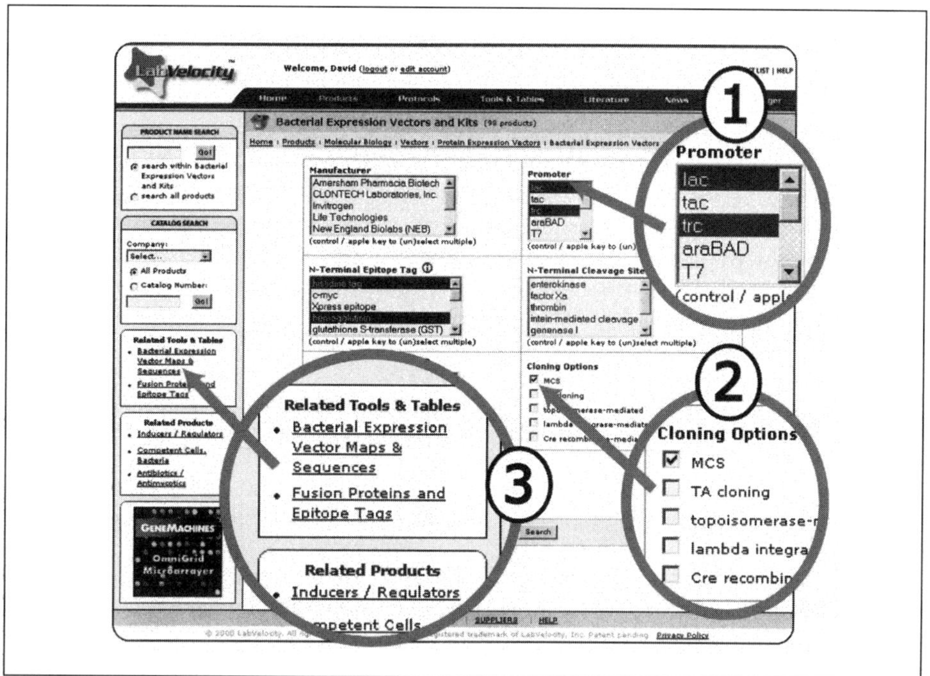

Figure 1. Customizing a LabVelocity parametric search to find bacterial expression vectors having (no. 1) *lac* or *trc* promoters and (no. 2) multiple cloning sites; (no. 3) each parametric search is linked to useful technical information and related products.

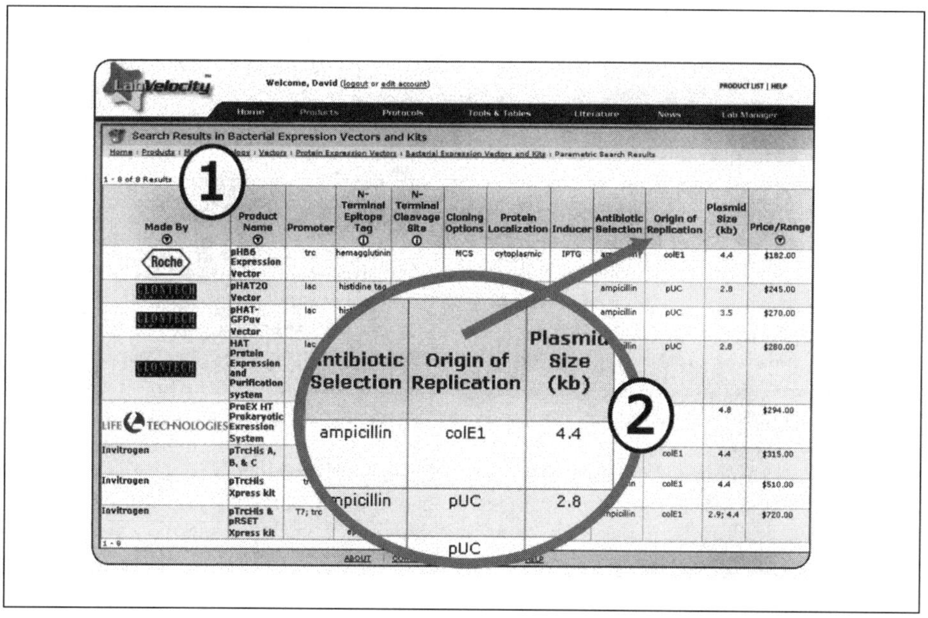

Figure 2. Following a completed search, (no. 1) product tables allow (no. 2) attribute-based product comparisons.

DISCUSSION

Life Science Product Searching

By 1999, the world's 550 000 life scientists could already choose from more than 8000 scientific suppliers offering more than two million products (1). With the vast potential for new therapeutic targets mined from the Human Genome, the life science industry is predicted to experience double-digit expansion each year throughout the next decade (http://www.biotechsage.com/outlook.htm). As the life science industry grows and becomes increasingly specialized, so will the product choices offered to its researchers. Today's scientist needs online product databases and search engines that are able to handle the sophisticated queries necessary to locate increasingly specialized commercial life science materials.

Unfortunately, many life science Web sites have product databases that can only be searched by keyword search engines, a technology designed to search text and word-processed documents. Keyword search engines are insufficient for all but the simplest research. While they may produce perfect accuracy—in the narrow sense that they find all the occurrences of a specified query—keyword searches are frustratingly simple-minded, finding precisely what researchers ask for, but not necessarily what researchers need. They lack the intuition of a talented researcher, and their pattern matching is best suited to retrieving indexed documents. Web sites that rely on keyword searching force researchers to guess which combination of the right kind of words (e.g., full text, abbreviations, Greek or Latin characters) and Boolean linkers (e.g., AND, OR, NOT) will result in a productive search. Compounding researcher uncertainty, many of these Web sites only search their product database by product titles or catalog numbers, for which there are no universal nomenclatures.

To examine the effectiveness of keyword searches for complex life science queries, we searched online for bacterial expression vectors by the Boolean keyword construction, "vector AND expression AND bacteria AND lac AND promoter AND histidine AND tag AND N-Terminal AND Epitope AND multiple AND cloning AND sites." This keyword search yielded no results at several life science Web sites with large universal product databases, while successful truncated searches yielded many hundreds, even thousands, of results. Internet-wide Web portals with more advanced keyword search engines (e.g., Google™) performed better with long Boolean constructions, but the identity, source, and quality of each link from their hit lists had to be tested individually. In addition to products, their hit lists included links to non-product information such as lecture texts, press releases, and documents composed entirely of lists of keywords. While some of this information may be useful, this low signal-to-noise ratio can be frustrating when researchers are looking only for products. Additionally, useful product comparisons were difficult because each supplier Web site presented its product data in a different format. Clearly, the lengthy Boolean constructions necessary to ensure specific results from keyword searches are not a practical or efficient way to conduct online scientific information gathering.

Parametric searching, originally developed to identify complex technical products in engineering catalogs, is an improvement over keyword searches because it allows researchers to find products based on their attributes when the title, supplier, and catalog number of a product are unknown (Reference 2 and http://idm.internet.com/text/features/cadis-1.shtml). At LabVelocity, parametric searches are combined with a user-friendly interface to make it easy for researchers to quickly discover and compare product information. The catalog is organized by discipline and technique into a tree-like hierarchy, within which

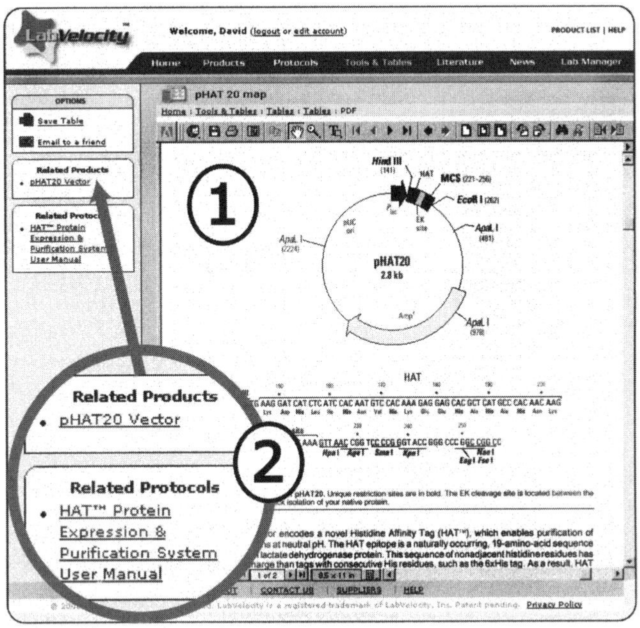

Figure 3. Each LabVelocity table and protocol (no. 1) can be saved online and e-mailed and is linked to (no. 2) recommended products, related protocols, and other technical references.

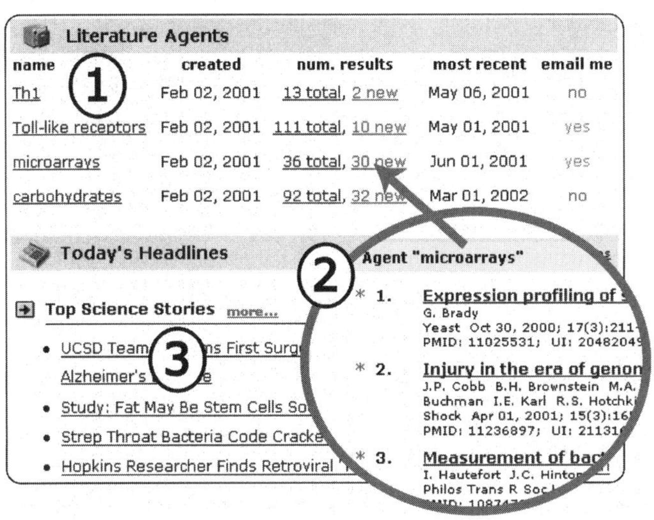

Figure 4. A customized LabVelocity home page displays (no. 1) active Literature Agents, which automatically search the MEDLINE database multiple times each week for (no. 2) new abstracts of interest and (no. 3) personalized daily science news.

researchers select categories (e.g., molecular biology) and subcategories (e.g., vectors) appropriate to their product to reach one of LabVelocity's 600 individually designed parametric searches (e.g., for bacterial expression vectors). Each parametric search can be easily tailored to find only products with very specific attributes (e.g., *lac* promoters) simply by choosing from options menus. This design eliminates keyword and Boolean-operator construction guesswork.

For instance, with a few mouse clicks, researchers can customize LabVelocity's "Bacterial Expression Vectors and Kits" parametric search to find only those bacterial expression vectors with *lac* or *trc* promoters, histidine or hemagglutinin N-terminal epitope tags, multiple cloning sites, and cytoplasmic protein localization (Figure 1)—a combination of product attributes that would be cumbersome to search for by keyword.

This particular search yields a comparison table of matching vectors from multiple manufacturers. Using this table, researchers can compare vectors based not only on the previously selected search attributes but also on information such as a vector's inducer, antibiotic selection, origin of replication, plasmid size, and price (Figure 2). Each vector is linked to its manufacturer's Web site for further information and purchasing. Researchers can add any desired product from this and other searches to LabVelocity product lists. They can e-mail these product lists to their purchasing department or save them for future reference.

In addition to finding a suitable bacterial expression vector, researchers can also customize: the "Antibodies" search to find antibodies based on their antigen, clone name, species specificity, host species, isotype, and conjugation; the "Affinity for Fusion Proteins/Epitope Tags" search to find only those bulk and column resins that affinity purify His-tagged proteins; and the "PCR/Thermostable DNA Polymerases and Kits" search to find only *Taq* and *Pfu* polymerases suitable for high-fidelity and hot-start reactions. With hundreds of other customizable product-specific searches covering a wide range of techniques and services (biochemistry/protein chemistry, cell biology/immunology, chemicals, chromatography, custom services, electrophoresis/blotting, lab equipment/instrumentation, molecular biology, premade buffers and solutions, software, and supplies/accessories), researchers now have user-friendly tools sophisticated enough to find increasingly complex commercial life science products easily and quickly.

LabVelocity's product catalog and parametric searches are valuable laboratory resources. Researchers can discover novel products and suppliers from search results. Researchers who need help with antibody production and other animal work, DNA and peptide purification and sequencing, and gene array synthesis can use the "Custom Services" searches to make well-informed decisions about these and many other labor-saving services. A researcher who is setting up a new laboratory would benefit from the site's many "Lab Equipment" searches, where instruments (e.g., automated synthesizers and sequencers, thermal cyclers, centrifuges, electrophoresis systems, fluorometers, luminomoters, plate readers) can be searched and compared using multiple combinations of selected equipment options. Researchers may also discover new solutions for experimental problems when using parametric search menus; for instance, they could discover new bacterial expression vector promoters or cloning options from that search interface's "Promoter" and "Cloning Options" menus (Figure 1).

Integrated Protocols and Technical Tables

LabVelocity has extensive libraries of technical tables (e.g., pathway diagrams and vector maps) and over 2500 protocols covering a wide range of techniques—all tested by

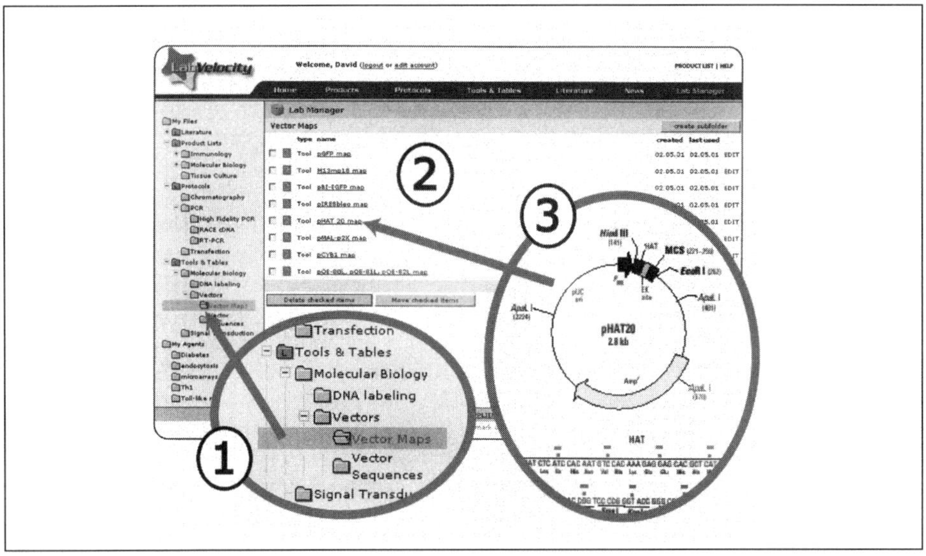

Figure 5. Using the Lab Manager to view a vector map from an online technical library. Clicking a personalized folder named (no. 1) "Vector Maps" within the Lab Manager's folder tree displays (no. 2) a list of its saved vector maps. A saved map (no. 3) can then be viewed, printed, and e-mailed at any time by clicking its name within this list.

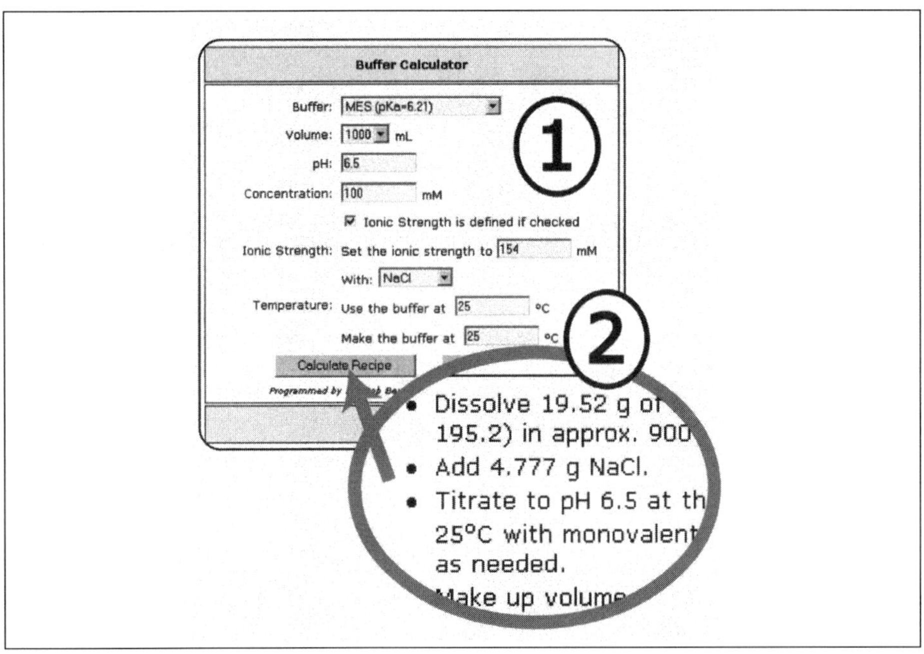

Figure 6. LabVelocity's (no. 1) Buffer Calculator can quickly create (no. 2) accurate recipes for dozens of different buffers at any scientist-selected volume, pH, concentration, ionic strength, and temperature.

commercial suppliers and research institutions, such as Clontech Laboratories and the Joint Genome Institute. Using the full organizational and interactive power of the Internet to streamline experimental planning, LabVelocity integrates these protocol and technical table databases with its product database (Figure 3). This integration allows scientists to quickly collect and save on-site all the information necessary to carry out an experiment without having to perform multiple searches. Because only very basic knowledge is required for information retrieval from multiple databases, this design is especially suited to novice researchers and to those with more experience but who want to venture outside their field of expertise. For example, from a vector product search, researchers only have to click sidebar links to view, save, and e-mail vector maps and sequences, fusion and epitope tag tables, protocols, and related products such as inducers, bacteria, and antibiotics (Figure 1)—without having to search separate databases for each additional ingredient. Individual products are similarly linked to resources specific for their use (e.g., a specific vector's map, sequence, and protocol).

Literature Agents

To further aid researchers, LabVelocity offers customizable literature services. LabVelocity hosts the National Library of Medicine's MEDLINE database, which can be searched on-site by keyword, article title, author name, publication date and type, journal, language, country, institutional affiliation, and MeSH headings. Researchers can also create personalized literature agents that automatically search the MEDLINE database several times each week and can e-mail alerts when new abstracts of interest arrive. Once created, active literature agents are displayed on a registered user's home page (Figure 4). Abstracts from MEDLINE search results are linked to the publishing journal's Web site when available.

The Lab Manager and Institutional Memory

All of the information collected at LabVelocity—product lists, protocols, technical references, and MEDLINE abstracts—can be saved and organized online using a section of the site called the Lab Manager. An unlimited number of maps and sequences of favorite vectors (and their associated protocol and product information) can be stored within personalized Lab Manager folders and then viewed, printed, and e-mailed at any time with just a click of the mouse. Many such folders could be part of a customized online technical library (Figure 5). Laboratories, which often experience high levels of researcher turnover, can use these libraries to create "institutional memory": a centralized archive of life science information specific to a laboratory or institution. Information in online archives can be accessed, updated, and shared by authorized laboratory users at anytime and from anywhere, offering significant intra-institutional communication and labor-saving advantages over paper-based methods of record keeping. All information archived in the Lab Manager is protected by user-created passwords and by a firewall.

A Valuable Research Tool

LabVelocity provides many other labor-saving services: a collection of interactive tools that perform common laboratory calculations [e.g., buffer recipes (Figure 6), DNA and protein conversions]; up-to-date, annotated bioinformatics, genomics, and software-download links; and customizable daily science news. These services—added to automated MEDLINE agents, to the efficiency and accuracy of parametric searching, and to the power of linking attribute-based product information to protocols and technical references—significantly streamline the scientific information-gathering process. The ease of gathering information from integrated life science databases is especially suited to novice scientists, to experienced researchers who want to venture outside their field of expertise, and to those researchers who are setting up new laboratories. Researchers can create personalized libraries of all the technical information found on the site using the Lab Manager's password-protected file system, and laboratories can use the Lab Manager to create "institutional memory" to reduce information loss caused by researcher turnover. All of these features—available for free—make LabVelocity a valuable online resource for the working life scientist.

REFERENCES

1. **Davis, J.** 1999. Chemdex's New Supply-Side Economics. Business 2.0. March:4.
2. **Frederickson, G.N.** 1997. Parametric search and locating supply center in trees. Proceedings of the second Workshop on Algorithms and Data Structures. *519*:299-319.
3. **McEntyre, J.** 2000. Is publishing to perish? A sea of change afoot for life science publishing. New Technologies for Life Scientists: A Trends Guide *12*:3-6.
4. **Studt, T.** 2000. Researchers integrate Internet tools into their work. Res. Dev. Tech. Rep. *42*:E13-E14.

Update to: LabVelocity™: Online Tools for Life Science Products, Protocols, Technical Information, MEDLINE® Searches, and Laboratory Calculations

David A. Riethof and Rama Balakrishnan

LabVelocity, Inc., San Francisco, CA, USA

Since the publication of this paper in June 2001, there have been many enhancements and new features added to the online information services LabVelocity™ provides scientists. While a number of enhancements and new features are freely available, several are available only through a subscription service called ResearchLink. Additionally, all features are available in a new location (http://researchlink.labvelocity.com).

Resource Centers, one of LabVelocity's new free features, allow researchers to access many different types of information related to a technique or application from a single Web page (Figure 1). For example, from the chromatography Resource Center, scientists can find chromatography-related products, protocols, and reference tables, without having to search separate sections of the site. Current Resource Centers include polymerase chain reaction (PCR), nucleic acid cloning, protein expression, chromatography, and cell and/or tissue culture.

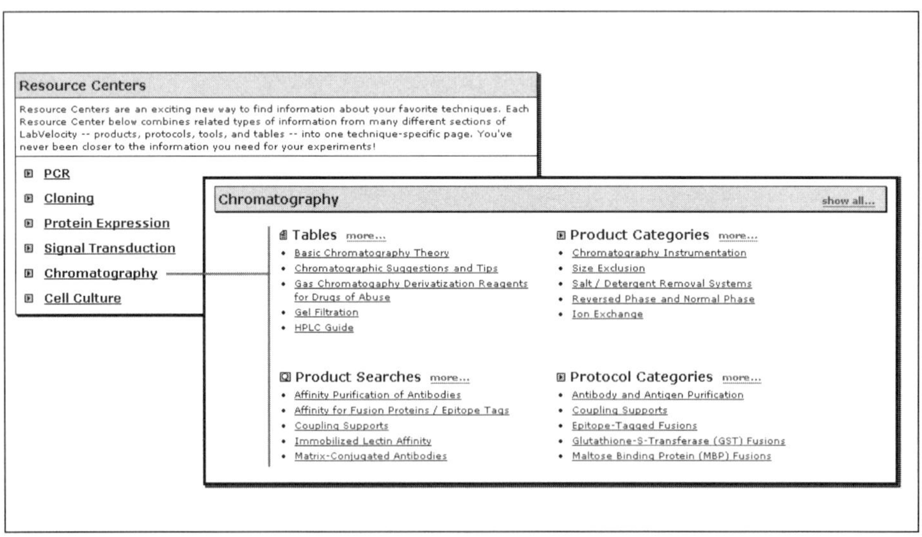

Figure 1. LabVelocity's new Resource Centers save time by providing easy access to products, protocols, or reference tables relating to various laboratory procedures.

New features available only with a subscription include: (*i*) enhancements to literature searching; (*ii*) enhancements to file sharing for scientific collaborations; and (*iii*) the gene analysis software package Jellyfish.

ResearchLink's enhanced literature features include on-line access to full-text articles and the ability to purchase hard-to-find articles on-line using the Infotrieve service. Automated literature searching is now available only with a ResearchLink subscription. Manual literature searching is still free at ResearchLink.LabVelocity.Com.

ResearchLink's file management system, My Files (formerly the LabManager), has been enhanced with Workgroup functions to facilitate communication among colleagues. Workgroups allow researchers to securely share scientific information, such as product ordering lists, purchase orders, saved literature abstracts, protocols, and ordering histories within a laboratory or with collaborators miles away. Workgroups also eliminate repetitive searching and provide convenient access to routinely used products, protocols, and other laboratory information.

Jellyfish (Figure 2) is an affordable gene analysis software package that includes all the functions commonly required for nucleic acid or protein sequence analysis. Researchers can import GenBank® sequences, design primers, create restriction maps, and perform Basic Local Alignment Search Tool (BLAST) analyses using a convenient graphical user interface. Jellyfish is available as a download from LabVelocity on an annual subscription basis. Researchers can download a 30-day free trial of Jellyfish from jellyfish.labvelocity.com.

For more information on these and other LabVelocity services, please visit home. labvelocity.com.

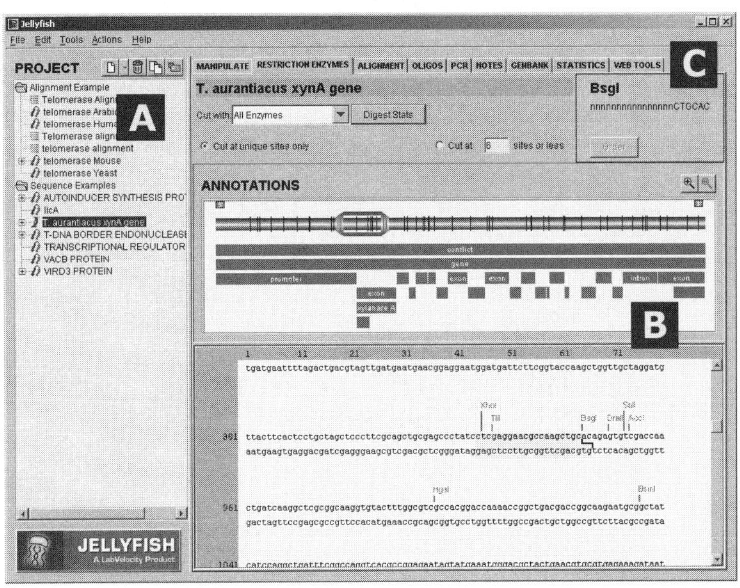

Figure 2. Jellyfish's user-friendly interface brings together all the essential sequence analysis functions: a convenient project management filing system (A), visual displays of the genetic structures (B), easy-to-use functions (C), and more.

CHAPTER 55

Using a World Wide Web Server as a Local Organizer for Protein and DNA Sequences

Richard Atwell, Frank Gibbins, and Chris Upton
University of Victoria, Victoria, British Columbia, Canada

Reprinted from BioTechniques 19:966-970 (December 1995)

ABSTRACT

We have used a local World Wide Web (WWW) server to organize protein and DNA sequences that are used frequently in our laboratory. WWW server programs are available for most computer platforms and are easily set up with minimal computer skills. This approach allows for the easy retrieval of sequence data, which can then be used as input for other analysis programs. This format is especially simple to use in conjunction with WWW database searches. The sequence files may be served to the "public" Internet or kept private by requiring a password for access. Other advantages are (i) *sequences can be accessed from multiple computer platforms using the appropriate WWW-browser;* (ii) *files can be accessed remotely from any computer on the Internet;* (iii) *only a single sequence format is used, simplifying the updating and archiving of data; and* (iv) *links to remote files can also be served in addition to local files.*

INTRODUCTION

We are in danger of being swamped by the data from the various "Genome Projects". This rapid and exponential growth in the acquisition of DNA and protein sequence information provides a number of challenges not only for the maintainers of the various databases but also for the individual laboratories that must use this information on a daily basis. Thus, even small laboratories are finding themselves in an "information-overload" situation. The sequences generated or used frequently "in-house" must be organized efficiently in a user-friendly manner, and there is an increasing need to search the most current DNA and protein sequence databases more frequently. These problems may be compounded by the use of several analysis programs and multiple computer platforms, which require that sequence data be available in various formats. Although some analysis pro-

BioComputing: Computer Tools for Biologists
Edited by Stuart M. Brown

Table 1. List of URLs to Link Remote Computers

Number	URL (Uniform Resource Locator)
1	http://www.ncbi.nlm.nih.gov/Recipon/bs_seq.html
2	http://www.blocks.fhcrc.org
3	http://psort.nibb.ac.jp
4	http://expasy.hcuge.ch
5	http://www.public.iastate.edu/~pedro/research_tools.html
6	http://home.netscape.com
7	http://www.biap.com
8	http://www.ncsa.uiuc.edu/General/Internet/WWW/HTMLPrimer.html
9	http://dragon.acadiau.ca/~giles/HTML_Editor/Documentation.html
10	http://www.nih.gov/pdb/pdb
11	http://www.prosci.uci.edu/kinemages/kinpage.html
12	http://hermes.bioc.uvic.ca
13	http://atlas.nlm.nih.gov:5700/Entrez/index.html
14	http://hermes.bioc.uvic.ca/Poxvirus/vaccinia/VCOP-IND.HTML

grams can read multiple formats and other programs are available to interconvert formats, this constant switching between formats usually leads to multiple copies of sequence files. This in turn results in file-naming difficulties and may have serious consequences if care is not taken to ensure the data of the original files remain unaltered.

Access to the World Wide Web (WWW) is almost universal in research institutions, and the result has been an explosion in the amount of information and the number of resources available to molecular biologists. Examples include performing database searches with BLAST (Reference 1; Table 1, #1); using BLOCKS to search for protein motifs (Reference 3; Table 1, #2); predicting protein localization sites (Reference 5; Table 1, #3); and protein analysis with programs of the ExPASy site at the University of Geneva, Geneva, Switzerland (Table 1, #4). An excellent catalog of such resources is provided by Pedro Coutinho (Table 1, #5). Indeed, one of the simplest ways of running DNA or protein database searches is to first retrieve a sequence from DNA and protein databases through the WWW and then use remote computers to perform these computationally intensive searches.

However, while "WWW forms" have simplified the process of submitting database queries, it is still necessary to type or "paste" the query sequence into the "form". Our own situation exemplifies some of these difficulties. We are interested in proteins that are involved in the virulence of ectromelia virus, a mouse poxvirus. Poxviruses are large viruses with DNA genomes in excess of 180 kb, and they encode more than 200 proteins. Two viral genomes (2,4) have been completely sequenced, and there is a large amount of sequence data from other viruses available. Since the function of many of the viral proteins is currently unknown, we frequently wish to compare these proteins to new sequences in the database. Even with viral proteins of known activity, it is equally important to find newly sequenced homologs since comparison of multiple sequences helps define the most conserved amino acids. In addition to WWW tools, we use a variety of programs on different computer platforms to analyze the sequences. Thus, our goal was to create a system to help organize these relatively large DNA sequences and several hundred protein sequences in a simple and intuitive way that would overcome the problems discussed previously. Following the successful example of using the WWW to provide access to the entire protein and DNA database, we decided to establish a local WWW server to provide similar functions with our own customized data set. Through this paper, we wish to demonstrate the utility of a simple local WWW server in organizing protein and DNA sequence data for an individual research laboratory. Potential users of such a system should not be put off by the apparent complexity of some WWW sites on the Internet.

MATERIALS AND METHODS

A computer with an Internet connection is required to access the WWW. A number of WWW-browser programs are available, each with versions for most computer platforms. These are used to browse files made available by computers acting as WWW servers. We have used Netscape™ (Table 1, #6) as a browser and MacHTTP™ (Table 1, #7) as the server running on a Macintosh® Centris 650 computer (Apple Computer, Cupertino, CA, USA) with an Ethernet connection to the University of Victoria's computer network. It is not necessary to dedicate a computer to the WWW server function since the program runs in the background.

The HyperText Transport Protocol (HTTP) specifies methods of communication between computers on the WWW but is of little concern to the average user. Instead, one uses HyperText Markup Language (HTML) to organize the format of a text document and define HyperText links within or between documents. HTML uses a simple set of commands, and a variety of information is available for the beginner (Table 1, #8). Although HTML documents are simple text files, a number of programs are also available to edit and check formatting of HTML commands. HTML Editor™ is a semi-what-you-see-is-what-you-get (WYSIWYG) editor for the Macintosh (Table 1, #9). More advanced features provided by the HTTP server include graphics, sound and computer-movies. It is also a simple process to serve Protein Data Bank (Table 1, #10) or kinemage (Table

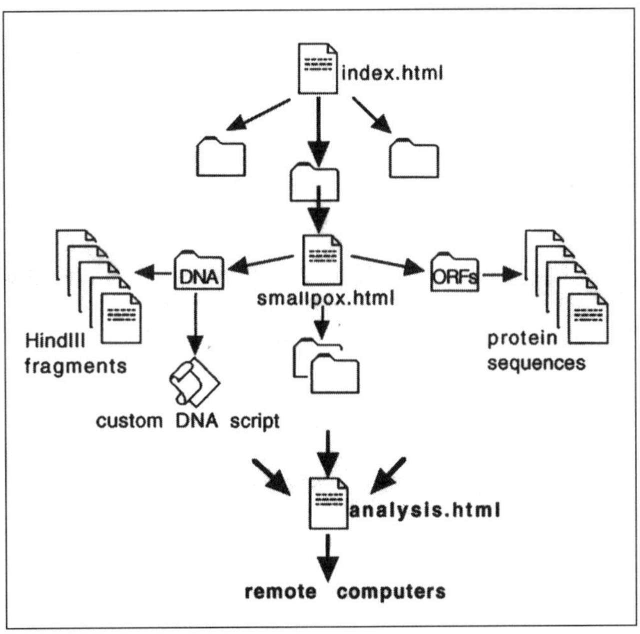

Figure 1. Outline of the organization of files on our WWW server. The first file sent to a user by the WWW server is the default.html page. Clicking on Poxvirus Section takes the user to the index.html page; it presents basic information and the options available. Only the file path relating to the DNA and protein sequences of the smallpox virus is shown in detail.

1, # 11) files and have a WWW browser automatically start programs to view protein or DNA structures.

The script, used by the WWW server for retrieving user-specified nucleotide ranges from large DNA files, was written in Applescript™ (Apple Computer).

RESULTS

The key feature of WWW documents is the use of HyperText, which allows the linking of text or graphics in one file to other parts of the same file or to other files. In WWW nomenclature, files are frequently referred to as "pages" and usually have the extension .html. The links can be used as menus to allow users to navigate quickly to the data they require. The organization of our WWW server (Table 1, #12), in a slightly simplified version, is shown in Figure 1. Upon initial connection, the "home page" of the WWW server is presented (default.html). One choice is the Poxvirus section, which is the gateway to the files dealing with the organization of the DNA and protein sequences of a smallpox virus (variola major, strain Bangladesh). Simply clicking on the highlighted HyperText **Poxvirus Section** commands the WWW server to send the index.html file (Figure 1), and clicking on **Smallpox virus** brings up information specific to this virus (Figure 1). A *Hin*dIII restriction map of the viral genome, saved as a GIF file, is presented along with a HyperText index of the fragments (Figure 2A; HyperText links are double-underlined in figures), which link to other regions of the same document containing lists of the individual open reading frames (ORFs; Figure 2B). Some data are presented with each ORF name, which is also a HyperText link to a file containing the relevant amino acid sequence (Figure 2C). For simplicity, each protein sequence is contained in a separate file, which also contains a HyperText link to an "analysis page", and HyperText links to references for that protein. The "analysis page" itself contains HyperText links to remote WWW services for the analysis of DNA and protein sequences.

The DNA sequence of variola major can be accessed in three ways. First, the DNA sequence of any *Hin*dIII fragment can be requested by clicking on the name from an index. Second, we have written a small script to retrieve user-specified regions of the genome. The user specifies the range of nucleotides to be returned and can specify if the sequence is to be reserved and complemented. Finally, the complete GenBank® entry for the viral genome can be requested.

All information is returned to the screen of the WWW browser, and it is a simple matter to SELECT-COPY-SAVE or SELECT-COPY-PASTE the material into other applications. This process is especially easy when preforming WWW-based database searches. Table 2 illustrates an example of the steps required to submit a protein search to BLAST. The process takes considerably less than a minute. An alternative connection to the databases for retrieval of sequences is by using the network or WWW versions of ENTREZ (Table 1, #13) produced by the National Center for Biotechnology (NCBI). Using ENTREZ, with *variola* and *Q2L* as the search keywords, results in five matches. Each is correct in that a variola virus Q2L protein is returned, but initially, it is not obvious which virus strain the protein is from, and sequences must be requested and the annotations checked for this information. However, when using a local WWW server, one has total control over the information that can be customized as required, and if preferred, rather than storing all sequences locally, just the WWW HyperText links to specific sequences in the databases can be presented by the local server. A further advantage of using a local WWW server is that a variety of information can be accessed; this could include unpublished sequences, laboratory protocols or plasmid maps.

We have also put, albeit in a simplified form, the 200 ORFs of vaccinia virus on the WWW server (Table 1, #14). One file contains an index of the viral proteins that consists of HyperText links to the location of the sequence data in a single file. While this is easier to set up, a large file must be accessed for each request. Since vaccinia and variola viruses are closely related, if one wished, each protein sequence could be associated with a HyperText link to its counterpart in the other virus. However, it is necessary to balance the effort required and the benefit gained from this work.

Table 2. Steps to Submit Protein Search to BLAST

1) Link directly to WWW server Poxvirus Page
[the WWW browser bookmark feature allows any link to be saved for direct access]
2) Click on Smallpox virus
3) Click on Hind III fragment Q from index
[this link jumps to another part of the document]
4) Click on Q2L from ORF list
[this link jumps to another file containing the amino acid sequence of Q2L]
5) Select and copy amino acid sequence
6) Click on Analysis Page
[this links to a page containing HyperText links to remote computers]
7) Click on BLAST Search Submission Form
8) Set BLASTP parameters for protein query and elect database
[these options are presented as buttons and menus]
9) Paste amino acid “query sequence” into search form
10) Click on “Perform Search” button

As with any experiment it is of great benefit to plan in detail the setting up of a local WWW server. Anyone with a little experience storing files in directories or folders on computers should be able to do this with ease. Formatting files is quite simple, and many shortcuts are obvious once the outline of the server has been established. For example, since we used a single analysis page, exactly the same link is in each ORF file. Similarly, many other links are identical except for the name of the ORF they point to. At the start, one should have a good idea of what information needs to be accessed, and this should be tested on a subset of the data. While it is simple to modify the files of the server, it is time-consuming to add a single HyperText link to each of 200 protein sequence files when it could have been incorporated in a template file at the outset.

DISCUSSION

The number of WWW server sites is growing as fast as the DNA sequence databases; many are clearly for entertainment or business purposes while others present valuable information of a general nature. However, we wish to demonstrate the utility of running a WWW server locally within our own laboratory. The WWW server has solved a major problem of organizing large numbers of DNA and protein sequences into an easily accessible format. The data involved are from the publicly available databases, and thus we have made our WWW server accessible to the Internet and other members of the poxvirus community. However, if the data must be protected, the WWW server can be customized in several ways to restrict access from outside users.

There are a number of advantages to this method of organizing DNA and protein sequences. The identical data are available to multiple computers in the laboratory without the many problems associated with networking different computer platforms. Indeed, the WWW server may be accessed from any computer on the Internet with a WWW browser. This may reduce computer-access problems in a laboratory. Since these data are consolidated on a single computer, in a single format, it is a simple process to update and back up the sequence files. Also, due to rapid data retrieval from the local WWW server and because the text format allows easy incorporation into WWW forms and other programs, the retrieved sequences may be considered to be "disposable". Therefore, the users are more likely to access the WWW server rather than store copies in several formats that may become mislabeled or corrupted. Finally, it is possible to store HyperText links to infrequently used DNA or protein sequences in the public databases. This again simplifies data management and makes the information available to all who need it in a form that also speeds access to the sequences.

ACKNOWLEDGMENTS

R.A. was supported by the Apple Educational Consultant Program. F.G. was supported by the British Columbia Student Assistance Program. This work was funded by the Natural Sciences and Engineering Research Council of Canada Operating Grant to C.U. We are grateful to Mike Reddy and Paula Davies for testing the WWW server and to those who have made the WWW into a valuable resource for the Sciences.

REFERENCES

1. **Altschul, S.F., W. Gish, W. Miller, E.W. Myers and D.J. Lipman.** 1990. Basic local alignment search tool. J. Mol. Biol. *215*:403-410.
2. **Goebel, S.J., G.P. Johnson, M.E. Perkus, S.W. Davis, J.P. Winslow and E. Paoletti.** 1990. The complete DNA sequence of vaccinia virus. Virology *179*:247-266.
3. **Henikoff, S. and J.G. Henikoff.** 1994. Protein family classification based on searching a database of blocks. Genomics *19*:96-107.
4. **Massung, R.F., L. Liu, J. Qi, J.C. Knight, T.E. Yuran, A.R. Kerlavage, J.M. Parsons, J.C. Vente and J.J. Esposito.** 1994. Analysis of the complete genome of smallpox variola major virus strain Bangladesh-1975. Virology *201*:215-240.
5. **Nakai, K. and M. Kanehisa.** 1992. A knowledge base for predicting protein localization sites in eukaryotic cells. Genomics *14*:897-911.

Section VII

Lab Automation and Miscellaneous Software

Lab automation is a classic venue for the do-it-yourself scientist. Many of these efforts involve using equipment in novel ways, and often involve the development of custom computer applications. Similar to the sections in this volume that describe applications built with macros and scripting languages, these articles are useful both for people who wish to automate the exact same process and for others who can make use of the software as a model to build a solution for their own unique laboratory situation.

CHAPTER 56

An Introduction to Molecular Visualization: Seeing in Stereo with RasMol

Robert M. Horton and Remington J.S. Stone

Department of Dermatology, Biophysics Graduate Program, University of California, Davis, Davis, CA, USA

Reprinted from BioTechniques 22:660-662 (April 1997)

Just as protein and nucleic acid sequences are collected in databases accessible on the Internet, so too are three-dimensional structures from X-ray crystallography and nuclear magnetic resonance studies. To appreciate the insights this data provide into macromolecular function, we need to visualize it structurally. Programs that translate mathematical coordinate values into interactive images on a computer screen can help us understand the relative positions and conformations of active, reactive or interactive sites. And besides that, molecules are really pretty.

Downloading the Software

This article focuses on RasMol (RASter MOLecules), a well-established molecular visualization program that runs on personal computers (1). Its development is currently supported by Glaxo-Wellcome Research and Development (Greenford, England, UK), and it can be downloaded at no cost from their Web site (http://www.glaxowellcome.co.uk/netscape/software/). More information and links, as well as a reference manual, can be found on the RasMol Home Page at http://www.umass.edu/microbio/rasmol. It is available for many platforms, including Windows®, Macintosh®, Unix and even the Acorn Archimedes.

RasMol can run on older computers, including 286-based IBM compatibles, or 680x0 Macintoshes, but on a newer Pentium® or Power Macintosh, it can really spin molecules around. Here we will demonstrate some useful features and tricks that can be done with Version 2.6b2, which allows stereo viewing. This is a "beta" version, so it is technically still in testing phase. While there are still a few bugs in the features that distinguish Version 2.6 from Version 2.5 (i.e., stereo), the program is quite useable.

BioComputing: Computer Tools for Biologists
Edited by Stuart M. Brown

Obtaining Structure Files

Molecular coordinates are archived at the Protein Data Bank (PDB)of Brookhaven National Laboratory at http://www.pdb.bnl.gov/cgi-bin/browse. A simplified interface is found at http://molbio.info.nih.gov/cgi-bin/pdb. The most basic way to use this data is to save a PDB structure file as text on your hard drive, then open it from within the RasMol program. Alternatively, your Web browser can be configured to automatically launch RasMol as a "helper application" whenever you download a file of the appropriate type. In Netscape, go to the options menu, choose "general preferences" and select the "helpers" tab. You need to create a new application MIME type with sub-type "x-rasmol". Set it to work with the extension "pdb" and to launch the RasMol application.

Setting the Viewing Variables

A variety of display modes can be chosen from the RasMol menu bar. For example, the molecule can be represented as "spacefilled" spheres, "wireframes" or "ribbons". Other options let you display the secondary structures or chains of a protein in different colors. The "specular" option (from the Latin specularis, meaning "of a mirror") adds (virtual) reflections of light from the molecule's surface.

Many additional commands and options can be typed in the command line window (This window may not be obvious; look for a "RasMol Command ..." button on your task bar in Windows 95® or a desktop icon in Windows 3.x. The command line window should be open behind the display window on a Macintosh.) For example, the command *set specpower n* (where n is a number between 1 and 100) changes the shininess of the molecule's surface. The command select *hydrophobic* selects all hydrophobic amino acids (aa), and the command *select (within(4.0,75-80)) and (not 75-80)* picks all atoms within the specified distance of aa 75–80, but not aa 75–80 themselves. The command *color blue* colors currently selected atoms blue. Information about commands can be found under "help" on the menu bar or in the RasMol manual.

Stereo Viewing

Clicking on "stereo" under "options" from the menu bar causes two similar pictures of the molecule to be displayed side by side. If the molecule appears off-center, it can be brought back to the middle of the viewing area by typing in the command *translate x -30.* Depending on the molecule, you might use a number from -25 to -35.

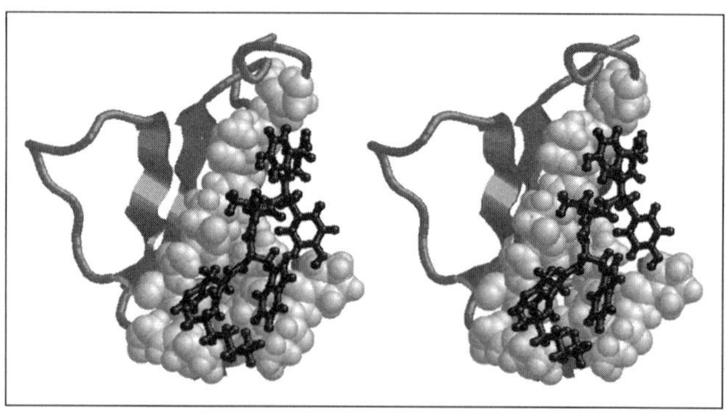

Figure 1. NMR structure of a snake neurotoxin, α-bungarotoxin, bound to a peptide from a nicotinic acetylcholine receptor (Reference 2; PDB ID code "1abt"). The peptide is shown in a modified ball-and-stick representation, and the amino acids from the toxin that interact with the peptide are shown in spacefill. The remainder of the toxin molecule is shown in "cartoon" format, with beta sheets depicted as broad arrows.

There are two basic

methods for stereo viewing without special glasses: cross-eyed viewing and parallel viewing. Both rely on altering the normal relationship between the convergence of your eyes and the focal distance. Information on stereo viewing and tutorials to help you learn the methods can be found at http://www.vision3d.com/3views.html. By default, RasMol's "stereo" menu option gives you stereoscopic pairs set up for cross-eyed viewing, which means your right eye should see the left image, while your left eye sees the right image. Many people (including these authors) find the parallel method easier. To generate stereo pairs for parallel viewing, enter *stereo -8* on the command line. Again, the parameter -8 is an approximation; some people may be more comfortable with -4 or -10.

For parallel viewing, it is particularly important that the images not be set too far apart; you can change the size and shape of the display window by dragging the lower right corner with the mouse cursor. For parallel viewing, unfocus your eyes—imagine you are looking at something very far away—and allow the images to overlap. If you find one method of viewing is too difficult or causes too much eye strain, try the other. If you get the molecule in focus and it looks "inside out", your eyes are focused for the other viewing method, so "set stereo" appropriately. Figure 1 should be viewed by the parallel method.

Command Scripts

RasMol commands can be stored in a file and run as a batch. Once you have your molecule displayed the way you like it, use the command *write script figure.txt* to automatically create a script file (in this case named figure.txt). Then, the current display configuration can be recreated by running the script with the command *script figure.txt.* Script files can be modified using a word processor, and saved as plain ASCII (American Standard Code for Information Interchange) text.

Exporting Images

RasMol can export images in several formats, including Windows bitmap (BMP, which can be labeled and edited with the Windows Paint program), and vectored postscript (for high-resolution printing). Note that vectored postscript works for some display modes, like ball-and-stick and spacefill, but not for others, like ribbons and cartoon. Also, it does not produce stereo images. This is, after all, a beta test version of the software.

Images can also be exported in Compuserve graphics interchange format (GIF) format, to use on a Web page. We have put several examples, including an animated stereo molecular movie, on our Web page at http://128.120.169.130/ Fairclough/Btx.html.

The Internet Spirit

Another noteworthy thing about RasMol is that the source code is freely available, so other people can make modified versions or use parts of it in their own programs. For example, there is a University of California, Berkeley version of RasMol (you can find the link from the RasMol home page above) that allows you to load more than one molecule at a time, which is potentially useful for visualizing molecular interactions. The authors of the Netscape plug-in "Chemscape Chime" have also used some RasMol code to produce a remarkable piece of software for use with Web pages. This open-minded sharing of code exemplifies what many people feel is the true spirit of the Internet. These and other freely available molecular visualization programs are briefly reviewed at http://www.umass.edu/microbio/rasmol/othersof.html, and a more comprehensive list of molecular modeling software is available at http://cmm.info.nih.gov/modeling/software_list/sw_index_document.html.

REFERENCES

1. **Sayle, R.A. and E.J. Milner-White.** 1995. RasMol: Biomolecular graphics for all. Trends Biochem. Sci. *20*:374-376.
2. **Basus, V.J, G. Song and E. Hawrot**. 1993. NMR solution structure of an alpha-bungarotoxin/nicotinic receptor peptide complex. Biochemistry *32*:12290-12298.

CHAPTER 57

The Primer Generator: A Program that Facilitates the Selection of Oligonucleotides for Site-Directed Mutagenesis

Alexander Turchin and Joseph F. Lawler, Jr.

Johns Hopkins University School of Medicine, Baltimore, MD, USA

BioTechniques 26:672-676 (April 1999)

ABSTRACT

Site-directed mutagenesis is a powerful tool that has enabled molecular biologists to perform functional analysis of altered nucleic acids and proteins. Newer PCR-based mutagenesis techniques have reduced the process of mutagenesis to as little as one day. While each technique has its advantages, both require a strategy to isolate the desired clone from a population that contains mutagenized and wild-type genes. In this report, we describe a World Wide Web-based computer program that facilitates the design of mutagenic primers such that successfully mutagenized clones can be identified by the presence or absence of a unique restriction site.

INTRODUCTION

Many site-directed mutagenesis (SDM) protocols call for the use of two oligonucleotides (2,5). The mutagenic oligonucleotide introduces the desired mutation, and a selection oligonucleotide increases the fraction of successfully mutagenized clones and facilitates their subsequent identification. The selection oligonucleotide does improve the fraction of mutagenized clones but it is not 100% effective (3). Clones to which the selection oligonucleotide has annealed but are not annealed to the mutagenic oligonucleotide, are only identified with subsequent DNA sequencing.

If the mutagenic primer introduces or deletes a restriction site, then it can aid in the subsequent identification of mutagenized clones. In effect, the mutagenic primer can be carefully designed to also be a selective primer. The Primer Generator automates the

BioComputing: Computer Tools for Biologists
Edited by Stuart M. Brown

selection of a mutagenic oligonucleotide that meets these criteria. This technique, of carefully selecting a mutagenic oligonucleotide to facilitate subsequent mutant identification, is a useful and cost-effective method for optimizing the procedure of SDM.

MATERIALS AND METHODS

Software

The following software is used: Perl (10) with a CGI module (9) and National Center for Supercomputing Applications (NCSA) HTTPd server (6).

Algorithm

Figure 1 shows an outline of the program. Initially, the investigator inputs the nucleotide sequence that encodes the current protein fragment and the protein sequence desired following mutagenesis. The latter sequence is reverse-translated using the redundancy of the genetic code to produce a number of DNA sequences that could potentially encode the given protein sequence.

Following reverse-translation, the sequences are interrogated for restriction enzyme sites using data from the Restriction Enzyme Database (REBASE) (http://www.neb.com/rebase/rebase.html). The results of these searches are independently compiled for the current and all possible target nucleotide sequences. Potential restriction sites in the existing sequence are subtracted from restriction sites in the mutagenized sequence, thereby generating a list of restriction sites unique to the mutagenized product. Similarly, restriction sites in the mutagenized sequence are subtracted from restriction sites in the existing sequence, which identifies sites that were deleted by the mutagenesis procedure.

From these operations, a list is compiled of unique restriction sites in the mutagenized sequence and of the oligonucleotide probes needed to generate them. In addition, restriction sites that are deleted from the existing sequence through the mutagenesis procedure are also listed. The mutagenic primer output is ordered by an increasing number of base pairs that differ from the original sequence. These base pairs are boldfaced. A useful feature of the Primer Generator is the ability to limit the output by the maximal number of substitutions that will be tolerated in the mutagenic primer. It can be set for 1 through 10 or for no limit. This feature facilitates selection of a mutagenic primer that necessitates minimal alteration of the existing sequence. To decrease computation time, especially for longer sequences, it is also possible to limit the number of restriction endonucleases to only the most commonly used enzymes. These enzymes are likely to be readily available in most laboratories.

Introducing or deleting a unique restriction site may necessitate mutagenizing more than the minimal number of residues required to elicit the desired change in the primary sequence. The mutagenesis procedure is however sufficiently robust to facilitate changing more than one base pair. A mutagenic oligonucleotide can also be chosen such that it introduces a unique restriction site in the mutagenized sequence, deletes a specific restriction site from the existing sequence or produces some combination of the two. A number of clones from the mutagenesis procedure can then be screened in a short amount of time with a restriction digest before being sequenced.

The restriction enzymes' names in the output are linked to the corresponding entries in the REBASE database, which provides the user with additional information about the enzyme. Included among this information is the enzyme type, the microorganism from which it was isolated, the prototype of the enzyme (the isoschizomer that was discovered first), the recognition sequence, the sites at which the DNA strand is cleaved, commercial sources of the enzyme and references pertaining to the discovery and characterization of the enzyme.

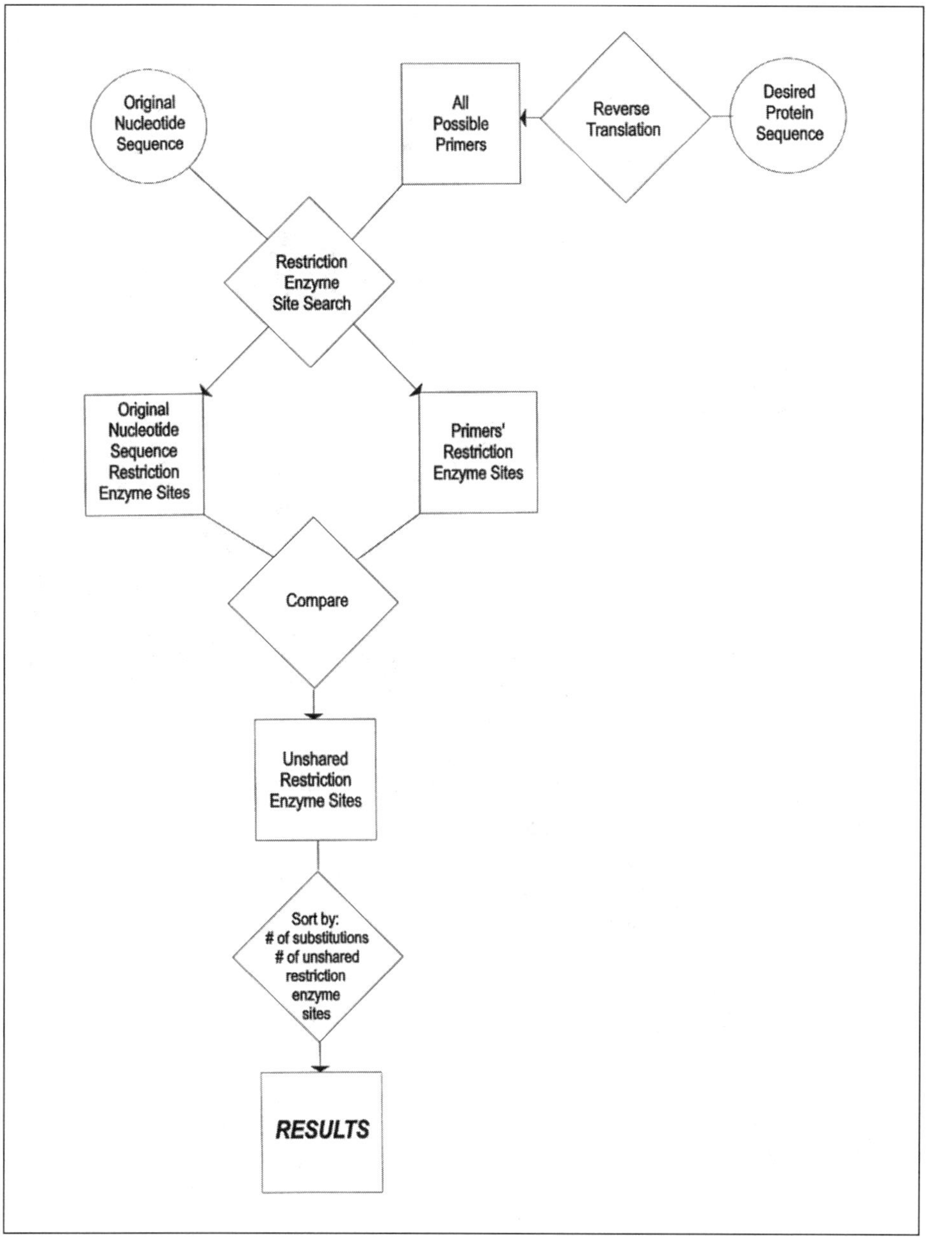

Figure 1. Flowchart of the Primer Generator program.

RESULTS

To demonstrate the effectiveness of this algorithm, we programmed the software to develop primers that mutagenize a critical serine residue in the rapamycin and FKBP12 target protein 1 (RAFT1) to arginine. When administered to cells, rapamycin forms a complex with the immunophilin FKBP12. This complex subsequently interacts with RAFT1 (8)/FRAP (1). This FKBP12-rapamycin-RAFT (1) interaction is dependent on the presence of a serine residue at position 2035 in RAFT1/FRAP (1). RAFT1/FRAP mutants that do not complex with FKBP12 in the presence of rapamycin are proving instrumental in elucidating the signaling capacity of this protein. Figure 2 summarizes the output of the program.

The Primer Generator

Original nucleotide sequence	GAGGCCTCTCGCTTG
Original protein sequence	EASRL
Desired amino acid sequence	EARRL
Use of restriction enzymes with site size less than 5 bp	No
Use of only most common restriction enzymes	No
Maximum number of substitutions compared to original sequence	3

The following primers have been generated:

- Click on enzyme name to access corresponding entry in The Restriction Enzyme Database (REBASE).
- Positions of cut sites are calculated relative to the beginning of the primer.
- Differences between primer sequence and original nucleotide sequence are **bolded**

Primer Sequence	Deleted Restriction Sites	Cut Positions	New Restriction Sites	Cut Positions	Number of Substitutions
GAGGCCC**CG**TCGCTTG	AatI	4	None		2
	Eco147I	4			
	Pme55I	4			
	SseBI	4			
	StuI	4			

Primer Sequence	Deleted Restriction Sites	Cut Positions	New Restriction Sites	Cut Positions	Number of Substitutions
GAGGCCC**CGG**CGCTTG	AatI	4	AsuC2I	6, 7	3
	Eco147I	4	BcnI	6, 7	
	Pme55I	4	Bsp143II	12, 8	
	SseBI	4	BstH2I	12, 8	
	StuI	4	HaeII	12, 8	
			NciI	6, 7	

Figure 2. Partial representation of primers constructed by the Primer Generator program. For each primer, the following data are presented: (*i*) nucleotide sequence of the primer,(*ii*) list of restriction endonucleases that cleave the WT sequence but not the mutagenized sequence, (*iii*) number of residues from the beginning of the primer that the endonuclease cleaves, (*iv*) list of restriction endonucleases that cleave the mutagenized sequence but not the WT sequence, (*v*) number of residues from the beginning of the primer that the endonuclease cleaves and (*vi*) number of substitutions that the primer introduces into the WT sequence. The primers are sorted in order of decreasing number of changes introduced in the WT sequence. Underlined text represents hyperlinks.

The output in Figure 2 is only a representative selection of the entire output of the program, which included 15 different mutagenic oligonucleotides. The mutagenic oligonucleotides (limited to the maximum of three substitutions compared to the original sequence) are capable of introducing 14 unique restriction sites, deleting the five existing sites and allowing the user to select from combinations thereof. Alternatively, with settings changed to limit restriction endonucleases used in analysis to only the most common enzymes and the maximum number of substitutions raised to four, we are still able to introduce six new and remove two old restriction sites (data not shown).

DISCUSSION

The Primer Generator has a number of uses. For example, an approach that has been used to define the binding specificities of modular protein-protein interaction domains (such as the phosphotyrosine-directed SH3 domain) is to independently vary the amino acid residue at a single position to the other 19 amino acids (7). This method has been successful in defining the binding characteristic of a single position (hydrophobic, basic, etc.) in the target peptide. By using the Primer Generator, one could order a single mutagenic primer that contains three N's in place of the codon to be mutated and contains an N in the wobble position of the adjacent invariant codons (over a thousand different sequences). A single mutagenesis reaction would be performed, and a few well-chosen restriction digests would rapidly identify the desired mutants, while preventing the sequencing of redundant mutants.

In addition to its use for mutagenesis, the Primer Generator can also be of use in clinical settings. A number of genetic diseases such as cystic fibrosis are characterized by a specific mutation that accounts for >70% of all cases (11). If one wished to screen polymerase chain reaction (PCR) products generated by primers flanking the region most commonly mutated, they need only input the wild-type (WT) and mutant sequences into the program to find restriction enzymes that can be used to distinguish between the two.

Computers are assuming an ever increasing role in molecular biology. The repertoire of computer-based tools and utilities available to investigators continues to grow. Here, we report an algorithm that allows one to choose a mutagenic oligonucleotide such that successfully mutated DNA can readily be identified by a restriction digest. The flexible, user-friendly interface offers the user extensive control of the program and its output. The hot links provide the user with additional information about the enzyme chosen. The availability of site-specific, hybrid restriction endonucleases is expected to increase the number of enzymes available for use by this program (4). A carefully chosen mutagenic oligonucleotide can save time and money by reducing the number of primers used and clones sequenced. The Primer Generator can be accessed at the Web site: (http://www.med.jhu.edu/medcenter/primer/primer.cgi).

ACKNOWLEDGMENTS

The authors wish to thank Dr. Robert Cottingham and Dr. Peter Li for constructive feedback and technical comments. We are also indebted to the Division of Biomedical Information Sciences of the Johns Hopkins University School of Medicine for allotting the project free space on the JHMI Affiliates Server.

REFERENCES

1. **Chen, J., X. Zheng, E.J. Brown and S.L. Schreiber.** 1995. Identification of an 11-kDa FKBP12-rapamycin-binding domain within the 289-kDa FKBP12-rapamycin-associated protein and characterization of a critical serine residue. Proc. Natl. Acad. Sci. USA *92*:4947-4951.
2. **Hemsley, A., N. Arnheim, M.D. Toney, G. Cortopassi and D.J. Galas.** 1989. A simple method for site-directed mutagenesis using the polymerase chain reaction. Nucleic Acids Res. *17*:6545-6551.
3. **Hutchinson, C.A., M. Phillips, M.H. Edgell, S. Gillam, P. Jahnke and M. Smith.** 1978. Mutagenesis at a specific position in a DNA sequence. J. Biol. Chem. *253*:6551-6560.
4. **Kim, Y., J. Cha and S. Chandrasegaran.** 1996. Hybrid restriction enzymes: zinc finger fusions to Fok I cleavage domain. Proc. Natl. Acad. Sci. USA *93*:1156-1160.
5. **Kunkel, T.A.** 1985. Rapid and efficient site-specific mutagenesis without phenotypic selection. Proc. Natl. Acad. Sci. USA *82*:488-492.
6. **NCSA HTTPd Development Team.** 1998. NCSA HTTPd v. 1.5.2a. Available from: **http://hoohoo.ncsa.uiuc.edu/docs/Overview.html** [Internet].
7. **Pawson, T.** 1995. Protein modules and signalling networks. Nature *373*:573-580.
8. **Sabatini, D.M., H. Erdjument-Bromage, M. Lui, P. Tempst and S.H. Snyder.** 1994. RAFT1: a mammalian protein that binds to FKBP12 in a rapamycin dependent fashion and is homologous to yeast TORs. Cell *78*:35-43.
9. **Stein, L.D.** 1998. CGI.pm v. 2.42. Available from: **http://www-genome.wi.mit.edu/ftp/pub/software/WWW/** [Internet].
10. **Wall, L., T. Christiansen and R. Schwartz.** 1996. Programming Perl, 2nd ed. O'Reilly & Associates, Sebastopol, CA.
11. **Welsh M.J., L.C. Tsui, T.F. Boat and A.L. Beaudet.** 1993. Cystic fibrosis, p. 3799-3875. *In* C.R. Scriver, A.L. Beaudet, W.S. Sly and D. Valle (Eds.), The Molecular and Metabolic Basis of Inherited Disease, 7th ed. McGraw Hill, New York.

CHAPTER 58

Computer Program for Calculating the Melting Temperature of Degenerate Oligonucleotides Used in PCR or Hybridization

Haoyuan Chen and Guan Zhu[1]

Emory University, Atlanta, GA and [1]New York State Department of Health, Albany, NY, USA

BioTechniques 22:1158-1160 (June 1997)

ABSTRACT

Degenerate primers or probes have been widely used in molecular cloning, but the calculation of their melting temperatures could not simply be done using thermodynamic parameters because of degeneracy and the lack of a computer program. We present here a simple computer program named dPrimer for the calculation of melting temperature of degenerate oligonucleotides based on the nearest-neighbor model. The program was written in C++ computer language and implemented in Macintosh® with a Symantec® C++ compiler. The degenerate sequencing data were read into a graph data structure. All possible oligonucleotide sequences were then determined by a depth-first search algorithm. Their melting temperature (T_m) values were individually calculated, and output was given as T_m range, mean and standard deviation. These data could help one in the selection of PCR annealing and hybridization temperatures as well as in the design of degenerate oligonucleotides with a desired range of T_m.

INTRODUCTION

Degenerate primers or probes, usually designed from partially sequenced peptides or conserved regions on the basis of comparison of several proteins, have been widely used in the polymerase chain reaction (PCR), the DNA library screening or Southern blot analysis (9,10). It is necessary to estimate the melting temperature (T_m) of a primer or probe so that one can select reasonable annealing or hybridization temperatures.

The nearest-neighbor model calculates the free energy of the double helix for an

BioComputing: Computer Tools for Biologists
Edited by Stuart M. Brown

oligonucleotide based on the sequence, thus providing an accurate estimation of the T_m of the oligonucleotide (1–3,6,7). Although the algorithm of the model is not complicated, a computer program may be needed in the calculation of T_m of oligonucleotides because of the number of parameters and steps involved in the calculation (6,7). A degenerate primer or probe is actually a number of mixed oligonucleotides. The number of oligonucleotides is determined by the degeneracy, which ranges from at least 2 to several thousand. Since each oligonucleotide has its own T_m, a degenerate primer or probe actually has a range of T_m values. When using the nearest-neighbor model, it is not practical for one to dissect the degeneracy and individually calculate the T_m value of all possible oligonucleotides.

A popular algorithm could easily be applied to calculate T_m of a degenerate oligonucleotide:

$$T_m\,(^\circ C) = 2 \times AT + 4 \times GC \qquad \text{[Eq. 1]}$$

where AT and GC are the numbers of AT and GC base pairs, respectively (8). Based on previous tests with oligonucleotides whose experimental T_m values were known, this does not seem to be an accurate method for either PCR or hybridization (6,7). An empirical way to determine the optimal annealing or probing temperature is to test a series of temperatures in a set of experiments, which is a time-consuming process, especially when several pairs of degenerate primers or several probes are used in one experiment. Therefore, a convenient computer program for the calculation of T_m values of degenerate primers or probes using the nearest-neighbor model would greatly help researchers in the selection of an annealing or hybridization temperature or in the design of degenerate oligonucleotides with a desired range of T_m values. Hence, we were motivated to design the present computer program to fulfill the need.

MATERIALS AND METHODS

There have been several reported algorithms for the determination of T_m of an oligonucleotide duplex, but the most accurate are based on the nearest-neighbor interaction model (1,3,6,7). If the oligonucleotide is not self-complementary, the T_m can be calculated as:

$$T_m\,(^\circ C) = \Delta H/(\Delta S + R \times \ln(c/4)) - 273.15 + 16.6 \times \log\,[salt] \qquad \text{[Eq. 2]}$$

where ΔH and ΔS are the enthalpy and entropy for the helix formation of an oligonucleotide duplex, respectively; R is the molar gas constant (1.987 cal/°C × mol); c is the total concentration of the annealing oligonucleotide; and [salt] is the potassium ion (K^+) concentration in the PCR or sodium ion (Na^+) in the hybridization of a blot. The effect of magnesium ions on the T_m of PCR primers has not been considered here because of the lack of published thermodynamic data. In a standard protocol, $K^+ = 50$ mM in PCR, while $Na^+ \approx 1.0$ M in hybridization of a blot, in which 6× standard saline citrate (SSC) or sodium chloride sodium phosphate EDTA (SSPE) buffer is usually used for both hybridization and washings (5). We also empirically assigned c = 50 nM or 0.1 nM to Equation 2 for the calculation of T_m values of degenerate oligonucleotides used in PCR or hybridization, which were in good agreement when those oligonucleotides with published thermodynamic data were tested (6,7).

The program, named dPrimer, was written in C++ computer language and implemented in Macintosh® with a Symantec® C++ compiler (Symantec, Cupertino, CA, USA).

The degenerate DNA sequencing data were read into a graph data structure. All possible oligonucleotide sequences were then determined by a depth-first search algorithm (4). Individual T_m values of all oligonucleotides were calculated with Equation 2 using the nearest-neighbor thermodynamic data determined by Breslauer et al. (2). The output was given as a range of T_m values, the mean of T_m values and the standard deviation (SD).

RESULTS AND DISCUSSION

The launched dPrimer program provides two small windows for the input of either amino acids or nucleotides. A brief instruction is also given for the use of one-letter symbols. The amino acids can be converted to degenerate oligonucleotides by checking the 'A.A. to DNA' button. The oligonucleotides can also be converted to complementary and reverse sequences by checking the 'complement & reverse' button. The default maximum degeneracy is set at 5000 but can be adjusted accordingly. The thermodynamic parameters for both PCR and hybridization have been preset in the program, and a user can select one of them by checking the corresponding button. The program will begin the calculation if the 'Process' button has been checked. All results will then be shown in the large window on the right. Although the dPrimer program is designed for degenerate oligonucleotides, the T_m of a regular oligonucleotide can also be calculated with a degeneracy of 1.

The depth-first search algorithm was applied in the program to search for all possible sequences. T_m values of all possible oligonucleotides in a degenerate primer or probe were then calculated, and output was given as a T_m range (minimum and maximum T_m) with the mean T_m values and SD. The length of the degenerate oligonucleotides could easily be adjusted in the input windows for quick calculations if a desired range of T_m is required.

For example, one of the degenerate PCR primers used in the molecular cloning of the β-tubulin gene of *Eimeria tenella* has an amino acid sequence of DNEALY (9). The dPrimer program will convert it into a degenerate primer sequence as GAY-AAY-GAR-GCN-YTN-TA and calculate the degeneracy (256), T_m range (30.81°–51.24°C), mean of T_m (40.89°C) and SD (4.01°C). If this primer is used for hybridization, the range of T_m values, the mean of T_m values and the SD would be 41.78°–61.83°, 51.75° and 4.00°C, respectively. The two sets of data reflect the difference in the stability of the DNA duplex of oligonucleotides, which is caused by the change of concentrations of salt and oligonucleotides (1).

Considering that most degenerate oligonucleotides are designed for PCR amplification, DNA library screenings or Southern blot analysis, we include only thermodynamic data for the formation of a DNA duplex in this program. Therefore, the T_m obtained using the program is only accurate when the degenerate oligonucleotides are annealing or hybridizing to DNA templates.

Although dPrimer will give users a T_m range covering all individual oligonucleotides in a degenerate primer or probe, the T_m of the oligonucleotide that actually forms the DNA duplex is not clear. Previous studies have shown that, besides the T_m of primers, other factors such as the length and T_m of PCR product, should also be considered when determining an optimal annealing temperature for PCR (7). In the hybridization of a membrane blot, the hybridization temperature may be much lower than the T_m of an oligonucleotide. The first two washings are usually done at room temperature, while the final washing is conducted for 5–15 min at the T_m of oligonucleotide (5,6). Degenerate primers have been intensively used in our laboratory where we generally begin with an annealing temperature between the mean and maximum T_m of primers for PCR. In the case of hybridization, temperature for the final washing is usually chosen between the minimum and mean T_m of the probe, which may be increased if higher stringency is required. Using this strategy, we have successfully amplified desired genomic DNA fragments of the β-tubulin gene in *Eimeria tenella* (9) and a P-ATPase gene in *Cryptosporidium parvum* (G. Zhu, unpublished) by PCR with degenerate primers. We have also cloned other types of P-ATPases in *C. parvum* by screening a genomic DNA library with a degenerate probe (G. Zhu, unpublished).

The dPrimer Version 1.0 for Macintosh computer is available upon request.

REFERENCES

1. **Breslauer, K.J.** 1995. Extracting thermodynamic data from equilibrium melting curves for oligonucleotide order-disorder transitions. Methods Enzymol. *259*:221-242.
2. **Breslauer, K.J., R. Frank, H. Blšcker and L.A. Marky.** 1986. Predicting DNA duplex stability from the base sequence. Proc. Natl. Acad. Sci. USA *83*:3746-3750.
3. **Freier, S.M., R Kierzek, J.A. Jaeger, N. Sugimoto, M.H. Caruthers, T. Neilson and D.H. Turner.** 1986. Improved free-energy parameters for predictions of RNA duplex stability. Proc. Natl. Acad. Sci. USA *83*:9373-9377.
4. **Horowitz, E., S. Sahni and D. Mehta.** 1995. Fundamentals of Data Structures in C++, p. 345-346. Computer Science Press, New York.
5. **Rychlik, W., L.L. Domier, P.R. Gardner, G.M. Hellmann and R.E. Rhoads.** 1987. Amino acid sequence of the mRNA cap-binding protein from human tissues. Proc. Natl. Acad. Sci. USA *84*:945-949.
6. **Rychlik, W. and R.E. Rhoads.** 1989. A computer program for choosing optimal oligonucleotides for filter hybridization, sequencing and in vitro amplification of DNA. Nucleic Acids Res. *17*:8543-8551.
7. **Rychlik, W., W.J. Spencer and R.E. Rhoads.** 1990. Optimization of the annealing temperature for DNA amplification in vitro. Nucleic Acids Res. *18*:6409-6412.
8. **Suggs, S.V., T. Hirose, E.H. Miyake, M.J. Kawashima, K.I. Johnson and R.B. Wallace.** 1981. Use of synthetic oligodeoxyribonucleotides for the isolation of specific cloned DNA sequences, p. 683-693. *In* D.D. Brown (Ed.), ICN-UCLA Symposium: Development Biology Using Purified Genes. Vol. 23. Academic Press Inc., New York.
9. **Zhu, G. and J.S. Keithly.** 1996. The beta tubulin gene of *Eimeria tenella*. Mol. Biochem. Parasitol. *76*:315-319.
10. **Zhu, Y.Y., M.J. Fillenwarth, D. Crabb, L. Lumeng and R.C. Lin.** 1996. Identification of the 37-kD rat liver protein that forms an acetaldehyde adduct in vivo as delta(4)-3-ketosteroid 5-beta-reductase. Hepatology *23*:115-122.

CHAPTER 59

Software to Determine Optimal Oligonucleotide Sequences Based on Hybridization Simulation Data

D. Hyndman[1], A. Cooper[1], S. Pruzinsky[1], D. Coad[1] and M. Mitsuhashi[1,2]

[1]Advanced Gene Computing Technologies, Inc., Irvine, CA, and [2]Hitachi Chemical Research Center, Irvine, CA, USA

BioTechniques 20:1090-1097 (June 1996)

ABSTRACT

*In the design of oligonucleotide sequences for targeting DNA or RNA sequences, it can be difficult to identify sequences that will hybridize only to the intended target. The term "sequence-specific" or "sequence-nonspecific" is often used to describe the interactions of an oligonucleotide with a mixture of DNA or RNA. Our new computer program, HYB*simulator™ *(formerly OligoProbe DesignStation), creates a set of candidate oligonucleotides from a target gene. For each of the candidate oligonucleotides, a large sequence database is searched for sequences that will hybridize to the oligonucleotide. This is referred to as computer hybridization simulation (CHS). Using the nearest-neighbor model, the HYB*simulator *takes into account mismatches in hybridization and calculates the melting temperature (T_m) or free energy for hybridization to all sequences in a database. The specificity of each oligonucleotide is then quantified by the number of genes that may hybridize and the predicted T_ms or free energies of hybridization to those genes. The CHS data are used to select oligonucleotides based on their specificity with respect to a database.*

INTRODUCTION

Oligonucleotides are essential components for various types of biological experiments. They can be used for sequencing reactions, gene amplification procedures, specific gene detection (Northern/Southern blot experiments or in situ hybridization), detection of point mutations or the inhibition of gene expression with antisense oligonucleotides. Because they can be affordably synthesized, oligonucleotides are now frequently used in a wide range of basic and applied sciences, including the food, pharmaceutical and cosmetic industries. A major problem encountered when using oligonucleotides is the difficulty in obtaining specific sequences for each application. The terms "sequence-specificity" or

BioComputing: Computer Tools for Biologists
Edited by Stuart M. Brown

"sequence-nonspecificity" are often used, but there is no consensus as to how to define and quantify sequence specificity. Previously (14,15), we introduced a new concept for designing oligonucleotides based on a technique for calculating and representing specificity of oligonucleotide sequences. Using HYB*simulator*™, which runs on an IBM®-compatible PC or Power Macintosh®, we have adapted our technology for the design of oligonucleotides for many different applications. Here we discuss the capabilities and applications of HYB*simulator* for the design of oligonucleotide sequences.

LIMITATIONS OF CURRENTLY AVAILABLE SOFTWARE

Programs for designing polymerase chain reaction (PCR) primer pairs are available from various commercial and non-commercial sources (11,16). These programs select primer pairs that satisfy user-defined parameters, such as length, secondary structure of primers, G/C content of primers and amplicons, stability of primers, length of amplicons, etc. Because such software does not provide information regarding the specificity of the oligonucleotides, users cannot determine if particular primers will hybridize to other sequences in the genome. Another deficiency is that such software sometimes produces a long list of candidate primer pairs, making it arduous to initially select the "best" primers. These programs are also not suitable for antisense oligonucleotides.

Homology analysis programs (2,10,12,17,19,23) and multiple alignment programs (5,9) are sometimes used to identify specific oligonucleotides. These programs compare the similarity of long sequences and are designed for identifying homologous genes in a database. Because they compare the entire sequence, they are not well-suited for identifying short specific oligonucleotide sequences (Figure 1, I). Also, if multiple copies of a short sequence are present within the same gene, a multiple alignment program will only identify one of those sequences, since the entire sequence is being compared (Figure 1, II). Furthermore, because some sequences in GenBank® contain introns (genomic DNA), and some sequences for the same gene may not contain introns (cDNA or RNA), a homology analysis often will not identify similar genes (Figure 1, III). Lastly, "homology" implies sequence similarity, which is not necessarily related to hybridization strength. If a sequence is found to be the most "homologous" to another sequence, it is not necessarily the sequence that is the most favorable for hybridization ("hybridizable") (Figure 1, IV).

Another way to estimate the specificity of particular sequences in the genome is to estimate the frequency of a given oligonucleotide based on the frequencies of the constituent dimer, trimer, tetramer, pentamer and hexamers of the oligonucleotide (7). While such a technique may identify oligonucleotides containing hexamers, which are present in the genome in high frequency, it cannot be expected to accurately predict the presence or absence of an actual 20-mer.

PRINCIPLE OF HYB*simulator*

The principle of HYB*simulator* is to simulate actual hybridization experiments between probes and sequences in a database, then select probes based on the data as outlined in Figure 2. Hybridization simulation requires three files: a model, a ProbeSet and a sequence library.

Model

HYB*simulator* calculates melting temperature (T_m) and free energy of hybridization using the nearest-neighbor model. The nearest-neighbor model calculates the total

I. Difficulty in finding short oligonucleotide sequences.

```
AGCTAGCATTAGCTAGCTAGGGTGCCATGGCAGG
: ::  :: :  ::: :::  : ::  : :: ::
ACCTTTCAATGACTACCTACCGAGCATTAGCGGG
::::  :: :  ::: :::  :  :  :   :::
ACCTAGCATTAGCTAGCTAGGGTTCCATGAAGGG
```

II. Difficulty in finding multiple repeats.

```
ACCTAGCATTAGCTACCTACCGAGCATTAGCGGG
:::::::::::::: :::   :  :  :   :::
ACCTAGCATTAGCTAGCTAGGGTTCCATGAAGGG
```

III. Interference of homology by introns.

```
GGACGATGCAGGAAC
:::::::       :
GGACGATAGCTTGAGATGCAGCAGGAAC
```

IV. Sequence similarity does not correlate with hybridization strength.

```
AATTATTTTAAGCTGAACCTGCCCGCGGCACTTA
::::::::::   : :    :::::::      :
AATTATTTTATTGTTATTGAGCCCGCGCGTGGTT
```

Homology=100% Homology=70%

ΔG=-12.6 Kcal/mol ΔG=-17.8 Kcal/mol

Figure 1. Summary of current problems of homology/multiple alignment programs for designing optimal oligonucleotide sequences.

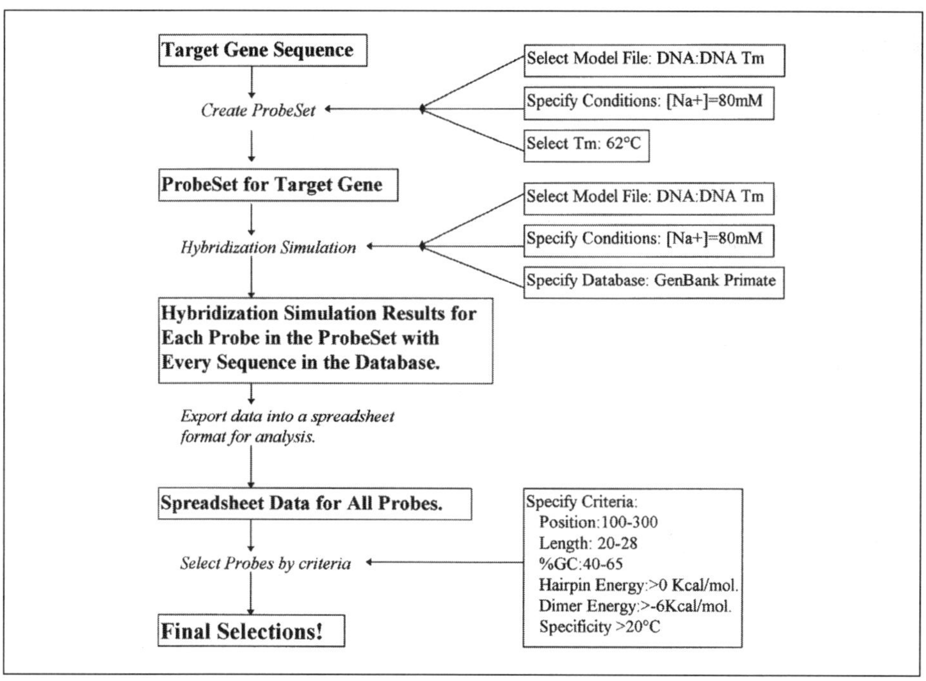

Figure 2. HYB*simulator* strategy for design of oligonucleotides.

enthalpy or entropy of hybridization based on contributions of every two consecutive base pair combination. T_m or free energy is determined from the total entropy and enthalpy (3,6). The parameters for the contributions of the different nearest-neighbor combinations are contained in the model files. The parameters for DNA:DNA (3), RNA:RNA (6,22) or RNA:DNA (21) are different and are specified by the appropriate model file. When new parameters become available (i.e., for phosphorothioate oligonucleotides or peptide nucleic acid), they can be put into a new model file for use by HYB*simulator*. The model files also contain values that are used for calculating the T_m or free energy for hybridization with mismatches (1,8,18,20).

ProbeSet

A ProbeSet is a set of candidate oligonucleotides. A ProbeSet created by HYB*simulator* contains all possible oligonucleotides derived from the target sequence that fit a chosen specification. ProbeSets can be created with a specified length, T_m or free energy of hybridization. HYB*simulator* creates a candidate oligonucleotide from each position in the target gene sequence that is closest to fitting the specified input criteria (i.e., T_m, free energy or length). For example, if the target gene is 1000 bp, approximately 1000 candidate oligonucleotides will be created and stored in the ProbeSet. Once a ProbeSet is created, the user can view the sequence of each oligonucleotide, edit the sequences, complement the entire ProbeSet or add a tag sequence to each oligonucleotide entry. Also, a ProbeSet can be created by inputting oligonucleotide sequences directly.

Hybridization

It is from a ProbeSet that the designed oligonucleotides are selected for a particular application. The specificity of probes in the ProbeSet is determined based on computer hybridization simulation (CHS) data. CHS simulates hybridization of each probe in the ProbeSet with every sequence in a specified GenBank database. For each sequence to which a particular probe may hybridize, the binding strength (as T_m or free energy) is calculated using parameters provided by the selected model file. In addition to the conventional calculation of T_m or free energy for the entire length of the probe, HYB*simulator* performs multiple calculations for the possible sub-sequences, then selects the most favorable value, as shown in Figure 3, I. This calculation will yield the value for the most stable hybridization of a probe to a mismatched sequence (Figure 3, II). CHS for a 1000-bp gene with the GenBank primate database takes approximately 4 h with typical hardware (Pentium® PC with 32 Mbytes random-access memory [RAM]).

I. HYB*simulator* ™ (subsequence analysis):

```
Probe*:CATCCAGCTACGAATCTCCG   ΔG (Kcal/mol)
gene  :AGCCCAGCTACGGATCTTGG   -12.0
          ---------G----TG-   -13.0
          ---------G----      -13.4 (HYBsimulator™'s choice)
          ---------           -10.0
```

*:actual probe in experiments should be complement

II. Effect of subsequence analysis:

	ΔG (Kcal/mol) subsequence analysis without	ΔG (Kcal/mol) subsequence analysis with**
Probe* :CATCGAGCCGACGACGACCG		
gene1 :AGC-C------------TG-	-15.8	-18.3
gene2 :AGC------C-------TG-	-15.8	-17.4
gene3 :AGC---------C----TG-	-15.8	-17.4
gene4 :AGC------------C-TG-	-15.8	-18.3

*:actual probe in experiments should be complement
**: Values are derived from HYB*simulator* ™

Figure 3. Comparison between conventional nearest neighbor thermodynamic calculation and HYB*simulator* in mismatched nucleotide sequences.

For CHS, HYB*simulator* accepts the GenBank Flat File CD-ROM (National Center for Biotechnology Information [NCBI], Bethesda, MD, USA). Also, HYB*simulator* can use a researcher's own sequence database if the data are in the GenBank ASCII format.

VIEWING AND ANALYSIS OF HYB*simulator* RESULTS

Hybridization Window

The *Hybridization* window in HYB*simulator* displays CHS data (Figure 4). The left half of the window contains a graphical representation of all of the CHS data generated. The right half of the window displays specific data for individual probes. Each position on the X-axis of the graph designates a single probe. As you move the cursor horizontally across the graph, the probe data on the right half of the window change to correspond to the selected probe position.

The right half of the *Hybridization* window contains the found hybridizable sequences for individual probes. In the top box, the name (including position within the target), T_m and sequence of the selected probe are displayed. The lower box contains the found hybridizable sequences. For each found sequence, the following is displayed:

- The GenBank entry that contains the sequence ("Locus").
- The position within the entry where the hybridizable sequence was found ("Pos").
- The T_m or ΔG of hybridization of the probe to the hybridizable sequence ("T_m").
- The nucleotide sequence of the hybridizable sequence. Identical nucleotides are displayed as "-", and mismatches are displayed as the actual nucleotide of the sequence ("Sequence").

The graph is a plot of the T_m or ΔG of all of the found hybridizable sequences in the results file for each of the probes. The T_m of the probe is

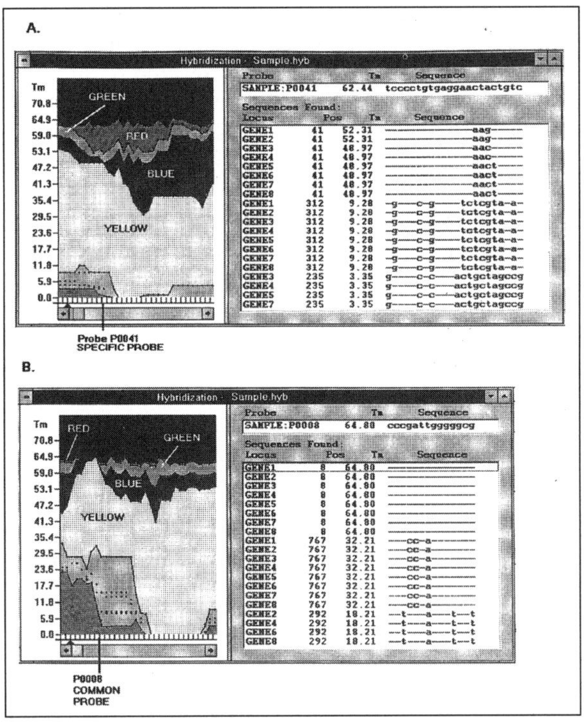

Figure 4. The *Hybridization* Window: identifying a specific or a common probe. Each X-axis position in the graph represents a different probe. Moving the cursor across the graph causes the data window on the right of the graph to display the data for the selected probe. (A) Probe:P0041 is selected. For each probe, the T_m of the probe hybridizing to its target is graphed in red. The T_m of the first hybridization is graphed in green. The T_m of the third hybridization is graphed in blue, and the T_m of the 7th hybridization is graphed in yellow. Probe:P0041 is specific because the distance between the red graph and the green graph (i.e., the specificity) is large. (B) Probe P0008 is selected. Probe:P0008 is identified as specific because the yellow graph reaches the level of the red graph, indicating that at least 7 sequences have exact matches to Probe:P0008. The data to the right of the graph verifies that 8 P0008 have 8 exact matches.

graphed in red and will always be the highest graphed point. The T_m of the first hybridization (the sequence to which the probe hybridizes with the highest T_m) for each probe is graphed in green. The distance from the top of the red graph to the top of the green graph indicates the difference between the T_m of the probe and the first hybridization. More specific probes will have greater differences in these T_ms. Since the difference of T_m between the hybridization to the target and the first hybridization to nonspecific sequences is a measurement of a probe's specificity, we refer to this value as "*Specificity*". Probe P0041, indicated in Figure 4A, is the most specific probe of those in the graph since it has the highest *Specificity*. The third hybridization T_ms are graphed in blue, and the seventh hybridization T_ms are graphed in yellow. If the yellow graph reaches the height of the red graph, the probe has at least 7 exact matches (Probe P0008 in Figure 4B). By scanning the graph, the user can identify areas of particular specificity or areas in which the probes are common to other sequences.

The *Library* window contains the sequence database that is used for CHS. The user can view information about the hybridizable genes in the *Hybridization* window. If the user selects the gene name in the right side of the CHS results window, the *Library* window updates to display the selected sequence. The user can then examine the sequence and read the details regarding the sequence.

Spreadsheet

The data generated by CHS can be exported into a spreadsheet format, which can be opened and analyzed with most commercially available spreadsheet programs. This format allows easy manipulation and selection of candidate oligonucleotides based on the many different parameters.

The spreadsheet created by HYB*simulator* contains a row for each oligonucleotide in the ProbeSet. The columns contain data, including CHS results data, that can be used to select the best oligonucleotides.

CHS data are represented by three different methods: *(i)* the T_m of the 1st, 2nd, 4th, 8th etc. saved hybridizable sequences found for each probe; *(ii)* the number of hybridizable and cross-hybridizable sequences at various ΔT_m or $\Delta\Delta G$ relative to the T_m or ΔG of the probe hybridizing to its target; and *(iii)* The difference of T_m or ΔG between the binding strength of the probe to its target and the closest non-exact match ("*Specificity*").

In addition to the CHS data, the following data are also provided for each probe: *(i)* length: the number of nucleotides in the oligonucleotide; *(ii)* GC content: the percentage of guanine or cytosine in the oligonucleotide; *(iii)* terminal GC content: the percentage G/C of the 3′ or 5′ end of the nucleotide. For PCR primers, because of the weaker binding of G/C base pairs, a low G/C content on the 3′ terminal may provide higher specificity; *(iv)* hairpin energy: the free energy of formation of a hairpin structure by the oligonucleotide (in Kcal/mol). Larger negative values indicate a more stable hairpin structure, which will prevent hybridization with the target and is therefore undesirable; and *(v)* dimer energy: the free energy of formation of a dimer by the hybridization of an oligonucleotide with the same oligonucleotide. Larger negative values indicate a more stable dimer, which will compete with the hybridization with the target and can adversely affect PCR.

In a spreadsheet program, the best oligonucleotides can be selected based on all the available data with criteria specific to the application (13). Additionally, other data, such as secondary structure of the target site or estimated frequency of the oligonucleotides based on subsequences (7), can be added to the spreadsheet for selection. Also, in most commercial spreadsheet programs like Microsoft® Excel™, users can write macros (small programs that run within the spreadsheet program) that can further derive data from the probes, which may be useful for some specific type of application or for automated selection.

APPLICATIONS OF HYB*simulator*

Basic Probes

To design optimally specific DNA probes for dot blots, Southern blots, Northern blots, etc., the user first creates a candidate ProbeSet from the target gene sequence. After selecting the model file (either for DNA:DNA, RNA:RNA or DNA:RNA), the user specifies salt concentration, oligonucleotide concentration and finally the desired T_m. The ProbeSet is then created containing all possible probes with the approximate T_m specified. Then, CHS is executed for the ProbeSet against the GenBank database of interest. For example, if the project is to identify a viral DNA sequence in a human specimen, GenBank viral and human databases would be used. When the CHS is complete, the results are saved in a spreadsheet file, and the optimal or best probes are selected from the candidate probes based on the data in the spreadsheet. Typical specifications include: G/C content between 40% and 65%; hairpin formation energy greater than 0.0 Kcal/mol (more negative values indicate more hairpins); dimer formation energy above -6.0 Kcal/mol (more negative values indicate more stable dimers); and the highest specificity possible to each of the databases.

Common or Consensus Probes

It is often desirable to detect the same gene in related organisms or related genes in the same organism. It is also sometimes desirable to try to detect unknown varieties of a gene family by using a probe that identifies all known types of a particular gene family. This can be accomplished by selecting common probes to all the targets of interest.

The design of common probes is very similar to the design of basic probes. To design common probes, in addition to the steps explained for the design of basic probes, users perform an additional CHS against a small database comprised of all the target genes. The spreadsheet data for this CHS are then pasted beside the spreadsheet for the other CHS's as described for basic probes. The user first selects for the number of exact hybridizations ($\Delta T_m = 0$) to the multiple target gene database to equal the number of target sequences. From this selection, the remaining candidates are all possible common probes. From these remaining candidates (i.e., the possible common probes), the user selects based on the remaining criteria as described for basic probes.

Table 1. Example of Oligonucleotide Sequences Data

Sequence (5′–3′)	Hybridization Strength (ΔG, Kcal/mol)			Hairpin Dimer (Kcal/mol)		Number of Matches in GenBank Primate Database at $\Delta\Delta G$		
	DNA:DNA	RNA:RNA	DNA:RNA			≤2	≤3	≤4
Δ tgcagtcatagaac	14.8	20.5	13.2	4.1	-6.4	11	24	51
∇ cagtcatagaacatg	15.8	20.9	14.2	5.3	-3.0	1	1	2

Basic PCR

The design of PCR primers is similar to the design of probes with the following additional considerations.

Terminal G/C content. It is common practice to select primers with a low G/C content at the 3′ end (i.e., the leading edge of replication). If the 3′ end of the primer is A/T rich, the weaker binding of A/T base pairs may put more importance on complete hybridization of the primer to prime polymerization. A low G/C content on the 3′ end of a PCR primer will therefore increase the specificity of the PCR.

Alternate binding within the target gene. It is important to ensure that primers are not going to bind at alternate sites within the target sequence. Such alternate binding can interfere with elongation or cause aberrant-sized bands. HYB*simulator* can determine the most stable alternate site binding energy for each primer. Primers can be selected for minimum hybridization to alternate sites within the target.

Compatibility of primers. PCR requires two primers, and the two primers should be compatible. Compatibility of PCR primers means that both primers have very similar T_ms, and they do not hybridize with each other. If the T_ms are different, at any chosen annealing temperature, either one of the primers is not annealing or the other primer is annealing to nonspecific sequences by allowing mismatches. If the primers are taken from a ProbeSet created with a fixed T_m, then the two primers will have the same T_m.

To select compatible primer pairs, the user first selects several good sense primers and then several good antisense primers. Typical criteria for selection include: G/C content between 40% and 65%; hairpin formation energy greater than 0.0 Kcal/mol (more negative values indicate more hairpins); dimer formation energy above -6.0 Kcal/mol (more negative values indicate more stable dimers); 3′ end G/C content below 50%; minimum hybridization to alternate sites within the target sequence and the highest specificity possible to each of the databases. From the selected sense and antisense primers, HYB*simulator* will then calculate the free energy of heterodimer formation between each combination of sense with antisense primer, and the best PCR primer pair is chosen based on minimum interaction between primers.

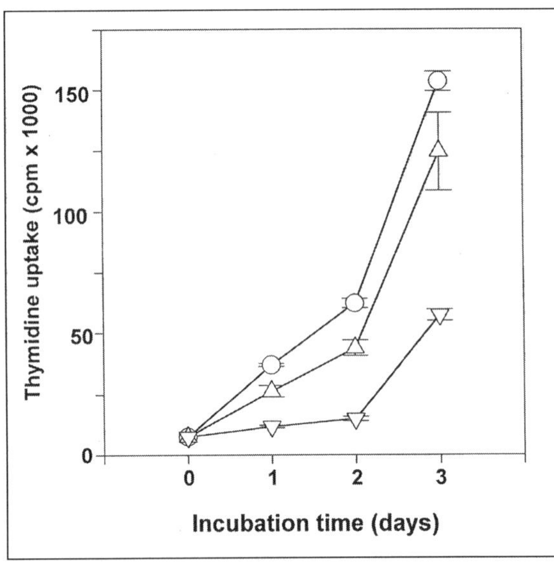

Figure 5. Inhibition of cell growth by antisense oligonucleotides. Human U937 mononuclear cells were seeded at a density of 2500 cells per well in 100 µL RPMI containing 0.5% fetal calf serum and incubated with (Δ,∇) or without (O) 10 µM of antisense phosphorothioate oligonucleotides against the human c-*jun* oncogene. After 24 h of incubation, cells were stimulated with 10% serum for 1–3 days. Cells were exposed to ^{3}H-thymidine (0.5 µCi per well) for the last 6 h, then incorporated radioactivity was determined in a liquid scintillation counter. Each data point is the mean ± standard deviation from triplicate determinations. Two oligonucleotide sequences [low specificity (Δ) and high specificity (∇)] were selected by HYB*simulator* from the start codon region as shown in Table 1.

Antisense Oligonucleotides

For antisense oligonucleotide sequence design, in addition to the basic oligonucleotide characteristics of G/C content, length and hairpin energy, and the specificity of the oligonucleotide with respect to the database, secondary structure must also be considered. Unlike hybridization experiments, the conditions of an antisense experiment cannot be changed, as they are the intracellular conditions. Therefore, more attention should be paid to the secondary structures of both the oligonucleotide and the target gene. HYB*simulator* can calculate the most stable free energy of secondary structure for the target site of each probe and add the data to the spreadsheet output. Based on this information, oligonucleotides can be chosen that will hybridize outside a region containing secondary structure.

The specificity of the oligonucleotides is crucial for effective antisense inhibition. It has been reported that 9-mer phosphorothioate antisense oligonucleotides can inhibit gene function effectively and specifically when microinjected into oocytes (4). Although 9-mer oligonucleotides can be effective antisense agents, a 9-mer is more likely to hybridize to unrelated genes; therefore, extra care must be taken to select specific oligonucleotides. HYB*simulator* can select candidate oligonucleotides not only by minimizing cross-hybridization against a genomic database in GenBank, but also by placing more emphasis on hybridizability against an mRNA or coding sequence database extracted from GenBank.

Figure 5 shows the results of an antisense experiment with two different antisense oligonucleotides to the c-*jun* oncogene and the data for the oligonucleotides.

DISCUSSION

There is no universal way to design the best oligonucleotide sequences, and design philosophies are varied for different applications. Therefore, we developed HYB*simulator* to provide as much information as possible on each candidate oligonucleotide. In addition to basic oligonucleotide characteristics such as G/C content, hairpin formation energy and dimer formation energy, we provide a quantitative means to select for specific oligonucleotides with respect to a particular database. Because of the simple spreadsheet format of the output of HYB*simulator* data, users can easily add additional data that may be important for their particular application. Furthermore, because of the flexibility of HYB*simulator* to use different thermodynamic models and sequence libraries, it will be useful as sequence information grows and additional thermodynamic nearest-neighbor parameters are developed.

One limitation of CHS is the incompleteness of current GenBank sequence data. Clearly, as GenBank becomes more complete, the correlation of CHS to the actual hybridization experiment will become closer. We believe, though, that the existing sequence data are a valid representation of the complete genome. Therefore, specificity of oligonucleotides with respect to the existing data can be a useful indication of the specificity of the oligonucleotide in the actual experiment. Moreover, if the users know particular sequences that they specifically do not want their oligonucleotide to hybridize to, CHS data will closely represent the actual experiment.

The HYB*simulator* runs on an IBM compatible 486 or higher (with Windows NT™) or a Power Macintosh with 16–32 Mbytes of RAM, depending on the application. Because of the increasing popularity of these personal computers, we believe that HYB*simulator* will be accepted as a standard tool for designing oligonucleotides in the future.

ACKNOWLEDGMENTS

We would like to thank Dr. J. SantaLucia (Wayne State University, Detroit, MI) for providing new nearest-neighbor coefficients to us, and Dr. D.G. Payan (Khepri Pharmaceuticals, South San Francisco, CA) for his critical review of the manuscript.

REFERENCES

1. **Aboul-ela, J., D. Koh, I. Tinoco and F.H. Martin.** 1985. Base-base mismatches. Thermodynamics of double helix formation for dCA3XA3G + cDT3YT3G (X,Y=ACGT). Nucleic Acid Res. *13*:4811-4824.
2. **Altschul, S.F., W. Gish, W. Miller, E.W. Myers and D.J. Lipman.** 1990. Basic local alignment search tool. J. Mol. Biol. *215*:403-410.
3. **Breslauer, K.J., R. Frank, H. Blocker and L. Marky.** 1986. Predicting DNA duplex stability from the base sequence. Proc. Natl. Acad. Sci. USA *83*:3746-3750.
4. **Fakler, B., S. Herlitze, B. Amthor, H.P. Zenner and J.P. Ruppersberg.** 1994. Short antisense oligonucleotide-mediated inhibition is strongly dependent on oligo length and concentration but almost independent of location of the target sequence. J. Biol. Chem. *269*:16187-16194.
5. **Feng, D.F. and R.F. Doolittle.** 1987. Progressive sequence alignment as a prerequisite to correct phylogenetic trees. J. Mol. Evol. *25*:351-360.
6. **Freir, S.M., R. Kierzek, J.A. Jaeger, N. Sugimoto, M.H. Caruthers, T. Neilson and D.H. Turner.** 1986. Improved free-energy parameters for predictions of RNA duplex. Biochemistry *83*:9373-9377.
7. **Han, J., Z. Zhu, C. Hsu and W.H. Finley.** Selection of antisense oligonucleotides on the basis of genomic frequency of the target sequence. Antisense Res. Dev. *4*:53-65.
8. **He, L., R. Kierzek, J. SantaLucia, A.E. Walter and D.H. Turner.** Nearest-neighbor parameters for G-U mismatches. Biochemistry *30*:1124-1132.
9. **Higgins, D.G. and P.M. Sharp.** 1988. CLUSTAL: a package for performing multiple sequence alignment on a microcomputer. Gene *73*:237-244.
10. **Landau, G.M., U. Vishkin and R. Nussinov.** 1990. Fast alignment of DNA and protein sequences. Methods Enzymol. *183*:487-502.
11. **Lowe, T., J. Sharefkin, S.Q. Yang and C.W. Dieffenbach.** 1990. A computer program for selection of oligonucleotide primers for polymerase chain reactions. Nucleic Acids Res. *18*:1757-1761.
12. **Marizel, J.V. and R.P. Lenk.** 1981. Enhanced graphic matrix analysis of nucleic acid and protein sequences. Proc. Natl. Acad. Sci. USA *78*:7665-7669.
13. **Mitsuhashi, M.** Technical Report: oligonucleotide probe design series: basic requirements for designing optimal oligonucleotide probe sequences. J. Clin. Lab. Anal. (In press).
14. **Mitsuhashi, M., A. Cooper, M. Ogura, T. Shinagawa, K. Yano and T. Hosokawa.** 1994. Oligonucleotide probe design—a new approach. Nature *367*:759-761.
15. **Mitsuhashi, M. and T. Hosokawa.** 1994. Novel computerized method for designing nucleotide sequence used for DNA probes and PCR primers. Nippon Rinsho *52*:530-541.
16. **Montpetit, M.L., S. Cassol, T. Salas and M.V. O'Shaughnessy.** 1992. OLIGSCAN: a computer program to assist in the design of PCR primers homologous to multiple DNA sequences. J. Virol. Methods *36*:119-128.
17. **Pearson, W.R. and D.J. Lipman.** 1988. Improved tools for biological sequence analysis. Proc. Natl. Acad. Sci. USA *85*:2444-2448.
18. **SantaLucia, J., R. Kierzek and D.H. Turner.** 1990. Effects of GA mismatches on the structure and thermodynamics of RNA internal loops. Biochemistry *29*:8813-8819.
19. **Smith, T.F. and M.S. Waterman.** 1981. Comparison of bio-sequences. Adv. Appl. Math. *2*:482-489.
20. **Sugimoto, N., R. Kierzek, S.M. Freier and D.H. Turner.** 1986. Energetics of internal GC mismatches in ribooligonucleotide helix. Biochemistry *25*:5755-5759.
21. **Sugimoto, N., S. Nakano, M. Katoh, A. Matsumura, H. Nakamuta, T. Ohmichi, M. Yoneyama and M. Sasaki.** 1995. Thermodynamic parameters to predict stability of RNA/DNA hybrid duplexes. Biochemistry *34*:11211-11216.
22. **Turner, D.H., N. Sugimoto and S.M. Freier.** 1988. RNA structure prediction. Annu. Rev. Biophys. Chem. *17*:167-192.
23. **Wilbur, W.J. and D.J. Lipman.** 1983. Rapid similarity searches of nucleic acid and protein data banks. Proc. Natl. Acad. Sci. USA *80*:726-730

DN*Assist*, a C++ Program for Editing and Analysis of Nucleic Acid and Protein Sequences on PC-Compatible Computers Running Windows 95, 98, NT4.0 or 2000

Hugh G. Patterton and Steven Graves[1]

University of Cape Town, Rondebosch, South Africa, and
[1]Los Alamos National Laboratories, Los Alamos, NM, USA

Reprinted from BioTechniques 28:1192-1197 (June 2000)

ABSTRACT

*The size of sequence databases, the rapid rate of generating sequence data and the need to routinely construct recombinant DNA molecules in molecular biological research necessitate the frequent handling and analysis of nucleotide and protein sequences on a computer. Although several console or Web-based utility programs are available, the application of these programs generally requires reformatting the data when proceeding from one such program to the next. The acquisition of elaborate, integrated program suites is financially prohibitive to smaller laboratories. Here, we report the development of DN*Assist*, a shareware program for editing and analysis of nucleic acid and protein sequences. It was developed as a multiple-document interface program—similar to a word processor—where sequences are entered, edited and analyzed in a single integrated environment. DN*Assist *can calculate the physicochemical properties of a sequence, convert between nucleic acid and protein sequences, translate DNA in multiple frames, identify open reading frames and locate ambiguous sequence patterns allowing gaps and mismatches. DN*Assist *also performs restriction enzyme and transcription factor-binding site analyses of DNA sequences, the multiple alignment of nucleic acid and protein sequences and the analysis of DNA sequences for nucleosome positioning sites.*

INTRODUCTION

Research in molecular biology generally requires microcomputers to edit, manipulate and analyze nucleotide and protein sequences. The availability of the sequence of the entire genome of some organisms (3,20), the rapid generation of sequence data and the require-

BioComputing: Computer Tools for Biologists
Edited by Stuart M. Brown

ment to design and construct recombinant DNA molecules in research have made it necessary for almost all molecular biological laboratories to store and analyze sequences on computers. Common analyses include the identification of specific and functionally significant sub-sequences, the conversion of one sequence type to another, the location of consensus sites such as those for restriction enzymes or for transcription factors and the multiple alignment of sequences. Often the physicochemical properties such as molecular weight, melting temperature, isoelectric point, molar absorption coefficient at a given wavelength and the composition of polynucleotides and polypeptides are also required.

These analyses can be performed with individual utility programs that are often console-based (19), form-based Web interfaces to server-side Perl or C scripts (22) or with larger program suites such as GCG (7) that require mainframe or workstation computers running UNIX® or Solaris® operating systems. However, when performing a series of analyses with separate utility programs, the data are generally required or generated in specific formats and need to be reformatted in a text-based editor when proceeding from one analysis program to the next. The acquisition of sophisticated program suites that offer an integrated environment for analysis is often financially prohibitive to many smaller research laboratories. Here, we report the development of DN*Assist*, a shareware C++ program designed with a multiple-document interface (MDI) with integrated editing and extensive analysis capabilities that can be used on a PC-compatible computer running the Microsoft® Windows® 95, 98, NT® version 4.0 or 2000 operating system.

MATERIALS AND METHODS

Program Design

DN*Assist* is coded in C++ (1,18) and the graphic user interface developed using version 5.02 of the Borland® Objects Windows Library (Inprise, San Jose, CA, USA). The code was compiled with version 5.02 of the Borland® C++ compiler (Inprise) for Microsoft Windows 95, 98, NT4.0 or Windows 2000, and was optimized for execution speed; C++ exceptions were enabled. The code is object orientated, with classes designed for sequence storage, management, manipulation, analysis and presentation (a paper detailing the design and relationship between classes will be published elsewhere). The proper allocation and de-allocation of memory blocks by the program was tested with version 2.0 of Borland® Codeguard® (Inprise), and the program was subjected to three beta testing phases in unrelated molecular biology laboratories. DN*Assist* requires an Intel Pentium MMX or related CPU with 32 MB RAM and at least 15 MB available disk space for installation; it can be downloaded from http://www.dnassist.com.

RESULTS AND DISCUSSION

General Program Layout

DN*Assist* is designed as an MDI, as shown in Figure 1. The user is presented with a familiar environment, similar to that of a word processing program, in which sequences are managed and edited and analyses performed. All file and windows management, sequence editing and analysis operations may be selected from the menu, where relevant methods and commands are logically grouped under the File, Edit, Convert, Analyze,

Options, Windows and Help menu items. The more frequently used commands are also accessible from buttons on a movable toolbar (Figure 1) that displays hints when the mouse pointer is moved over individual buttons. All the analysis methods can be selected from the Analysis menu or from a context-sensitive pop-up menu. When a method is selected, it is directly applied to a displayed sequence or sequence selection in the editor window with input focus. The user can analyze multiple types of sequences in a myriad of ways within one easy-to-use integrated environment.

The user can open several windows simultaneously in which sequences (editor windows) or the results of analyses (result windows) are displayed. The windows can be resized and arranged using the mouse pointer or the Windows menu, which also lists all open editor and result windows (Figure 1).

Sequence Editor

There are four different editor-type windows in which to display and edit DNA or degenerate DNA, RNA or protein sequences. New editor windows can be opened from the File menu or from the toolbar. The sequence editor window depends on the type chosen by the user from the menu, in the case of a new editor window, or on the sequence type of a file opened from a disk. For sequence files that do not specify a sequence type,

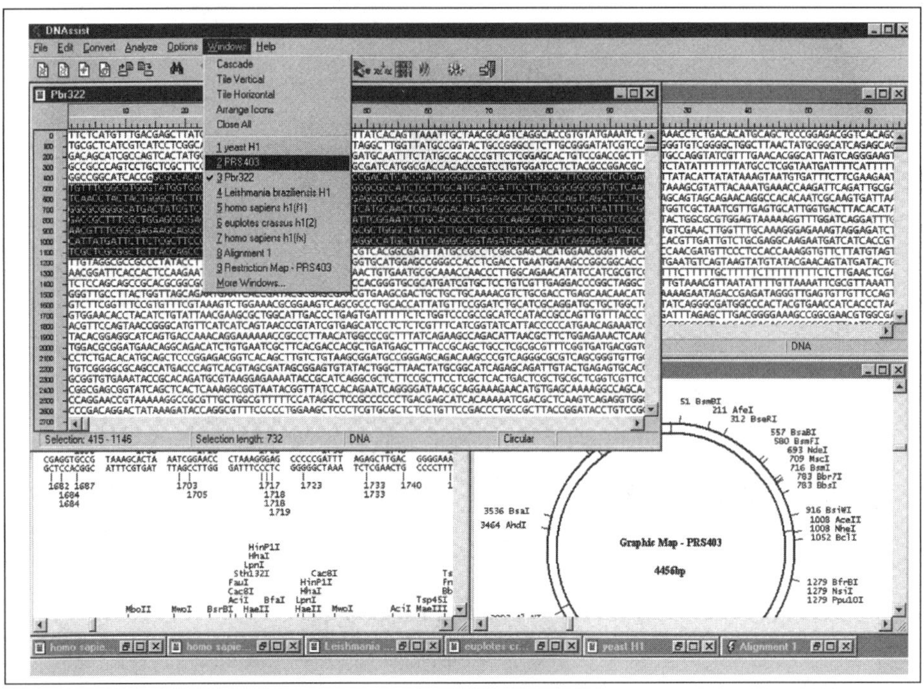

Figure 1. The DN*Assist* sequence editing and analysis environment. DN*Assist* is designed as an MDI program that allows the user to have multiple-sequence editing and analysis-result windows open simultaneously. These windows can be resized, arranged or minimized. All analysis methods can be selected from the menu and are directly applied to the sequence or sequence selection in the editor window with input focus. Frequently used actions are also accessible from the toolbar and from a context sensitive pop-up menu that is displayed by clicking the right mouse button.

such as text or FASTA (15) format files, DN*Assist* assigns a type that recognizes the largest percentage of characters in the sequence file.

The user can enter a sequence from the keyboard or load a sequence from a disk if the file is in text, FASTA (15), GenBank® (16), GCG (7), DNA Strider (13) or DN*Assist* format. The editor window responds to all uppercase or lowercase character keys that represent legitimate symbols for the window sequence type, general editing keys such as Delete and Backspace and the navigation arrow buttons. In addition, several shortcut keystrokes for common editing actions such as Copy, Cut and Paste are also defined. Alternatively, sequences can be dragged and dropped within (or between) compatible editor window types or pasted from the clipboard into a compatible window. Selections in a sequence can be made using the keyboard or the left button on the mouse and dragging the pointer within an editor window. The user can specify the font face and size of installed fixed pitch fonts, as well as the line length displayed in an editor window from the Options menu. All editor and analyses settings specified by the user are saved in a unique subkey of HKEY_CURRENT_USER in the Registry when the user exits the program, a feature that allows DN*Assist* to be used in a multi-user environment.

A sequence displayed in an editor window is manipulated in memory as a continuous series of eight-bit character data types and is managed by a specialized C++ class. This class will dynamically expand or shrink the allocated memory block in which the character array is placed in response to additions or deletions to the sequence. This allows optimum memory use and the manipulation of sequences with a size constrained only by the physical configuration of the computer.

Sequences can be saved to disk in ASCII text or DN*Assist* format. The format file of the latter contains a file header of five semicolon-delimited fields, in which information regarding the stored sequence is given and is followed by a series of bytes that correspond to the ASCII character values of the nucleotides or amino acid residues in the sequence.

Sequence Analysis

The analysis methods that can be selected from the main menu or the context-sensitive pop-up menu depend on the sequence type of editor window with input focus. The menu entries and toolbar icons of analysis methods that cannot be applied to a given sequence type are dimmed to present the user with a more intuitive graphic interface.

The result of an analysis is displayed in a result MDI window. In the case of a text-based result, the content of the result window can be saved to a disk or copied to the clipboard in rich-text format (RTF), from which it can be opened or pasted into other applications that recognize this format. Similarly, graphic results can be saved or copied and imported into other applications that recognize the enhanced Windows metafile format. The user has extensive control over the format in which text-based results are displayed. In the case of sequence listings, one or both DNA strands can be shown as a continuous sequence or grouped into sequence blocks with a block size and block spacing that may be specified. Sequence numbering can be displayed above, below or to the right of such listings.

Sequence Properties

DN*Assist* can calculate and display the general physicochemical properties of nucleic acid and protein sequences, plus the hydrophobicity (12), hydrophilicity (11) and antigenicity (21) profiles of protein sequences. The following are displayed for nucleic acid sequences: the number of each of the four nucleotides, G+C%, A+T%, sequence length, melting temperature calculated by the nearest neighbor method (6), annealing temperature, molecular weight, molar absorption coefficient and $E_{0.1\%}$ at 260 nm (2). Displayed for protein sequences are the number and mole percentage of each of the 20 amino acid residues, sequence length, molecular weight, pI (4), molar absorption coefficient and $E_{0.1\%}$ at 280 nm (14). These properties can be displayed for the entire sequence or for only a selected subset of the sequence.

Sequence Reversals, Conversions and Translations

A DNA sequence can be transformed between a linear and circular configuration or converted to its complementary sequence in the 5′ to 3′ orientation. This conversion takes place as a combined reverse-complement operation. A nucleic acid sequence can also be converted to a protein sequence, DNA and RNA sequences can be interconverted (Figure 2A) and a DNA sequence can be translated to a protein sequence in one, three or all six frames (Figure 2B). In addition, a protein sequence can also be reverse translated to a degenerate DNA sequence. Nucleic acid to protein conversions and translations are performed according to the entries in the codon table selected from the Options menu.

Codon tables generated by the GCG program, Codonfrequencies (7), are compatible with DN*Assist*. A translated sequence is displayed as a nucleotide sequence listing aligned with the single-letter amino acid abbreviation. Stop codons are indicated by asterisks. If an in-frame stop codon is encountered during a nucleic acid-to-protein conversion, the user is alerted to the truncation of the protein at the given position.

Finding ORFs and Patterns

All the sequence searches in DN*Assist* make use of a class implementation of the Boyer-Moore algorithm (5). Apart from routine substring searches from the Edit menu, DN*Assist* also allows searching of a sequence for ORFs and sequence patterns. The ORF search function uses the start and stop codons in the selected codon table to search for continuous sequence stretches that may code for proteins. The pattern-search function searches for a defined sequence pattern, allowing ambiguous nucleotides or amino acid residues in the pattern and a user-defined number of sequence gaps and mismatches. The sequence of an identified ORF or pattern match is highlighted in the editor window.

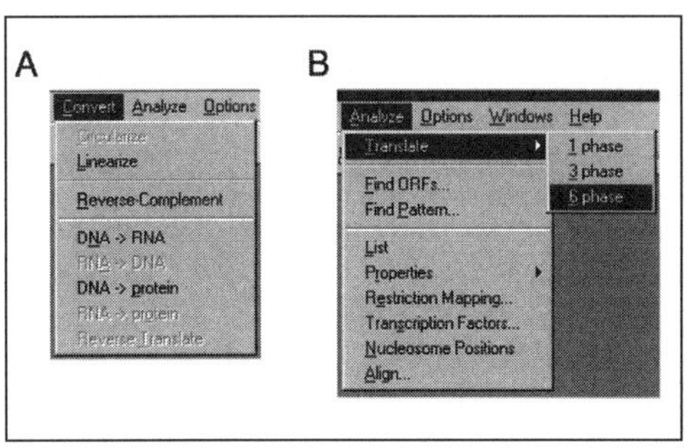

Figure 2. Sequence conversions and translations in DN*Assist*. (A) Nucleic acid sequences can be reversibly interconverted or converted to a protein sequence from the Convert menu. (B) DNA sequences can be translated in 1, 3 or all 6 reading frames from the Analyze menu.

Restriction Enzyme Analysis

DN*Assist* performs a restriction enzyme analysis using restriction enzyme data from a Rebase database file (17). Updated versions of this enzyme data file can be downloaded from the Rebase Web site at http://rebase.neb.com/rebase/rebase.files.html. All or a selection of the restriction enzymes in the file can be used (Figure 3), and enzymes leaving a specific type of cleaved terminus can also be specified from this set. The result of a restriction enzyme analysis can be displayed as a graphic map (Figure 4), a sequence listing and in tabulated format. The tabulated restriction enzyme analysis lists the enzymes used in the analysis, the number of times that each enzyme cleaves, the cleavage positions and fragment sizes in descending order and a list of restriction enzymes that do not have recognition sequences in the analyzed sequence.

Predicting Nucleosome Rotational Positions

The propensity of a DNA sequence to accommodate a rotationally positioned nucleosome is calculated with the Drew-Calladine algorithm (8) and is displayed as a plot of the probability of finding a rotationally positioned nucleosome at every position of the analyzed sequence.

Sequence Alignments

DN*Assist* conducts pairwise and multiple alignments of DNA, degenerate DNA, RNA and protein sequences by constructing a phylogenetic tree that is based on a matrix of all pairwise sequence similarity scores, as used in CLUSTALW (10,19). The alignment allows ambiguous nucleotides or amino acid residues. Aligned sequences are displayed as sequence lists with identical nucleotides and amino acid residues or as conserved amino acid residues, indicated in specified font foreground and background colors (Figure 5). From the Options menu, the user can specify the colors and the minimum level of sequence identity or conservation required before residues are indicated as such. The user can also specify Gap extension, opening penalties and the minimum level of divergence before alignment is delayed. DNA sequence alignments can be performed with a user-selected DNA weight matrix, and the penalty associated with a nonidentical purine-purine

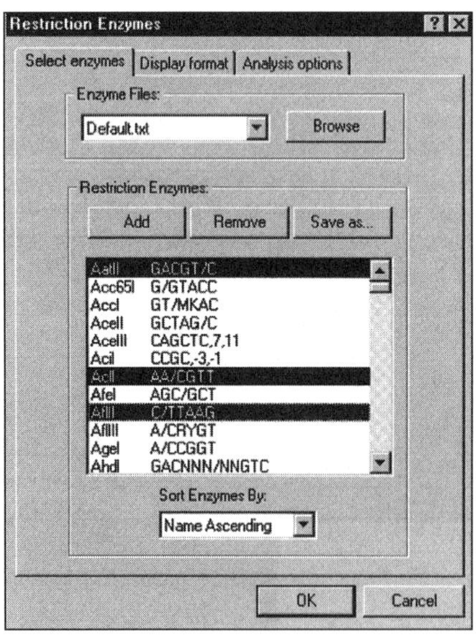

Figure 3. Restriction enzyme analysis. All restriction enzymes in a specified file are listed in the Restriction Enzymes dialog box. Individual restriction enzymes can be deleted, new enzymes added and the modified file saved to disk. The user can also select specific restriction enzymes from the enzyme list.

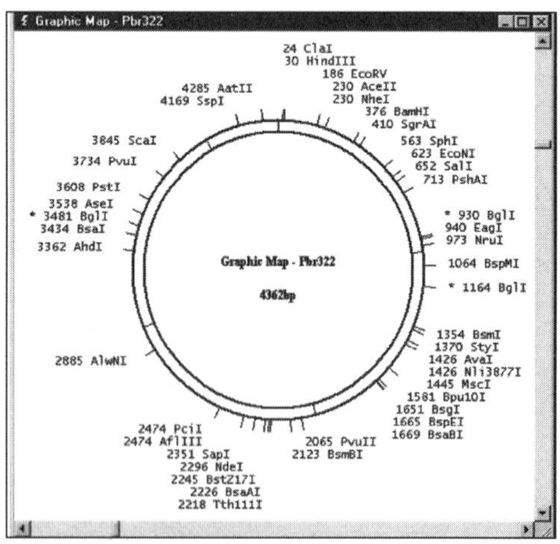

Figure 4. The display of a restriction enzyme analysis. A graphic map of a restriction enzyme analysis will display all unique sites and, if specified by the user, will indicate the presence of a cleavage site for a selected restriction enzyme, with an asterisk.

or pyrimidine-pyrimidine alignment can be set. For protein sequences, the protein weight matrix and penalties for opening gaps at specific residues, sequence termini or hydrophilic regions are also user selectable.

Promoter Analysis for Transcription Factor Binding Sites

A DNA sequence can be analyzed for the presence of transcription factor binding sites listed in the Transfac database (9). DN*Assist* lists transcription factors in the database in a hierarchic manner, grouped by kingdom, species and factor. These groups are displayed in list boxes with a master-slave relationship. For example, the user can select all vertebrate factors or only the factors identified in *Homo sapiens*. DN*Assist* issues a local structured query language (SQL) query against the Transfac database and displays only the factors returned for the selected kingdom(s) and species. The user can then choose to perform the analysis with all or only a selection of displayed transcription factors. Updated versions of the Transfac database can be downloaded from http://transfac.gbf.de/TRANSFAC/index.html. The result of a transcription factor analysis can be displayed as a graphic map, sequence listing or in tabulated form, where the name and binding position of the selected transcription factors are given.

SUMMARY

DN*Assist* is an excellent tool set for the molecular biologist while remaining affordable for the research scientist. As such, it is a marked advance over other such tools on the market that cost as much as 10 times more. In today's atmosphere of increasingly tight funding, DN*Assist* is an essential and cost-effective program for almost all molecular biology researchers.

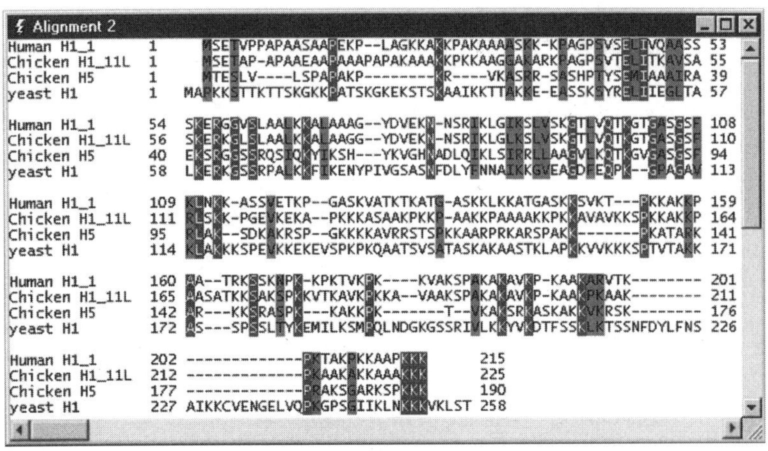

Figure 5. Multiple alignments in DN*Assist*. A multiple alignment is displayed with conserved and identical amino acid residues or identical nucleotides, as shown in a selected font foreground and font background color. The minimum number of nucleotides or amino acid residues that must be identical or conserved to be indicated as such can also be specified.

ACKNOWLEDGMENTS

The authors thank Andrew Flaus and Allison Johnson for their many helpful suggestions.

REFERENCES

1. **American National Standards Institute.** 1998. Information Technology-Programming Languages-C++. Document number ISO/IEC 14882-1998. American National Standards Institute, Bethesda, MD.
2. **Ausubel, F., R. Brent, R.E. Kingston, D.D. Moore, J.G. Seidman, J.A. Smith and K. Struhl.** 1995. Short Protocols in Molecular Biology. Third Edition, John Wiley and Sons, New York.
3. **Blattner, F.R., G. Plunkett, C.A. Bloch, N.T. Perna, V. Burland, M. Riley, J. Collado-Vides, J.D. Glasner et al.** The complete genome sequence of *Escherichia coli* K-12. Science *277*:1453-1474.
4. **Bjellqvist, B., B. Basse, E. Olsen and J.E. Celis.** 1994. Reference points for comparisons of two-dimensional maps of proteins from different human cell types defined in a pH scale where isoelectric points correlate with polypeptide compositions. Electrophoresis *15*:529-539.
5. **Boyer, R.S. and J.S. Moore.** 1977. A fast string searching algorithm. Commun. ACM. *20*:762-772.
6. **Breslauer, K.J., R. Frank, H. Blocker and L.A. Marky.** 1986. Predicting DNA duplex stability from the base sequence. Proc. Natl. Acad. Sci. USA *83*:3746-3750.
7. **Devereux, J., P. Haeberli and O. Smithies.** 1984. A Comprehensive Set of Sequence Analysis Programs for the VAX. Nucleic Acids Res. *12*:387-395.
8. **Drew, H.R. and C.R. Calladine.** 1987. Sequence-specific positioning of core histones on an 860-base-pair DNA. Experiment and theory. J. Mol. Biol. *195*:143-173.
9. **Heinemeyer, T., X. Chen, H. Karas, A.E. Kel, O.V. Kel, I. Liebich, T. Meinhardt and I. Reuter.** 1999. Expanding the TRANSFAC database towards an expert system of regulatory molecular mechanisms. Nucleic Acids Res. *27*:318-322.
10. **Higgins, D.G. and P.M. Sharp.** 1988. CLUSTAL: a package for performing multiple sequence alignment on a microcomputer. Gene *73*:237-244.
11. **Hopp, T.P. and K.R. Woods.** 1981. Prediction of protein antigenic determinants from amino acid sequences. Proc. Natl. Acad. Sci. USA *78*:3824-3828.
12. **Kyte, J. and R.F. Doolittle.** 1982. A simple method for displaying the hydropathic character of a protein. J. Mol. Biol. *157*:105-132.
13. **Marck, C.** 1988. DNA Strider: a C program for the fast analysis of DNA and protein sequences on the Apple Macintosh family of computers. Nucleic Acids Res. *16*:1829-1836.
14. **Pace, C.N., F. Vajdos, L. Fee, G. Grimsley and T. Gray.** 1995. How to measure and predict the molar absorption coefficient of a protein. Protein Sci. *4*:2411-2423.
15. **Pearson, W.R.** 1990. Rapid and sensitive sequence comparison with FASTP and FASTA. Methods Enzymol. *183*:63-98.
16. **Read, R.L., D. Davison, J.E. Chappelear and J.S. Garavelli.** 1992. GBPARSE: a parser for the GenBank flat-file format with the new feature table format. Comput. Appl. Biosci. *8*:407-408.
17. **Roberts, R.J. and D. Macelis.** 1999. REBASE-restriction enzymes and methylases. Nucleic Acids Res. *27*:312-313.
18. **Stroustrup, B.** 1997. The C++ Programming Language. Third Edition. Addison-Wesley Publishing Company, Reading, MA.
19. **Thompson, J.D., D.G. Higgins and T.J. Gibson.** 1994. CLUSTAL W: improving the sensitivity of progressive multiple sequence alignment through sequence weighting, position-specific gap penalties and weight matrix choice. Nucleic Acids Res. *22*:4673-4680.
20. **Walsh, S. and B. Barrell.** 1996. The *Saccharomyces cerevisiae* genome on the World Wide Web. Trends Genet. *12*:276-277.
21. **Welling, G.W., W.J. Weijer, R. van der Zee and S. Welling-Wester.** 1985. Prediction of sequential antigenic regions in proteins. FEBS Lett. *188*:215-218.
22. **Zhang, J. and T.L. Madden.** 1997. PowerBLAST: a new network BLAST application for interactive or automated sequence analysis and annotation. Genome Res. *7*:649-656.

Update to: DN*Assist*, a C++ Program for Editing and Analysis of Nucleic Acid and Protein Sequences on PC-Compatible Computers Running Windows 95, 98, NT4.0 or 2000

Hugh G. Patterton[1] and Steven Graves[2]

[1]*University of Cape Town, Rondebosch, South Africa, and* [2]*Los Alamos National Laboratories, Los Alamos, NM, USA (patterh@science.uct.ac.za)*

Version 2.0 of DN*Assist* is now available at the new DN*Assist* Web home at (http://www.dnassist.org). Version 2.0 of the program includes several improvements suggested by users, as well as minor bug fixes. Among the new features included is a reverse translation option, which is selectable from the Convert menu item. This allows the user to convert a protein sequence to a degenerate DNA sequence. This feature is particularly useful for the design of degenerate oligonucleotides for the polymerase chain reaction (PCR) amplification of the gene encoding a protein for which the amino acid sequence is known. The reverse translation of the sequence is performed according to the entries in the codon table, which can be selected from the Options menu item.

Previously, DN*Assist* assigned a sequence type to an ambiguous sequence text file according to the highest percentage of characters in the file that could be matched to a specific sequence type. In version 2.0, DN*Assist* allows the user to assign a sequence type in the case of ambiguity. This allows the user to open a text file containing the sequence GATC, as a protein sequence, if, indeed, the user intended for the sequence to be a polypeptide.

Version 2.0 of DN*Assist* includes a proprietary database of transcription factors to perform transcription factor binding site analyses of putative promoter sequences. DN*Assist* can also use the TransFac database (1). However, the text file version of the TransFac database that is used by DN*Assist* to generate a Paradox format database is no longer available to academic institutions for free, but is sold by Research Group Bioinformatics/AG Bioinformatik (RGB/AGB). Users who wish to make use of the TransFac database with DN*Assist* will need to purchase the database from RGB/AGB (http://transfac.gbf.de). Updated restriction enzyme files in DN*Assist* format are still available free of charge at (http://rebase.neb.com).

The Help files supplied with DN*Assist* 2.0 have been updated and converted to hypertext markup language (HTML) format. This allows the user to view help topics in a Help Viewer or Internet browser such as Internet Explorer. The help files can also be searched for specific words or phrases that may not be listed in the index. Version 2.0 of DN*Assist* Help contains 90 topics with 328 links and 27 graphic images.

DN*Assist* version 2.0 is available as a shareware program from (http://www.dnassist.org). To allow the use of security features in the registry, DN*Assist* is no longer supported on Windows® 95. Help with installation or use of DN*Assist* can be obtained by e-mail (help@dnassist.org). Users are requested to register DN*Assist* for a fee of $200 after a 90-day evaluation period. Enquiries regarding registration can be sent to (sales@dnassist.org). User comments and suggestion are very welcome, and can be e-mailed to (support@dnassist.org).

REFERENCE

1. **Wingender, E., X. Chen, R. Hehl, H. Karas, I. Liebich, V. Matys, T. Meinhardt, M. Prüß, I. Reuter, and F. Schacherer.** 2000. TRANSFAC: an integrated system for gene expression regulation. Nucleic Acids Res. *28*:316-319.

CHAPTER 61

Laboratory Voice Data Entry System

Jeremy L. Praissman and John C. Sutherland

Brookhaven National Laboratory, Upton, NY, USA

Reprinted from BioTechniques 27:1202-1208 (December 1999)

ABSTRACT

We have assembled a system using a personal computer workstation equipped with standard office software, an audio system, speech recognition software and an inexpensive radio-based wireless microphone that permits laboratory workers to enter or modify data while performing other work. Speech recognition permits users to enter data while their hands are holding equipment or they are otherwise unable to operate a keyboard. The wireless microphone allows unencumbered movement around the laboratory without a "tether" that might interfere with equipment or experimental procedures. To evaluate the potential of voice data entry in a laboratory environment, we developed a prototype relational database that records the disposal of radionuclides and/or hazardous chemicals. Current regulations in our laboratory require that each such item being discarded must be inventoried and documents must be prepared that summarize the contents of each container used for disposal. Using voice commands, the user enters items into the database as each is discarded. Subsequently, the program prepares the required documentation.

INTRODUCTION

The storage and analysis of digital experimental data acquired by computer-controlled instruments is playing an ever increasing role in scientific laboratories. Computer database programs are also being used to organize the information needed to run a laboratory (1). In addition, experimental protocols and other documents that are generated by people, rather than instruments, are being prepared in digital form with increasing frequency, typically using a word processor or a spreadsheet program.

One area that has been relatively slow to convert to a digital format is the preparation or modification of documents in which data entry or annotation must be performed during the course of some other activity, such as the preparation of samples. It occurred to us that recently introduced software that permits computers to recognize and act on human speech might facilitate the greater use of digital data during activities in which

BioComputing: Computer Tools for Biologists
Edited by Stuart M. Brown

the use of a keyboard or pointing device is awkward or distracting. Thus, we assembled a computer system to perform speech recognition at the same time the user moves freely around the laboratory. We also developed a prototype database to perform functions while workers enter data without shifting their hands from experimental samples or apparatus to traditional digital input devices. Here, we describe the system for speech recognition that we assembled and the database application that we wrote to take advantage of voice data entry.

MATERIALS AND METHODS

Computer and Audio Systems

The components required for the voice data entry system are shown in Figure 1. We used a generic personal computer with a 300 MHz Intel (Santa Clara, CA, USA) Celetron processor, 128 Mb of random access memory (RAM); the Microsoft (Redmond, WA, USA) NT operating system (version 4, service pack 4) equipped with Microsoft Office 97 Professional Edition; Dragon NaturallySpeaking® Preferred version 3.52 software (Dragon Systems, Newton, MA, USA) and a Sound Blaster Pro audio interface system (Creative Labs, Milpitas, CA, USA). The microphone headset for input to the audio system supplied with the voice recognition software was replaced by the base unit of an FM Wireless Microphone System (Tandy/Radio Shack, Fort Worth, TX, USA). This system operates in the 6 m fixed/mobile communications band. The microphone and battery-powered transmitter of the wireless microphone system are worn by the person using the data entry system. The transmitter, which measures 60 × 90 × 18 mm and weighs 130 g, can be kept in the pocket of a laboratory coat or attached to the belt. The microphone is connected to the transmitter by a 700 mm long lightweight insulated wire cable, which also serves as the unit's antenna. When attached to the collar of a shirt or blouse, the microphone is positioned close to the user's throat. In some experiments, we replaced this microphone and transmitter with a radio transceiver (21-410 or 21-407; Tandy/Radio Shack) that operates on one of five predetermined frequencies in the same band as the wireless microphone system. The 21-407 transceiver is equipped with a headset that positions a small headphone over one ear and the microphone on a boom near the wearer's mouth. In contrast, the 21-410 combines the functions of a microphone and speaker in a single transducer that is mounted in an earplug worn in the user's ear. Unlike the transmitter that is supplied with the wireless microphone system and broadcasts continuously, both transceivers broadcast only on command in either a "push-to-talk" or a "voice-activated" mode. They can easily be modified to transmit continuously, albeit with reduction in battery life. Both transceivers connect to their audio transducers and attach to the user's clothing in a manner similar to the supplied transmitter, but are slightly larger and heavier.

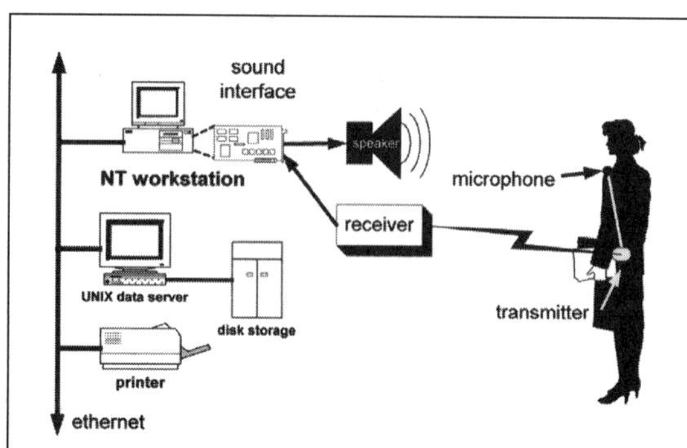

Figure 1. Diagram of the voice-controlled data entry system.

The personal computer used in this application is connected to a local-area network (LAN) and functions as part of a system for data acquisition, storage and analysis, which has been described elsewhere (2). Thus, data files can be stored on a server and later recalled by any person with proper authorization and access to the LAN or the Internet. Various components attached to the network, such as laser printers, are used by the system described here for the generation of reports.

Radiological Waste Inventory Data System

To demonstrate the use of voice recognition in a laboratory environment, we developed a system to track radiological wastes that were discarded during the preparation of experimental samples. We implemented the waste inventory system as a relational database using Microsoft Access 97, which is a component of Microsoft Office 97 Professional Edition. The scheme of the database is shown in Figure 2. The central table of this database records transactions in which a worker discards one or more items of a particular type into a particular container. The other tables in the database contain supporting information about the properties of the objects that can be discarded, the people performing the action, the containers in use, the types of containers available, the types of materials that can be placed in different types of containers, etc.

The various fields available in each table are also shown in Figure 2, as are the dependencies between tables. The transaction form for entering the information required to fill out one row in a Radioactive Waste Inventory Sheet (RWIS) is shown in Figure 3. (In the lexicon of Access and other database management systems, a form is a collection of information formatted primarily for display on the computer screen, while a report is a collection of information formatted primarily for printing. Because the paper-based waste dis-

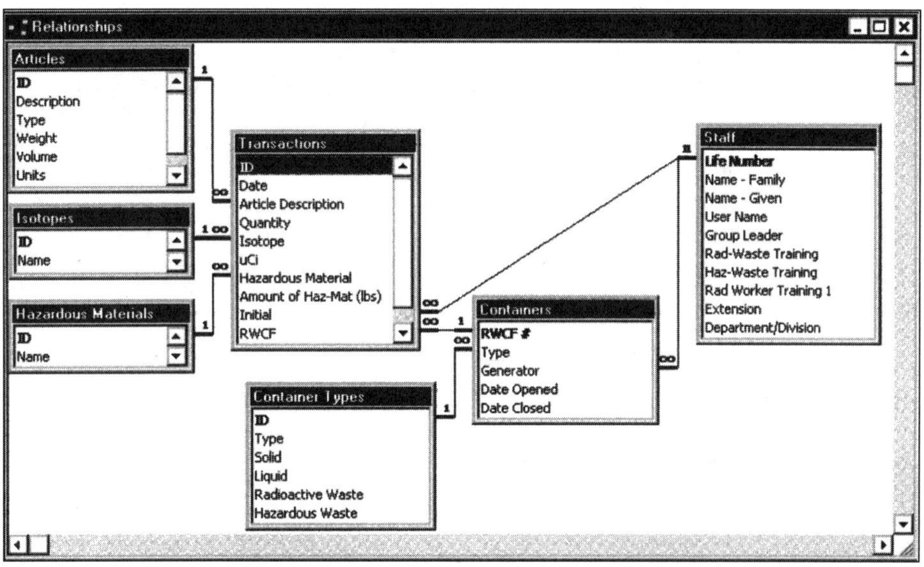

Figure 2. Scheme of the waste inventory database. Data are stored in 7 tables, which are related to one another as indicated by the lines. The names of the tables are indicated at the top with the list of fields given below. The relations are all one-to-many, indicated by the symbols 1 and ∞, respectively. For example, each staff member may be responsible for several containers. Five of the tables hold information used to provide the options available for completing each field of a single transaction in which one or more items are discarded (i.e., a single row of the transactions table).

posal system of our laboratory does not make this distinction, the piece of paper called a Radioactive Waste Control Form [RWCF] is generated as a report in the data system.)

Large fonts are used so that operators can confirm entries easily at some distance from the computer display. The operator can use voice commands to move the focus of the computer screen to various objects (controls) on the form, select items from pick lists that some controls contain or enter data when the range of acceptable values are not known a priori, and to execute other commands. All such functions can also be performed using a keyboard or pointing device, which are provided. Other forms are used when containers are logically opened and closed and to perform other functions.

Table 1 lists the system's forms and reports. Logically, closing a container automatically generates a report that provides a summary of its contents and contains the information that must accompany it through the waste management system. Figure 12 shows a portion of such a report, which is formatted to duplicate the paper RWCF.

DISCUSSION

Our experiments demonstrate that voice data entry in a laboratory environment is possible using readily available hardware and software and offers significant advantages for certain activities. We have also identified factors that should be considered in evaluating the use of voice entry of laboratory data.

Speech Recognition System

Within the past few years, several vendors have introduced modestly priced voice recognition software. Voice entry also requires audio input hardware (i.e., sound cards) that are being supplied as standard components with many personal computers. Likewise, the radio system that permits free movement around the laboratory is designed for the consumer market and is thus reasonably priced. Paradoxically, the most significant cost in equipping a computer system for voice data entry is not in any specialized hardware or software but in the computer itself. Speech recognition is a computationally

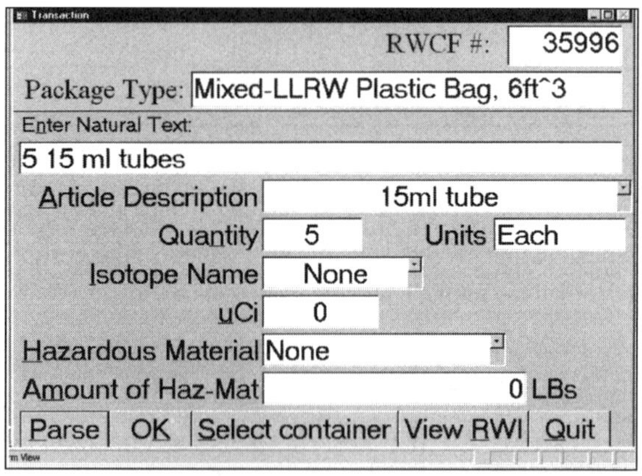

Figure 3. The form used to enter individual transactions into the waste inventory database. The name of the person using the database is supplied to the program from information acquired by the operating system when the person logged in. The type of items that can be discarded are provided in pick lists. Other tables in the database (Figure 8) record information about the discardable items such as the units in which they are measured, their average unit weight and class (paper, plastic, metal, etc.). Such information is required to provide summaries of the contents of filled containers, as required by existing regulations. The date and time of each transaction are entered automatically when the transaction is completed using data from the computer's clock. The user supplied data (Article Description, Quantity, etc.) can be entered by moving from one control to the next. However, the default data entry method is for the user to speak a phrase such as "numeral 5 fifteen mL tubes". After the voice recognition software interprets these words, the phrase appears in the Enter Natural Text control. If the phrase was recognized correctly, the user activates the Parse control (Parse is the default entry on this form, so one option is to say "new line"), thus activating a code module that parses the phrase and loads the results into the relevant text controls. Both the recognized phrase and the resulting text control entries are shown in the figure above. After checking the entries, the user completes the transaction by activating the OK control (one option is to say "type alt K").

Table 1. Forms and Reports Used in the Waste Inventory Data System

Name	Type	Description
Main Menu	form	Allows selection of the other forms and termination of the program.
Select Container	form	Allows the users to select the container into which items will be discarded from the list of presently open containers. Also provides for the logical closing of a filled container, or opening the New Container Form.
New Container	form	Allows a user to logically open a new container entry in the database. A number corresponding to an Radiological Waste Control Form (RWCF) identifies each such container.
Transaction	form	Contains controls for entering data describing an item or items being discarded. The information from this form constitutes one record in a Radioactive Waste Inventory Form.
Add Article	form	Adds a record describing a new class of items that can be discarded into the Articles table.
Close Check	form	Confirms that the user has specified that a particular container is to be logically closed.
Generate Report	form	Initiates generation of a Radioactive Waste Inventory report or an RWCF, which are described below.
Radioactive Waste Inventory	report	Lists all of the records of items placed in a container plus header information such as the type of container and the name of the person who generated the waste. A portion of a Radioactive Waste Inventory prepared by the Waste Inventory Data System is shown in Figure 5.
RWCF	report	Summarized the contents of each container. An example is shown in Figure 6.
Articles	report	Lists the contents of the Articles table (i.e., the classes of articles that can be entered into the database as having been discarded). This report is generated by a control located on the Add Articles Form.

demanding process and requires both a fast processor and a lot of RAM. The system we assembled frequently requires a few seconds to recognize some words and phrases, a delay that can be distracting during an experimental procedure. Thus, we recommend that voice data entry systems should be based on the fastest available microprocessor. Some recently introduced microprocessors have special on-chip hardware that facilitates computationally intensive multimedia tasks. Presumably, such systems would be worth the additional cost in voice data entry applications.

The microphones we tested represent a range of trade offs. The headset supplied with the voice recognition software appeared to provide the highest quality sound and would be adequate in applications in which the operator remains seated in one location (e.g., while operating a microscope). However, the connections to the computer are too restrictive if the operator must move around the laboratory. The clip-on microphone supplied with the FM wireless microphone system is unobtrusive, but when attached to a person's collar, it is positioned near the wearer's throat rather than near the side of the mouth, as recommended by the speech recognition software documentation. The FM radio transceiver with a headset positions the microphone properly, but is intrusive. The transceiver with the microphone in an earplug provides adequate audio quality with minimum bulk and reduced the effects of room noise. However, some users found the earplug uncomfortable. All of these transmitters provided signals that the speech recognition software was able to process correctly. The software documentation recommends retraining the program whenever the microphone or other components of the audio system are changed.

The accuracy of speech recognition software is influenced by background noise, thus a quiet environment is desirable. We should have liked to use the text-to-speech capability of the software to provide feedback to the user indicating the data was entered, but this capability was not available for use with Access in our version of the software. However, we use large fonts in the transaction form to permit operators to check the data even when they are located some distance from the computer. We also programmed the computer to generate a bell-like sound to confirm that a record has been entered. Text-to-speech using the speakers of the sound system would prevent more than one person from using voice entry in a given area. However, multiple users could be accommodated, if the speaker system was replaced by a radio link from the computer to the personal headphones of the user. Our experience with the radio transceivers suggests that using the same frequency for both directions would not be desirable. Pressing the "push-to-talk" button of the transceiver defeats the goal of hands-free data entry, while the "voice-operated-transmission"

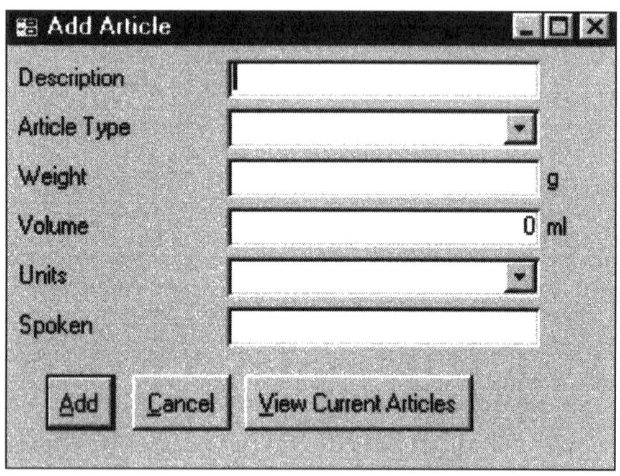

Figure 4. Add Articles form used to add a new class of discardable items to the data base. This form allows end users to add new classes of articles to the Articles table (Figure 2), which provides the pick list for the Article Description control on the Transaction form (Figure 3). Records can also be added to or removed from the Articles table and the five other supporting tables shown in Figure 2 by opening the table in Datasheet View using the Access programming environment, but such procedures are normally performed only by the database manager.

mode causes the user to preface most phrases with “aggggah” or some similar sound to initiate a transmission, thereby confusing the speech recognition program. Thus, user-specific voice feedback would be most effective if separate radio frequencies were used in each direction, i.e., full-duplex mode.

Speech recognition is a statistical process, so recognition errors at some level are unavoidable. Effective operation of the speech recognition software requires significant training of both the operator and the speech recognition program. The speech recognition program is “trained” by having each user read text displayed by the program on the computer screen. The system maintains a separate voice recognition resource file for each user. There is also a provision for building application-specific vocabularies by providing files that contain technical terms and by responding to questions from the program regarding words not in its dictionary. Such training is also specific to each user.

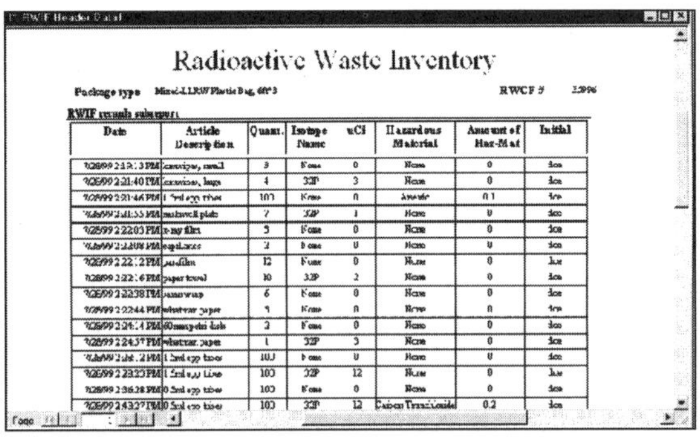

Figure 5. Radioactive Waste Inventory. This report is generated when the container designated by the indicated RWCF is logically closed. It lists all of the items placed in the container including data automatically added by the computer system such as the date and time of entry and the person recording the transaction, plus “header” information such as the type of the container and the RWCF number. It can be printed or displayed on a computer screen, as shown here.

Voice recognition is, in some ways, like a “command line” computer interface, in that the operator needs to know about a dozen spoken commands that are used to control the computer. These requirements mean that voice data entry is best suited for applications in which a small number of people frequently perform repetitive tasks that benefit significantly from capturing data in digital form as the work proceeds. An example of such a program is described below.

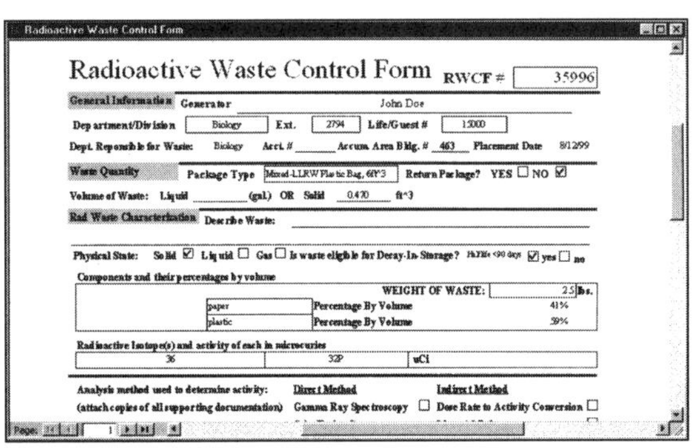

Figure 6. Part of a report generated by the waste inventory database system summarizing the contents of a specific container. The summary information is generated automatically, saving the user the time and effort required to compile such information by hand. It can be printed or displayed on the screen, as shown here.

Laboratory Applications of Speech Recognition

Our prototype Waste Inventory Data System database application has many of the attributes appropriate for voice data entry in a laboratory environment. Experiments involving the handling of radionuclides take place in reserved areas that can be entered only by authorized individuals, hence reducing interference from background noise. Workers handling radionuclides must complete several types of training, so the additional investment in learning the use of the voice data entry system is a small fraction of the total training load. The requirement that items discarded during such experiments must be inventoried, whether or not they are contaminated, places a premium on capturing such data as the items are discarded. The types of items discarded during an experiment are usually limited (tubes, paper wipes, pipet tips, etc.) and thus can be easily entered in lists from which the user can make a selection. New items can be added to the list using the "Add Articles" form, which also records information on the items' weight, type (paper, plastic, metal, etc.), state (solid or liquid) and the name that will be used in voice recognition, as shown in Figure 4. The major advantage of the voice data entry system is realized when a container is "closed" using the Select Container Form. This step calls a code module (written in Access Visual Basic for Applications) that moves the entry for each item in the transactions table to the Radioactive Waste Inventory report, as shown in Figure 5. The information in this report is also condensed to produce an RWCF, which summarizes the contents of the container. Figure 6 shows an example of this report, which can be printed and attached to the container when it is delivered to the waste management system. The code module performs a variety of required calculations, such as determining the fraction of the container composed of paper, plastics, metals, etc., hence eliminating considerable time that a person would take to perform these calculations.

An advantage of implementing the radiological inventory system using a relational database as opposed to, for example, a spreadsheet, is the ease of checking that organizational rules are followed. For example, entering a transaction in which a liquid sample is reported as being placed in a type of container not designed for liquids generates a warning message. We also check that the person generating the waste is currently certified to handle the materials being discarded. This check is implemented using the various training fields in the Staff Table (see Figure 2), which contain the date through which a staff member is authorized to perform a given function. At present, such data must be maintained by the local database administrator, but could be obtained by our database program by submitting a query over the network to the laboratory-wide training database. Similarly, the data listing the contents of each container and the various summaries contained in the RWIS and RWCF, respectively, could be transmitted directly in digital form to a central database.

CONCLUSION

Speech recognition software can be used to enter data during an experiment without diverting the user from the task at hand. Replacing the microphone supplied with speech recognition programs permits the user to move freely around the laboratory. While the cost of the auxiliary hardware and software to implement such a system is modest, the computationally demanding nature of speech recognition requires a fast microprocessor and a large quantity of RAM to provide acceptable performance. Effective use of speech recognition in a laboratory is facilitated by a quiet environment and requires significant training of both the people using the system and the speech recognition software. Thus, speech recognition is most useful for applications in which real-time capture of data during a procedure provides substantial benefits. The recording of items discarded during work involving radioactive or hazardous samples is such an application because the database program, rather than a staff member, prepares the documentation required by waste management regulations.

ACKNOWLEDGMENTS

We thank Denise C. Monteleone and John G. Trunk for assistance in configuring the computer system, Paula V. Bennett and Robert L. Colichio for explaining the operation of the radiological waste protocols and evaluating the operation of our prototype radioactive waste database and Ann Emrick for assistance in preparation of the manuscript. J.L.P., a computer science major at Carnegie-Mellon University, was the recipient of an Energy Research Undergraduate Laboratory Fellowship from the United States Department of Energy. Our research was supported by the Office of Biological and Environmental Research, Office of Science, United States Department of Energy.

REFERENCES

1. **Nayler, O. and S. Stamm.** 1999. ScienceLabDatabase: a computer program to organize a molecular biology laboratory. BioTechniques *26*:1186-1191.
2. **Sutherland, J.C., D.C. Monteleone and B.M. Sutherland.** 1997. Computer network for data acquisition, storage, and analysis. J. Photochem. and Photobiol. *B 40*:14-22.

Update to: Laboratory Voice Data Entry System

John C. Sutherland

Brookhaven National Laboratory, Upton, NY and East Carolina University, Greenville, NC, USA (sutherlandj@mail.ecu.edu)

Voice recognition, combined with database technologies, offers the potential of increased efficiency in the capture and reporting of a variety of data in the laboratory environment. It is particularly valuable in situations in which repetitive and, hence, predictable operations must be performed and where ordered and predictable data must be stored for future reference. The gains in productivity that can be achieved in such environments can justify the costs associated with developing voice recognition database systems. In appropriate circumstances, increases in the ease of data entry outweigh the inherent loss in flexibility that such systems impose.

The technologies underlying voice data entry systems has improved significantly in the short time since we described this system. Voice recognition technology, the most application-specific enabling technology, has shown significant improvement with several vendors offering excellent systems. Dragon NaturallySpeaking® voice recognition software is currently shipped as version 6, compared to the version 3.5 product that we used. Among the most significant improvements are the ability of the current version to correctly recognize words spoken as a part of a continuous stream, compared to the "___distinct___word___pattern___required___in___nineteen___ninety___nine." The available tools for developing voice recognition applications have also improved.

There have been important advancements in several other supporting technologies. Computer performance has forged ahead, as we have come to expect. The rapid penetration of mobile phones in the communications marketplace has brought with them a wide assortment of improved audio input/output gear—headphones and microphones—which are both more comfortable and offer improved fidelity at modest cost. Computer displays represent a less obvious, but perhaps more important, area of improvement. The cost of flat screen monitors has dropped sharply in the last few years, and these devices are far less obtrusive and safer in the laboratory workspace than their cathode-ray tube-based predecessors.

Despite these significant improvements, voice recognition is still only a niche form of human–computer interface. More training and practice are required to enable effective use of voice recognition input compared to the ubiquitous keyboard and mouse. Thus, voice recognition is still most likely to be successful in situations in which laboratory personnel regularly perform generally similar operations in reasonably isolated and quiet areas.

Voice data entry will be most likely to return benefits greater than cost when it is part of a comprehensive systems approach to facilitating laboratory operations, rather than an independent addition to existing procedures. Adding voice data entry to existing procedures is particularly problematic if the organizational entity developing the new capability is not in complete control of the entire process. For example, developing a database to fill out one or more forms or reports should not be attempted unless the authority to specify the format of the report rests with the same entity that develops the data system. Otherwise, the effort expended in implementing the data system may be nullified by changes in the underlying process or the reporting requirement.

CHAPTER 62

ScienceLabDatabase: A Computer Program to Organize a Molecular Biology Laboratory

Oliver Nayler and Stefan Stamm

Max-Planck Institute of Biochemistry and Neurobiology,
Martinsried, Germany

Reprinted from BioTechniques 26:1186-1191 (June 1999)

ABSTRACT

A description of a novel laboratory management software is provided. ScienceLabDatabase (SLD) offers a useful platform to organize a molecular biology research laboratory. The program manages stocks of biological samples, including plasmids, antibodies and cell lines, laboratory protocols and addresses, and it includes an easy ordering and funds managing system. Preformed data sheet templates help to store and maintain valuable information on samples or reagents and to facilitate the transfer of accurate information between researchers. Password protection regulates access, and simple button functions allow the use of this system without prior database knowledge. SLD is based on FileMaker Pro Version 4.0, which allows easy customization and import of preexisting data from other applications. The SLD program was successfully tested in several independent research groups and proved a useful tool to efficiently organize a molecular biology laboratory.

INTRODUCTION

The handling of rapidly increasing amounts of data poses a considerable problem in different areas of industry and academia. However, the development of database software greatly improves data management for a variety of applications, which increases efficiency of daily routines and significantly reduces cost. Today, relational databases help to organize material stocks, streamline purchasing and handle information. They are therefore widespread platforms to organize and maintain any form of business enterprise. However, maximum efficiency is only achieved when the specific needs of a customer are met. Surprisingly, to our knowledge, no available commercial program has been designed to manage a small-to-medium scale molecular biology research laboratory. Current alternatives seem to be listing information in a spreadsheet program, such as Excel®, or collecting data in printed form.

BioComputing: Computer Tools for Biologists
Edited by Stuart M. Brown

A scientific research laboratory does not greatly differ from any commercial business in that it needs to control costs and to manage information and material. In our own laboratories, we were increasingly struggling to keep up with the management of plasmid clones, antibody stocks, cell lines, etc., and we spent a considerable amount of time filling out data sheets and searching for lost samples. Furthermore, it is not only important to be able to track down valuable samples in an overflowing freezer, but it is equally important to have adequate and reliable documentation to share with colleagues or to use for patenting purposes. The costs of constructing a customized database have been relatively prohibitive and many require, once established, full-time administrators and programmers. However, the release of FileMaker Pro Version 4.0 enabled the development of an affordable and customized database solution. FileMaker Pro 4.0 offers both power and flexibility—a database can be maintained on either a PC or Macintosh®-based computer system and can be accessed by either system over a network. Furthermore, database files are compatible across computer platforms, and customized record entry forms can be created easily with little computing knowledge. Using FileMaker Pro 4.0, we initiated the development of the ScienceLabDatabase (SLD), and we hope that it will find interest in a wide community. Since SLD was developed by molecular biologists, it incorporates in-depth knowledge of a research laboratory and accounts for the specific everyday laboratory routine, its personnel, equipment and frequently used labware. In addition, to organize stocks of valuable biological material or consumables, SLD features an easy ordering system and provides a complete overview of expenses and grants. It furthermore serves as a database to store addresses and to accumulate experimental protocols. Together, SLD offers an affordable and easy-to-install database system, and it offers an efficient and user-friendly way to organize a molecular biology laboratory.

MATERIALS AND METHODS

ScienceLabDatabase is a FileMaker Pro 4.0 (Claris Corporation, Santa Clara, CA, USA) based program and is used on Macintosh computers (Apple Computer, Cupertino, CA, USA) or on a standard PC. ScienceLabDatabaseMac and ScienceLabDatabasePC are available as programs using FileMaker Pro 4.0 or as stand-alone programs, which do not require FileMaker Pro 4.0. The stand-alone program does not allow the user to customize the software or change passwords, and the program cannot be accessed simultaneously by multiple users over a network. A minimum of 36 Mbyte random-access memory (RAM) and a 4 Gbyte hard drive are recommended (depending on the planned size of the database). The programs run on all systems released following Macintosh System 6.0 (Apple Computer) or Windows 95® (Microsoft, Redmond, WA, USA). See Availability for information about obtaining the programs.

DISCUSSION

SLD is a hierarchically organized relational database for the use in molecular biology research laboratories. It is divided into several sub-databases (Figure 1) and helps to manage data concerning common biological samples (plasmids, antibodies, primers etc.), general administrative data (ordering, inventory) and addresses. A separate section is devoted to laboratory protocols. Password protection is included to recognize different user categories before entering the program's home page. Currently, three password tiers have been set: *(i)* administrator, *(ii)* laboratory scientist and *(iii)* guest. Each of these tiers contain different permissions regarding data entry, editing and viewing. Additional features such as grant and cost controls are only accessible to the administrator or respective group leader, whose password recognition permits the user to modify the current program (not in stand-alone version) or to change passwords, if necessary. The different databases are activated through buttons, and additional help functions provide useful information on each subject. Furthermore, since SLD can import data from Excel spread sheets (Microsoft), a preexisting database is quickly converted.

Database Entry

This section describes the procedure to enter a plasmid into the plasmid database. This process is identical for all the other databases such as antibodies, cell lines and so on. New entries are made on preformed data sheets (exemplified by the plasmid data sheet in Figure 2), and a numbering system allocates a new identification (ID) number to each entry. To ensure that each item is unmistakably recognized, the ID number will not be reused even after deleting an old entry. Each data sheet is designed to contain most of the commonly used information, and we have found that this system encouraged users to enter complete data. However, since SLD is constructed with FileMaker Pro 4.0 software, additional entry fields or modification of existing SLD forms can be easily customized to suit specific needs (not in stand-alone version).

The plasmid entry form contains useful cross-links. Similarly, other sub-databases are cross-linked if they contain complementary information; for example, entries from the **Users** and the **Suppliers** databases appear in all other databases in pull-down menus in the *Made by* and *Supplier* field. The **DNA Analysis** button starts up a preferred program, such as DNAstrider or DNAsis® (Hitachi Software Engineering America, San Bruno, CA, USA), and copies the contents of the *Sequence* field into the DNA program for further manipulation, such as restriction digest maps. A link to a program of your choice is easily established (described in the **Help** function in SLD). Because each field can hold up to

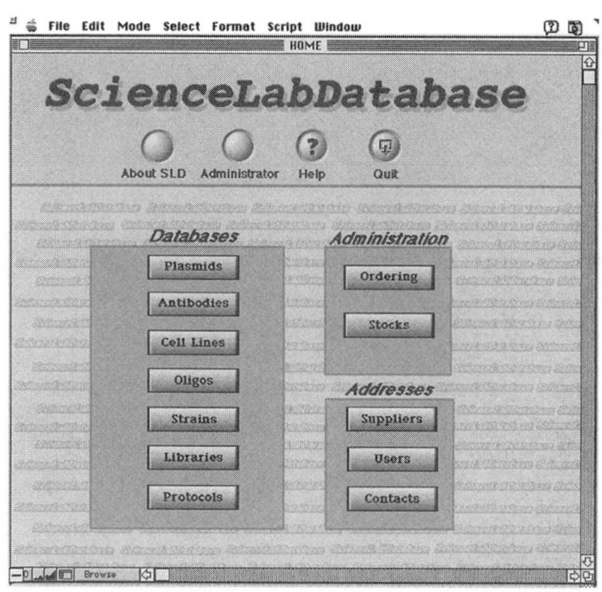

Figure 1. Home page of the SLD. A screen shot of the home page shows the different sub-databases, which are activated by clicking on the individual database buttons. General information and explanations on the use of the different databases is accessible with the **Help** button.

65 000 characters, the entire DNA sequence information can be stored. The plasmid map can be made with a simple drawing program that automatically starts by pressing the **Draw Map** button. Complete maps that are created by more sophisticated programs can be simply copied and pasted into the *Map* field. FileMaker Pro 4.0 can import a variety of Macintosh and PC graphic formats, making it simple to insert scanned or internet-downloaded maps.

All the important information for a particular plasmid is displayed in a data sheet using the *browse* mode, whereas an overview of all plasmids is found in a list display. The power of an electronic database usually resides in its *sort* and *find* features. Every database can be sorted according to any feature, such as date, name, etc.. With a *find* option, a particular entry or a group of entries containing a common feature can be rapidly picked out. So far, the entry of a record into the plasmid entry form has been described. Many of these described features can also be found in **Antibodies**, **Oligos**, **Cell Lines**, **Strains**, **Libraries** and **Protocols**. A particular advantage of the SLD lies in its capability to orga-

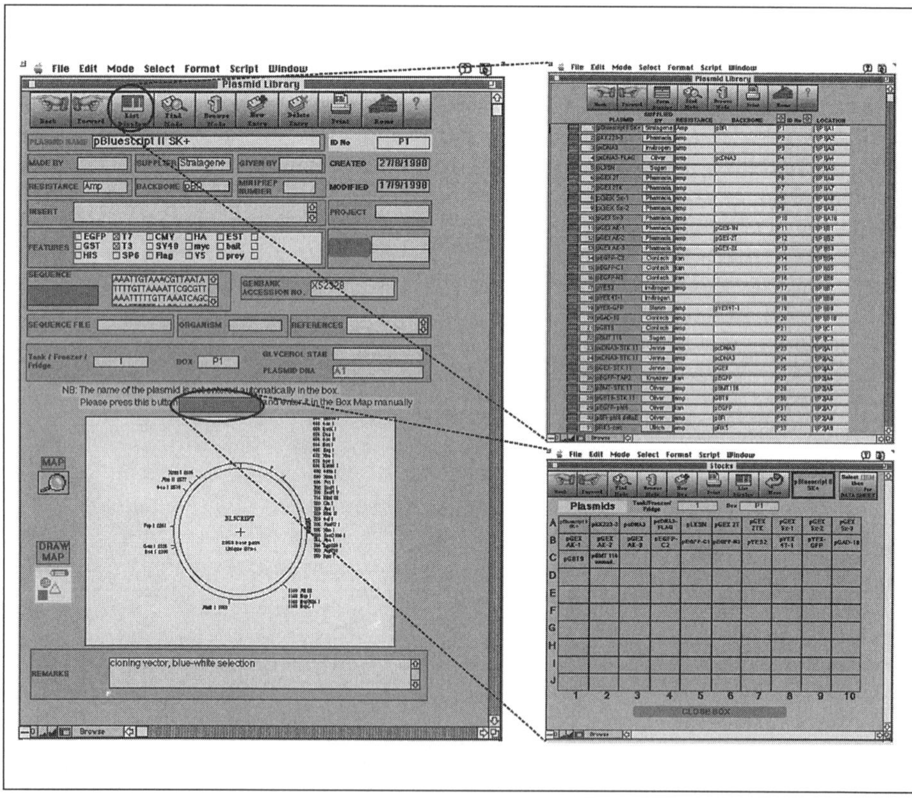

Figure 2. Preformed data entry sheet of the Plasmid sub-database. The screen shot of the **Plasmid** database displays the data entry form in the *browse* mode (left). The entry of the pBluescript® plasmid illustrates the information that can be stored. Buttons are used to open a DNA analysis program or a **Draw map** program. The **List Display** button is circled. This button displays lists of all plasmids entered, indicating their name, the person that made or provided the plasmid, the resistance, the unique identification number and the location (top right). The circled **Box it** button opens the stocks database (bottom right). The name of the plasmid indicated on the top right is transferred to its precise location in this virtual box with a **Drag&Drop** function.

nize sample storage. Items are currently stored in standard polypropylene freezer boxes using a 10 × 10 space format (e.g., Nunc®; Nalge Nunc International). Although different box formats are not featured here, these could be easily customized. After entering data, for example into the **Plasmid** database, the sample is manually allocated to a specific freezer and box. The **Box it** command is used to open a graphical display of the chosen box, and the sample is entered (Figure 2). All boxes can be viewed in the **Stocks** database, which is a virtual representation of the laboratory freezers. Furthermore, to identify a sample in the box, simply clicking on it and then pressing the **MENU** button takes you to the original entry in the correct database.

Ordering

The ordering of reagents or equipment is an almost daily routine in most laboratories. However, we have found that it is often difficult to control outgoing and incoming orders and to maintain an overview of available reagents to avoid redundancies or erroneous orders. Furthermore, looking up addresses or phone numbers of suppliers, especially for

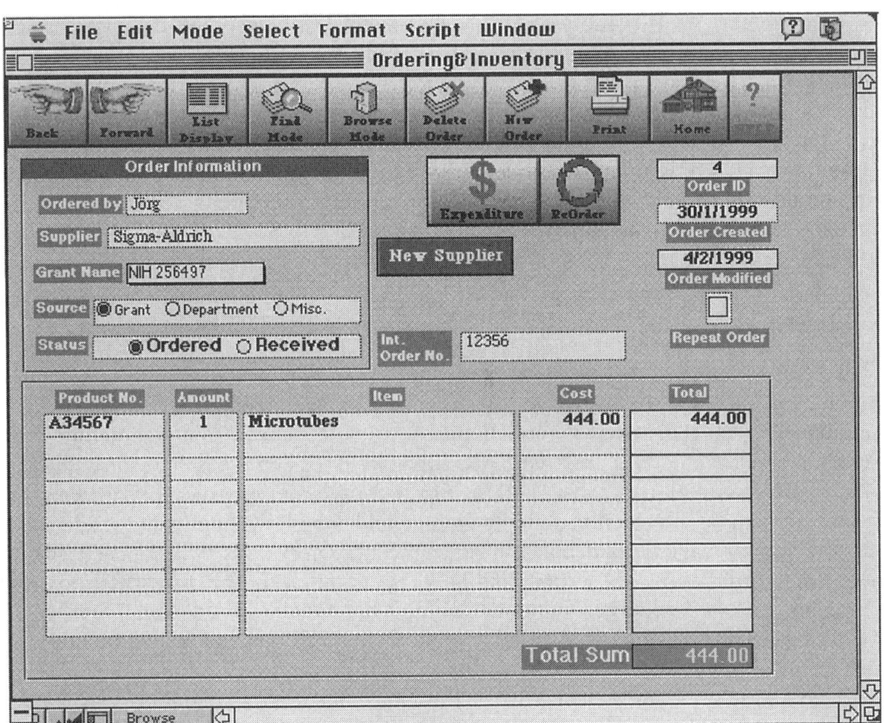

Figure 3. Preformed ordering data sheet. The preformed data sheet is used to place an order and to allocate the purchase to a particular fund and grant name. A balance of accounts is easily acquired with the **Expenditure** button. Frequent orders are simplified using a repeat order status. The **ReOrder** button creates a new order form from a previously used entry bearing a repeat order status.

frequent reorders, can be unnecessarily time-consuming. Several aspects of a regular ordering process, including supplier addresses, details of the purchase, etc. are combined in a preformed order data sheet in SLD (Figure 3). A **ReOrder** button allows the instant creation of a new order form from a previously used order. In addition, all orders can be allocated to a particular grant or money source, so that an instant balance of the accounts on a monthly or yearly basis can be viewed. A specific search feature also enables the user to view expenditure according to, for example, supplier or user. The expenditure information is again only available to the individual with the administrator permission. The **Ordering** database program incorporates all received orders and provides an accurate overview of all available reagents. Thus, redundancies of orders and reagents can be effectively eliminated.

Administrative Functions and Address Books

After SLD has been opened with the administrator's password, all program features, such as layouts, passwords and so on can be changed (not in the stand-alone version). Also, several additional features are available in the program. Clicking of the **Administrator** icon in the home page (Figure 1) opens a dialogue that allows the definition of projects that will appear as pull-down menues in the various database sheets. In addition, this page allows the entry of fund names and the amount of money allocated to a particular grant. Clicking on *Expenditure* in the order menu results in a sub-menu that gives a weekly, monthly or yearly expenditure report. With the administrator's password, new members of the laboratory can be added, including their user names, phone numbers and addresses. This user name will appear in all data sheets as a pull-down list under **Made By** or **Ordered By**. Finally, only the administrator can delete entries by pushing the **Delete Entry** button. In contrast, the laboratory scientist password allows entries of supplier and contact names, which are used to store information about vendors and fellow scientists, respectively, providing reagents.

SUMMARY

We describe some of the central features of ScienceLabDatabase, molecular biology laboratory management software specifically developed by molecular biologists using FileMaker Pro 4.0. This database meets the specific needs of a small-to-medium scale research laboratory and will greatly help to organize stocks of valuable reagents, to store and maintain information on those items and to simplify ordering processes. Furthermore, as all orders can be allocated to different grants and funds, a group leader maintains an overview over all reagents and consumables in the laboratory and has instant access to the balance of a project's money sources and how they were used. The power and flexibility of FileMaker Pro 4.0 with the SLD does not limit itself to one laboratory. Because of its networking potential and its cross-platform capability, this database could be initialized on an institute level.

We have successfully introduced the SLD management system in our own laboratories and found that it greatly facilitates many aspects of our daily routines. Furthermore, since it is very easy to use, it is appreciated by our laboratory members as an easy information retrieval system. Based on our own experience, it is thus conceivable that ScienceLab-Database could become a widespread laboratory management program.

AVAILABILITY

The ScienceLabDatabase is made available at a reasonable fee. For further information, visit the authors' home page (http://www.neuro.mpg.de/stamm.htm).

ACKNOWLEDGMENT

This work was supported by Axel Ullrich (O.N.) and the Max Planck Society (S.S).

CHAPTER 63

Emulation of an ELISA Reader and Quantitative Image Analysis Using the Computer Program Digital-OD

Stephan Jansen, Bernd Kriegesmann and Bertram Brenig
Tierärztliches Institut der Universität Göttingen, Göttingen, Germany

Reprinted from BioTechniques 24:998-1001 (June 1998)

ABSTRACT

Here, we present a Microsoft® Windows® application that has been written to emulate a microplate reader. After scanning a standard 96-well microplate, color gradations in the slots can be analyzed quantitatively and be used for the estimation of enzyme kinetics, binding assays or concentration determination. In a test, series accuracy of the program was determined. We show that the returned optical densities correlated highly with the true concentrations (r >0.97). Thus, the software is useful for a wide number of applications. Further, the program is easy to handle and is available free of charge.

INTRODUCTION

In molecular biology and biochemistry, microplate-based assays are commonly used for measuring concentrations of various substances for the determination of binding capacities or for the estimation of enzyme kinetics. In most cases, the final visualization of the results is done with a dye reaction, in that the color of the dye or the intensity of the dye or both is changed. Usually, this difference is analyzed spectrophotometrically in sophisticated devices like enzyme-linked immunosorbent assay (ELISA) plate readers. Also, the visualization of results can be achieved using a carrier material, and intensity nuances are measured with a densitometer. In this context, it has been shown that in addition to fairly expensive equipment being able to detect the differences in density or color, commercially available image-digitizing instruments like digital charge-coupled device (CCD) cameras (2) can also be used. Even flatbed scanners have been proposed for densitometric analysis or the quantification of immunoblots (5). Besides, software has been demonstrated to solve tasks like quantitative image analysis (3,4). Although no special software is

BioComputing: Computer Tools for Biologists
Edited by Stuart M. Brown

available, reading microplates using a desktop scanner with the appropriate computer program should also be possible. Here, we present a computer application that has been particularly written for this purpose. The Digital-OD program is supposed to emulate a microplate reader and allows quantitative analysis of reactions in the wells. The software is a Microsoft® Windows® application and is available free of charge. The required files can be downloaded from the author's web site (**http://www.gwdg.de/ ~sjansen/**).

DESCRIPTION OF THE PROGRAM

A bit-map file is originated by an image-digitizing procedure (e.g., by a scanner) and loaded into the program. This image serves as a source for the subsequent analysis. Uncompressed bit maps were chosen because virtually all scanning programs are able to generate this format. By default, the size of the picture under investigation is scaled to the size of the window; however, the program tries to maintain the proportions of microplates. Otherwise, the picture can be horizontally and vertically rescaled.

The *Reader functions* entry in the menu bar lists procedures necessary for a simple run (Figure 1). In microplate readers, a wavelength must be chosen from a set of filters for analysis. In the Digital-OD program, these filters are emulated electronically. A red, yellow, green and blue filter that reflects wavelengths at approximately 700, 585, 520 and 435 nm, respectively, has been implemented. Furthermore, it is possible to determine only the gray values of certain areas on the screen. Also, for special applications or for optimization, the filter dialog permits a switch to *Advanced settings*. In this dialog, filter colors and wavelengths other than the pre-set parameters can be imitated. Instead of the subtractive color scheme, the additive color scheme was chosen for filter-color generation. This was done on the observation that it is far easier to add the colors red, green and blue than to delineate a wavelength that results from subtraction of cyan, magenta and yellow. In the additive color scheme for instance, red and green makes yellow or green and blue makes a greenish blue. In both examples, the visual appearance of the merged colors correlate with their physical wavelengths. The only exception that would not work in this way is to merge red and blue. This would give a sort of violet in the additive color scheme but not give a short wave filter. Another alternative would have been a direct input of the desired wavelength. Unfortunately, exact generation of a filter depends specifically on the graphic card present in the computer. So a fine tuning would have been obligatory in any case. And finally, the additive color scheme harmonizes best with the computer hardware and abates inaccuracies.

The principle of action when measuring the color gradation in the slots of the microplate is to determine the digital presence of each of the colors in the additive red, green and blue (RGB) system. This is different from ELISA plate readers. In these plate readers, maximum contrast is achieved by eliminating unwanted wavelengths with optical filters, and the specific absorption of a probe at a certain wavelength is measured. In the Digital-OD program, the opposite color of the digital filter is taken into account. The program measures which color is there and with what intensity. Unlike ELISA readers, this means that measurement occurs positively and not according to that what has been extincted. When comparing the program with an ELISA reader however, there is no difference in terms of the choice of the optimal filter. This means that in the measuring process, red gradations are read best with a green filter or yellow gradations are determined best with a blue filter and vice versa. This is very similar to the procedure when using pure optical systems.

In the program, there are two procedures implemented to read a plate: *(i)* either single areas can be examined (like the "color temperature tool" in photo-publishing programs) or *(ii)* all slots of an ELISA plate can be measured simultaneously. In the latter procedure, only the slots A1, A12 and H12 of the plate have to be indicated. The other slots of a standard 96-well plate will be aligned automatically. In both procedures, areas of 5×5 pixels are checked, and the average of the color values within this field is determined. Blanks can be set by clicking on the menu entry *Consider blanks*. The average of the respective blank values will then be automatically subtracted from the other slots to be measured. To ease typical ELISA applications, an upper and a lower threshold can be fixed. When measuring a plate using the thresholds, the output file does not contain raw data in terms of optical densities. Depending on the threshold settings, positive, negative or ambiguous reactions are given instead. All results can then be stored in a file and printed out or subjected to further mathematical analysis. Existing data files will not be overwritten, but the results will be appended in the file.

VALIDATION OF THE PROGRAM

To estimate the accuracy of the results obtained by the program, two series of dilutions were made. The first was a dilution series of picrinic acid, which serves otherwise as a control substance to calibrate microplates. The picrinic acid dilution shows a yellow gradation and was diluted from 3.2 optical density $(OD)_{600}$ down to 0.025 OD_{600}. The other dilution that served as a reference for testing was made from a protein concentration gradient. Protein concentration was measured using the Bradford assay (Reference 1; Bio-Rad, Hercules, CA, USA). A typical color gradient from a dark brown to a deep blue was formed. ELISA plates were then measured in a THERMOMax® ELISA Reader (Molecular Devices, Sunnyvale, CA, USA) at 405 and at 600 nm to obtain best results for the picrinic acid dilution and the protein concentration gradient, respectively. The plate was then scanned using an AGFA Studio Scan II Flatbed Scanner (Agfa Germany, Cologne, Germany) at 300×300 dots per inch (dpi) and a color resolution of 2^{30} colors. Analogous to the procedure with the microplate reader, the yellow (585 nm) and the blue filters (435 nm) were applied to measure the corresponding dilution series. The video

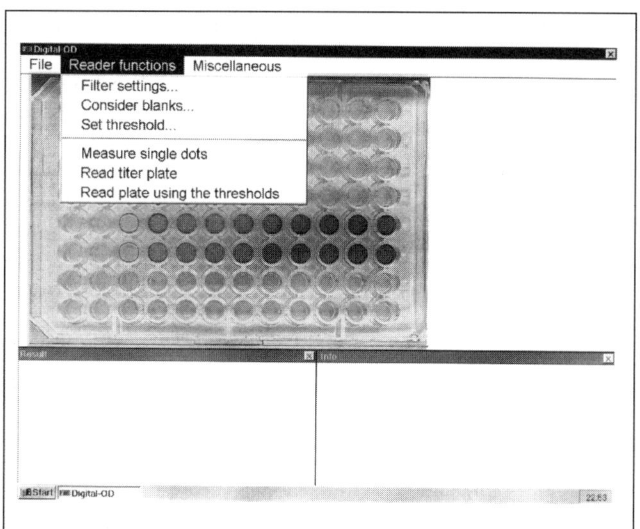

Figure 1. Screenshot of the Digital-OD program. Available functions to emulate a microplate reader are listed under the menu entry *Reader functions*.

graphics adapter (VGA) card resolution of the computer used for this investigation was 1024 × 768 pixel with 2^{24}-bit color resolution. After measuring of the plate in both systems, data were collected and subjected to regression analysis with the true concentration.

With the ELISA reader, both the picrinic acid dilution and the protein concentration gradient returned results that matched with the true concentration (r >0.99). An only slightly less accurate estimation of the true concentrations were returned by the scanning procedure with subsequent computer analysis. The picrinic acid dilution showed a correlation of $r = 0.97$ with the true concentration, while the estimation of protein concentration returned a value of $r = 0.99$. This indicates that the ELISA reader returns the most exact results; nevertheless, for most applications these minor deviations are not crucial. The most important difference concerning the interpretation of results is that in microplate readers the relation between the extinction and the true concentration of the substance is linear. In the computer-based procedure using the Digital-OD program, the digitally estimated optical density and the true concentration follow a logarithmic order (Figure 2). Calibration of the whole system occurs in an identical manner as with microplate readers, apart from the fact that the logarithmic slope of the curve has to be considered. After calibration, even quantitative assays are possible, as shown with the determination of the protein concentration.

As can be seen with the picrinic acid dilution, the detection range spans from OD = 3.2–0.2. Measuring higher or lower optical densities is not recommended, and substances to be investigated need either dilution or concentration. Other limiting factors are the quality of the scanner and the graphic card. Although the computer program also works well with a lower pixel density, to avoid dithering and to ensure a smooth color gradation, the color resolution should be not less than 2^{24} bits. Conversely, a higher color resolution would not improve results because the program itself cannot resolve more colors. The quality of the monitor is not important, but in some instances it causes suboptimal settings of the graphic card. This may affect the quality of the image analysis. Additionally, a Model UC1200 SE Scanner (UMax Technologies, Willich, Germany) with a transillumination gear has been tested. Although the image of the microplate is excellent, one has to take care that illumination intensity is not too high. This can cause problems when the dyes themselves are very light. Other digitizing instruments like CCD cameras can possibly be used for analysis. Even a simple hand-held scanner worked well and gave satisfying results, provided that the microplates were placed

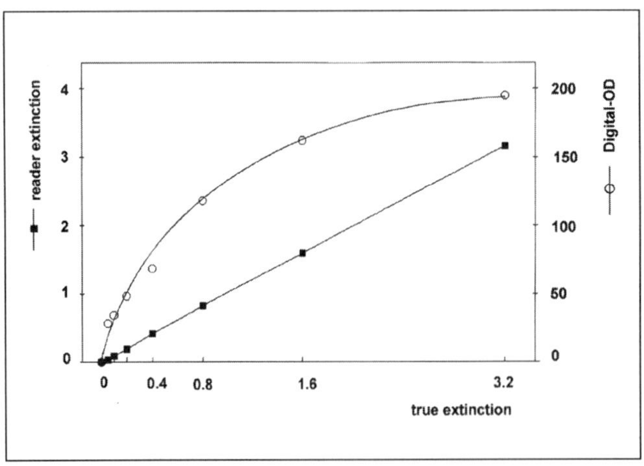

Figure 2. Correlation of the given, true concentration with the extinction returned from the reader (squares) and the optical density (circles). The main difference between an ELISA reader and the Digital-OD program can be seen in the different slopes of the curves. In the Digital-OD program, the slope follows a logarithmic order. For quantitative analysis, the logarithmic order needs to be considered.

on a white reflecting surface. Although the program presented here is principally designed to analyze images of microplates, there are further applications like the quantification of any kind of dot-blot assays. The accuracy of the image-digitizing process has been shown to be high and to fit the needs of wide number of applications (2,5). Since the prices for high-performing scanners and high-quality graphic cards have dropped notably, the program presented here is useful to set up a low-cost and versatile alternative to expensive ELISA readers.

REFERENCES

1. **Bradford, M.M.** 1976. A rapid and sensitive method for the quantification of microgram quantities of protein utilizing the principle of dye binding. Anal. Biochem. *72*:248-254.
2. **Malek, A.M., S. Izumo and S.L. Alper.** 1997. Quantitative densitometric analysis using a commercially available handheld CCD digital camera. BioTechniques *22*:1150-1153.
3. **Shea, T.B.** 1994. An inexpensive densitometric analysis system using a Macintosh™ computer and a desktop scanner. BioTechniques *16*:1126-1128.
4. **Van de Lest, C.H.A., J.H. Veerkamp and T.H. van Kuppevelt.** 1995. ImageCalc: a Microsoft® Windows™ application for quantitative image analysis and comparison BioTechniques *18*:1050-1055.
5. **Velleman, S.G.** 1995. Quantifying immunoblots with a digital scanner. BioTechniques *18*:1056-1058.

Update to: Emulation of an ELISA Reader and Quantitative Image Analysis Using the Computer Program Digital-OD

Stephan Jansen[1], Bernd Kriegesmann[2], and Bertram Brenig[2]

[1]Berufliche Schule am Schützenpark, Kiel, and [2]Tierärztliches Institut der Universität Göttingen, Göttingen, Germany (jansen-grossenaspe@t-online.de)

We received numerous responses to the digital-OD program and were happy to learn that the application is used for many routine and scientific purposes. Fortunately, no severe bugs were reported. The main inquiries, however, were concerned with the quality of the acquired digital image, and this seems to be a general problem with flat bed scanners. As a solution, some users of the digital-OD program proposed other image digitizing instruments than scanners. In fact, we can confirm that charge-coupled device (CCD) cameras, video-cameras with a digitizer, and even Web-cameras provide better results than some scanners do. Especially in scanners that lack a transparency adapter, a disturbing shadow is thrown during the run. The shadow is caused by the titer plate itself under illumination. Unfortunately, there is no other way than to fiddle around and to find an appropriate position so that the shadow of the titer plate does not spoil the image to be investigated. However, this remains mainly a specific problem with these scanners.

When using still, video, or even Web cameras, other aspects have to be considered. First, the illumination of the titer plate should be optimized, and second, the angular deflection in cameras has to be regarded. To minimize the occurrence of any shadows, the titer plate should be placed on a white transilluminator or, as a low cost version, the titer may be put on a white piece of paper. In the latter case, the paper should be illuminated by lamps that produce an evenly dispersed white light like fluorescent lamps do. To enhance the uniformity of the illumination, we recommend to adjust at least two fluorescent lamps in a top light position. To estimate the success of an uniform illumination, we recommend to measure the whole respective surface with the digital-OD program using greyscales as a filter setting. Having done this procedure for standardization once, even small differences in concentrations can be measured with high accuracy using the digital-OD program.

Under the assumption that the camera's optical system is focused on the middle of the titer plate, a certain angular deflection from the center to the edge of the plate is normal and cannot be avoided. Hence, the light to be measured travels a slightly longer way through the (colored) liquid in the wells that are located at the edge of the plate than it does in the more central wells. This would cause darker wells at the edge and is a systematic error, and as a consequence, a higher concentration is pretended in the wells at the edges. However, this bias can be easily reduced when the picture is taken from a greater distance and when a long focal length of the optical system is used.

Taken together, the quality of the measurement depends mainly on the quality of the optical system that is employed to generate the digitized image. For low cost purposes, Web cameras have proven to be sufficient, but we also learned that many sophisticated CCD cameras are in use together with the digital-OD program. The application is still free of charge and is available from the site (http://jansen-grossenaspe.bei.t-online.de).

INDEX

B

C

D

E

F

G

J

K

L

M

N

O

Q

S

T

U

V

W

Z